Encyclopedia of
Plant Physiology

New Series Volume 18

Editors

A. Pirson, Göttingen
M.H. Zimmermann, Harvard

Higher Plant Cell Respiration

Edited by
R. Douce and D.A. Day

Contributors

T. ap Rees M. Chauveau S. Delorme P. Dizengremel
I.B. Dry G. Ducet G.E. Edwards D. Falconet P. Gardeström
J.B. Hanson J.L. Harwood H. Lambers C. Lance B. Lejeune
C.A. Mannella A.L. Moore M. Neuburger J.M. Palmer
F. Quétier P.R. Rich M. Steup M. Stitt J.A. Ward
J.T. Wiskich

With a Foreword by H. Beevers

Springer-Verlag
Berlin Heidelberg NewYork Tokyo

Dr. ROLAND DOUCE
Centre d' Etudes Nucléaires et Université Scientifique
et Médicale de Grenoble
Département de Recherche Fondamentale
Laboratoire de Physiologie Cellulaire Végétale
UA CNRS N° 576
85X
F-38041 Grenoble Cedex
France

Dr. DAVID A. DAY
Botany Department
The Australian National University
Box 475 P.O.
Canberra City, ACT 2601
Australia

With 81 Figures

ISBN 3-540-13935-4 Springer-Verlag Berlin Heidelberg New York Tokyo
ISBN 0-387-13935-4 Springer-Verlag New York Heidelberg Berlin Tokyo

Typesetting, printing and bookbinding: Universitätsdruckerei H. Stürtz AG, Würzburg
2131/3130-543210

Foreword

I am honored by the editor's invitation to write a Preface for this volume. As a member of an older generation of plant physiologists, my lineage in plant respiration traces back to F.F. BLACKMAN through the privilege of having M. THOMAS and W.O. JAMES, two of his "students," as my mentors. How the subject has changed in 40 years!

In those dark ages B.^{14}C. most of the information available was hard-won from long-term experiments using the input–output approach. Respiratory changes in response to treatments were measured by laborious gas analysis or by titration of alkali from masses of Pettenkofer tubes; the Warburg respirometer was just beginning to be used for plant studies by pioneers such as TURNER and ROBERTSON. Nevertheless the classical experiments of BLACKMAN with apples had led to important results on the relations between anaerobic and aerobic carbohydrate utilization and on the climacteric, and to the first explicit concept of respiratory control of respiration imposed by the "organization resistance" of cell structure. THOMAS extended this approach in his investigations of the Pasteur effect and the induction of aerobic fermentation by poisons such as cyanide and high concentrations of CO_2. JAMES began a long series of studies of the partial reactions of respiration in extracts from barley and YEMM's detailed analysis of carbohydrate components in relation to respiratory changes added an important new dimension. Much emphasis was placed on the soluble oxidases, phenol oxidase, ascorbate oxidase, and peroxidase, those distinctive plant enzymes which seemed so inevitably associated with their respiration. The work of HILL, GODDARD, and OKUNUKI on the plant cytochromes refocused attention to where it belonged, as expected from considerations of comparative biochemistry. JAMES' important book on *Plant Respiration* was an authoritative survey of the achievements up to 1953, and the various editions of *Plant Physiology* by Thomas and his colleagues provided extensive coverage of this subject.

The early 1950's saw a transition in the study of plant respiration that was shared by most other fields of biological research. The increasing availability of research funds, more sophisticated instruments, labeled substrates, coenzymes and other biochemicals, were major contributing factors. The investigation of plant respiration proceeded along several lines. The new instruments, O_2-electrodes, and infra-red analyzers allowed rapid and continuous recording of gas exchanges and the Warburg and other respirometers became generally available. High speed centrifuges allowed the separation of organelle fractions, and spectrophotometers made possible precise investigations of individual enzyme reactions.

The enzymatic reactions of glycolysis were firmly established during this period, and mitochondria from plants, first isolated from mung bean by BONNER's group at Caltech in 1950, became the focus of attention from many groups around the world. The work of LATIES and others established the central importance of the TCA cycle in plant respiration. In retrospect it is ironic that many of the acids were first discovered in plants long before they were so elegantly enshrined in the TCA cycle by KREBS. Cyanide-resistant respiration in plant tissues and mitochondria engaged attention, and AXELROD's investigations led to the establishing of the oxidative pentose phosphate pathway as a constitutive part of respiration in plants. Volume XII of the first edition of the Encyclopedia, edited by J. WOLF, appeared in 1960 and dealt comprehensively with *Plant Respiration* in over 2000 pages, and my own book on *Respiratory Metabolism* in 1961 summarized these developments rather more briefly. By that time the dual role of respiration in providing intermediates for all kinds of synthetic reactions, as well as the reducing equivalents and ATP required for their accomplishment, had been demonstrated, although not precisely quantified. The overall rate of normal respiration had been shown to be due not to the limited capacity of some key enzymatic reaction, but was governed instead by intracellular control mechanisms. The ways in which the pace of reactions driven by respiration determined its rate, and particularly the role of ATP turnover in the control of aerobic and anaerobic respiration were being recognized.

Since that time many of the respiratory enzymes have been prepared in highly purified form by TURNER and others, and much has been learned about the details of regulation imposed by allosteric effectors. Progress has been forthcoming on the mechanisms underlying respiratory changes in aging of tissue slices and in ripening. Purified mitochondria, those marvelous machines of ATP generation, have been intensively investigated and the discussion of the many aspects of this subject occupy, appropriately, a large part of the present volume. The discovery of photorespiration and the role of organelles, including mitochondria in this process, was not anticipated in 1960. The impressive progress recorded in this volume concerning all aspects of the biochemistry of plant respiration has made possible the investigation of linked processes such as ion and metabolite transport within and between cells, and more definitive estimates of the respiratory costs of dependent processes in the life of the plant. The efficiency with which respiratory breakdown of carbohydrate is linked with synthesis of cell constituents, and with repair and maintenance of cell structures is now being investigated, and models indicating how respiratory energy is apportioned to various functions have been generated.

These developments in the understanding of the biochemical and physiological aspects of plant respiration represent important achievements since the coverage on the topic in the first edition of the Encyclopedia. The authors of the various chapters in the present volume are among those who have contributed most significantly, and appropriately reflect the present international character of research on this important subject. We can only marvel at the progress that has been made and wonder what new developments the next 20 years will bring. Certainly the questions now being asked are very different from those that intrigued us 40 years ago, when even the main pathways were not established. A major challenge now is to understand more precisely how the process of respiration is regulated and integrated into the overall functioning of the plant as a whole.

HARRY BEEVERS
Biology Department
University of California, Santa Cruz

Contents

3 Plant Mitochondrial Lipids: Structure, Function and Biosynthesis
J.L. Harwood (With 9 Figures)

4 Plant Mitochondrial Cytochromes
G. Ducet (With 2 Figures)

5 The Outer Membrane of Plant Mitochondria
C.A. Mannella (With 10 Figures)

8 The Cyanide-Resistant Pathway of Plant Mitochondria
C. LANCE, M. CHAUVEAU, and P. DIZENGREMEL (With 10 Figures)

9 Membrane Transport Systems of Plant Mitochondria
J.B. HANSON (With 5 Figures)

10 The Tricarboxylic Acid Cycle in Plant Mitochondria: Its Operation and Regulation

J.T. WISKICH and I.B. DRY (With 8 Figures)

11 Leaf Mitochondria ($C_3 + C_4 + CAM$)

P. GARDESTRÖM and G.E. EDWARDS (With 3 Figures)

12 Starch and Sucrose Degradation
M. Stitt and M. Steup (With 5 Figures)

13 The Organization of Glycolysis and the Oxidative Pentose Phosphate Pathway in Plants
T. AP REES (With 4 Figures)

14 Respiration in Intact Plants and Tissues: Its Regulation and Dependence on Environmental Factors, Metabolism and Invaded Organisms
H. LAMBERS (With 6 Figures)

List of Contributors

T. ap REES
  Botany School
  University of Cambridge
  Downing Street
  Cambridge, CB2 3EA/United Kingdom

M. CHAUVEAU
  Laboratoire de Biologie Végétale IV
  Université Pierre et Marie Curie
  12, Rue Cuvier
  F-75005 Paris/France

S. DELORME
  Laboratoire de Biologie Moléculaire
  Végétale Associé au CNRS
  BAT 430 – Université Paris Sud
  F-91405 Orsay/France

P. DIZENGREMEL
  Laboratoire de Biologie Végétale IV
  Université Pierre et Marie Curie
  12, Rue Cuvier
  F-75005 Paris/France

I.B. DRY
  Botany Department
  University of Adelaide
  Adelaide, S.A. 5001/Australia

G. DUCET
  Laboratoire de Physiologie cellulaire
  Faculté des sciences de
  Marseille-Luminy
  F-13288 Marseille Cedex 09/France

G.E. EDWARDS
  Department of Botany
  Washington State University
  Pullman, WA 99164-4230/USA

D. FALCONET
  Laboratoire de Biologie Moléculaire
  Végétale Associé au CNRS
  BAT 430 – Université Paris Sud
  F-91405 Orsay/France

P. GARDESTRÖM
  University of Umeå
  Department of Plant Physiology
  S-901 87 Umeå/Sweden

J.B. HANSON
  Department of Plant Biology
  University of Illinois
  505 So. Goodwin
  Urbana, IL 61801/USA

J.L. HARWOOD
  Department of Biochemistry
  University College
  Cardiff, CF1 1XL/United Kingdom

H. LAMBERS
  Department of Plant Ecology
  University of Utrecht
  Lange Nieuwstraat 106
  NL-3512 PN Utrecht/The Netherlands

C. LANCE
  Laboratoire de Biologie Végétale IV
  Université Pierre et Marie Curie
  12, Rue Cuvier
  F-75005 Paris/France

B. LEJEUNE
  Laboratoire de Biologie Moléculaire
  Végétale Associé au CNRS
  BAT 430 – Université Paris Sud
  F-91405 Orsay/France

C.A. MANNELLA
  Wadsworth Center for Laboratories
  and Research
  New York State Department of Health
  Albany, NY 12201/USA

A.L. MOORE
  Department of Biochemistry
  University of Sussex
  Falmer, Brighton BN1 9QG
  United Kingdom

M. Neuburger
  Laboratoire de
  Physiologie Cellulaire Végétale
  U.A. CNRS N° 576
  C.E.N.G. – 85X
  F-38041 Grenoble Cedex/France

J.M. Palmer
  Department of Pure and Applied
  Biology
  Imperial College of Science and
  Technology
  Prince Consort Road
  London SW7 2BB/United Kingdom

F. Quetier
  Laboratoire de Biologie Moléculaire
  Végétale Associé au CNRS
  BAT 430 – Université Paris Sud
  F-91405 Orsay/France

P.R. Rich
  Department of Biochemistry
  University of Cambridge
  Tennis Court Road
  Cambridge, CB2 1QW/United Kingdom

M. Steup
  Botanisches Institut der
  Westfälischen Wilhelms-Universität
  Schloßgarten 3
  D-4400 Münster/FRG

M. Stitt
  Institut für Biochemie
  der Pflanze
  Untere Karspüle 2
  D-3400 Göttingen/FRG

J.A. Ward
  Department of Pure and Applied
  Biology
  Imperial College of Science and
  Technology
  Prince Consort Road
  London, SW7 2BB/United Kingdom

J.T. Wiskich
  Botany Department
  University of Adelaide
  Adelaide, S.A. 5001/Australia

Introduction

D.A. DAY and R. DOUCE

One of the major functions of mitochondria from all organisms is to provide ATP as the principal energy source for the cell. This is true also of plant mitochondria and it is therefore no surprise that many basic features of mitochondrial membranes have been conserved between animals and plants despite a billion years of divergent evolution. Thus the morphology of plant mitochondria closely resembles that of their animal counterparts (the basic features are depicted in Fig. 1), as do their cytochrome chain, ATPase complex, energy conservation (H^+ ejection) mechanisms, and membrane phospholipid composition. The latter point is illustrated by the data in Table 1: the pattern of phospholipids in membranes from both photosynthetic and nonphotosynthetic plant mitochondria and from human heart mitochondria are virtually identical. Presumably these basic features of mitochondrial membranes are essential for their functioning in energy transduction.

Nonetheless, there are many unique features of plant mitochondria which, one assumes, reflect their functioning in autotrophic metabolism, and it is on these aspects that most of the articles presented in this volume focus attention. Since respiration is an important parameter in the growth and yield of plants (see LAMBERS, Chap. 14, this Vol.), it is equally important that we understand the unique properties of plant mitochondria (and respiratory metabolism in general) and the way they are integrated into the general metabolism of the plant cell. As editors, we hope that this volume will contribute to that understanding.

The unique features of plant mitochondria include the following:
1. Cyanide- and antimycin A-insensitive respiration (Chaps. 8 and 14).
2. Respiratory-linked oxidation of external NADH and rotenone-insensitive oxidation of internal NADH (Chap. 7).
3. Matrix-located malic enzyme (Chaps. 10 and 11).
4. Oxaloacetate and amino acid transport (Chap. 9).
5. Carrier-mediated transport of nucleotides and other cofactors (Chaps. 1 and 9).
6. Rapid glycine oxidation by leaf mitochondria (Chap. 11).
7. The size and complexity of their DNA (Chap. 2).

In addition, the rate of O_2 consumption on a protein basis is much higher than that in animal mitochondria (see IKUMA 1972), while fatty acid oxidation is either very low (THOMAS and WOOD 1982) or not detectable (GERHARDT 1983, MACEY 1983) in plant mitochondria (the bulk of fatty acid oxidation in the plant cell being confined to microbodies).

It is now generally agreed that the mitochondrial genome of higher plants is the largest one identified to date. Most animal mtDNA have only some

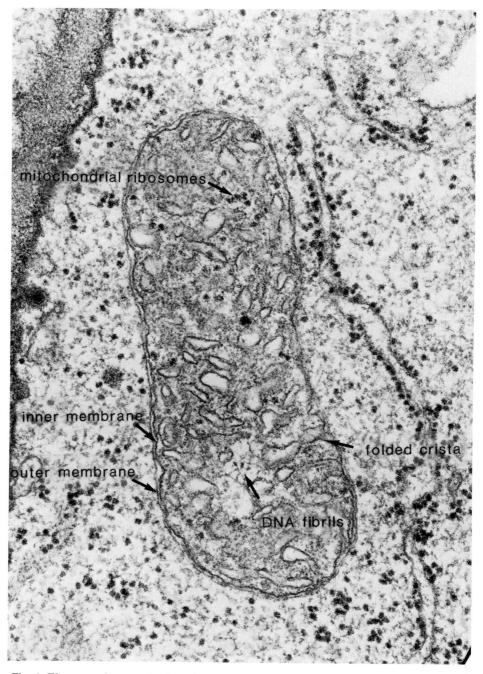

Fig. 1. Electron micrograph of a mitochondrion within a leaf cell of *Spinacia oleracea* L. Between the inner and outer membranes is the intermembrane space. The inner membrane bounds a soluble matrix in which are localized the Krebs cycle enzymes. (Photograph courtesy of J.P. Carde, Bordeaux, France)

Table 1. Phospholipid composition of mitochondrial membranes

Phospholipid	Percent Composition (w/w)		
	Pea leaf[a]	Sycamore cell[b]	Human heart[c]
Phosphatidylcholine (PC)	40	43	43
Phosphatidylethanolamine (PE)	37	35	34
Diphosphatidylglycerol (DPG)	15	13	18
Phosphatidylinositol (PI)	6	6	5
Phosphatidylglycerol (PG)	2	3	n.d.

[a] DAY et al. (1984),   [b] BLIGNY and DOUCE (1980)   [c] ROUSER et al. (1968)

15,000 nucleotides, but in yeast the mtDNA is five times as long, and in plants it is as much as five times longer still. There are at least two explanations for the enormous sequence complexity of plant mitochondrial DNA; (a) there are many more genes and/or regulatory sequences in plant mtDNA than in animal mtDNA; (b) there is far more noncoding DNA in the mitochondria of plants than in those of animals. In addition, the possibility remains that real differences in the organization of the mitochondrial genome may exist in different plants. This raises the important question of how such heterogeneity is transmitted from generation to generation with regard to biological aspects of plant mtDNA replication. The recent discovery by STERN and LONSDALE (1982) that chloroplast and mitochondrial DNA share a common sequence provides a new way of thinking about the maintenance of these genomes and the possible communication between plastids, mitochondrial and nuclear genomes during the coordinated changes in cellular function associated with the development and differentiation of the plant cell. These questions serve as a reminder that some of the most fascinating problems of plant mitochondrial biogenesis are only now appearing.

With the exception of the DNA, most of the above-mentioned differences between animal and plant mitochondria have been known for many years (see IKUMA 1972) but were often attributed to artifacts generated by damage or contamination of the mitochondria during their isolation from plant tissues (PACKER et al. 1970, CHANCE et al. 1968), and until the mid-1970's plant mitochondria were usually considered to be essentially similar to animal mitochondria. This early pre-occupation with the criteria of "normal" mitochondrial functioning established with animal tissues has led to a considerable body of work on the isolation, purification, and integrity of plant mitochondria, and it is now possible to obtain preparations of the latter organelles which display better than 95% intactness and which are virtually completely free of enzymes and membranes originating from other cell compartments (see NEUBURGER, Chap. 1, this Vol.). Hence, after many years of rigorous testing, we can now be sure that the observed differences in plant mitochondria listed above are, for the large part, representative of real (and often unique) properties of fully functional organelles. For example, CHANCE et al. (1968) in an early review stated that "if the rate of malate oxidation is accelerated by the addition of

NAD$^+$, the mitochondria have been damaged during isolation and have lost endogenous pyridine nucleotide". However, we now know that the active uptake of NAD$^+$ via a specific transporter is a constitutive property of intact mitochondria (NEUBURGER and DOUCE 1983, TOBIN et al. 1981), and recent evidence shows that both influx and efflux of NAD$^+$ from mitochondria can occur via this carrier and that variations in the endogenous NAD$^+$ content of potato and cauliflower mitochondria occur in vivo (NEUBURGER et al. 1985).

In vitro, plant mitochondria prove to be biochemically very flexible organelles. One example of this is glycine metabolism in leaf mitochondria, where the reoxidation of NADH produced in the matrix upon operation of the glycine decarboxylase complex can be reoxidized equally well by either the respiratory chain (in which case glycine decarboxylation will be linked to the energy status of the mitochondria) or substrate shuttles (in which case the NADH is transferred out of the mitochondria and glycine decarboxylation will not be directly linked to the energy status of the mitochondria). One is faced with another choice at the level of the respiratory chain, since electron transport to O$_2$ can be either phosphorylating (via cyt oxydase) or nonphosphorylating (via the alternative oxidase). Which of these mechanisms operates in vivo is of vital importance to our understanding of the photorespiratory cycle (GARDESTROM and EDWARDS, Chap. 11, this Vol.). Another example of the flexibility and complexity of plant mitochondrial metabolism can he found in the oxidation of malate. This occurs via the action of two enzymes, malate dehydrogenase and NAD-linked malic enzyme, both of which are localized in the matrix and which interact via oxaloacetate and NADH, resulting in very complex patterns of respiration in vitro (see WISKICH and DRY, Chap. 10, this Vol.). This complexity is compounded by the fact that malic enzyme, in particular, can be regulated by a wide range of extramitochondrial conditions and metabolites in vitro. It is only very recently that we have begun to understand the details of malate oxidation by isolated mitochondria, and the situation in vivo remains a puzzle. In addition to possessing carriers similar to those found in mitochondria from other sources for the uptake of malate and pyruvate, and an additional malate-oxidizing enzyme (malic enzyme), plant mitochondria also possess a specific carrier with a high affinity for oxaloacetate (DAY and WISKICH 1981, CHEN and HELDT 1984) and the plant cytoplasm contains substantial activity of phosphoenolpyruvate carboxylase (ap REES, Chap. 13, this Vol.). Thus, in addition to the conventional operation of the TCA cycle, utilizing pyruvate from glycolysis as the sole carbon source, one can imagine situations arising in vivo where malate (produced by the combined operation of phosphoenolpyruvate carboxylase and malate dehydrogenase in the cytosol) is the end product of glycolysis and provides the TCA cycle with both pyruvate and oxaloacetate after transport into the matrix. Alternatively, carbon input to the TCA cycle could occur in the form of cytosolic oxaloacetate, with pyruvate being provided either by the action of pyruvate kinase in the cytosol or by operation of malic enzyme in the matrix utilizing malate generated in the cycle. The latter mechanism may be important in C$_3$ plants under conditions where adenylate control of glycolysis and the cyt chain is imposed but the alternative path operates, since operation of malate dehydrogenase in the matrix will probably be impaired under these conditions by the mitochondrial NADH/NAD ratio, thus disrupting conventional operation of the TCA cycle (PALMER and WARD, Chap. 7. this Vol.). The former two alternative mechanisms may be important when TCA cycle intermediates such as α-ketoglutarate leave the mitochondrion for use elsewhere in the cell. To date, no experiments have been devised to distinguish between these possibilities. Ironically, the very flexibility of isolated plant mitochondria makes the task of placing them in a physiological context even more difficult, and this task remains one of the main challenges to researchers in plant respiration.

Despite the fact that plant respiration is a very active field of research, many questions remain unanswered and some controversies are not yet resolved. Since it was impossible to completely avoid overlap between authors when compiling this volume, readers will find that certain topics are dealt with in more than one chapter and often very different opinions are expressed. This is particularly true of malate oxidation and its involvement with nonphosphorylating electron transport pathways, which is discussed in Chaps. 7, 8, 10, and 11. As editors, we can only suggest that readers seeking information on this topic study each discussion to obtain a balanced overview. It is our hope that the presentation of different points of view will stimulate further critical research, which will help to resolve the controversies.

Finally, we wish to express our thanks to our friends and fellow scientists who so willingly contributed to this volume and to FRANÇOISE BUCHARLES for her invaluable secretarial assistance.

References

Bligny R, Douce R (1980) A precise localization of cardiolipin in plant cells. Biochim Biophys Acta 617:254–263

Chance B, Bonner WD, Storey BT (1968) Electron transport in respiration. Annu Rev Plant Physiol 19:295–320

Chen J, Heldt HW (1984) Oxaloacetate translocator in plant mitochondria. Proc 6th Int Congr Photosynthesis (in press)

Day DA, Wiskich JT (1981) Effect of phthalonic acid on respiration and metabolite transport in higher plant mitochondria. Arch Biochem Biophys 211:100–107

Day DA, Neuburger M, Douce R (1985), unpublished data

Gerhardt B (1983) Localization of β-oxidation enzymes in peroxisomes isolated from non-fatty plant tissues. Planta 265:1–9

Ikuma H (1972) Electron transport in plant respiration. Annu Rev Plant Physiol 23:419–436

Macey MJK (1983) β-oxidation and associated enzyme activities in microbodies from germinating peas. Plant Sci Lett 30:53–60

Neuburger M, Douce R (1983) Slow passive diffusion of NAD^+ between intact isolated plant mitochondria and suspending medium. Biochem J 216:443–450

Neuburger M, Day DA, Douce R (1984) Transport of NAD^+ in Percoll purified potato tuber mitochondria: Inhibition of NAD^+ influx and efflux by N-4-azido-2-nitro-phenyl-4-aminobutyryl-3'-NAD^+. Plant Physiol (in press)

Packer L, Murakami S, Mehard CN (1970) Ion transport in chloroplasts and plant mitochondria. Annu Rev Plant Physiol 21:271–304

Rouser G, Nelson GJ, Fleisher S, Simon G (1968) In: Chapman D (ed) Biological mem-
    branes. Academic Press, London New York, pp 5–69
Stern DB, Lonsdale DM (1982) Mitochondrial and chloroplast genomes of maize have
    a 12-kilobase DNA sequence in common. Nature (London) 299:698–702
Thomas DR, Wood C (1982) Oxidation of acetate, acetyl CoA and acetylcarnitine by
    pea mitochondria. Planta 154:145–149
Tobin A, Djerdjour B, Journet E, Neuburger M, Douce R (1980) Effect of NAD$^+$
    on malate oxidation in intact plant mitochondria. Plant Physiol 66:225–229

1 Preparation of Plant Mitochondria, Criteria for Assessement of Mitochondrial Integrity and Purity, Survival in Vitro

M. Neuburger

1 Introduction

Plant mitochondria are small organelles of the cytoplasmic compartment of the cell, involved in respiration and providing energy (ATP) and metabolic precursors for the rest of the cell. The mitochondria consist of a folded inner membrane involved in different processes (electron transport, metabolite transports via selective carriers, ATP synthesis, etc.) surrounding a matrix space containing all the tricarboxylic acid (TCA) cycle enzymes as well as a genetic system (DNA, RNA and ribosomes) for protein synthesis. The inner membrane is protected from the cytosol by an outer limiting membrane, the role of which remains uncertain. However the presence of b-type cytochrome (Douce et al. 1973), porine (Zalman et al. 1980) and hexokinase (Tanner et al. 1983) demonstrates that the external membrane must perform several specific functions as well as a mechanical role. In addition, little is known about the presence of soluble enzymes in the intermembrane space and their possible functions.

Isolation of fully competent mitochondria presenting the same morphological structure as shown in vivo and displaying high rates of TCA cycle substrates oxidations, high respiratory control ratios and optimum P/O values requires methods which not only avoid rupture of both membranes, but also protect isolated mitochondria from harmful products from other cell compartments capable of damaging membranes and leading to a functional impairment of the mitochondria. Furthermore, purity of the isolated plant mitochondria should be a prerequisite for all studies on mitochondrial composition, metabolism, transport processes and biogenesis.

2 General Considerations for the Isolation of Intact Mitochondria

A common feature of all plant cells is the presence of a rigid cell wall which must be disrupted to liberate cytoplasmic organelles into the grinding medium. The mechanical resistance of the cell wall generally necessitates the use of shearing forces which lead inevitably to the rupture of the ubiquitous vacuole of the cell and are known to contain harmful products (Matile 1976) such as hydrolytic enzymes, phenolic compounds, tannins, alkaloids, and terpenes. These products released in the grinding medium inevitably interact with mitochondrial membranes, and care must be taken to avoid or minimize such interactions by protective agents added to the grinding medium and able to strongly chelate inhibiting substances or block the functioning of hydrolytic enzymes. In addition, it must be kept in mind that strong and prolonged grinding can also mechanically injure mitochondria, resulting in a mixed population of intact and damaged mitochondria (envelope-free mitochondria and mitochondria which have resealed following rupture and loss of matrix content) in the final preparation.

Phenolic compounds, as well as their oxidation products (quinones), are highly reactive and interact strongly with membrane proteins in different ways (Loomis 1974). Such interactions with mitochondrial membranes lead to a strong inhibition of electron transport through the respiratory chain (Ravanel et al. 1982) or uncoupling of oxidative phosphorylation (Ravanel et al. 1981). For this purpose, defatted bovine serum albumin (BSA) is routinely added to the grinding medium as it binds not only free fatty acids but also quinones (Weinbach and Garbus 1966). Furthermore, additional reductants in the medium are useful to prevent quinone interactions (Anderson and Rowan 1966). For example, cysteine (free base), β-mercaptoethanol or isoascorbate are convenient protective agents used, since the initial work from Wiskich and Bonner (1963). In addition, the nontoxic high molecular weight polyvinylpyrrolidone (PVP) has been shown to improve the quality of the mitochondria extracted from tissues with high phenolic content (Hulme et al. 1964) as it acts as a strong scavenger of phenols and tannins (Loomis and Battaile 1966). Addition of PVP to the grinding medium is also beneficial for obtaining well-coupled mitochondria from green leaves (Douce et al. 1977, Neuburger and Douce 1977, Arron and Edwards 1979, Day and Wiskich 1980). It is also recommended to keep the pH of the extracting medium within the range of 7.2–7.5 by including a strong buffer in the medium because alkaline pH will increase phenol autooxidation, while acidic pH will increase interactions between phenolics and protein functional groups (Loomis 1974). In addition, buffers counterbalance the acidity of the vacuolar sap. The acidity of the sap is mainly due to organic acids (such as TCA-cycle intermediates) or inorganic acids (such as phosphate). These compounds released in the medium, following rupture of vacuoles occurring during the grinding step, would lead to a dramatic shift of the pH toward acidic values in the absence of a strong additional buffer in the grinding medium. Buffers such as 3-(N-morpholino)propanesulfonic acid (MOPS), N-(2-hydroxy-ethyl) piperazin-N'-2-ethanesulfonic acid (HEPES) or sodium pyrophosphate are highly convenient for this purpose.

Another troublesome problem arises from various lipolytic acylhydrolases present in all plant tissues (Galliard 1980) and released in the medium during the grinding step. Such hydrolases, probably of vacuolar origin (Matile 1976, Nishimura and Beevers 1978), may produce rapid hydrolysis of membrane phospholipids and dramatically impair the mitochondrial functioning (Douce and Lance 1972, Bligny and Douce 1978). Free fatty acids thus formed, mainly linoleic and linolenic acids, are the pre-eminent substrates for lipoxygenases (Galliard and Chan 1980). These enzymes rapidly catalyze oxidation of polyunsaturated fatty acids to fatty acid hydroperoxides in the presence of molecular oxygen. Hydroperoxides thus formed may react with membrane proteins such as cytochromes (HAUROWITZ et al. 1941, DUPONT et al. 1982). The chelation of free fatty acids with defatted serum albumin (hydrophobic interactions) is the most efficient to prevent their deleterious effects. Furthermore, addition of chelating agents such as EDTA or EGTA which chelate Ca^{2+}, a cofactor of some lipolytic acylenzymes (BLIGNY and DOUCE 1978, GALLIARD 1980), is recommended.

The use of a very cold grinding medium (2–4 °C) also minimizes enzymatic activity. Finally, keeping the ratio of grinding medium to tissues very high will help to decrease interactions during the grinding step.

Tissue homogenization is a crucial step in the isolation of fully intact mitochondria. As strong mechanical forces are generally needed for physical disruption of the rigid cell wall, this step is routinely carried out with electrical blenders such as a Moulinex mixer, Polytron, or Waring blender for large-scale preparations of mitochondria. Whatever the mechanical device used for the grinding, this step has to be reduced to a minimum. First, strong vortex effects produced during the blending can injure mitochondria. Second, prolonged grinding promotes phenol oxidation and increases contact between mitochondria and the harmful products outlined above. Since short grinding times lead to a low yield of mitochondria extracted, large quantities of tissues are needed to isolate sufficient quantities of highly intact mitochondria.

3 Large-Scale Preparation of Washed Mitochondria

"Washed" mitochondria (a crude pellet containing mitochondria) can be routinely prepared from nongreen (tubers, hypocotyls, etc.) or green leaf tissues (spinach, pea leaves) by the general method of BONNER (1967) modified by DOUCE et al. (1972).

3.1 Reagents

Grinding Medium. 0.3 M mannitol (or sorbitol); 20–50 mM MOPS (or HEPES or tetrasodium pyrophosphate); 1 mM EDTA (or EGTA); 0.1–0.2% (w/v) cysteine, free base, (or 2 mM β-mercaptoethanol or 10 mM isoascorbate); 0.1% (w/v) BSA. Adjust pH to 7.5 with NaOH or with HCl (when pyrophosphate is used). Note that for green tissues, addition of polyvinylpyrrolidone (PVP) as a soluble form [0.5% (w/v) PVP 350 or PVP 40] or insoluble form [0.8% (w/v) polyclar AT] is strongly recommended.

Wash Medium. 0.3 M mannitol; 10 mM potassium phosphate; 1 mM EDTA; 0.1% defatted BSA. Adjust pH to 7.2.

It is recommended to keep both mediums at 0°–4 °C.

3.2 Procedure for Potato Tuber Mitochondria

Tubers (3–4 kg) previously kept overnight in the cold (5 °C) are cut in to small pieces. The material is homogenized in three successsive batches in a 1-gallon Waring blender at low speed for 2 s (about 1.5 kg tubers for 3 l of grinding medium). It is desirable to have the extraction medium as a semifrozen slush. The brei is squeeze-filtered through six layers of muslin and a 50-µm nylon net. The filtered solution is then centrifuged at 1,500 g for 15 min (Rotor GS3, RC5 Sorvall centrifuge) to remove starch and cell debris. The supernatant is centrifuged at 10,000 g for 20 min.

The pellets obtained are suspended in a small volume of wash medium by using a homogenizer of the Potter-Elvehjem type with a loose fitting pestle. The mitochondrial fraction is centrifuged at 1,500 g for 5 min (SS34 Rotor, RC5 Sorvall centrifuge). The small pellet, which consists of aggregated membranes and starch grains, is discarded and the supernatant centrifuged at 10,000 g for 15 min. The final pellet thus obtained is resuspended in a small volume of wash medium (30–40 mg protein ml^{-1}).

The technique described for the isolation of potato mitochondria is also suitable for extracting mitochondria from other storage tissues, etiolated tissues, and green leaves, insofar as the medium contains PVP (DOUCE et al. 1977, NEUBURGER and DOUCE 1977, JACKSON et al. 1979, ARRON et al. 1979, DAY, NEUBURGER and DOUCE, unpublished data). However, some modifications have been brought to this basic procedure to isolate mitochondria more rapidly from the homogenate (PALMER and KIRK 1974).

All the preparations thus obtained and termed "washed mitochondria" inevitably contain some damaged organelles due to the mechanical grinding process and harmful compounds released from other cell compartments. It is therefore essential to check the degree of intactness of isolated mitochondria before the following purification step because this latter step slightly improves the quality of mitochondria but, of course, cannot restore damaged mitochondria. This may also prevent misleading interpretations with regard to enzyme localizations, transport processes, and the real oxidative capacities of plant mitochondria.

4 Assessment of Mitochondrial Integrity

Simple and sensitive tests for mitochondrial integrity have been described by DOUCE et al. (1972) and NEUBURGER et al. (1982).

4.1 Spectrophotometric Assay for Succinate: Cytochrome c Oxidoreductase (DOUCE et al. 1972)

Reaction medium: 0.3 M mannitol; 10 mM phosphate buffer, pH 7.2; 5 mM MgCl$_2$: 0.5 mM ATP; 50 µM cytochrome c; 1 mM KCN and 0.1–0.5 mg mitochondrial protein.

Electron transport through the respiratory chain is initiated with 10 mM succinate in the presence of ATP to activate the succinate dehydrogenase. The rate of cytochrome c reduction is followed at 550 nm ($\varepsilon_{550} = 21 \cdot 10^3$ mol^{-1} cm^{-1}). When the mitochondria are fully intact, no reduction of cytochrome c occurs because the outer mitochondrial membrane is impermeable to cytochrome c, thus preventing it from interacting with endogenous cytochrome c on the inner membrane (WOJTCZAK and ZALUSKA 1969, DOUCE et al. 1972). When the experiment is repeated with mitochondria in the reaction medium deprived of mannitol (which leads to osmotic swelling and rupture of the outer membrane), the rate of cytochrome c reduction increases considerably as cytochrome c has access to the inner membrane (this rate correspond to 100% damaged mitochondria). Comparison between the rate of cytochrome c reduction in normal medium and the rate observed after bursting the mitochondria gives a rapid estimation of the percentage of mitochondria with intact outer membranes.

4.2 KCN-Sensitive Ascorbate-Cytochrome c-Dependent O_2 Uptake
(NEUBURGER et al. 1982)

This assay is based on the measurement of KCN-sensitive ascorbate-cytochrome c-dependent O_2 uptake, with an oxygen electrode (Fig. 1). Reaction medium: 0.3 M mannitol; 10 mM phosphate buffer, pH 7.2; 5 mM MgCl$_2$; 8 mM ascorbate and 0.2–0.4 mg mitochondrial protein. Oxygen consumption is initiated with 30 μM cytochrome c. After a linear rate of O_2 uptake has been recorded, the reaction is blocked by addition of 200 μM KCN. The small residual O_2 uptake corresponds to a nonenzymatic oxidation of reduced cytochrome c by molecular oxygen, while the KCN-sensitive O_2 uptake is due to the cytochrome oxidase functioning. If the mitochondria are fully intact, the rate of O_2 uptake is entirely KCN-insensitive because reduced cytochrome c has no access to the inner membrane. When the experiment is repeated with burst mitochondria, the rate of KCN-sensitive oxygen uptake increases (corresponding to 100% damaged mitochondria). Rupture of the mitochondria is again achieved by osmotic swelling. A convenient way of achieving this is to incubate the same

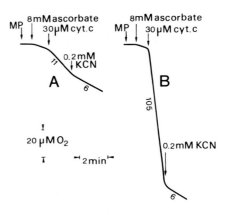

Fig. 1A, B. Ascorbate: cytochrome c dependent O_2 uptake of Percoll-purified mitochondria. A intact mitochondria; B burst mitochondria (see text). The basic medium (1 ml) contains 0.2 mg protein, 8 mM ascorbate and 30 μM cytochrome c. Cyanide (0.2 mM) is added as indicated. Numbers on the traces refer to nmol O_2 consumed min^{-1}. MP purified mitochondria

quantity of mitochondria in distilled water (in the O_2 electrode vessel) for 10 s and then adding double-strength medium to restore isotonic conditions following the osmotic shock. Therefore it should be possible to estimate very rapidly the proportion of intact mitochondria in a given preparation from the relative rate of KCN sensitive cyt c-dependent O_2 uptake by untreated and by osmotically shocked mitochondria in the same manner as for succinate-dependent cyt c reduction (see above).

Measurements of respiratory control and ADP/O ratios can also be used to give an indication of the quality of the mitochondria in terms of the ability of the inner membrane to maintain ion gradients (Hanson and Day 1980). However, these parameters are extremely variable, depending on the source of mitochondria and the choice of substrate. For example, mitochondria which have an active alternative pathway (see Lance et al., Chap. 8, this Vol.) may display low respiratory control and ADP/O ratios but still be highly coupled, while use of malate alone as substrate may lead to very high apparent respiratory control because of accumulation of oxaloacetate (Palmer 1976, Lance et al. 1967, Day et al. 1984).

5 Control of Mitochondrial Purity

Although crude mitochondrial preparations from different etiolated tissues have been largely used to study the components of the respiratory chain (e.g., cytochromes) or respiratory metabolism, it is obvious that preparations obtained by simple differential centrifugation are generally more or less contaminated by other intact or damaged organelles and aggregated membranes deriving from ruptured cell compartments. The presence of contaminants in a mitochondrial preparation is of importance for scientists involved not only in mitochondrial metabolism, but also in other fields such as protein, nucleic acid, and lipid biosynthesis.

Depending on the origin of the tissues used, microbodies (peroxisomes, glyoxysomes) or proplastids frequently contaminate crude mitochondrial preparations. For example, mitochondria from storage tissues (e.g., potato tubers) are mostly contaminated by peroxisomes and amyloplast membranes, but mitochondria from etiolated tissues (e.g., cauliflower buds) by plastids. In addition, membrane vesicles of different origin are frequently observed by electron microscopy in crude mitochondrial preparations. Assays for enzymatic activities known to be clearly associated with the different cell compartments allow a rapid estimation of contaminants present in the preparations. Quail (1979) has reviewed most of the enzymatic markers which can be used with confidence for this purpose. For instance, catalase and glycolate oxidase are representative of a peroxisomal contamination (Tolbert 1982). Glyoxysomes, which are essentially present in tissues with large fatty acid reserves, such as seedlings, contained not only catalase but also enzymes from the glyoxylate cycle, in particular isocitrate lyase or malate synthetase, which are specific markers of that type of organelles (Beevers 1982).

The use of enzymes from the plastidial glycolytic sequence, such as triose-P-isomerase or from the pentose phosphate pathway (latent 6-phosphogluconate dehydrogenase or glucose 6-P-dehydrogenase), provides highly sensitive assays for plastidial contamination. However, as isoenzymes of both pathways are also present in the cytosol, further analyses related to galactolipid synthesis (DOUCE 1974) or direct analysis of other specific constituents of plastid membranes such as carotenoids and galactolipids (DOUCE and JOYARD 1979) are accurate indicators of plastid contaminants. In addition, "microsomes", a general term which defines a heterogenous membranous fraction which could be sedimented at high speed (FLEISCHER and KERVINA 1974), are characterized by an antimycin A- insensitive NADPH cytochrome c reductase activity.

Finally, attention must be drawn to some harmful enzymes deriving from the vacuole which often contaminate washed mitochondria; this is the case for lipoxygenase and lipolytic acyl hydrolases which can be easily detected, as described respectively by SIEDOW and GIRVIN (1980) and BLIGNY and DOUCE (1978).

Consequently, a careful purification of washed mitochondria is imperative to avoid misinterpretation on composition and physiological capacity of plant mitochondria.

6 Purification of Plant Mitochondria

Purification is usually achieved with density gradient centrifugation but other methods have also been employed, particularly for mitochondria from green leaves, and the reader is referred to GARDESTRÖM and EDWARDS (Chap. 11, this Vol.).

6.1 Purification on Sucrose Gradients

Purification on a step density gradient of sucrose is the main technique so far used to remove most of the contaminants from washed mitochondria obtained from different etiolated tissues. The following method has been introduced by BAKER et al. (1968) to purify avocado fruit mitochondria, and modified by DOUCE et al. (1972) for large-scale preparation of pure mitochondria from etiolated mung bean hypocotyls, white potato, and Jerusalem artichoke tubers.

Methods. Washed mitochondria (30–50 mg protein) suspended in 2 ml of wash medium (see above) are layered on top of a discontinuous gradient of sucrose consisting of four layers of 1.8 M, 1,45 M, 1.2 M and 0.6 M sucrose, each containing 10 mM phosphate buffer, pH 7.2, and 0.1% defatted bovine serum albumin. The sample is then centrifuged at 40,000 g for 60 min in a swinging bucket rotor (SW27, BECKMAN). Mitochondria are recovered as a sharp band at the interface of the 1.2 M and 1.45 M sucrose layers.

The mitochondrial fraction is collected and diluted very carefully to isoosmotic conditions with 10 mM phosphate buffer, pH 7.2, containing 0.1% defatted bovine serum albumin. The dilution step is critical and must be done very slowly. The mitochondria are recovered from the gradient in a medium of high osmotic strength (roughly 1.4 M sucrose) in a contracted form probably due to a slow sucrose diffusion inside the matrix space occurring during centrifugation (Sitaramam and Sambasiva 1984). Consequently, a too rapid dilution leads to a fast expansion of the inner membrane and damages the mitochondria (Douce et al. 1972).

Although this technique for the preparation of mitochondria is reasonably successful with etiolated tissues, it fails to purify mitochondria from green leaves (Jackson et al. 1979). However, using a continuous sucrose gradient Nash and Wiskich (1983) have succeeded in purifying pea leaf mitochondria which are substantially devoid of chlorophyll and peroxisomes.

6.2 Purification on Percoll Gradients

A new technology for mitochondrial purification under isotonic conditions emerged recently with the introduction of a nontoxic silica sol Percoll (Pharmacia Fine Chemicals). Unlike sucrose, the very low osmolality of Percoll allows the formation of density gradients which are isoosmotic throughout, provided that a suitable osmoticum (sucrose, mannitol, sorbitol or raffinose) is included in the gradient mixture. Unfortunately, there is no general standard method of purification on Percoll gradient due to the diversity of contaminants present in washed mitochondria from different tissues. In addition variations of the size and the protein/lipid ratio may slightly influence the apparent density of plant mitochondria in a Percoll medium. Thus the various parameters which influence the shape of the Percoll gradient (Percoll concentration, rotor geometry, diameter of the tubes, centrifugal force and time) must be carefully determined for each type of mitochondrial sample which has to be purified. We have also observed that a careful choice of the osmoticum (mannitol, sucrose, or raffinose at room temperature) which modifies slightly the density and the viscosity of the medium can improve the separation of mitochondria from contaminating membranes and particles (Neuburger et al. 1982, Neuburger and Douce 1983). For details of the characteristics and behavior of Percoll, the reader is directed to the Pharmacia booklet *Percoll Methodology and Applications*.

Consequently, three strategies have been used for mitochondrial purification in Percoll solution:

Step Density Gradient of Percoll. A nonlinear gradient is made with layers of Percoll solutions of different concentrations. The mitochondria are layered on top of the gradient and centrifuged. Centrifugation in a swinging bucket rotor will preserve the step gradient, while centrifugation in a fixed angle rotor will modify the gradient. This method has been successfully used to purify spinach leaf mitochondria (Jackson et al. 1979, Nishimura et al. 1982). Starting from the same tissue, mitochondria have been obtained using combined methods involving partition in aqueous polymer two-phase system (dextran-polyethyleneglycol) and a final purification on a step Percoll gradient (BERGMAN et al. 1980, see also GARDESTROM and EDWARDS, Chap. 11, this Vol.).

Linear Percoll Gradient. The purification step consists in layering washed mitochondria on top of a pre-made linear Percoll gradient followed by centrifugation in a swinging-bucket rotor. Pre-made gradients have been used by GOLDSTEIN et al. (1981) to purify wheat seed mitochondria free from lipoxygenase activity and by METTLER and BEEVERS (1980) to separate glyoxysomes and mitochondria from castor bean endosperm.

Self-Generated Gradient. This method takes advantage of the ability of Percoll to self-generate a density gradient during the course of the centrifuging step. It has been successfully used to purify washed mitochondria from *Neurospora crassa* (SCHWITZGUEBEL et al. 1981), potato tubers (NEUBURGER et al. 1982), avocado fruits (MOREAU and ROMANI 1982) and more recently from pea leaf (DAY, NEUBURGER and DOUCE, unpublished data). Details of this technique are given below for the purification of potato mitochondria and pea leaf mitochondria.

6.2.1 Purification of Mitochondria from Potato Tubers

Reagents

– Percoll (Pharmacia Fine Chemicals)
– Sucrose solution: 0.6 M sucrose containing 20 mM phosphate buffer, pH 7.2; 2 mM EDTA and 0.2% (w/v) bovine serum albumin.
– Mannitol solution: 0.6 M mannitol containing 20 mM phosphate buffer, pH 7.2; 2 mM EDTA and 0.2% (w/v) bovine serum albumin.
– Wash medium: 0.3 M mannitol containing 10 mM phosphate buffer, pH 7.2; 1 mM EDTA and 0.1% (w/v) bovine serum albumin.
– Medium A: mix 28 ml of stock Percoll with 50 ml of sucrose solution and adjust to 100 ml with distilled water; final pH 7.2.
– Medium B: mix 28 ml of stock Percoll with 50 ml of mannitol solution and adjust to 100 ml with distilled water; final pH, 7.2.

Procedure

Two centrifuge tubes (40 ml capacity) are filled with 36 ml of medium A. Aliquots (3-ml sample, 50–60 mg protein) of the mitochondrial suspension are then layered on the Percoll medium. The Percoll gradient is self-generated by centrifuging at 40,000 g for 30 min using a fixed angle rotor (Sorval SS34 rotor). Mitochondria are recovered beneath a yellow layer at the top of the gradient (Fig. 2), while peroxisomes are recovered as a sharp band sitting above a clear colorless layer of Percoll at the bottom of the tube. The mitochondrial fraction is collected with a pipette, diluted tenfold with wash medium and centrifuged for 15 min at 10,000 g (Sorvall, SS34 rotor). The pellet is taken up in approximately 2 ml of wash medium, layered on top of 36 ml medium B in a 40-ml capacity tube and centrifuged at 40,000 g for 30 min (Sorvall SS34 rotor). Mito-

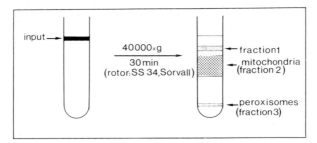

Fig. 2. Purification of potato tuber mitochondria. Schematic representation of the sucrose Percoll gradient fractionation of the washed mitochondria (*input*) in order to obtain intact mitochondria (*fraction 2*) and microbodies (*fraction 3*). The rotor used is an angle head rotor (SS34, Sorvall). For experimental details see text

Table 1. α-Ketoglutarate oxidation and measurements of various extramitochondrial markers in washed and purified mitochondria isolated from potato tubers. Washed mitochondria were purified by centrifugation (40,000 g, 30 min; rotor SS34 Sorvall) in a Percoll gradient (28% (v/v) Percoll; 0.3 M sucrose; 10 mM phosphate buffer, pH 7.2; 1 mM EDTA; and 0.1% (w/v) bovine serum albumin)

	Washed mitochondria	Purified mitochondria
State 3 rate of α-ketoglutarate oxidation (nmol of O_2 min^{-1} mg^{-1} protein)	153	216
Catalase (μmol of O_2 min^{-1} mg^{-1} protein)	12	0.2
Lipoxygenase (μmol of O_2 min^{-1} mg^{-1} protein)	3.5	ND
Carotenoids (μg mg^{-1} protein)	0.1	0.008

Note: if the purification is repeated once [i.e., after pelleting mitochondria and layering the suspending pellet on top of a new Percoll gradient: 28% (v/v) Percoll/0.3 M mannitol/ 10 mM-H_2KPO_4 (pH 7.2) 0.1% (w/v) bovine serum albumin/1 mM EDTA], mitochondrial catalase activity can be reduced to less than 0.02 μmol of O_2 min^{-1} mg^{-1} protein. ND, not detected.

chondria are recovered in a sharp band close to the bottom of the tube. The mitochondrial fraction is removed, diluted tenfold with wash medium and centrifuged at 10,000 g for 10 min. The final mitochondrial pellet is resuspended in a small volume of wash medium (60–80 mg protein ml^{-1}). These mitochondria are generally free from contamination by other membrane particles (Table 1).

6.2.2 Purification of Mitochondria from Pea Leaves

Reagents

– Percoll (Pharmacia Fine Chemical)
– Medium A: 0.3 M sucrose, 28% (v/v) Percoll; 10 mM phosphate buffer, pH 7.2; 1 mM EDTA and 0.1% (w/v) bovine serum albumin.

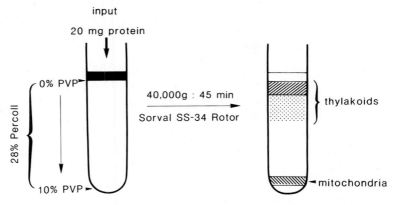

Fig. 3. Purification of pea leaf mitochondria. Schematic representation of the separation of thylakoids and mitochondria after centrifugation of crude mitochondria on a 0–10% (w/v) PVP-25 gradient in 28% Percoll in a fixed angle rotor (Sorvall SS34), for 45 min at 40,000 g (for details see text)

Table 2. Purification of pea leaf mitochondria. Crude mitochondria were purified by differential centrifugation and subsequently purified on a Percoll/PVP gradient as described in the text. Succinate oxidase was measured as O_2 consumed in the presence of 10 mM succinate and 1 mM ADP and glycolate oxidase as O_2 consumed in the presence of 10 mM glycolate and 0.3 mM KCN. A and B refer to separate experiments

		Chlorophyll	Glycolate oxidase	Succinate oxidase
		$\mu g\ mg^{-1}$ protein	$nmol\ O_2\ min^{-1}$	mg^{-1} protein
Washed mitochondria	A	47	142	95
	B	39	200	140
Purified mitochondria	A	0.3	35	176
	B	0.4	95	197

Note: if the purification procedure is repeated once (see text) mitochondrial glycolate oxidase is reduced to less than 5 nmol O_2 min^{-1} mg^{-1} protein, chlorophyll and galactolipids are barely detectable.

— Medium B: 0.3 M sucrose, 28% (v/v) Percoll, 10% (w/v) soluble polyvinyl pyrrolidone (PVP-25, Serva); 10 mM phosphate buffer, pH 7.2; 1 mM EDTA and 0.1% (w/v) bovine serum albumin.

Procedure

Using a gradient maker, 18 ml of medium A are mixed with 18 ml of medium B in a Sorvall SS34 tube (40 ml capacity), leading to a linear gradient of 0 to 10% (w/v) PVP-25 (top to the bottom). Washed mitochondria obtained by the method of DOUCE et al. (1977) and free from intact chloroplasts, but containing thylakoid membrane fragments are suspended in approximately 2 ml of wash medium, layered on top of the Percoll/PVP solution and centrifuged for

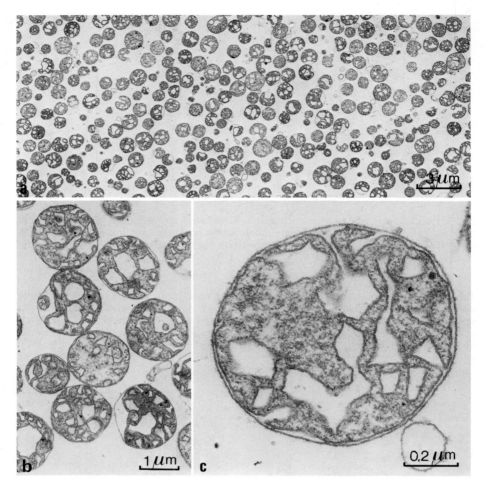

Fig. 4a–c. Electron micrographs of Percoll-purified potato tuber mitochondria. Note that mitochondria exhibit intact membranes and a dense matrix

45 min at 40,000 g (Sorvall SS34 rotor). The mitochondria are found in a light white band near the bottom of the tube, while the thylakoid membranes remain near the top of the tube (Fig. 3). The mitochondrial fraction is removed, diluted tenfold with wash medium and centrifuged at 12,000 g for 15 min. Mitochondria are resuspended in a small volume of cold wash medium (15–30 mg protein ml^{-1}).

The mitochondria thus obtained generally contain less than 0.3 µg chlorophyll mg^{-1} protein and are only slightly contaminated by peroxisomes. However, if mitochondria are repurified in the same condition except for Percoll concentration, which is increased up to 35%, most of the peroxisomal contamination is removed from the mitochondrial fraction which is recovered in the lower part of the Percoll/PVP gradient (Table 2) (Day, Neuburger and Douce, unpublished data).

Fig. 5. Oxidation of succinate, NADH and α-ketoglutarate by potato tuber mitochondria. *Upper traces* washed mitochondria (*MW*); *lower traces* Percoll-purified mitochondria (*MP*). The concentrations given are the final concentrations in the reaction medium. The numbers on the traces refer to nmol O_2 consumed min^{-1} mg^{-1} protein. Note that purified mitochondria showed a markedly higher rate of O_2 uptake in state 3 than did unpurified mitochondria

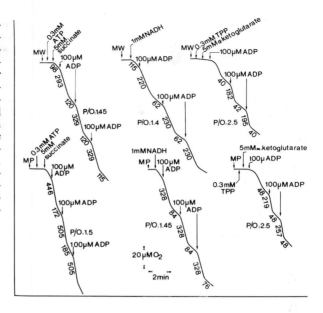

Fig. 6. Succinate reduced minus oxidized difference spectra of potato tuber mitochondria at liquid nitrogen temperature (77 K). Mitochondrial protein concentrations were 3.3 mg ml^{-1} for washed mitochondria (*MW*) and purified mitochondria (*MP*). Optical path: 2 mm. In this particular experiment concentrations of cyt aa$_3$ in MW and MP were respectively 0.27 and 0.46 nmol mg^{-1} protein. Expressed on a relative basis, taking the protein concentration as unity, the succinate oxidation rates in MW and MP were, respectively 273 and 460 nmol O_2 consumed min^{-1}

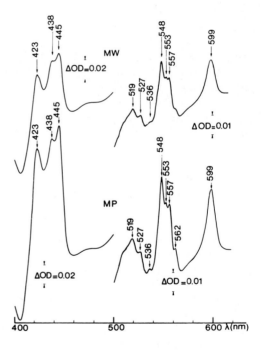

6.2.3 Properties of Percoll-Purified Mitochondria

Isolated mitochondria purified on Percoll density gradient appear mainly in the dense configuration and show no contamination with intact plastids or microbodies (Fig. 4). In addition, mitochondria thus obtained generally exhibit

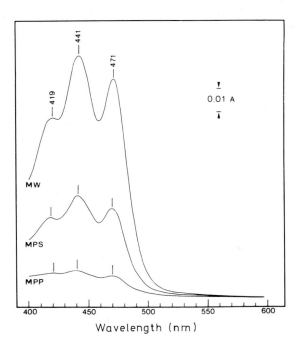

Fig. 7. Absorption spectra of ethanol extracts of potato tuber mitochondria. *MW* washed mitochondria (3.2 mg protein; 0.32 μg carotenoid); *MPS* mitochondria purified on a sucrose gradient (3.2 mg protein; 0.14 μg carotenoid); *MPP* mitochondria purified on a Percoll gradient (3.2 mg protein; 0.03 μg carotenoid)

intact membranes and a dense matrix. In several experiments, the percentage intactness, determined by enzymatic assays (KCN-sensitive ascorbate-cyt c-dependent O_2 uptake or succinate:cyt c-oxidoreductase, see above), ranged from 80 to 98% with a mean value of 95%. Due to the removal of proteinous contaminants, Percoll purified mitochondria show a markedly higher rate of O_2 uptake in state 3 than do unpurified mitochondria (Fig. 5). In the same way the concentration of the different cytochromes when expressed in terms of nanomoles per milligram of mitochondrial protein is greater in purified than in unpurified mitochondria (Fig. 6). In addition, it has been shown (NEUBURGER et al. 1982) that, on a cytochrome oxidase basis, the rate of succinate oxidation by unpurified mitochondria was equal to that recorded for Percoll-purified mitochondria. Thus, it is clear that physiological integrity of the mitochondria is maintained during the Percoll purification process.

Main contaminants normally found in unpurified mitochondria are removed after passage through Percoll density gradient, which appears more effective than purification on sucrose gradient (Fig. 7). In good agreement with previous results (McCARTY et al. 1973), mitochondria purified on Percoll are devoid of galactolipids as well as carotenoids (Fig. 7, NEUBURGER et al. 1982). Complete separation of mitochondria from microbodies is also achieved as catalase and glycolate oxidase activities are almost completely removed from the mitochondrial preparation (NEUBURGER et al. 1982, DAY, NEUBURGER and DOUCE, unpublished data). It is also interesting to note that again Percoll gradient appears much more effective than sucrose gradient to remove hydrolases such as lipolytic acyl hydrolases (NEUBURGER et al. 1982) or lipoxygenase (GOLDSTEIN et al. 1981,

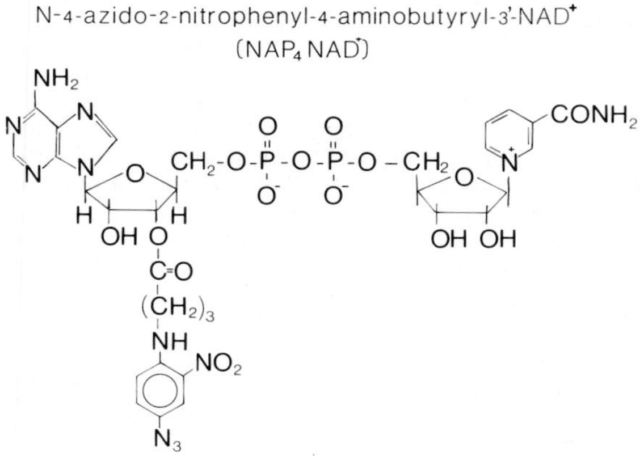

Fig. 8. Structure of N-4-azido-2-nitrophenyl-4-aminobutyryl-3′-NAD$^+$ (NAP$_4$NAD$^+$), a specific inhibitor of NAD$^+$ transport in plant mitochondria

N-4-azido-2-nitrophenyl-4-aminobutyryl-3′-NAD$^+$
(NAP$_4$ NAD$^-$)

NEUBURGER et al. 1982, DUPONT et al. 1982), and it seems likely therefore that the bound PVP (coating the silica particles) may act beneficially by binding the hydrolases. Interestingly, mitochondria carefully purified on Percoll gradients and devoid of hydrolase activities are remarkably stable. When stored on ice in a concentrated suspension, in wash medium, the mitochondria retained rapid coupled rates of O$_2$ uptake with succinate and NADH for up to 30 h. However, when NAD-linked substrates were used, O$_2$ uptake became increasingly dependent on added NAD$^+$. This phenomenon is due to slow leakage of NAD$^+$ from the mitochondrial matrix (NEUBURGER and DOUCE 1983). Subsequent uptake of NAD$^+$ via a specific carrier fully restores electron transport (NEUBURGER and DOUCE 1983). This leakage of NAD$^+$ is not due to damage of mitochondrial membranes because the mitochondria retained their impermeability to cytochrome c over the storage period. Both the uptake of added NAD$^+$ by depleted mitochondria and the efflux of endogenous NAD$^+$ during storage were inhibited by incubating the mitochondria with N-4-azido-2-nitrophenyl-4-aminobutyryl-3′-NAD$^+$ (Fig. 8), a specific inhibitor of the NAD$^+$ carrier (NEUBURGER and DOUCE 1983).

7 Concluding Remarks

This chapter has dealt only briefly with the general requirements for isolation of intact and pure mitochondria, and has focused on specific procedures for particular tissues such as potato tubers and pea leaves, which have been developed in our laboratory. These techniques are expected to be applicable for a wide range of plant tissues but may need to be modified slightly according to the source of mitochondria. This is particularly true for the concentration of Percoll, the choice of the osmoticum and the method of centrifugation employed. Density-gradient centrifugation in silica sol (Percoll) is at present the best means of preparing pure, active mitochondria from higher plants. Gradient-purified mitochondria retain at least as much biological activity as crude mitochondria prepared by classical differential centrifugation.

References

Anderson JW, Rowan KS (1966) Extraction of soluble leaf enzymes with thiols and other reducing agents. Phytochemistry 6:1047–1056

Arron GP, Edwards GE (1979) Oxidation of reduced nicotinamide adenine dinucleotide phosphate by plant mitochondria. Can J Biochem 57:1392–1399

Arron GP, Spalding MH, Edwards GE (1979) Isolation and oxidative properties of intact mitochondria from the leaves of *Sedum praealtum.* Plant Physiol 64:182–186

Baker JE, Elfvin LG, Biale JB, Honda SI (1968) Studies of ultrastructure and purification of isolated plant mitochondria. Plant Physiol 43:2001–2022

Beevers H (1982) Glyoxysomes in higher plants. Ann NY Acad Sci 386:243–253

Bergman A, Gardeström P, Ericson I (1980) Method to obtain a chlorophyll-free preparation of intact mitochondria from spinach leaves. Plant Physiol 66:442–445

Bligny R, Douce R (1978) Calcium-dependent lipolytic acyl-hydrolase activity in purified plant mitochondria. Biochim Biophys Acta 529:419–428

Bonner WD Jr (1967) A general method for the preparation of plant mitochondria. In: Estabrook RW, Pullman ME (eds) Methods in enzymology, vol X. Academic Press, London New York, pp 126–133

Day DA, Wiskich JT (1980) Glycine transport by pea leaf mitochondria. FEBS Lett 112:191–194

Day DA, Neuburger M, Douce R (1984) Activation of NAD-linked malic enzyme in intact plant mitochondria by exogenous coenzyme A. Arch Biochem Biophys 231:233–242

Douce R (1974) Site of galactolipid synthesis in spinach chloroplasts. Science 183:852–853

Douce R, Joyard J (1979) Structure and function of the plastid envelope. In: Woolhouse HH (ed) Advances in botanical research, vol VII. Academic Press, London New York, pp 1–116

Douce R, Lance C (1972) Alteration des activités oxydatives et phosphorylantes des mitochondries de chou-fleur sous l'action de phospholipases et du vieillissement. Physiol Vég 10:181–198

Douce R, Christensen EL, Bonner WD (1972) Preparation of intact plant mitochondria. Biochim Biophys Acta 275:148–160

Douce R, Manella CA, Bonner WD (1973) The external NADH dehydrogenases of intact plant mitochondria. Biochim Biophys Acta 292:105–116

Douce R, Moore AL, Neuburger M (1977) Isolation and oxidative properties of intact mitochondria isolated from spinach leaves. Plant Physiol 60:625–628

Dupont J, Rustin P, Lance C (1982) Interaction between mitochondrial cytochromes and linoleic acid hydroperoxide. Possible confusion with lipoxygenase and alternative pathway. Plant Physiol 69:1308–1314

Fleischer S, Kervina M (1974) Long-term preservation of liver for subcellular fractionation. In: Fleischer S, Packer L (eds) Methods in enzymology, vol 31/A. Academic Press, London New York, pp 3–41

Galliard T (1980) Degradation of acyl lipids: hydrolytic and oxidative enzymes. In: Stumpf PK, Conn EE (eds) The biochemistry of plants, vol IV. Academic Press, London New York, pp 85–116

Galliard T, Chan HW-S (1980) Lipoxygenases. In: Stumpf PK, Conn EE (eds) The biochemistry of plants, vol IV. Academic Press, London New York, pp 131–161

Goldstein AH, Anderson JO, McDaniel RG (1981) Cyanide-insensitive and cyanide-sensitive O_2 uptake in wheat. Plant Physiol 67:594–596

Hanson JB, Day DA (1980) Plant mitochondria. In: Stumpf PK, Conn EE (eds) The biochemistry of plants, vol I. Academic Press, London New York, pp 315–358

Haurowitz F, Schwerin P, Yenson MM (1941) Destruction of hemin and hemoglobin by the action of unsaturated fatty acids and oxygen. J Biol Chem 140:353–359

Hulme AC, Jones JD, Wooltorton LSC (1964) Mitochondrial preparations from the fruit of the apple. Preparation and general activity. Phytochemistry 3:173–188

Jackson C, Dench JE, Hall DO, Moore AL (1979) Separation of mitochondria from contaminating subcellular structures utilizing silica sol gradient centrifugation. Plant Physiol 64:150–153

Lance C, Hobson GE, Young RE, Biale JB (1967) Metabolic processes in cytoplasmic particles of the avocado fruit. IX the oxidation of pyruvate and malate during the climacteric cyle. Plant Physiol 42:471–478

Loomis WD (1974) Overcoming problems of phenolics and quinones in the isolation of plant enzymes and organelles. In: Fleischer S, Packer L (eds) Methods in enzymology, vol 31/A. Academic Press, London New York, pp 528–544

Loomis WD, Battaile J (1966) Plant Phenolic compounds and the isolation of plant enzymes. Phytochemistry 5:423–438

Matile Ph (1976) Vacuoles. In: Bonner J, Varner JE (eds) Plant biochemistry, 3rd edn. Academic Press, London New York, pp 189–224

McCarty RE, Douce R, Benson AA (1973) The acyl lipids of highly purified plant mitochondria. Biochim Biophys Acta 316:266–270

Mettler IJ, Beevers H (1980) Oxidation of NADH in glyoxysomes by a malate-aspartate shuttle. Plant Physiol 66:555–560

Moreau F, Romani R (1982) Preparation of *Avocado* mitochondria using self generated Percoll density gradients and changes in buoyant density during ripening. Plant Physiol 70:1380–1384

Nash D, Wiskich JT (1983) Properties of substantially chlorophyll-free pea leaf mitochondria prepared by sucrose density gradient separation. Plant Physiol 71:627–634

Neuburger M, Douce R (1977) Oxydation du malate, du NADH et de la glycine par les mitochondries des plantes en C_3 et C_4. CR Acad Sci 285:881–884

Neuburger M, Douce R (1983) Slow passive diffusion of NAD^+ between intact isolated plant mitochondria and suspending medium. Biochem J 216:443–450

Neuburger M, Journet EP, Bligny R, Carde JP, Douce R (1982) Purification of plant mitochondria by isopycnic centrifugation in density gradients of Percoll. Arch Biochem Biophys 217:312–323

Nishimura M, Beevers H (1978) Hydrolases in vacuoles from castor bean endosperm. Plant Physiol 62:44–48

Nishimura M, Douce R, Akazawa T (1982) Isolation and characterization of metabolically competent mitochondria from spinach leaf protoplasts. Plant Physiol 69:916–920

Palmer JM (1976) Electron transport in plant mitochondria. Annu Rev Plant Physiol 27:133–157

Palmer JM, Kirk BI (1974) The influence of osmolarity on the reduction of exogenous cytochrome c and permeability of the inner membrane of Jerusalem artichoke mitochondria. Biochem J 140:79–86

Quail PH (1979) Plant cell fractionation. Annu Rev Plant Physiol 30:425–484

Ravanel P, Tissut M, Douce R (1981) Effects of flavone on the oxidative properties of intact plant mitochondria. Phytochemistry 20:2101–2103

Ravanel P, Tissut M, Douce R (1982) Uncoupling activities of chalcones and dihydrochalcones on isolated mitochondria from potato tubers and mung bean hypocotyls. Phytochemistry 21:2845–2850

Schwitzguebel JP, Møller IM, Palmer JM (1981) Change in density of mitochondria and glyoxysomes from *Neurospora crassa*: a re-evaluation utilizing silica sol gradient centrifugation. J Gen Microbiol 126:289–295

Siedow JN, Girvin ME (1980) Alternative respiratory pathway. Plant Physiol 65:669–674

Sitaramam V, Sambasiva R (1984) The intermembranous space and the structure of mitochondrial membrane in oxidative phosphorylation. Trends Biochem Sci 9:222–223

Tanner GJ, Copeland L, Turner JF (1983) Subcellular localization of hexose kinases in pea stems: mitochondrial hexokinase. Plant Physiol 72:659–663

Tolbert NE (1982) Leaf peroxisomes. Ann NY Acad Sci 386:254–268
Weinbach EC, Garbus J (1966) Restoration by albumin of oxidative phosphorylation
    and related reactions. J Biol Chem 241:169–175
Wiskich JT, Bonner WD (1963) Preparation and properties of sweet potato mitochondria.
    Plant Physiol 38:594–603
Wojtczak L, Zaluska H (1969) On the permeability of the outer mitochondrial membrane
    to cytochrome c.I. Studies of whole mitochondria. Biochim Biophys Acta 193:64–72
Zalman LS, Nikaido H, Kagawa Y (1980) Mitochondrial outer membrane contains a
    protein producing nonspecific diffusion channels. J Biol Chem 255:1771–1774

2 Molecular Organization and Expression of the Mitochondrial Genome of Higher Plants

F. Quetier, B. Lejeune, S. Delorme, and D. Falconet

1 Introduction

The existence of extranuclear genetic information in the cytoplasm of higher plants was discerned long ago by formal genetics. However, it was not until the early 1960's that direct evidence was obtained of the isolation of both chloroplastic (ct) DNA and mitochondrial (mt) DNA.

During a first period, preceding 1974, a few groups determined the physico-chemical parameters of both ct- and mtDNA's in various higher plants and accordingly the structure and organization of these organelle genomes seemed quite conventional. Only two parameters were controversial for the mt genome and did not fit well with a simple organizational pattern: the K_2 plot of reasso-ciation kinetics of the mtDNA isolated from potato tubers was not linear, as it should be for an homogeneous DNA, and direct length measurement of mtDNA molecules showed a marked dispersion (Vedel and Quetier 1974). It should be emphasized that all experiments carried out at this time used rather drastic DNA extraction procedures.

A second period opened with the use of restriction endonucleases; whereas restriction analysis rapidly confirmed and considerably hastened the elucidation of the organization of the chloroplast genome, it revealed a very complex and surprising situation among mitochondrial genomes of higher plants (Quetier and Vedel 1978). The complex mt restriction patterns prompted a series of reinvestigations on mtDNA size and configuration by electron microscopy, which turned out to corroborate this molecular heterogeneity; no explanation was available, however.

The advent of cloning experiments certainly constitutes the most reliable approach to the molecular organization of the mt genome of higher plants during the past few years. All recent results obtained through cosmid cloning confirm a complex organization in which recombination events play an impor-tant role.

The first 20 years of research in this field resulted in a serious controversy; namely, the molecular heterogeneity of higher plant mt genomes. During this period, very different approaches were used, based on various parameters which overlap only partially since they involve different aspects of the genome com-plexity. These difficulties were enhanced by several unreproducible results and changing interpretations. It is only since 1983 that a consensus scheme has been reached which reconciles most of these controversies.

2 Physicochemical Characterization

2.1 Buoyant Density and Melting Point

Whereas the buoyant density of animal mtDNA's varies between 1.684 and 1.711 g ml^{-1} (BORST 1972), mtDNA's isolated from widely dispersed plants, including both monocotyledones and dicotyledones, band in the very narrow range of 1.705–1.707 g ml^{-1} (WELLS and INGLE 1970). Only one exception is known with *Oenothera* whose mtDNA bands at 1.710 g ml^{-1} (BRENNICKE 1980). This remarkable constancy, which indicates a conserved G + C content, should be related to peculiarities in mutational/evolutional processes. A study of a possible compositional heterogeneity was carried out on the potato mtDNA by analytical CsCl ultracentrifugations of molecules sheared to different molecular weights (VEDEL and QUETIER 1974): the standard deviations in density distribution did not reveal such heterogeneity.

This remarkable constancy of the G + C content of the higher plant mtDNA has been also evidenced by thermal denaturation curves (T$_m$ values), with only 2% variations (WELLS and BIRNSTIEL 1969, KOLODNER and TEWARI 1972, VEDEL and QUETIER 1974, WARD et al. 1981). The T$_m$ of *Oenothera* mtDNA is higher and fits well the G + C% obtained from its buoyant density. Likewise, melting curves of potato mtDNA molecules sheared to various molecular weights rule out major internal heterogeneity, as reported for *Drosophila* mtDNA (BULTMANN and LAIRD 1973).

The two strands of plant mtDNA's cannot be separated in alkaline CsCl (potato mtDNA, VEDEL and QUETIER 1974), indicating that the purine/pyrimidine ratio is equivalent for both strands, in contrast to several animal mtDNA's (BORST 1972). The intramolecular heterogeneity has been also studied by Cs$_2$SO$_4$ − Ag$^+$ gradients which were used to fractionate cucumber mtDNA into several bands (VEDEL et al. 1972) whose nature was not studied; on the other hand, the in vitro binding of synthetic homopolyribonucleotides to potato mtDNA revealed the presence of d-A- and d-C-rich clusters (QUETIER and VEDEL 1973).

2.2 Direct Observation of mtDNA Molecules by Electron Microscopy

The first observations of higher plant mtDNA by electron microscopy revealed only linear molecules: WOLSTENHOLME and GROSS (1968) found only linear molecules ranging from 1 to 62 µm with a mean value of 19.5 µm (mung bean); MIKULSKA et al. (1970) also observed only linear molecules, with a mean value of 10 µm, in pea; linear molecules from 1 to 28 µm were also observed for potato (VEDEL and QUETIER 1974). The subsequent use of mild isolation procedures allowed the observation of circular molecules but all authors have reported a very low proportion of circular molecules (5%), with the exception of KOLODNER and TEWARI (1972), who claimed to observe up to 55% of circles. This last result was not reproducible in an attempt by SEVIGNAC and QUETIER (unpublished).

Electron microscopic observations have shown a heterogeneous distribution in the size of the circular molecules. LEVINGS et al. (1979) reported three major discrete classes at 16, 22, and 30 µm in maize; SYNENKI et al. (1978) found seven apparent classes in soybean at 5.9, 10, 12.9, 16.6, 20.4, 24.5, and 29.9 µm. SEVIGNAC (1980) scored 200 circles ranging from 0.5 to 29 µm in Virginia creeper liquid-cultured cells and 268 circles ranging from 0.5 to 28 µm in wheat; no discrete classes were apparent. More recently, FONTARNAU and HERNANDEZ-YAGO (1982) have also reported a heterogeneous size distribution for the circular mtDNA molecules isolated from four species of the genus *Citrus* which occurred from 0.3 to 20 µm without discrete classes. Only two reports presented homogeneous distribution in size for circular molecules; pea mtDNA appeared homogeneous at 30 µm to KOLODNER and TEWARI (1972) but recent observation (SEVIGNAC and QUETIER, unpublished) showed well-dispersed sizes. For *Oenothera*, BRENNICKE (1980) reported first a homogeneous size at 33 µm, but more recently isolated different circular mtDNA species (BRENNICKE and BLANZ 1982). The general situation now is that the size distribution of circular molecules is heterogeneous regardless of whether discrete classes can be discerned. In all the species studied, small molecules from 0.3 to several µm have been observed and some of these may be related to cytoplasmic male sterility (maize, KEMBLE and BEDBROOK 1980, WEISSINGER et al. 1982; surgarbeet, POWLING 1981; sorghum, DIOXON and LEAVER 1982; Brassicaceae, PALMER et al. 1983; *Vicia faba*, NEGRUK et al. 1982, BOUTRY and BRIQUET 1982).

2.3 C_0t Curves and Kinetic Complexity

The complexity of a DNA can be also approached by measuring the reassociation kinetics of the denatured molecules under standardized conditions (C_0t curves). This efficient method has also led to unexplained and varying results. The first value, determined for lettuce mtDNA by WELLS and BIRNSTIEL (1969), was 210 kb and a heterogeneous curve was observed. KOLODNER and TEWARI (1972) found no heterogeneity in lettuce, spinach, bean, and tobacco; they estimated the pea mtDNA at 110 kb. A value of 150 kb was calculated for potato mtDNA (VEDEL and QUETIER 1974) but a second-order kinetics was observed for only 75% of the DNA. WARD et al. (1981) determined the kinetic complexity for musk melon (2,400 kb), cucumber (1,500 kb), zucchini squash (840 kb), watermelon (330 kb), pea (360 kb) and corn (480 kb). A heterogeneous C_0t curve was observed for corn mtDNA and sequences reiterated 10 to 50 times represented from 5 to 10% of these genomes.

2.4 Detection of Discrete Circular mtDNA Molecules by Gel Electrophoresis

Two groups succeeded in detecting the presence of discrete classes of circular mtDNA molecules. Supercoiled molecules have been isolated from in vitro cultured cells (*Nicotiana, Datura, Solanum, Phaseolus, Zea*, DALE et al. 1981) or tissue cultures (*Oenothera*, BRENNICKE and BLANZ 1982) and were analyzed by agarose gel electrophoresis. Heterogeneous patterns have been observed and the smallest bands have been extracted and characterized; their sizes range from 1.5 to 28.8 kb. It was concluded from DNA × DNA cross-hybridizations that the different circular mtDNA molecules of a plant share common sequences (tobacco, maize, Dale 1981) or correspond to unique sequences (*Oenothera*, Brennicke and Blanz 1982).

As mentioned in Sect. 2.2, the small DNA molecules which are probably related to cytoplasmic male sterility have also been detected by gel electrophoresis.

3 Restriction Analysis and Molecular Cloning

3.1 Restriction Patterns

This approach above all others has revealed the heterogeneity of the population of mtDNA molecules in higher plants (Quetier and Vedel 1978). Brassicaceae excepted, all higher plants so far studied have exhibited very complex patterns sharing two peculiarities: (1) the number of bands is considerably larger than that obtained for mtDNA of other origins or for chloroplastic DNA (enzymes which usually give few cuts, like Sal 1 and Kpn 1, generate 30 to 40 bands)[1]; (2) the stoichiometry is complicated since both very intense bands and faint bands are present. In addition, band multiplicity involves several noninteger numbers; some of them are <1 and have remained unexplained until recently. Accordingly, the assignment of a given band to be the unit for the stoechiometry calculations is arbitrary and leads to unreliable estimations of the genome complexity.

Table 1. Genome size as estimated from restriction patterns

Plant	Exact value (kb)	Reference
Virginia creeper	165[a]	Quetier and Vedel (1978)
Cucumber	120[a]	Quetier and Vedel (1978)
Wheat	140[a]	Quetier and Vedel (1978)
Wheat	590[c]	Delorme et al. (1984)
Oenothera	190[a]	Brennicke (1980)
Zucchini squash	750[b]	Ward et al. (1981)
Watermelon	350[b]	Ward et al. (1981)
Pea	350[b]	Ward et al. (1981)
Corn	180[a]	Spruill et al. (1980)
Corn	600[b]	Ward et al. (1981)
Corn	320[a]	Borck and Walbot (1982)
Corn	470[a]	Borck and Walbot (1982)
Corn	2,000[b]	Borck and Walbot (1982)
Nicotiana	400–450[a]	Belliard et al. (1979)
Brassicaceae	110–140[b]	Lebacq and Vedel (1981)

[a] Stoichiometry not taken into account
[b] Stoichiometry taken into account; reference fragment arbitrarily chosen
[c] Sum of all the different Sal I fragments (52) identified by molecular cloning

---

1 Numerous restriction endonucleases, such as Sal 1, Kpn 1 or BamH 1 (for restriction enzymes nomenclature, see Roberts 1980) have been used to analyze mtDNA

Table 1 shows the values reported for various plants and ranging from 190 to 2,000 kb. Some of them represent minimal estimates since the stoichiometry has not been taken into account; others have been calculated after having assigned one band to be present once per genome. Even in the latter case, different values were found for the same plant according to different restriction nucleases and authors. These complex restriction patterns have been shown not to originate from a partial methylation (BONEN and GRAY 1980).

Brassicaceae have been found to give less complex mtDNA restriction patterns (LEBACQ and VEDEL 1981); the complexity ranged from 150 to 200 kb, but a molecular heterogeneity has nevertheless been evidenced recently in these plants (PALMER and SHIELDS 1984; see Sect. 3.3).

3.2 Molecular Cloning

A partial BamH 1 library of maize mtDNA was constructed in the plasmid pBR322 by SPRUILL et al. (1980). Twenty-eight clones were recovered which represented about one-third of the library. Several bands were shown to correspond to different comigrating fragments. BORCK and WALBOT (1982) cloned 20 EcoR 1 fragments of maize mtDNA and obtained similar results. A complete Sal 1 library of the wheat mtDNA was constructed in plasmid pBR322 and in cosmid pHC79 (DELORME et al. 1984); it contained 52 different fragments ranging from 0.4 kb to 29 kb and the genome size amounted to 590 kb when each fragment was counted once per genome.

An interesting feature was revealed when these cloned fragments were hybridized back to restricted mtDNA: whereas a part of the cloned fragments hybridized only to the expected bands, others hybridized to multiple bands. SPRUILL et al. (1980) identified five such fragments out of their 28 clones and BORCK and WALBOT (1982) three such fragments out of 20 clones (partial libraries). DELORME et al. (1984) analyzed the complete Sal 1 library of the wheat mtDNA and 27 of the 52 fragments were shown to share common sequences with other fragment(s). Several repeated sequences have also been evidenced in the maize mt genome (LONSDALE et al. 1984).

3.3 Physical Map(s) of mtDNA

The use of plasmids does not allow the insertion of fragments longer than 15 kb and physical mapping can only be approached by using cosmids, as evidenced by LONSDALE. Such vehicles accommodate 35 to 45 kb inserts which can be prepared either by a partial restriction hydrolysis or by a mechanical shearing of the genomic DNA. Each recombinant cosmid thus contains several restriction fragments whose arrangement can be easily determined. The comparison of overlapping inserts allows "walking" along the genome. Large stretches of mtDNA have thus been mapped somewhat easily, but difficulties arose with recombinant cosmids exhibiting an unexplainable overlapping, named multiple possibilities. An example is represented in Fig. 1.

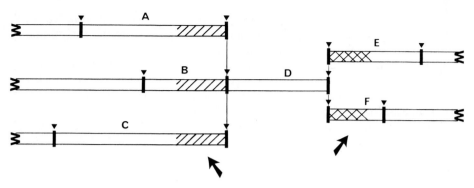

Fig. 1. Example of multiple arrangements as found by overlapping cosmid inserts. The sequences *ADE, BDE, BDF, CDE,* and *CDF* have been cloned in cosmids and lead to multiple possibilities on both sides of fragment *D*. *Arrows* indicate the flanking fragments. The left flanking fragments *A, B,* and *C* share a common sequence (*dashed*) as well as the right flanking fragments *E* and *F* (*hatched*)

Considered first as artifacts, these recombinant cosmids have been analyzed for the homology of the flanking fragments. In every case this homology was found and involved the repeated sequences mentioned in Sect. 3.2. It was then clear that these different arrangements corresponded to the results of *recombination events*. The molecular organization of the mt genome of higher plants can be thus represented by a collection of molecules which share various homologies and originate from one or several "master" molecules which contain several recombination points (cf. Fig. 2).

LONSDALE et al. (1983) predicted two possibilities for the maize mt genome which would correspond either to a huge circular molecule of 1,200 kb or to a collection of three circular molecules respectively 600, 250 and 350 kb in size. In the latter case, the 600 kb circle contains most of the sequences, whereas the 250 kb and the 350 kb arise by internal recombination of the 600 kb circle at a 3 kb repeated site.

PALMER and SHIELDS (1984) predicted a tripartite structure of the Chinese cabbage and turnip mt genome. A master circle, 218 kb in size, contains the entire set of sequences and two smaller circles of 135 kb and 83 kb arise by internal recombination at a 2 kb repeat site. Additional recombinations are likely to occur and to generate other circles, since weak homologies were detected which could correspond to potential recombination points.

LEJEUNE et al. (1984) found for the wheat mt genome a master circle of 430 kb containing all the unique sequences and several repeated stretches as well. These repeated sequences comprise seven pairs of repeats and two sets of three repeats each, the latter concerning the mt-rRNA genes. A large number of various sized molecules can be generated by internal recombinations. The restriction fragments which are present on patterns but which are not included in the master circle are generated through these recombinations: it was shown

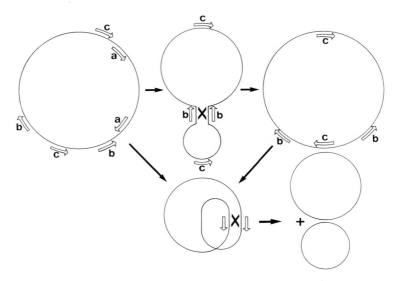

Fig. 2. Generation of different molecules by reciprocal recombinations. The circular molecule represented on the *left* contains a pair of direct repeated sequences (*a*) and two pairs of inverted repeated sequences (*b* and *c*). A pair of direct repeated sequences (*a*) can generate two different molecules after a reciprocal recombination as shown at bottom. A pair of inverted repeated sequences (*c*) can be converted into a pair of direct repeated sequences through a reciprocal recombination involving another pair of inverted repeated sequences (*b*); this flip-flop mechanism gives in a first step the circular molecule which is represented on the *right* side (and which differs from the left one); in a second step, the repeated sequences *c* which are now in a direct position can generate two different molecules by the same mechanism as that mentioned for *a*. This model can also work with linear molecules. It can explain the observed heterogeneity of the population of mtDNA molecules in a higher plant

that in this genome a recombination between two partially homologous fragments generates two "new" fragments, which effectively correspond to restriction fragments of the pattern (cf. Fig. 3).

    This structure appears to be a good compromise since it reconciles the major discrepancies mentioned above, namely
– the heterogeneity in size distribution of the mtDNA molecules observed by electron microscopy
– the complex restriction patterns which contain nonstoichiometric bands (related to the stoichiometry of the different circles) and very faint bands (junction fragments related to the recombinations).

    Whether all the native mtDNA molecules are circular or not remains still unsolved. Since supercoiled megaplasmids > 500 kb have been evidenced in *Rhizobium* on agarose gels, factors other than the large size of the mtDNA molecules from higher plants should affect their conformational state.

    It should be noticed that such recombinational events had already been substantially described in yeast (DUJON 1982), *Podospora anserina* (CUMMINGS

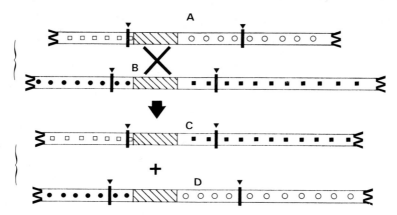

Fig. 3. Effects of a reciprocal recombination on the restriction fragments. The restriction fragments *A* and *B* share a common sequence (*hatched box*) leading to a recombination. As a result, the restriction fragments *C* and *D* are generated. The stoichiometry of fragments *A*, *B*, *C* and *D* reflects the stoichiometry of the different kinds of mtDNA molecules

et al. 1979) and trypanosoma (HOEIJMAKERS and BORST 1982) mtDNA's. The chloroplast genome consists of two molecular forms which can be generated by a recombination (flip-flop mechanism) at the level of the rRNA repeats (*Cyanophora*, BOHNERT and LÖFFELHARDT 1982; *Phaseolus*, PALMER 1983). The case of hybrids is also very significant since mtDNA's isolated from plants regenerated after protoplast fusion display several bands specific of hybrids and absent in the two parents (BELLIARD et al. 1979, GALUN et al. 1982, NAGY et al. 1981). It is likely that such recombinations are involved in the cytoplasmic male sterility (LONSDALE et al. 1983); they would also explain the differences observed by DALE et al. (1981) between cells isolated from plants and from tissue cultures or liquid-cultured cells.

4 Identified Mitochondrial Genes

4.1 rRNA Genes

BONEN and GRAY (1980) first identified the fragments coding the 26S, 18S and 5S mt-rRNA's in wheat mitochondria. The 18S and 5S rRNA genes are closely linked and well apart from the 26S rRNA gene. This feature is broadly distributed among higher plants (HUH and GREY 1982, STERN et al. 1982). Mapping and cloning experiments revealed that four different Sal 1 fragments each contain a copy of the 18S–5S tandem in wheat (FALCONET et al. 1984); at least two different arrangements of the 26S rRNA gene are present (FALCONET 1983).

4.2 tRNA Genes

Mitochondria encode 30 mt-tRNA's (GRIENENBERGER et al. 1982; see also LEAVER et al. 1983). An initiator methionine tRNA gene is separated by only one base pair from 5′ end of the 18S rRNA gene in wheat (GRAY and SPENCER 1983).

4.3 Protein Genes

By using yeast probes, several higher plant mt genes have been identified: cytochrome c oxidase subunit I and II (see LEAVER et al. 1983), α subunit of F1-ATPase (HACKE and LEAVER 1983), ATPase subunit 6 (DELORME, unpublished), ATPase subunit 9 (LEAVER et al. 1983) and apocytochrome b (WILSON et al. 1983). An intron has been found in the gene encoding the cytochrome c oxidase subunit II in maize (FOX and LEAVER 1981); recent results indicate that the homologous gene in *Oenothera* contains no intron (HIESEL and BRENNICKE 1983).

5 Concluding Remarks

The multipartite structure of the mt genome in higher plants raises several questions. A major problem lies in the level of autonomy of the different circles predicted from cosmid mapping and DNA × DNA hybridizations. The identification of replication origin(s) will certainly give the key to the inheritance mechanism; it is expected that the occurrence of only one replication origin will involve the transmission of the master circle to the progeny, and that further internal recombinations will reconstruct the collection of the different molecules. An alternative mechanism is that each type of molecule contains a replication origin conferring an autonomy. It should be kept in mind that mtDNA restriction profiles appear quite stable through sexual crosses and that mtDNA displays the same GC content whatever the level of evolution of the different plants so far studied (*Oenothera* excepted). The second important problem is that of the transcriptional activity of the different copies of the rRNA genes or of other repeated sequences.

On the other hand, the amount of unique sequences among these very large genomes remains unexplained. "Foreign" genes have recently been evidenced in mt genomes of higher plants. A 12 kb sequence is common to the chloroplast genome in maize and contains a copy of the 16S rRNA and a copy of two tRNA chloroplastic genes (STERN and LONSDALE 1982). More recently, LONSDALE et al. (1983) have shown that the maize mt genome also contains a sequence homologous to the RuBPcase large subunit of chloroplastic DNA. In addition, several copies of the S1 and S2 mitochondrial molecules associated with the cytoplasmic male sterility in maize have been found in the nucleus of this plant (KEMBLE et al. 1983). All these recent experiments reveal the important role played by recombinations in the complex structure of the higher plant genome.

References

Belliard G, Vedel F, Pelletier G (1979) Mitochondrial recombination in cytoplasmic hybrids of *Nicotiana tabacum* by protoplast fusion. Nature (London) 281:401–403

Bohnert HJ, Löffelhardt W (1982) Cyanelle DNA from *Cyanophora paradoxa* exists in two forms due to intramolecular recombination. FEBS Lett 150:403–410

Bonen L, Gray MW (1980) Organization and expression of mitochondrial genome of plants. The genes for wheat mitochondrial ribosomal and transfer tRNA: evidence for an unusual arrangement. Nucleic Acids Res 8:319–335

Bonen L, Huh TY, Gray MW (1980) Can partial methylation explain the complex fragment patterns observed when plant mitochondrial DNA is cleaved with restriction endonucleases? FEBS Lett 111:340–346

Borck KS, Walbot V (1982) Comparison of the restriction endonuclease digestion patterns of mitochondrial DNA from normal and male sterile cytoplasms of *Zea mays*. Genetics 102:109–128

Borst P (1972) Mitochondrial nucleic acids. Annu Rev Biochem 41:333–376

Boutry M, Briquet M (1982) Mitochondrial modifications associated with cytoplasmic male sterility in Faba beans. Eur J Biochem 127:129–135

Brennicke A (1980) Mitochondrial DNA from *Oenothera berteriana*. Plant Physiol 65:1207–1210

Brennicke A, Blanz P (1982) Circular mitochondrial DNA species from *Oenothera* with unique sequences. Mol Gen Genet 187:461–466

Bultmann H, Laird CD (1973) Mitochondrial DNA from *Drosophila melanogaster*. Biochim Biophys Acta 299:196–209

Cummings DJ, Belcour L, Grandchamp C (1979) Mitochondrial DNA from *Podospora anserina*. II. Properties of mutant DNA and multimeric circular DNA from senescent cultures. Mol Gen Genet 171:239–250

Dale RMK (1981) Sequence homology among different size classes of plant mtDNA's. Proc Natl Acad Sci USA 78:4453–4457

Dale RMK, Duesing JH, Keene D (1981) Supercoiled DNA's from plant tissue culture cells. Nucleic Acids Res 9:4583–4593

Delorme S, Lejeune B, Sevignac M, Delcher E, LeHégarat JC, Quétier F (1984) An approach to the complexity of wheat mitochondrial DNA: characteristics of the Sal I library and occurrence of repeated sequences. (submitted)

Dixon LK, Leaver CJ (1982) Mitochondrial gene expression and cytoplasmic male sterility in sorghum. Plant Mol Biol 1:89–102

Dujon B (1982) Mitochondrial genetics and functions. In: Strathern JN, Jones EW, Broach JR (eds) The molecular biology of the yeast: metabolism and gene expression. Cold Spring Harbor Lab, Cold Spring Harbor, NY, pp 505–523

Falconet D (1983) Organisation du génome mitochondrial chez les plantes supérieures: cartographie des fragments de restriction Sal I codant les ARN ribosomiques 5S, 18S et 26S chez le blé *Triticum aestivum*. Thes Doct 3. Cycle, Univ Paris-Sud, Orsay

Falconet D, Lejeune B, Quétier F, Gray MW (1984) Evidence for homologous recombination between repeated sequences containing 18S and 5S ribosomal RNA genes in wheat mitochondrial DNA. EMBO J 3:297–302

Fontarnau A, Hernandez-Yago J (1982) Characterization of mitochondrial DNA in *Citrus*. Plant Physiol 70:1678–1682

Fox TD, Leaver CJ (1981) The *Zea mays* mitochondrial gene coding cytochrome oxidase subunit II has an intervening sequence, and does not contain TGA codons. Cell 26:315–323

Galun E, Arzee-Gonen P, Fluhr R, Edelman M, Aviv D (1982) Cytoplasmic hybridization in *Nicotiana*: mitochondrial DNA analysis in progenies resulting from fusion between protoplasts having different organelle constitutions. Mol Gen Genet 186:50–56

Gray MW, Spencer DF (1983) Wheat mitochondrial DNA encodes a eubacteria-like initiator methionine transfer RNA. FEBS Lett 161:323–327

Grienenberger JM, Jeannin G, Weil JH, Lejeune B, Quétier F, Lonsdale D (1982) Studies on plant mitochondrial tRNA's and tRNA genes. In: Cifferi O (ed) Structure and function of plant genomes. Plenum Press, New York London, p 142

Hacke E, Leaver CJ (1983) The alpha-subunit of the maize F_1-ATPase is synthesized in the mitochondrion. EMBO J 2:1783–1789

Hiesel R, Brennicke A (1983) Cytochrome oxidase subunit II gene in mitochondria of Oenothera has no intron. EMBO J 2:2173–2178

Hoejmakers JHJ, Borst P (1982) Kinetoplast DNA in the insect trypanosomes Crithidia luciliae and Crithidia fasciculata; sequence evolution of the minicircles. Plasmids 7:210–220

Huh TY, Gray MW (1982) Conservation of ribosomal RNA gene arrangement in the mitochondrial DNA of angiosperms. Plant Mol Biol 1:245–255

Kemble RJ, Bedbrook JR (1980) Low molecular weight circular and linear DNA molecules in mitochondria from normal and male-sterile cytoplasms of Zea mays. Nature (London) 284:565–566

Kemble RJ, Mans RJ, Gabay-Laughnan S, Laughnan JR (1983) Sequences homologous to episomal mitochondrial DNAs in the maize nuclear genome. Nature (London) 304:744–747

Kolodner T, Tewari KK (1972) Physicochemical characterization of mitochondrial DNA from pea leaves. Proc Natl Acad Sci USA 69:1830–1834

Leaver CJ, Dixon LK, Hack E, Fox TD, Dawson AJ (1983) Mitochondrial genes and their expression in higher plants. In: Cifferi O, Dure L (eds) Structure and function of the plant genome. Plenum Press, New York London, pp 347–361

Lebacq P, Vedel F (1981) Sal 1 restriction enzyme analysis of chloroplast and mitochondrial DNA's in the genus Brassica. Plant Sci Lett 23:1–9

Lejeune B, Delorme S, Falconet F, Quétier F (1984) Molecular organization of the wheat mitochondrial genome. (submitted)

Levings CS III, Shah DM, Hu WWL, Pring DR, Timothy DH (1979) Molecular heterogeneity among mtDNA's from different maize cytoplasms. In: Cummings D, Borst P, David I, Weissman S, Fox CF (eds) Extrachromosomal DNA, ICN-UCLA Symp Mol Cell Biol. Academic Press, London New York, pp 63–73

Lonsdale D, Hodge TP, Howe CJ, Stern DB (1983) Maize mitochondrial DNA contains a sequence homologous to the ribulose-1,5-bisphosphate carboxylase large subunit gene of chloroplast DNA. Cell 34:1007–1014

Lonsdale D, Hodge TP, Fauron CMR, Flavell RB (1984) A predicted structure for the mitochondrial genome from the fertile cytoplasm of maize. In: Goldberg RB (ed) UCLA Symp Mol Cell Biol, New Ser, vol 12. Academic Press, London New York (in press)

Mikulska E, Odintsova MS, Turischeva MS (1970) Electron microscopic observation of mitochondrial DNA from pea leaves. J Ultrastruct Res 32:258–267

Nagy F, Torok I, Maliga P (1981) Extensive rearrangements in the mitochondrial DNA in somatic hybrids of Nicotiana tabacum and Nicotiana knightiana. Mol Gen Genet 183:437–444

Negruk VI, Cherny DI, Nikiforova ID, Aleksandrov AA, Butenko RG (1982) Isolation and characterization of minicircular DNA found in mitochondrial fraction of Vicia faba. FEBS Lett 142:115–117

Palmer JD (1983) Chloroplast DNA exists in two orientations. Nature (London) 301:92–93

Palmer JD, Shields CR (1984) Tripartite structure of the Brassica campestris mitochondrial genome. Nature (London) 307:437–440

Palmer JD, Shields CR, Cohen DB, Orton TJ (1983) An unusual mitochondrial DNA plasmid in the genus Brassica. Nature (London) 301:725–728

Powling A (1981) Species of small DNA molecules found in mitochondria from sugarbeet with normal and male sterile cytoplasms. Mol Gen Genet 183:82–84

Quétier F, Vedel F (1973) Interaction of polyribonucleotides with plant mitochondrial DNA. Biochem Biophys Res Commun 54:1326–1333

Quétier F, Vedel F (1978) Heterogeneous population of mitochondrial DNA molecules in higher plants. Nature (London) 268:365–368

Roberts RJ (1980) Restriction and modification enzymes and their recognition sequences. Gene 8:329–343

Sevignac M (1980) Étude de l'héterogénéité de l'ADN mitochondrial des végétaux supérieurs en microscopie électronique. Dipl Ecole Pratique Hautes Etudes, Paris

Spruill WM, Levings CS III, Sederoff RR (1980) Recombinant DNA analysis indicate that the multiple chromosomes of maize mitochondria contain different sequences. Develop Genet 1:363–378

Stern DB, Lonsdale DM (1982) Mitochondrial and chloroplast genomes of maize have a 12 kb DNA sequence in common. Nature (London) 299:698–702

Stern DB, Dyer TA, Lonsdale DM (1982) Organization of the mitochondrial ribosomal RNA genes of maize. Nucleic Acids Res 11:3333–3337

Synenki RM, Levings CS III, Shah DM (1978) Physicochemical characterization of mitochondrial DNA from soybean. Plant Physiol 61:460–464

Vedel F, Quétier F (1974) Physicochemical characterization of mitochondrial DNA from potato tubers. Biochim Biophys Acta 340:374–387

Vedel F, Quétier F, Bayen M, Rode A, Dalmon J (1972) Intramolecular heterogeneity of mitochondrial and chloroplastic DNA. Biochem Biophys Res Commun 46:972–978

Ward BL, Anderson RS, Bendich AJ (1981) The size of the mitochondrial genome is large and variable in a family of plants (Cucurbitaceae). Cell 25:793–803

Weissinger AK, Timothy DH, Levings CS III, Hu WWL, Goodman MM (1982) Unique plasmide-like mtDNA's from indigenous maize races of Latin America. Proc Natl Acad Sci USA 79:1–5

Wells R, Birnstiel M (1969) Kinetic complexity of chloroplastal DNA and mitochondrial DNA from higher plants. Biochem J 116:777–786

Wells R, Ingle J (1970) The constancy of the buoyant density of chloroplast and mitochondrial DNA in a range of higher plants. Plant Physiol 46:178–179

Wilson AJ, Treick RW, Wilson KG (1983) Identification of the soybean mitochondrial gene coding for cytochrome b. In: Ahmed F, Downey K, Schultz J, Voellmy RW (eds) Advances in gene technology: molecular genetics of plants and animals. 15th Miami Winter Symp. Univ Miami, USA, p 77

Wolstenholme DR, Gross NJ (1968) The form and size of mitochondrial DNA of the red bean, Phaseolus vulgaris. Proc Natl Acad Sci USA 61:245–252

3 Plant Mitochondrial Lipids:
Structure, Function and Biosynthesis

J.L. HARWOOD

1 Introduction to Lipid Structures

The lipid content of mitochondrial membranes is relatively simple, as will be seen in the next section. Phosphoglycerides predominate but there are smaller, though very important, quantities of other lipids such as quinones. To help the reader who is unfamiliar with lipids, the structures and more important features of such compounds are discussed in this introduction.

The structures of important mitochondrial phosphoglycerides are shown in Fig. 1. The glycerol backbone of such molecules exhibits asymmetry at the central carbon. All naturally occurring phospholipids have the same stereochemical configuration. Early names for phospholipids followed DL terminology and the glycerol derivative was put into the same category as the glyceraldehyde molecule to which it would be converted by oxidation, without any alteration or removal of substituents. Phosphatidylcholine, for example, would have been called L-α-phosphatidylcholine. However, because the hydroxyl groups can be distinguished from one another, L-3-phosphatidylcholine is a clearer name. Recently, rules for the unambiguous numbering of glycerol carbon atoms have been proposed. Under this system, phosphatidylcholines are named 1,2-diacyl-sn-glycero-3-phosphorylcholines or, in short, 3-sn-phosphatidylcholines. The letters sn refer to stereochemical numbering and indicate that such a nomenclature is being used. (The nomenclature of lipids is detailed in IUB 1978). The phosphate group of phosphoglycerides may be substituted by various other molecules which give rise to the name of specific phosphoglycerides. Thus, if ethanolamine is esterified to the phosphate group, then the phosphoglyceride is named phosphatidylethanolamine [(3-sn-phosphatidyl)ethanolamine). Other important mitochondrial phospholipids, apart from phosphatidylcholine, are phosphatidylinositol [1-(3-sn-phosphatidyl)-L-myoinositol] and diphosphatidylglycerol (cardiolipin).

It is very important to remember that mention of a given phosphoglyceride in the singular is almost invariably incorrect. This is because a whole series of fatty acids can be substituted at the sn-1 and sn-2 positions. Thus, it is more correct to refer to phosphatidylcholine*s* or phosphatidylethanolamine*s*. The fatty acids, themselves, are usually straight-chain, even-numbered, saturated and unsaturated fatty acids. In most mitochondrial phosphoglycerides, the principal fatty acids are the saturated palmitic and stearic and the unsaturated oleic, linoleic and α-linolenic acids. The systematic names of these fatty acids are shown in Table 1. A short-hand nomenclature has come into use where the fatty acids are described by two numbers separated by a colon. The number

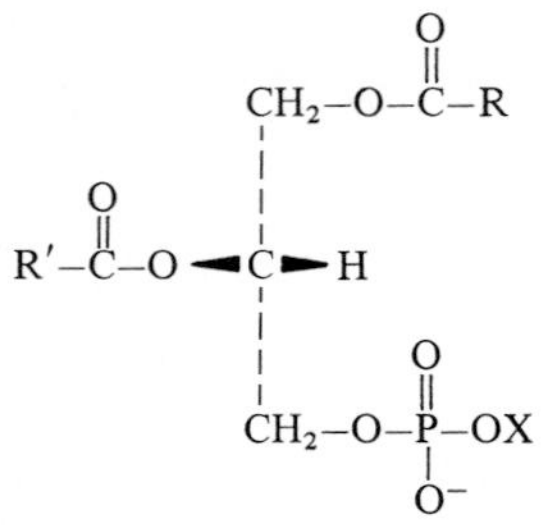

$$
\begin{array}{ll}
\underset{\displaystyle CH_2-O-\overset{\displaystyle O}{\overset{\displaystyle \|}{C}}-R}{} & \text{Position 1}
\end{array}
$$

$$
R'-\overset{\displaystyle O}{\overset{\displaystyle \|}{C}}-O \blacktriangleleft C \blacktriangleright H \qquad \text{Position 2} \qquad \textit{Basic structure}
$$

$$
CH_2-O-\overset{\displaystyle O}{\overset{\displaystyle \|}{P}}-OX \qquad \text{Position 3}
$$
$$
\underset{\displaystyle O^-}{}
$$

1,2–diacyl–sn–glycero–3–phosphoryl X

X moiety	Phosphoglyceride
X = H	Phosphatidic acid
X = $-CH_2CH_2\overset{+}{N}(CH_3)_3$	Phosphaticylcholine
X = $-CH_2CH_2NH_2$	Phosphatidylethanolamine
X = glycerol	Phosphatidylglycerol
X = myoinositol	Phosphatidylinositol

$$
\begin{array}{l}
X = -CH_2 \qquad\qquad H_2C-O-CO-R_4 \\
\qquad CHOH \quad O \quad HCOCO-R_3 \\
\qquad CH_2O-\overset{\displaystyle \|}{P}-O-CH_2 \\
\qquad\qquad\quad\; O^-
\end{array}
$$
        Diphosphatidylglycerol (cardiolipin)

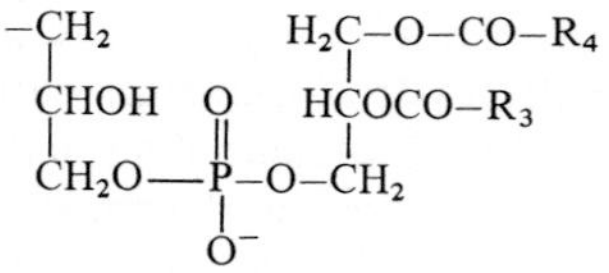

R = fatty acyl group

Fig. 1. Structures of phosphoglycerides

before the colon indicates the carbon chain length and the figure after shows the number of double bonds. Additional figures in parentheses show the position of double bonds and the letters c and t indicate whether the bond is cis-olefinic or trans-olefinic. The position of the double bonds is normally numbered from the carboxyl group but where it is more meaningful to number from the methyl end it is prefixed by ω. These abbreviations are also shown in Table 1 and will be seen in subsequent parts of this chapter. Most fatty acid analyses are performed by gas-liquid chromatography under conditions where separation is dependent on chain length and degree of unsaturation. Usually, conditions are not sufficient to separate positional isomers of, say, octadecenoic acid. The fact, therefore, that 18:1 may not just contain oleic acid (but may contain other isomers) is not always appreciated by either the authors or the readers of scientific papers!

The nature of the acyl groups and the head-group substituent molecule have a profound influence of the properties of the phosphoglyceride molecule.

Table 1. Structures of major mitochondrial fatty acids

Common name	Symbol	Structure	Systematic name
Palmitic	16:0	$CH_3(CH_2)_{14}COOH$	Hexadecanoic acid
Stearic	18:0	$CH_3(CH_2)_{16}COOH$	Octadecanoic acid
Oleic	18:1 (9c)	$CH_3(CH_2)_7CH \underline{=} CH(CH_2)_7COOH$	Octadecenoic acid
Linoleic	18:2 (9c, 12c)	$CH_3(CH_2)_4CH \underline{=} CHCH_2CH \underline{=} CH(CH_2)_7COOH$	Octadecadienoic acid
α-Linolenic	18:3 (9c, 12c, 15c)	$CH_3CH_2CH \underline{=} CHCH_2CH \underline{=} CHCH_2CH \underline{=} CH(CH_2)_7COOH$	Octadecatrienoic acid

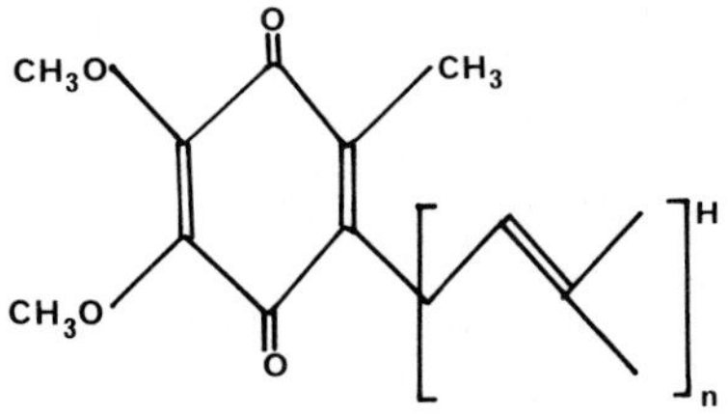

Fig. 2. The structure of ubiquinone

The type of fatty acid in the acyl groups determines their melting point and, hence, the production of the so-called liquid-crystalline state. The head groups influence the packing of the lipids, their interaction with other molecules such as proteins and the nature of their physical state. These topics are dealt with more fully later (see Sect. 4).

Lipids other than phosphoglycerides are also present in mitochondria. Of these, the electron transport component ubiquinone is particularly important (Fig. 2). The ubiquinones, like the plastoquinones of chloroplasts, contain a polyprenol side chain. The ubiquinone (Q) isoprenanalogue usually present is Q-9 or Q-10; i.e., the quinone residue with a side chain of nine (45C) or ten (50C) isoprene residues (THRELFALL and GRIFFITHS 1966).

2 Composition of Mitochondrial Membranes

2.1 Content of Acyl and Other Lipids

One of the most interesting aspects concerning the lipid content of naturally occurring membranes is the unique composition that many organelles possess. In the case of animal (PEARSONS et al. 1967) and yeast (HALLERMAYER and NEU-PERT 1974) mitochondria, it has long been recognized that the presence of di-

Table 2. Lipids of plant mitochondria

	% Total plant lipids (wt/wt)							Reference
	PC	PE	DPG	PI	PG	PS	PA	
Scorzonera tissue culture	45	21	3	15	2	–	10	Douce (1964)
Potato	43	30	8	7	3	3	5	Meunier and Mazliak (1972)
Potato	35	37	17	10	1	–	tr.	Journet, Harwood, Douce unpublished results (1982)
Cauliflower buds	44	34	11	7	4	–	–	Moreau et al. (1974)
Broad bean root	47	20	3	13	15	–	1	Oursel et al. (1973)
Lupin root	33	24	9	13	13	9	–	Oursel et al. (1973)
Castor bean endosperm	46	30	5	9	2	–	–	Ohmori and Yamada (1974)
Castor bean endosperm	37	40	10	4	2	2	a	Donaldson and Beevers (1977)
Sycamore tissue culture	43	35	13	6	3	–	–	Bligny and Douce (1980)
Mung bean hypocotyls	33	32	23	5	4	–	–	McCarty et al. (1973)
Mung bean hypocotyls	36	46	14	3	1	–	–	Bligny and Douce (1980)

[a] Included with PG

PC Phosphatidylcholine. PE Phosphatidylethanolamine. DPG Diphosphatidylglycerol (cardiolipin). PI Phosphatidylinositol. PG Phosphatidylglycerol. PS Phosphatidylserine. PA Phosphatidic acid

phosphatidylglycerol (cardiolipin) is a distinguishing feature. Mitochondria have been prepared from plant tissues by a number of different techniques. Unfortunately, the variations in separation methods may be reflected in the somewhat different lipid compositions reported (Harwood 1980a). In general, mitochondrial membranes contain a high concentration of phospholipids. Phosphatidylcholine and phosphatidylethanolamine are major components while phosphatidylinositol and phosphatidylglycerol, in addition to diphosphatidylglycerol, are present in significant amounts.

The phospholipid compositions of a number of different plant mitochondria are shown in Table 2. In most mitochondria, phosphatidylcholine is the major phosphoglyceride, representing up to half of the total phospholipids. However, in mung bean hypocotyls, phosphatidylethanolamine is the major phosphoglyceride and it always accounts for a quarter of the total phospholipids at a minimum (Table 2). The two types of root mitochondria were found to have rather high amounts of phosphatidylglycerol (Oursel et al. 1973). On average, phosphatidylinositol and diphosphatidylglycerol, the two other major acidic phospholipids each represent almost 10% of the total phosphoglycerides.

The presence or absence of glycosylglycerides in mitochondria has been a somewhat controversial issue. Because such lipids are the main components of plastid membranes, it is sometimes difficult to be certain that their presence in mitochondrial fractions is not due merely to contamination. Thus, Schwertener and Biale (1973) reported the presence of galactosylglycerides in mitochondrial preparations from cauliflower buds and avocado fruit. Mackender and Leech (1974) found galactosylglycerides in the mitochondrial fraction from broad bean leaves. They examined the fatty acid composition of these lipids and found that they were very different from that of the chloroplast galactosyl-

Table 3. Distribution of lipids between inner and outer membranes of plant mitochondria

Plant tissue	% Total phospholipids						
	PC	PE	DPG	PG	PI	PS	PA
Cauliflower buds[a]							
Outer Memb.	42	24	3	10	21	–	–
Inner Memb.	41	37	14	3	5	–	–
Castor bean endosperm[b]							
Outer Memb.	51	39	0	tr.	10		0
Inner Memb.	34	58	7	1	2		0
Sycamore tissue culture[c]							
Outer Memb.	54	30	0	5	11	–	–
Inner Memb.	41	37	15	3	5	–	–
Mung bean hypocotyl[c]							
Outer Memb.	68	24	0	2	5	–	–
Inner Memb.	29	50	17	1	2	–	–

[a] MOREAU et al. (1974).   [b] CHEESEBROUGH and MOORE (1980).
[c] BLIGNY and DOUCE (1980).
For abbreviations see Table 2

glycerides. They concluded that, in this case, cross-contamination could not account for all of the glycosylglycerides in the mitochondria. In contrast, other workers have shown that the presence of even trace amounts of galactosyglycerides is due to contamination from plastids (BEN ABDELKADER 1972, MCCARTY et al. 1973, OHMORI and YAMADA 1974, JOURNET E.-P., HARWOOD J.L. and DOUCE R. 1982, unpublished results).

Methods have been developed in several laboratories by which inner and outer mitochondrial membranes can be separated from each other. These include the use of osmotic shock, Yeda press treatment, or detergent treatment of isolated mitochondria. In an early study of potato tuber mitochondrial membranes (MEUNIER and MAZLIAK 1972), there were indications of an enrichment of diphosphatidylglycerol and other acidic phospholipids in the inner membrane, while the outer membrane contained high amounts of phosphatidylcholine. A later study of cauliflower bud mitochondria (MOREAU et al. 1974) demonstrated a great enrichment of diphosphatidylglycerol in the inner membrane, but other acidic phospholipids were concentrated in the outer membrane; phosphatidylcholine was evenly distributed (Table 3). Recently, there have been two very careful studies of the distribution of lipids within mitochondria. In both cases, tissues were specially chosen because they could be used to prepare extremely pure mitochondrial fractions. In these cases it was clear that the inner and outer mitochondrial membranes have a very different lipid composition. In castor bean (CHEESEBROUGH and MORE 1980), sycamore, and mung bean mitochondria (BLIGNY and DOUCE 1980), diphosphatidylglycerol was exclusively located in the inner membrane. In addition, the inner membrane was enriched in phosphatidylethanolamine while the outer membrane was relatively enriched in phosphatidylcholine and phosphatidylinositol (Table 3).

Table 4. The asymmetric distribution of phospholipids in membranes of castor bean endosperm mitochondria (Cheesebrough and Moore 1980)

Lipid class	Mitochondrial membrane	Distribution (mol %)		
		Outer leaflet	Inner leaflet	Inaccessible
Phosphatidylcholine	Outer	25	30	45
	Inner	21	20	59
Phosphatidylethanolamine	Outer	44	32	24
	Inner	70	18	12
Phosphatidylserine +	Outer	50	25	25
Phosphatidylinositol	Inner	0	100	0

Table 5. Fatty acid content of individual mitochondrial acyl lipids

Tissue	Lipid	Fatty acid content (% total)				
		16:0	18:0	18:1	18:2	18:3
Castor bean endosperm[a]	PC	17	7	18	52	4
	PE	18	8	9	59	5
	PI	39	11	9	37	2
	DPG	2	1	11	76	8
Sycamore tissue culture[b]	PC	25	2	4	46	23
	PE	29	2	3	51	15
	PI	50	2	3	38	7
	DPG	7	1	5	62	25
	PG	60	4	5	25	6
Mung bean hypocotyl[b]	PC	10	2	10	38	40
	PE	20	2	4	41	33
	PI	32	3	9	30	26
	DPG	3	1	7	39	50
	PG	63	4	6	11	15
Potato tuber[c]	PC	12	3	tr.	64	21
	PE	9	2	tr.	70	18
	PI	35	4	tr.	37	19
	DPG	5	1	1	65	37

[a] Donaldson and Beevers (1977).   [b] Calculated from Bligny and Douce (1980).
[c] Journet, Harwood, Douce, unpublished results (1982)
For abbreviations see Table 2

Cheesebrough and Moore (1980) have examined the sided-distribution of phospholipids in the outer and inner membrane of castor bean endosperm mitochondria (Table 4). The distributions were estimated by phospholipase A_2 digestion at several temperatures and, in the case of phosphatidylethanolamine, the sideness was checked with the chemical-labelling reagent TNBS (2,4,6-trinitrobenzenesulfonic acid). They found that the outer mitochondrial membrane showed little asymmetry in the distributions of phosphatidylcholine and phosphatidylethanolamine, but that the acidic lipids phosphatidylserine and phosphatidylinositol were concentrated in the outer leaflet. However, in all cases considerable quantities of these phosphoglycerides were inaccessible to digestion and, therefore, had an unknown distribution. By contrast, the inner mitochondrial membrane exhibited a high degree of asymmetry. Phosphatidylethanolamine was highly concentrated in the outer leaflet while the phosphatidylinositol and phosphatidylserine fraction was exclusively in the inner leaflet. It will be interesting to see if these distributions hold for mitochondrial membranes from other plant tissues.

There is some data on the fatty acid content on individual mitochondrial lipids. As noted many times in plant tissues (cf. HARWOOD 1980a), individual glycerides contain very different fatty acid compositions. While phosphatidylcholine and phosphatidylethanolamine had rather similar fatty acid compositions (with palmitate, linoleate and linolenate as major components), phosphatidylinositol was relatively enriched in saturated fatty acids (Table 5). These characteristics have been noted for the same three phosphoglycerides in other plant tissues and membranes (HARWOOD 1980a). Diphosphatidylglycerol was substantially enriched in polyunsaturated fatty acids. Phosphatidylglycerol contained a high level of palmitic acid. In addition, it has been reported that the fatty acids of individual phospholipids were more saturated in the outer membrane than in the inner membrane of mitochondria from cauliflower (MOREAU et al. 1974), mung bean and sycamore (BLIGNY and DOUCE 1980).

2.2 Comparison with Other Plant Membranes

It is of interest to compare the lipid content of mitochondrial membranes with those of other plant organelles. A few analyses are detailed in Tables 6 and 7. Like the other nonplastid plant membranes, those of the mitochondria are rich in phosphatidylcholine and phosphatidylethanolamine. However, they contain appreciably less phosphatidylinositol than the microsomal fractions or the

Table 6. The major glycerolipid classes of different plant membranes. (After HARWOOD 1980a)

Membrane	Percentage of total glycerolipids		
	Phospholipids	Glycolipids	Neutral lipids
Plasma	31– 65	9–24	26–48
Mitochondrial (total)	98–100	0– 2	tr.
Chloroplast (total)	18– 26	67–74	tr.
Chloroplast (lamellae)	9– 16	84–91	tr.
Etioplast	24– 42	58–76	tr.
Microsomal	47– 83	6–16	11–37
Nuclear	58	13	29

tr. trace

Table 7. The phosphoglyceride content of different plant membranes

Membrane	Plant	Percentage of total acyl lipids				
		PC	PE	PI	DPG	PG
Plasma	[a] Potato tuber	21	39	12	2	n.d.
Mitochondria	[b] Potato tuber	43	30	10[l]	12	[l]
	[c] Cauliflower	44	34	7	11	4
	[d] Sycamore	43	35	6	13	3
Peroxisome	[e] Potato tuber	61	20	4	0	15
	[f] Castor bean	51	27	9	2	3
	[g] Castor bean	48	36	8	0	4
Microsome	[h] Potato tuber	45	33	16	1	2
	[i] Pea seeds	67	14	19	0	tr.
Chloroplast lamellae	[j] Spinach	6	0	4	0	12
	[k] Broad bean	3	0	n.m.	0	6

[a] Demandre (1976). [b] Meunier and Mazliak (1972). [c] Moreau et al. (1974). [d] Bligny and Douce (1980). [e] Tchang (1974). [f] Donaldson et al. (1972). [g] Cheesebrough and Moore (1980). [h] Ben Abdelkader and Mazliak (1970). [i] Bolton and Harwood (1977). [j] Allen et al. (1966). [k] Mackender and Leech (1974.
[l] PI, PG and PS analyzed together. n.m. Not measured. n.d. None detected.

plasma membrane. Since it has, thus far, proved impossible to purify endoplasmic reticulum from plant microsomal fractions, we cannot compare endoplasmic reticulum with mitochondria. The chloroplast (and proplastid or etioplast) membranes are unique in containing little phospholipid. Their major lipid components are the glycosylglycerides. Furthermore, highly purified chloroplast membranes contain virtually no phosphatidylethanolamine and little phosphatidylinositol. Phosphatidylglycerol is the characteristic chloroplast phosphoglyceride (cf. Harwood 1980a). A glance at Table 7 will quickly show that diphosphatidylglycerol, a major mitochondrial phosphoglyceride, is only present in small amounts in other fractions. Indeed, it will be seen that in the more recent analyses, it could not be detected in membranes other than mitochondrial ones. In castor bean endosperm, for example, no diphosphatidylglycerol could be detected in endoplasmic reticulum or glyoxysome fractions (Cheesebrough and Moore 1980). The same was true of peroxisomal membranes from potato (Tchang F. 1974, unpublished results, quoted by Mazliak 1977). Bligny and Douce (1980) examined the question as to whether all disphosphatidylglycerol was present in the inner mitochondrial membrane. They found that the diphosphatidylglycerol/cytochrome a/a_3 ratio was the same in intact cells of sycamore as in the isolated mitochondria, suggesting that the lipid was only present in mitochondria. Furthermore, when outer and inner membranes of purified sycamore cell and mung bean hypocotyl mitochondria were separated, diphosphatidylglycerol was exclusively located in the inner mitochondrial membrane. It appears, therefore, that diphosphatidylglycerol can be used as a chemical marker for the inner membrane of plant mitochondria.

3 Metabolism

3.1 Sources of Precursors for Lipid Synthesis

It is now known that a number of subcellular fractions (and individual cell organelles) are capable of different reactions in lipid synthesis. Although technical problems have, so far, prevented the preparation of the complete array of uncontaminated subcellular fractions from a single plant tissue (although germinating castor bean has proved rather good in this regard), it seems most likely that cooperation between organelles takes place during lipid synthesis. The extent of this supposed cooperation and its importance has been the subject of several recent reviews both with regard to fatty acid synthesis (HARWOOD 1980b, STUMPF 1980, 1981) and acyl lipid formation (ROUGHAN and SLACK 1982).

The mitochondrion is certainly not self-autonomous with regard to its lipids. Indeed, only a few lipid synthetic reactions have been demonstrated there. The organelle is, therefore, to a large extent dependent on other parts of the plant cell for its supply of precursors.

Newly fixed photosynthate must be utilized by chloroplasts to make lipids directly or converted to other metabolites which can be readily transported to other parts of the cell. For a number of years the source of carbon for fatty acid synthesis has been a puzzle. Acetyl-CoA is clearly involved in de novo synthesis because carboxylation to malonyl-CoA is the first step. However, it has been noted by many workers that ^{14}C-acetate is actually a better precursor of fatty acids than ^{14}C-pyruvate (cf. HARWOOD 1979). This implied that acetate thiokinase was a more active enzyme than pyruvate decarboxylase (dehydrogenase) in plant tissues. The problem appears to have been solved by the following explanation (MURPHY and STUMPF 1981). Pyruvate, synthesized from dihydroxyacetone phosphate in the cytosol, enters the mitochondrion, where it is oxidized by an active pyruvate dehydrogenase to acetyl-CoA which can then be hydrolyzed to free acetate. The latter can then rapidly diffuse to and enter the chloroplast (Fig. 3) where it is activated by acetyl-CoA synthetase. This proposal is in keeping with the low levels of detectable plastid pyruvate dehydrogenase and the localization of a very active acetyl-CoA synthetase in spinach chloroplasts (KUHN et al. 1981). Comparisons of the relative rates of lipid synthesis from different precursors in *Spinacia oleracea* leaves have recently confirmed the importance of mitochondrially derived acetate for fatty acid formation (MURPHY and LEECH 1981). However, the detection of pyruvate dehydrogenase activity in plastids from several other species (ELIAS and GIVAN 1979, WILLIAMS and RANDALL 1979) may mean that phosphoglyceric acid can also be converted to acetyl-CoA by a pseudo-glycolytic pathway within the chloroplast. Furthermore, in pea leaves it seems that acetyl-CoA hydrolase activity is negligible in mitochondria and the great majority of the acetyl-CoA produced by pyruvate oxidation passes into the tricarboxylic acid cycle (GIVAN and HODGSON 1983). The general applicability of the cooperation of mitochondria with chloroplasts in the generation of acetyl-CoA as shown in spinach, therefore, remains to be demonstrated (cf. GIVAN 1983).

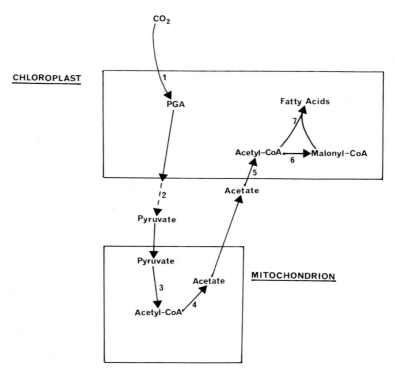

Fig. 3. The provision of acetyl-CoA for fatty acid synthesis in spinach leaves. Key processes involved: *1* photosynthetic fixation of CO_2 into phosphoglyceric acid; *2* glycolytic production of pyruvate in the cytosol; *3* mitochondrial pyruvate dehydrogenase; *4* mitochondrial acetyl CoA hydrolase; *5* chloroplastic acetyl-CoA synthetase; *6* chloroplastic acetyl CoA carboxylase; *7* chloroplastic fatty acid synthetase, palmitoyl-ACP elongase and stearoyl-ACP desaturase. (Murphy and Stumpf 1981)

In leaf tissue, most, if not all, de novo fatty acid synthesis takes place in chloroplasts (Stumpf 1980). The same situation is probably true of many other tissues which contain plastids of various types (cf. Weaire and Kekwick 1975). De novo synthesis produces palmitolyl-ACP as the product and this can be elongated to yield stearoyl-ACP which in turn can be desaturated to form oleate (cf. Stumpf 1980). All these reactions take place in isolated chloroplasts, as does the desaturation of linoleate to α-linolenate (Jones and Harwood 1980, Ohnishi and Yamada 1982). Whether plastids account for all of the oleate desaturation in plant cells is unknown at present. The balance of evidence favors at least some participation by extra-chloroplastic organelles, probably the endoplastic reticulum (cf. Harwood 1980b, Stumpf 1980, Roughan and Slack 1982). To summarize fatty acid synthesis, therefore, it is clear that the plastid plays a central role. Mitochondria hardly participate at all apart from providing a vital supply of precursor. Although plastids are extremely important, other cellular fractions undoubtedly contribute (for example in elongation) to the overall output of cellular fatty acids. Some examples of these reactions are given by HARWOOD (1979) and MOORE (1982).

Once fatty acids have been formed, they must be transferred to a complex lipid acceptor. These acylations (or transacylations) frequently involve phospholipids such as phosphatidylcholine in vivo (cf. ROUGHAN and SLACK 1982) and in vitro (SANCHEZ et al. 1983). De novo synthesis of phosphoglycerides, however, requires the acylation of glycerol-3-phosphate to form phosphatidic acid. These acylations have been reported in microsomal, mitochondrial and chloroplast fractions (HARWOOD 1979, MUDD 1980, SPARACE and MOORE 1979). Acylation of glyerol-3-phosphate can take place in the stroma of spinach chloroplasts, whereas the envelope fraction will acylate monoacylphosphatidic acid (lyso PA; JOYARD and DOUCE 1977). In castor bean, acylation of glycerol-3-phosphate was catalyzed mainly by the endoplasmic reticulum but some also by mitochondria; no such acylation occurred in the plastids (VICK and BEEVERS 1977). Thus, the key phospholipid precursor, phosphatidic acid, can be produced in several subcellular sites by acylation, as discussed more fully by MUDD (1980). In addition, it can also be synthesized by diacylglycerol kinase activity (DOUCE et al. 1968, DOUCE 1971).

In contrast to the formation of phosphatidic acid, its breakdown (by phosphatidic acid phosphatase) does not appear to be associated with mitochondria, at least in castor bean (MOORE et al. 1973). This phosphatase has been localized in the endoplasmic reticulum of castor bean (MOORE and SEXTON 1978) and mung bean (HERMAN and CHRISPEELS 1980), and an alkaline phosphatase (active with phosphatidic acid) has been reported in the chloroplast envelope of spinach (JOYARD and DOUCE 1977).

Phosphorylation of ethanolamine and choline are soluble activities and incorporation of the base phosphates into CDP-derivatives is partly soluble and partly membrane-bound. Mitochondria do not seem to be important here (cf. MOORE 1982). However, one major source of serine in the plant cell is the mitochondrion (MIFLIN and LEA 1977). Myo-inositol synthase has been located in both soluble and chloroplastic fractions from pea leaves and *Euglena gracilis* (LOEWUS and LOEWUS 1980).

It is apparent from the above discussion that the majority of the precursors for mitochondrial acyl lipid synthesis originate elsewhere in the cell. The contribution that mitochondria, themselves, can make is discussed in the next section.

3.2 Mitochondrial Phospholipid Synthesis

As discussed above, most of the precursors for acyl lipid synthesis by mitochondria come from other parts of the plant cell. However, the organelles are certainly able to carry out a number of steps involved in the formation of phospholipids. This particularly applies to reactions connected with phosphatidylglycerol and diphosphatidylglycerol synthesis. In the case of castor bean endosperm, where phospholipid synthesis has been studied in considerable detail, MOORE (1982) has compiled a summary of the synthetic reactions present in different organelles (Table 8). It will be seen that, with the exception of phosphatidyl-

Table 8. Compartmentation of phosphoglyceride synthesis in castor bean endosperm. (After Moore 1982)

Lipid formed	Reaction no. (cf. Fig. 4)	Percentage of total activity in isolated organelles				
		Soluble	Micro-somal	Mito-chondria	Plastids	Ref-erence
Fatty acids		[a]	[a]	0	100	1, 2
Phosphatidate	1	0	87	13	0	3, 4
Diacylglycerol	2	P	P	?	?	5, 6
Phosphatidylcholine	3	0	98	2	0	7
Phosphatidylcholine	5	0	90	10	0	8
Phosphatidylethanolamine	4	0	98	2	0	7
Phosphatidylethanolamine	7	?	P	?	?	9
Phosphatidylserine	6	0	100	0	0	9
CDP-diacylglycerol	8	0	75	25	0	3
Phosphatidylglycerol	9 + 10	0	50	50	0	10
Diphosphatidylglycerol	11	?	?	P	?	10
Phosphatidylinositol	12	0	100	0	0	11
Phosphatidylinositol	13	0	100	0	0	12

Reference: 1 = Vick and Beevers (1978); 2 = Zilkey and Canvin (1972); 3 = Sparace and Moore (1979); 4 = Vick and Beevers (1977); 5 = Moore et al. (1973); 6 = Moore and Sexton (1978); 7 = Sparace and Moore (1981); 8 = Moore (1976); 9 = Moore (1975); 10 = Moore (1974); 11 = Sexton and Moore (1978); 12 = Sexton and Moore (1981)

[a] Activity was localized in the plastid fraction with ^{14}C-acetate as precursor but with ^{14}C-malonyl-CoA some reactions (e.g., elongations) were present in the soluble and microsomal fractions (Harwood and Stumpf 1972).

P Present.

inositol, castor bean mitochondria are capable of synthesizing at least a proportion of their major phosphoglyceride components. A simplified scheme of phosphoglyceride synthesis is shown in Fig. 4 with relevant enzymes indicated therein.

The key intermediate of phosphoglyceride synthesis, phosphatidic acid, has been reported to be formed by a number of plant mitochondria. Early work with peanut cotyledons (Bradbeer and Stumpf 1960) and spinach leaves (Cheniae 1965) demonstrated phosphatidic acid synthesis in mitochondrial fractions prepared by differential centrifugation. With the increasing use of density gradient centrifugation, it became possible to prepare mitochondria with a high degree of purity. Douce et al. (1968) and Douce (1971) showed that such mitochondria, when prepared from cauliflower inflorescence, could incorporate $^{32}PO_4^{3-}$ from orthophosphate or ATP into phosphatidic acid. This incorporation was most probably catalyzed by diacylglycerol kinase, an enzyme not considered to be important compared to the acylation of glycerol-3-phosphate as a source of phosphatidic acid in spinach (Cheniae 1965). Vick and Beevers (1977) showed that castor bean mitochondria were capable of phosphatidate formation, although the majority of the synthesis was present in the endoplasmic reticulum fraction. Their results were confirmed and extended by Sparace and

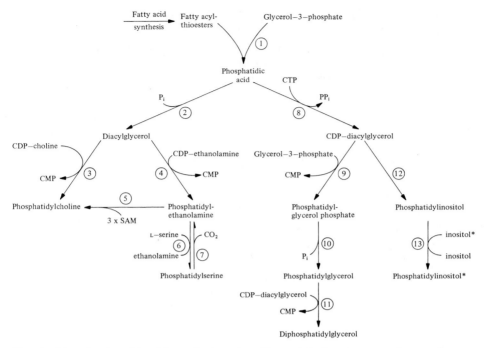

Fig. 4. Phosphoglyceride biosynthesis. *1* sn-Glycerol-3-phosphate acyl transferase; *2* phosphatidate phosphatase; *3* CDP-choline: 1,2-diacylglycerol cholinephosphotrans-ferase; *4* CDP ethanolamine: 1,2-diacylglycerol ethanolaminephosphotransferase; *5* S-Adenosyl-L-methionine: phosphatidylethanolamine N-methyl-transferase; *6* Phos-phatidylethanolamine: L-serine phosphotransferase; *7* Phosphatidylserine decarboxylase; *8* CTP: phosphatidate cytidyltransferase; *9* Glycerophosphate: CDP-diacylglycerol phosphatidyltransferase; *10* Phosphatidylglycerophosphate phosphohydrolase; *11* Phos-phatidylglycerol: CDP-diacylglycerol phosphatidyltransferase; *12* CDP-diacylglycerol: myo-inositol phosphatidyltransferase; *13* Phosphatidylinositol: myo-inositol phosphati-dyltransferase. (MOORE 1982)

MOORE (1979), who showed that phosphatidate would be formed by both the outer membrane and inner mitochondrial membrane fractions (Table 9). The total activities in the two fractions were quite similar in spite of a more than tenfold excess of protein in the inner membrane.

The synthesis of phosphatidate which was seen by DOUCE (1971) with cauli-flower could be augmented by the addition of CTP when some 7–8% of the radioactive lipid was CDP-diacylglycerol. Previously, DOUCE (1968) and SUMIDA and MUDD (1968) had both reported the synthesis of CDP-diacylglycerol in cauliflower by phosphatidate cytidyltransferase. SUMIDA and MUDD (1970) stud-ied some of the enzyme's characteristics using a 20,000 g particulate fraction. They found that activity could be increased by freezing and thawing or by the addition of 0.02% Triton X-100. Optimal reaction was obtained with 1 mM Mn^{2+} and at a pH of 5.6. In contrast, BAHL et al. (1970) found a pH optimum of 6.5 and that Mn^{2+} and Mg^{2+} were equally effective at 5 mM.

Table 9. Submitochondrial sites of phosphoglyceride synthesis in castor bean

Fraction	PG Synthesis[a]		CDP-diG Synthesis[b]		PA Synthesis[c]		PC Synthesis[d]		PC Synthesis[e]		PE Synthesis[f]	
	Total (%)	Sp. activ.	Total (%)	Sp. activ.	Total (%)	Sp. activ.	Total (%)	Sp. activ.	Total (%)	Sp. activ.	Total (%)	So. activ.
Intact mitochondria	100	365	100	0.077	100	60.4	100	22.1	100	1.33	100	0.66
Outer membrane	0.1	18.6	2.2	0.064	1.1	43.1	0.0	0	2.1	1.42	3.0	1.02
Ruptured inner membrane	10.8	515	227.4	3.691	3.3	34.1	0.7	5.8	6.0	0.65	6.2	0.33
Inner membrane	29.3	412	1561.2	5.693	1.6	4.3	4.4	4.1	3.0	0.21	5.4	0.19

Specific activity is quoted in nmol h^{-1} mg^{-1} protein. Figures are calculated from Sparace and Moore (1979, 1981). The enzymes assayed were [a] CDP diacylglycerol: glycerophosphate phosphatidyltransferase (EC 2.7.8.5) plus phosphatidylglycerophosphate phosphohydrolase (EC 3.1.3.27), [b] phosphatidate cytidyltransferase (EC 2.7.7.41), [c] sn-glycerol-3-phosphate acyltransferase (EC 2.3.1.15), [d] S-adenosyl-L-methionine: phosphatidylethanolamine N-methyltransferase (EC 2.1.1.17), [e] CDP-choline: 1,2-diacylglycerol cholinephosphotransferase (EC 2.7.8.2), [f] CDP-ethanolamine: 1,2-diacylglycerol ethanolamine phosphotransferase (EC 2.7.8.1).

They also showed that the K_m for CTP was 0.7 mM. Douce (1971) showed later that the reaction was stimulated by exogenous phosphatidic acid. Douce et al. (1972) showed that CDP-diacylglycerol formation was confined to the inner mitochondrial membrane in cauliflower and Sparace and Moore (1979) found a similar property in castor bean endosperm (Table 9).

During the above work, Douce (1968) and Sumida and Mudd (1970) found that the accumulation of radioactivity in CDP-diacylglycerol could be prevented by adding inositol or glycerol-3-phosphate (which react to produce phosphatidylinositol or phosphatidylglycerol phosphate). In castor bean endosperm, however, production of phosphatidylinositol by either the CDP-diacylglycerol: inositolphosphatidyltransferase or the base exchange reaction is exclusively localized in the endoplasmic reticulum (Moore 1982). In contrast, the CDP-diacylglycerol: glycerol phosphate phosphatidyltransferase in castor bean is equally distributed between mitochondria and endoplasmic reticulum (Moore 1974). The two fractions exhibited the same optimal requirements – 5 mM Mn^{2+}, 0.075% (w/v) Triton X-100 and a pH of 7.3. The reaction was inhibited by excess detergent and by $HgCl_2$, iodoacetate and NaF, but no build-up of phosphatidylglycerol phosphate could be seen, indicating that the phosphatidylglycerol phosphate phosphatase was active under all conditions. In contrast, Douce and Dupont (1969) found that 2 mM $HgCl_2$ could completely inhibit the phosphatase in cauliflower mitochondria. Sastry and Kates (1966) and Mashall and Kates (1972) observed some phosphatidylglycerol synthesis in spinach leaf mitochondria. The properties of the spinach enzyme were rather similar to the castor bean endosperm except that the affinity for glycerol-3-phosphate was much lower (K_m 0.2 mM versus 50 µM in castor bean). The apparent K_m for CDP-diacylglycerol was 2–3 µM in castor bean.

Further reaction of phosphatidylglycerol with CDP-diacylglycerol should yield diphosphatidylglycerol. Douce and Dupont (1969) found traces of the latter in their experiments with cauliflower mitochondria. Recently, the formation of diphosphatidylglycerol has been demonstrated in highly purified mitochondrial preparations (Hostetler K., Bligny R., Moore T.S. and Douce R. 1981, unpublished results) but other fractions were not tested (Moore 1982).

There are three other phospholipid synthetic enzymes which have been reported in mitochondrial fractions. Marshall and Kates (1974) and Macher and Mudd (1974) studied phosphatidylethanolamine synthesis by the Kennedy pathway in spinach leaves. Marshall and Kates (1974) showed that a significant portion of the total recovered activity of ethanolamine phosphotransferase was found in the 12,000 g pellet. However, the highest specific activity (with respect to protein) was in the microsomal pellet. The pH optimum was in the range 6.6–8.0 and Mg^{2+} or Mn^{2+} was required. Neither of the other two enzymes of the pathway (ethanolamine kinase, cytidyl transferase) were present in the mitochondrial fraction (Mudd 1980).

In castor bean endosperm, Lord (1975) concluded that the ethanolamine phosphotransferase was the same enzyme as the choline phosphotransferase. He also concluded that ethanolamine phosphotransferase was exclusively in the endoplasmic reticulum (Lord 1976). However, Sparace and Moore (1981) were able to show convincingly that a small amount (about 2%) of the total was localized in the mitochondria (Moore 1982). Neither of the other two alternative pathways for the formation of phosphatidylethanolamine (viz. phosphatidylserine decarboxylation and base-exchange) is present in plant mitochondria (Mudd 1980, Moore 1982).

Phosphatidylcholine can be formed by two different routes. The final enzyme of the Kennedy (CDP-base) pathway, choline phosphotransferase, was noted by Moore et al. (1970) and Devor and Mudd (1971) to be present in particulate fractions. The latter workers suggested that it was unlikely that all of the activity in their spinach leaf mitochondrial fractions was due to microsomal contamination. Johnson and Kende (1971) found the enzyme in a 44,000 g pellet from barley aleurone layers and noted that the total activity was greatly incrased following gibberellin exposure of the aleurone layer before fractionation.

In contrast to these studies, the enzyme in castor bean endosperm was reported to be localized exclusively in the endoplasmic reticulum fraction (Lord et al. 1972, Borden and Lord 1975, Moore 1976). However, in a careful recent study of purified castor bean mitochondria, Sparace and Moore (1981) demonstrated that a small amount of the total choline phosphotransferase was present in both outer and inner mitochondrial membranes (Table 9). As mentioned above, comparison of the characteristics of choline and ethanolamine phosphotransferases in spinach leaves and castor bean endosperm failed to demonstrate separate enzymes. However, as Mudd (1980) has discussed, it seems logical to expect two different enzymes to be present, if they can be purified, such as was shown for choline and ethanolamine kinases (Wharfe and Harwood 1979). Indirect evidence has been presented for separate phosphotransferase enzymes in germinating soybean, where the two phospholipids were synthesized from different molecular species of diacylglycerol (HARWOOD 1976). SPARACE and MOORE (1981) did, however, find some small differences in the properties of choline phosphotransferase from the mitochondrial and endoplasmic reticulum fractions from castor bean. The mitochondrial enzyme had a K_m for CDP-choline which was 8 µM instead of 10 µM, exhibited maximal activity with 10 mM Mg^{2+} instead of 4 mM and was less stable at 37 °C.

The alternative pathway for phosphatidylcholine formation is the "prokaryotic" one, involving successive methylation of phosphatidylethanolamine by S-adenosylmethionine. MARSHALL and KATES (1974) found that activity was present in the mitochondrial fraction of spinach leaves, although the highest specific activity was in the microsomal fraction. However, the in vitro system would not methylate phosphatidylethanolamine even though the methylation pathway was demonstrated in vivo. Phosphatidyldimethylethanolamine was the best acceptor in vitro. The methyl transferase enzyme has also been found in castor bean endosperm (MOORE 1976). The activity is dually compartmentated with about 10% of the total activity located in mitochondria. The castor bean enzyme would methylate phosphatidylethanolamine in contrast to spinach leaf enzyme, although phosphatidyldimethylethanolamine was still the best substrate. The castor bean enzyme had a pH optimum of 9.0, while that from spinach was optimally active at pH 8.0. Divalent cations were not required and detergents inhibited the activity in castor bean. The methyltransferase in mitochondria was only found in the inner membrane (SPARACE and MOORE 1979) (Table 9).

In summary, therefore, it is obvious that mitochondria have the capacity to carry out the final reactions in the assembly of most of their membrane lipids. They are dependent on a supply of substrates from other parts of the cell, however, and may also need to obtain significant portions of their phosphoglycerides from the endoplasmic reticulum. The quantification of the mitochondrion's capacity for self-autonomy in phosphoglyceride synthesis in vivo is an important research target for the future.

Mention of the necessity (at the very least for a supply of phosphatidylinositol) for an external source of mitochondrial phospholipids brings us, naturally, to the subject of phospholipid-exchange proteins. These proteins were shown first in rat liver by WIRTZ and ZILVERSMIT (1961). The phospholipid-exchange phenomenon was demonstrated between isolated plant organelles by BEN ABDELKADER and MAZLIAK (1969, 1970) and a partial purification of a soluble exchange protein made 3 years later (BEN ABDELKADER 1973). More recently, phospholipid-exchange proteins have been partially purified from potato tuber (KADER 1975), cauliflower inflorescence, Jerusalem artichoke tubers (DOUADY et al. 1978) castor bean endosperm (TANAKA and YAMADA 1979) and spinach leaves (TANAKA et al. 1980). These proteins are low molecular weight (about 20,000) proteins and they catalyze exchange between different isolated plant organelles. In most cases transfer is equally fast in both directions (e.g., TANAKA and YAMADA 1979), although MORGAN and MOORE (1978) have published a preliminary report of net transfer from labeled endoplasmic reticulum to mitochondria with exchange taking place. Certainly, free exchange between organelles would result in uniform membrane lipid compositions throughout the cell – which as Sect. 2.2 showed is definitely not the case. The lack of observed net transfer during incubations with purified exchange proteins does not necessarily preclude a role for them in vivo. All that would be required would be some mechanism to prevent complete accessibility of all mitochondrial lipids for exchange. This would favour transfer *to* but not *from* mitochondria regardless of the unrestricted rates of transfer. An analogous situation would be the operation of glycolysis in spite of the unfavorable equilibrium of the enolase reaction.

MAZLIAK and KADER (1980) have reviewed the occurrence, properties and possible function of phospholipid-exchange systems. In spite of the report of a phosphatidylcholine-exchange protein in spinach leaves (TANAKA et al. 1980), MURPHY and KUHN (1981) published evidence that such activity was due to the conductivity of salts in the same fraction. They proposed that spinach leaves contained no detectable lipid-exchange activity between microsomes (or liposomes) and chloroplasts, and that lipid transfer between those organelles was achieved by nonprotein-dependent means. This suggestion was quickly repudiated by JULIENNE et al. (1981), who purified a low molecular weight protein (20,000) from spinach leaves which would catalyze the transfer of phosphatidylcholine from liposomes to mitochondria and chloroplasts. Furthermore, YAMADA and co-workers (see TANAKA and YAMADA 1982) have purified several exchange proteins from castor bean endosperm and examined their properties. The importance of these proteins in supplying mitochondrial lipids in vivo is, however, still a matter of controversy.

3.3 Degradative Enzymes

In student textbooks, mitochondria are always portrayed as organelles which are principally concerned with oxidizing an array of food stuffs so as to produce a cellular supply of ATP. These degradative reactions include the β-oxidation of fatty acids. In plant cells, however, the importance of these organelles in the breakdown of fatty acids is much less clear. Indeed, in several tissues which have been studied in detail, β-oxidation is extramitochondrial!

There are a few reports of lipolytic enzymes in plant mitochondria. HUANG and MOREAU (1978) studied lipolytic activity in a range of germinating oil seeds. With the exception of castor bean, none contained acid lipase activity, but all had an alkaline lipase activity in their storage tissues (cf. GALLIARD 1980). Peanut mitochondria contained lipase activity towards triacylglycerols and monoacylglycerols (HUANG and MOREAU 1978) but, in other tissues, fractions such as the glyoxysomal were more important. A nonspecific acyl hydrolase, which had an absolute requirement for calcium, was reported in purified potato mitochondria (BLIGNY and DOUCE 1978). The enzyme was induced during aging of potato tuber tissue and caused severe functional impairment in the mitochondrial oxidative and phosphorylative properties. No monoacyl-intermediates were detected when either endogenous or exogenous phosphoglycerides were hydrolyzed. The enzyme was virtually inactive in the absence of added Ca^{2+} (Fig. 5)

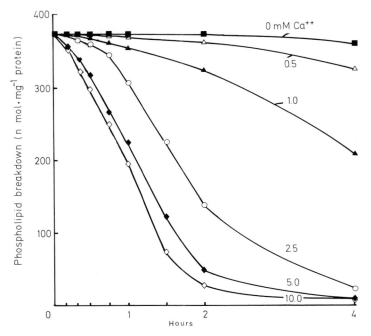

Fig. 5. The effect of Ca^{2+} concentration on the rate of endogenous phospholipid break-down during ageing of intact potato mitochondria. (Bligny and Douce 1978)

and was also detected in purified mitochondria prepared from cauliflower buds, mung bean hypocotyls and sycamore cells. Recently, Neuberger et al. (1982) tested mitochondria which had been purified by Percoll gradients for the same acyl hydrolase activity. In contrast to sucrose gradient purified organelles, these mitochondria showed little calcium-stimulated phospholipid degradation. It is possible, therefore, that the acyl hydrolase studied by Bligny and Douce (1978) was due to contamination. Neuburger et al. (1982) suggest that the hydrolase may originate from vacuoles and point out that sucrose-gradient-purified mito-chondria also contain small amounts of DNAase and RNAase. Alternatively, the Percoll treatment may be inhibitory although that seems less likely.

Early evidence for in vitro β-oxidation came from Stumpf and Barber (1956). Subsequent work on this process has been mainly carried out in germi-nating oil-rich seeds where fatty acid oxidation occurs at very high rates. How-ever, in such tissues it is the glyoxysome and not the mitochondrion which is important for β-oxidation (cf. Beevers 1980). Yamada and Stumpf (1965) estimated that only 3% of the total β-oxidation was mitochondrial in germinat-ing castor beans. In addition, in low-fat tissues, the oxidation of fatty acids by carefully purified mitochondria has been very difficult to demonstrate. Thus, Bligny and Douce (1978) showed that when the potato calcium-dependent lipase hydrolyzed phospholipids, the released fatty acids could be quantitatively accounted for in the unesterified fatty acid fraction (Fig. 6). This showed that

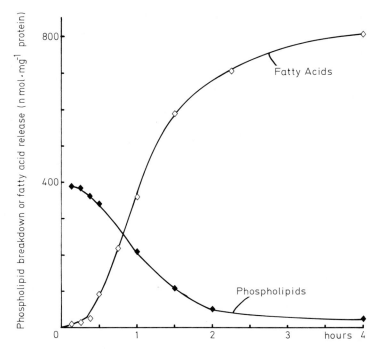

Fig. 6. The concomitant breakdown of endogenous phospholipids and release of nonesterified fatty acids during the aging of intact potato mitochondria induced by 5 mM Ca^{2+}. (BLIGNY and DOUCE 1978)

no β-oxidation was taking place and, indeed, these authors also reported that they had been unable to obtain fatty acid oxidation with any purified plant mitochondria (BLIGNY and DOUCE 1978).

THOMAS and MCNEIL (1976) purified mitochondria from pea cotyledons by differential centrifugation. They showed that maximal rates of oxidation of fatty acids required CoASH, ATP, Mg^{2+} and L-carnitine, i.e., the conditions known to be necessary for the oxidation of exogenous fatty acids by animal mitochondria. Oxidation took place with all saturated acids in the range C_4 to C_{16} with low activity for octadecanoate and tetradecanoate. The same authors (MCNEIL and THOMAS 1976) characterized the oxidation of palmitate, with the same mitochondrial preparations, in more detail. Unfortunately, since differential centrifugation does not yield uncontaminated subcellular fractions, it is difficult to know whether the β-oxidation demonstrated in the above pea cotyledon mitochondrial preparations was present in mitochondria per se. In a later study where sucrose density gradient centrifugation was used, palmitoylcarnitine synthesis appeared to be most active in etiochloroplasts from greening barley leaves (THOMAS et al. 1982).

Several workers have suggested that lipoxygenase activity is also present in mitochondria (e.g., BOUDNITSKAYA and BORISOVA 1978, DUPONT 1981). Lipoxygenase activity appears to be strongly resistant to cyanide and to be very sensitive to salicylhydroxamic acid and propyl gallate (Goldstein et al. 1980). It has been proposed that lipoxygenase could be the terminal oxidase of the cyanide-insensitive electron transport pathway or at least act in parallel with this oxidase and account for all or part of the cyanide-insensitive O_2 uptake in plant mitochondria (Dupont 1981, Parrish and Leopold 1978, Peterman and Siedow 1981, Shingles and Hill 1981). However, doubt has been cast recently on this interpretation, when Dupont et al. (1982b) showed that O_2 uptake could be observed which was due not to a lipoxygenase attack on linoleic acid but to oxidation of linoleic acid hydroperoxide already present in the substrate. In fact, potato mitochondria highly purified by Percoll gradients were devoid of lipoxygenase activity (Neuburger et al. 1982).

4 Functional Aspects of Lipids

4.1 Membrane Structure and Function

Any selected biological membrane, including those of plant mitochondria, contains an appreciable variety of different lipid types. In Sect. 2.1 it was seen that major components of plant mitochondrial membranes were the phosphoglycerides, phosphatidylcholine, phosphatidylethanolamine, phosphatidylinositol and diphosphatidylglycerol. Each of these lipids contains a mixture of fatty acids, differentially esterified to the sn-1 and sn-2 positions of glycerol. Thus, a huge mixture of lipid molecules will be present. The functional significance of such diversity is unknown, although the conservation of mitochondrial lipid composition with defined limits points to some functional advantage(s).

The usual model for a general membrane structure is that of Singer and Nicolson (1972). In this model amphiphilic acyl lipids such as the phosphoglycerides form a lipid bilayer and provide a fluid matrix in which integral membrane proteins can be embedded (Fig. 7). These proteins can either be localized on one side or traverse the membrane, thus being exposed at both surfaces. A vital aspect of membrane lipid mixtures is that, at physiological temperatures, there is sufficient hydrocarbon distortion for the hydrocarbon phase of the membrane to be fluid rather than crystalline. In addition, many membrane lipids are rather poor at forming bilayers in vitro, and it is believed that they may adopt a hexagonal-II configuration (Fig. 7) (Sen et al. 1982).

It is almost certainly too naive to consider mitochondrial lipids as though they were a simple mixture. Firstly, membrane asymmetry is the normal situation – as described for castor bean mitochondria in Sect. 2.1. Thus, there is a different lipid environment in each half of the lipid bilayer. Secondly, there may sometimes be a requirement for local lipid phase changes or separation to occur in response to rather small changes in temperature or other physical or chemical factors which are of physiological significance. Thirdly, specific lipids (and fatty acids) may be needed in membranes either for the function of a membrane protein or because their metabolism is specifically involved in a membrane function.

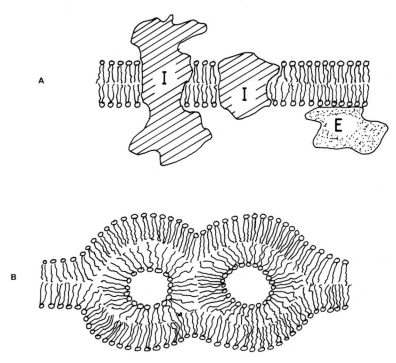

Fig. 7A, B. Possible arrangements of the lipid components in membranes. **A** The Singer-Nicholson fluid-mosaic model showing a lipid bilayer with intrinsic (*I*) and extrinsic (*E*) proteins. **B** Lipids in the hexagonal-II phase forming micelles which are embedded in a bilayer. (cf. SEN et al. 1982)

By the use of NMR and ESR it has been possible to gain an insight into the "microviscosity" of membrane lipids. Thus, the bilayer center is some 100–200 times less viscous than the surface. Within this broad statement are encompassed various extremes. For example, lipids which associate with integral membrane proteins ("boundary" lipids) may have the motions of their acyl chains severely impaired (cf. EDIDIN 1981). The diffusion rate of such lipids within the plane of the bilayer might also be restricted. The diffusion rates of lipids in various membranes have been calculated by the use of spin-labeled probes. Rates of around 10^{-8} cm^2 s^{-1} are typical (cf. RAISON 1980). In contrast, transbilayer movement (or flip-flop) is relatively slow. For artificial membranes, this type of movement has a half-time of many minutes or even hours, and was thought to be physiologically unlikely (McNAMEE and McCONNELL 1973) especially in view of the asymmetric distribution of lipids in membrane. However, it has recently been noted that transbilayer movement can be considerably increased when, for example, lipids are removed from one side of a bilayer. In these studies with phospholipase digestion, it was noted that the usual asymmetric proportions were maintained by movement of molecules from one side of the membrane to the other to replace similar lipids which have been degraded (op den Kamp 1981). The significance of this phenomenon in vivo is not known at present.

One of the most interesting aspects of mitochondrial lipid biochemistry, is the relationship of lipids with the activity of membrane proteins. The topic of membrane-bound enzymes has been the subject of many reviews (see Freed-man 1981 and Warren 1981). Lipid dependency has been demonstrated for a large number of enzymes, some of which can be reactivated by the addition of specific (or mixtures of) membrane lipids. Much of this research has been carried out with animal or bacterial systems but some experiments have utilized plant mitochondria. Studies on the E_a (the Arrhenius activation energy) and activity of succinate oxidase (Raison and Chapman 1976, Chapman et al. 1979) have provided good evidence that some membrane-associated enzymes from plants are modulated by the membrane lipid structure.

Mention was made above of the essential fluidity of the lipid bilayer. However, membrane lipids can undergo a liquid-crystalline to gel-phase transition within the temperature range of plant growth. Because membrane lipids consist of a large number of different molecules then their melting behaviour will reflect both inter- and intra-molecular mixing. Typical calorimetric melting curves for membrane lipids will, therefore, be broad. From the usual lipid compositions of mitochondrial membranes, it is likely that most of the lipids will be in the liquid-crystalline phase. Since such lipids occupy a larger area of the bilayer per molecule than rigid gel-phase lipids, it is unlikely that many, if any, of the latter will exist above 0 °C. It is, therefore, better to think of mitochondrial lipids as "ordered" and "disordered" to show our uncertainty as to their physical state (Raison 1980). Spin-labeled analogs of fatty acids can be added to mitochondrial membranes and their motion followed by ESR as the ambient temperature is altered. A typical result for mung bean mitochondria is shown in Fig. 8. Changes in the temperature coefficients of motion were observed at two temperatures, called T_f and T_s. Above T_f the lipids are relatively fluid and below T_s relatively solid. An alternative technique is to follow the fluorescence intensity and the fluorescence/polarization ratio of trans-parinaric acid in membrane lipids. These parameters show a change at a temperature where about 10% of solid lipid is present. For phospholipids from bean leaves this was about 15° (Sklar et al. 1975) – a temperature very similar to T_s for mung bean mitochondria (Fig. 8). In contrast, membranes of barley, wheat and other winter cereals do not undergo any major change in ordering in the range 0°–40 °C (Raison 1980).

The above changes in lipid fluidity obviously influence the functioning of membrane enzymes. Thus, such enzymes show an increase in E_a below both T_f and T_s. The combined effect of an increase in E_a for membrane-localized enzymes and a change in the permeability of the membrane can result in large changes in physiological processes at low temperatures (Lyons 1973). Plants which exhibit a lipid-ordering change in the temperature range 0°–15 °C also show considerable metabolic dysfunction when exposed to low, nonfreezing temperatures (Lyons 1973). In contrast, chilling-resistant plants do not show either a metabolic dysfuction or a lipid-ordering change in this temperature range.

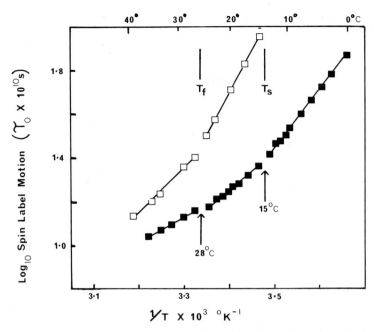

Fig. 8. Changes in the temperature coefficient of spin label motion of 12-nitroxide methyl stearate (□——□) and 16-nitroxide methyl stearate (■——■) with mitochondrial membranes of mung bean (*V. radiata*) as a function of temperature. Above T_f the lipids are relatively fluid and below T_s they are relatively solid. (After RAISON 1980)

For a number of plants, the temperature for the increase in E_a for succinate oxidase activity was the same as the temperature for a change in molecular ordering of the membrane lipids. For chilling-tolerant plants, the E_a of succinate oxidase showed no change in the 0°–30 °C range. The increase in E_a for this enzyme in mitochondria from chilling-sensitive plants was abolished by detergents which destroyed the phase transition (RAISON 1980). In a related study with mitochondria from Jerusalem artichoke (*Helianthus tuberosus* L.) an examination was made of the seasonal changes in the membranes of tubers. RAISON and co-workers were able to show that an increase in membrane fluidity (and an increase of E_a for succinate oxidase) occurs during winter, thus enabling this tissue to withstand freezing temperatures (RAISON 1980).

It should be borne in mind that correlations such as the above-named example do not necessarily provide a mechanism for enzyme activity modification or indeed chilling-damage at low temperatures. There are several reasons for this. For example, when a mixture of phospholipids is cooled, certain molecular species (those with the highest melting points) will solidify first and phase-separate from the remaining fluid lipids. There could, therefore, be considerable differences in the microenvironment of individual membrane proteins. Furthermore, a change in E_a at low temperatures does not mean that the reaction becomes rate-limiting, since other factors, such as supply of substrates, may be more important. This particularly applies to overall metabolic processes.

Many of the proteins of membranes are concerned with transport processes. In addition, passive diffusion of some molecules can also take place. Both of these processes are likely to be affected by the changes in lipid fluidity that are induced by temperature change. Measures of permeability of, for example, water through phosphoglyceride liposomes show that the rate of permeation decreases tenfold below the transition temperature for the lipids (BLOCK et al. 1975). Similar changes for molecules like erythritol and glycerol have been noted with plant membrane systems. Most of the work with isolated organelles has been carried out with chloroplasts (e.g., NOBEL 1974). There have also been studies on the leakage of ions, sugars and even proteins from the tissues of chilling-sensitive plants such as cucumber and *Passiflora* spp. In all of these tissues there was an increase in leakage at temperatures below the transition temperature of the membrane lipids.

The resistance of some plants to freezing temperatures can be neatly explained by reference to membrane lipid properties. Thus, plants which are sensitive to cooling have membrane lipids which become ordered ("solid") above 0 °C. This physical state slows water permeation and means that as ice forms outside the cell insufficient water can pass the plasma and other cell membranes to prevent the cytoplasm from freezing. In resistant plants, water permeability is maintained and the intracellular solute concentration can be increased sufficiently to prevent freezing (CHAPMAN et al. 1979).

In Fig. 8 it will be noted that spin-label experiments with the lipids of mung bean mitochondria show that a change in molecular ordering is observed at 28 °C. In general, the membrane lipids of plants undergo a change in molecular ordering between 22° and 34° (T_f values) (RAISON and CHAPMAN 1976, RAISON et al. 1977). Although decreases in E_a are sometimes seen for enzymes above T_f, there is no apparent correlation between T_f and the upper temperature limit (or optimal temperature) for plant growth (RAISON 1980).

4.2 Changes in Mitochondrial Lipids

There have been few studies on the effects of different conditions on mitochondrial lipids per se. However, there is much evidence that plant lipids are dependent on environmental growth conditions and the age of the plant and are altered by the application of chemicals such as herbicides and growth regulators. In these situations it is likely that the mitochondrial membranes are also altered in line with the changes for total tissue lipids.

One of the best-documented effectors is that of growth temperature. In general, as temperatures are lowered there is an increase in the content of unsaturated fatty acids, while the converse is true at higher temperatures (MAZLIAK 1977). This will cause a change in the transition temperature of the membrane lipids and, consequently, will alter the degree of disorder in the membrane. However, in certain cases changes may also occur in the chain length of the fatty acid moieties which sometimes neutralize the effect of increased desaturation (e.g., PATTERSON 1970).

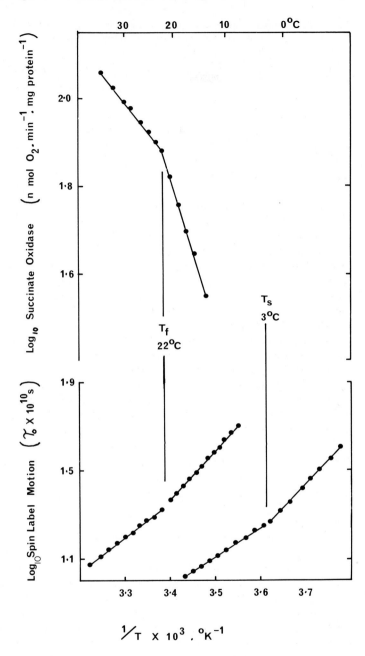

Fig. 9. Changes in the E_a of succinate oxidase activity and the temperature coefficient of motion for mitochondria from mature tubers of Jerusalem artichoke (*H. tuberosus* L.) as a function of temperature. The membranes were labeled with 12-nitroxide methyl stearate for the temperature range 8°–37 °C and with 16-nitroxide methyl stearate for the range −8°–28 °C. (After RAISON 1980)

Table 10. Changes in phospholipid content, choline incorporation and cholinephospho-transferase activity in rye roots at different temperatures. (After Kinney et al. 1982).

Root temp.	Phospholipid composition (nmol lipid P g^{-1} fr.wt.)						Choline incorp. into PC (pmol g^{-1} fr.wt. h^{-1})		Microsomal choline-phosphotransferase (pmol mg^{-1} protein h^{-1})	
	PC	PE	PG	PI	PA	Other	at 5 °C	at 20 °C	at 5 °C	at 20 °C
5 °C	296	365	39	145	57	44	38.4	90.1	200	257
20 °C	109	176	20	53	32	27	6.8	18.0	17	34

There has been some speculation as to the reason for an increase in unsaturation under certain conditions. Possibilities include the induction of specific enzymes, the modulation of desaturase activity by modification of the enzyme's environment or a change in the supply of one or more substrates. Mazliak (1977) has suggested that desaturase activity is increased at low temperatures, through an effect on membranes. However, Harris and James (1969a, b) had earlier suggested that the increase in unsaturation at lower temperatures could be attributed to an increase in dissolved O_2. More recently, Rebeille et al. (1980) reached a similar conclusion when considering temperature effects on the fatty acid composition of sycamore cells.

Physiological evidence suggests that ion transport in rye is greatly increased in roots that have been acclimatized at cool temperatures. This process is affected by the fluidity of the membrane and the environment of the transport proteins (see Sect. 4.1), so it is no surprise that fatty acid unsaturation (Clarkson et al. 1980) and lipid changes are seen. Rye roots grown at 5 °C had twice as much phospholipid on a fresh weight basis than those grown at 20°. Although all phospholipids were increased, the amounts of phosphatidylethanolamine (PE), phosphatidylcholine (PC) and phosphatidylglycerol were particularly so. In addition, the ratio of PE:PC in 5° roots was 0.81 whereas that in 20° roots was 0.62 (Table 10). It was suggested that the increase in PE and PC was due to an elevated activity of the CDP-base pathway enzymes (Kinney et al. 1982).

MacCarthy and Stumpf (1980a, b) examined the effect of growth temperature on the fatty acid composition of cells in tissue culture. Although fatty acid unsaturation increased on shifting the growth temperature from 25° to 15 °C, there were differences in detail, depending on the particular cell type. In *Catharanthus roseus,* they found an increase in oleate and linolenate proportions while in *Glycine max* and *Nicotiana tabacum,* increases in linoleate and linolenate were seen. When the metabolism of radioactive precursors such as ^{14}C-acetate, ^{14}C-oleate and ^{14}C-linoleate was followed, there was no consistent effect of temperature on fatty acid desaturation (MacCarthy and Stumpf 1980b). Similarly, Rebeille et al. (1980) failed to observe any temperature effect on the incorporation of radioactivity from ^{14}C-acetate into the fatty acids of sycamore cells, even though an increase in endogenous unsaturation could be seen.

In a study of mitochondrial lipids of mango fruits, MAZLIAK and KANE (1978) found that the molar ratio of palmitoleate/palmitate increased during ripening at 12° or 20 °C. The mango fruit is very chilling-sensitive and, when it was ripened at 4° or 8 °C, the ratio of palmitoleate/palmitate decreased along with the capacity of the isolated mitochondria to oxidize succinate.

Although it has been generally assumed that exposure of plants to lower temperatures leads to an increase in unsaturated fatty acids and that this was part of the mechanism for avoiding freezing injury, anomalous results have been obtained. Many plants will survive freezing temperatures without showing significant increases in unsaturation at lower temperatures (SIMINOVITCH et al. 1975, YOSHIDA and SAKAI 1973). In addition, membrane lipids from rye showed no significant lowering of their transition temperature even though their linolenic acid content was increased by low temperature growth (SINGH et al 1977). DE LA ROCHE et al. (1975) studied four varieties of wheat that different in their frost-hardiness. However, all four varieties showed the same increase in fatty acid unsaturation when they were grown at low temperatures. In addition, DALZIEL and BREIDENBACH (1982) examined two ecotypes of *Lycopersicon hirsutum*, which differed in their chilling sensitivity, but whose mitochondrial lipids showed identical physical properties as indicated by ESR and differential thermal analysis. These experiments shed some doubt on a simple relationship between fatty acid composition and membrane fluidity properties. Undoubtedly, some of the changes in fatty acid unsaturation are due to changes in oxygen solubility (HARRIS and JAMES 1969a, b, REBEILLE et al. 1980). In addition, it is quite possible that the changes in membrane lipids are too subtle for total lipid analyses to detect – for example, changes in molecular species or positional distributions of fatty acids.

The increase in total cell lipids which is seen on lowering growth temperatures (see above and GAWER et al. 1980) is also observed with water stress. Typically, there are large increases in phosphatidylcholine and phosphatidylethanolamine and the ratio of PE:PC goes up (HUITEMA et al. 1982), as seen for low temperature growth (e.g., KINNEY et al. 1982). Interestingly, droughtstressed plants show high frost resistance (HUITEMA et al. 1982, TYLER et al. 1981), although it has been noted that tolerant species are often more resistant to lipid changes than drought-susceptible species (PHAM THI et al. 1982). Since the changes noted in water stress and cool temperature growth apply to typical mitochondrial lipids, it would be interesting to know how much such changes are due to qualitative and quantitative alterations in mitochondria. Unfortunately, there has been little attention to this aspect.

In some cases, salt stress also leads to changes in plant lipids. ELLOUZE et al. (1982) found a decrease in lipid mass and linolenate percentage in sunflower leaves, while linoleate in the root tissue increased. In contrast, STUIVER et al. (1982) could show no effect of salt stress on the phospholipid content or composition of *Plantago* spp. In view of the conspicuous damage to mitochondria of the salt-sensitive ecotype of *Agrostis stolonifera* by salt stress (but not to the salt-tolerant ecotype) (SMITH et al. 1982) more research is clearly needed in this area. Both of the above species accumulated chloride to the same degree and with the same subcellular distribution, but no study was made of mitochondrial lipids (SMITH et al. 1982). Calcium has been reported to inhibit phospholip-id synthesis in a calcicolous plant, *Vicia faba* (OURSEL 1979), and to increase desaturation of ^{14}C-oleate in a calcifuge plant, *Lupinus luteus* (CITHAREL et al. 1982). Some of these changes are observed in subcellular fractions containing mitochondria, but more work is clearly required to characterize the effects of this organelle precisely.

An interesting possible role of mitochondria in calcium metabolism has been suggested by DUPONT (1978). He observed that purified mitochondria from cauliflower were able to accumulate calcium in the lipidic phase of their membranes. This accumulation was transient and concomitant with the appearance and disappearance of phosphatidic acid. A phospholipase D activity seemed to be responsible for the formation of phosphatidic acid. In further studies on the phenomenon, DUPONT and LANCE (1980) suggested that degradation of the phosphatidic acid was due to lipoxygenase attack of its acyl chains. However, lipoxygenase in potato tubers has been characterized as a nonmito-chondrial enzyme (DUPONT et al. 1982a). In view of the demonstration of a calcium-stimulated phospholipase (BLIGNY and DOUCE 1978; see Sect. 3.3), more work is needed to define the role of lipids in calcium transport and accumulation by plant mitochondria.

5 Conclusion

The lipid composition of mitochondria is unique amongst eukaryotic membranes. Although many of the constituents are also found in other membranes, the presence of components such as diphosphatidylglycerol and ubiquinones are characteristic. In recent years it has become increasingly clear that mitochondria are, themselves, capable of synthesizing many of their own lipid components. However, they are certainly not self-autonomous and rely on other parts of the cell for a supply of both precursors and also of fully formed lipids. The extent of mitochondrial self-autonomy with regard to lipids and the function of these lipids in the organelle's membranes are major areas for future research.

References

Allen CF, Good P, Davis HF, Fowler SD (1964) Plant chloroplast lipids. 1. Separation and composition of major spinach lipids. Biochem Biophys Res Commun 15:424–430

Allen CF, Hirayama O, Good P (1966) Lipid composition of photosynthetic systems. In: Goodwin TW (ed) Biochemistry of chloroplasts, vol I. Academic Press, London New York, pp 195–200

Bahl J, Guillot-Salomon T, Douce R (1970) Synthèse enzymatique du cytidine diphos-phate diglyceride dans les végétaux supérieurs. Physiol Veg 8:55–74

Beevers H (1980) The role of the glyoxylate cycle. In: Stumpf PK, Conn EE (eds). The biochemistry of plants, vol IV. Academic Press, London New York, pp 117–130

Ben Abdelkader A (1972) Biogenèse des lipides membranaires pendant la "survie" de tranches de tubercules de pomme de terre. Thes Fac Sci, Univ Paris

Ben Abdelkader A (1973) Isolement d'une fraction protéique stimulant les exchanges de phospholipids entre mitochondries et microsomes de tubercule de pomme de terre. CR Acad Sci 277:1455–1458

Ben Abdelkader A, Mazliak P (1969) Échange in vitro de phospholipides entre mitochondries, microsomes et surnageant cytoplasmique de cellules de pomme de terre. CR Acad Sci Ser D 269:697–700

Ben Abdelkader A, Mazliak P (1970) Échanges de lipides entre mitochondries, microsomes et surnageant cytoplasmique de cellules de pomme de terre ou de chou-fleur. Eur J Biochem 15:250–262

Bligny R, Douce R (1978) Calcium-dependent lipolytic acyl-hydrolase activity in purified plant mitochondria. Biochim Biophys Acta 529:419–428

Bligny R, Douce R (1980) A precise localisation of cardiolipin in plant cells. Biochim Biophys Acta 617:254–263

Block MC, Neut-Kok van der ECM, Deenen van LLM, Gier de J (1975) The effect of chain length and lipid phase transitions on the selective permeability properties of liposomes. Biochim Biophys Acta 406:187–196

Bolton P, Harwood JL (1977) Fatty acid biosynthesis by a particulate preparation from germinating pea. Biochem J 168:261–269

Borden L, Lord JM (1975) Development of phospholipid-synthesising enzymes in castor bean endosperm. FEBS Lett 49:369–371

Boudnitskaya EV, Borisova IV (1978) Investigation of lipoxygenase in chloroplasts and mitochondria from *Pisum sativum* seedlings. FEBS Lett 24:358–362

Bradbeer C, Stumpf PK (1960) Phosphatidic acid synthesis and diglyceride phosphokinase in mitochondria from peanut cotyledons. J Lipid Res 1:214–220

Chapman E, Wright L, Raison JK (1979) Seasonal changes in the structure and function of mitochondrial membranes of artichoke tubers. Plant Physiol 63:363–366

Cheesbrough TM, Moore TS (1980) Transverse distribution of phospholipids in organelle membranes from *Ricinus communis* L. var. Hale endosperm. Plant Physiol 65:1076–1080

Cheniae GM (1965) Phosphatidic acid and glyceride synthesis by particles from spinach leaves. Plant Physiol 40:235–243

Citharel B, Oursel A, Mazliak P (1982) Effect of calcium on the biosynthesis of linoleic and linolenic acids during the growth of a calcifuge plant (*Lupinus luteus* L.). In: Wintermans JFGM, Kuiper PJC (eds) Biochemistry and metabolism of plant lipids. Elsevier, Amsterdam, pp 39–42

Clarkson DT, Hall KC, Roberts JKM (1980) Phospholipid composition and fatty acid desaturation in the roots of rye during acclimisation of low temperature. Planta 149:464–471

Dalziel AW, Breidenbach RW (1982) Physical properties of mitochondrial lipids from *Lycopersicon hirsutum*. Plant Physiol 70:376–380

Demandre C (1976) Les membranes cytoplasmique de tubercules de pomme de terre: isolement, composition lipidique, biogenèse des lipides associés a ces membranes au cours de la survie de tranches de tubercle. Thes Doct 3. Cycle, Univ Paris

Devor KA, Mudd JB (1971) Biosynthesis of phosphatidylcholine by enzyme preparations from spinach leaves. J Lipid Res 12:403–411

Donaldson RP, Beevers H (1977) Lipid composition of organelles from germinating castor bean endosperm. Plant Physiol 59:259–263

Donaldson RP, Tolbert NE, Schnarrenberger C (1972) A comparison of microbody membranes with microsomes and mitochondria from plant and animal tissues. Arch Biochem Biophys 152:199–215

Douady D, Kader JC, Mazliak P (1978) Properties of plant phospholipid-exchange proteins. Phytochemistry 17:793–794

Douce R (1964) Identification et dosage de quelques glycérophosphatides dans des souches normales et tumorales de scorsonère cultivées in vitro. CR Acad Sci 259:2167–2170

Douce R (1968) Mise en évidence du cytidine diphosphate diglycéride dans les mitochondries végétales isolées. CR Acad Sci Ser D 267:534–537

Douce R (1971) Incorporation de l'acide phosphatique dans le cytidine diphosphate diglycéride des mitochondries isolées des inflorescences de chou-fleur. CR Acad Sci Ser D 272:3146–3149

Douce R, Dupont J (1969) Biosynthèse du phosphatidylglycerol dans les mitochondries végétales isolées: mise en evidence du phosphatidylglycerophosphate. CR Acad Sci Ser D 268:1657–1660

Douce R, Guillot-Salomon T, Lance C, Signol M (1968) Phospholipid composition of mitochondria and chloroplasts isolated from various plant tissues. Bull Soc Trans Physiol Veg 14:351–373

Douce R, Mannella CA, Bonner WD (1972) Site of the biosynthesis of CDP-diglyceride in plant mitochondria. Biochem Biophys Res Commun 49:1504–1509

Dupont J (1978) Phosphatidic acid formation linked to calcium accumulation by isolated plant mitochondria. In: Ducet G, Lance C (eds) Plant mitochondria. Elsevier, Amsterdam, pp 167–174

Dupont J (1981) Lipoxygenase-mediated cleavage of fatty acids in plant mitochondria. Physiol Plant 52:225–232

Dupont J, Lance C (1980) Calcium binding, phosphatidic acid formation and fatty acid breakdown in plant mitochondria. In: Mazliak P, Benveniste P, Costes C, Douce R (eds) Biogenesis and function of plant lipids. Elsevier, Amsterdam, pp 235–238

Dupont J, Rustin P, Lance C (1982a) Lipoxygenase in higher plants: a non-mitochondrial enzyme. In: Wintermans JFGM, Kuiper PJC (eds) Biochemistry and metabolism of plant lipids. Elsevier, Amsterdam, pp 293–296

Dupont J, Rustin P, Lance C (1982b) Interaction between mitochondrial cytochromes and linoleic acid hydroperoxide. Plant Physiol 69:1308–1314

Edidin M (1981) Molecular motions and membrane organisation and function. In: Finean JB, Michell RH (eds) Membrane Structure. Elsevier, Amsterdam, pp 37–82

Elias BA, Givan CV (1979) Localisation of pyruvate dehydrogenase complex in *Pisum sativum* chloroplasts. Plant Sci Lett 17:115–122

Ellouze M, Gharsalli M, Cherif A (1982) Effect of sodium chloride on the biosynthesis of unsaturated fatty acids of sunflower plants (*Helianthus annuus* L.). In: Wintermans JFGM, Kuiper PJC (eds) Biochemistry and metabolism of plant lipids. Elsevier, Amsterdam, pp 419–422

Freedman RB (1981) Membrane-bound enzymes. In: Finean JB, Michell RH (eds) Membrane structure. Elsevier, Amsterdam, pp 160–214

Galliard T (1980) Degradation of acyl lipids. In: Stumpf PK, Conn EE (eds) The biochemistry of plants Vol 4. Academic Press, New York, pp 85–116

Gawer M, Trapy F, Guern J, Mazliak P (1980) Lipid composition of sycamore cells cultivated at various temperatures. In: Mazliak P, Benveniste P, Costes C, Douce R (eds) Biogenesis and function of plant lipids. Elsevier, Amsterdam, pp 199–202

Givan C (1983) The source of acetyl-coenzyme A in chloroplasts of higher plants. Physiol Plant 57:311–316

Givan CV, Hodgson JW (1983) Formation of coenzyme A from acetyl-coenzyme A in pea leaf mitochondria: a requirement for oxaloacetate and the absence of hydrolysis. Plant Sci Lett 32:233–242

Goldstein AH, Anderson JO, McDaniel RG (1980) Cyanide-insensitive and cyanide-sensitive O_2 uptake in wheat 1. Gradient-purified mitochondria. Plant Physiol 66:488–493

Hallermayer G, Neupert W (1974) Lipid composition of mitochondrial outer and inner membranes of *Neurospora crassa*. Hoppe-Seylers Z Physiol Chem 355:279–288

Harris P, James AT (1969a) The effect of low temperatures of fatty acid biosynthesis in plants. Biochem J 112:325–330

Harris P, James AT (1969b) Effect of low temperature on fatty acid biosynthesis in seeds. Biochim Biophys Acta 187:13–18

Harwood JL (1976) Synthesis of molecular species of phosphatidylcholine and phosphati-dylethanolamine by germinating soya bean. Phytochemistry 15:1459–1463

Harwood JL (1979) The synthesis of acyl lipids by plant tissues. Prog Lipid Res 18:55–86

Harwood JL (1980a) Plant acyl lipids: structure, distribution and analysis. In: Stumpf PK, Conn EE (eds) The biochemistry of plants, vol IV. Academic Press, London New York, pp 1–55

Harwood JL (1980b) Fatty acid synthesis. In: Mazliak P, Benveniste P, Costes C, Douce R (eds) Biogenesis and function of plant lipids. Elsevier, Amsterdam, pp 143–152

Harwood JL, Stumpf PK (1972) Fatty acid biosynthesis by subcellular fractions of higher plants. Lipids 7:8–19

Herman EM, Chrispeels MJ (1980) Characteristics and subcellular localisation of phos-pholipase D and phosphatidic acid phosphatase in bean cotyledons. Plant Physiol 66:1001–1007

Huang AHC, Moreau RA (1978) Lipases in the storage tissues of peanut and other oil seeds during germination. Planta 141:111–116

Huitema H, Woltjes J, Vigh L, Hasselt Van P (1982) Drought-induced frost resistance in wheat correlates with changes in phospholipids. In: Wintermans JFGM, Kuiper PJC (eds) Biochemistry and metabolism of plant lipids. Elsevier, Amsterdam, pp 433–436

Huq S, Palmer JM (1978) The involvement and possible role of quinone in cyanide-resistant respiration. In: Ducet G, Lance C (eds) Plant mitochondria. Elsevier, Amsterdam, pp 225–232

IUB (International Union of Biochemistry) (1978) Biochemical nomenclature and related documents. Biochem Soc (London)

Johnson KD, Kende H (1971) Hormonal control of lecithin synthesis in barley aleurone cells: Regulation of the CDP-choline pathway by gibberellin. Proc Natl Acad Sci USA 68:2674–2677

Jones AVM, Harwood JL (1980) Desaturation of linoleic acid from exogenous lipids by isolated chloroplasts. Biochem J 190:851–854

Joyard J, Douce R (1976) L'enveloppe des chloroplastes est-elle capable de synthétiser la phosphatidylcholine? CR Acad Sci Ser D 282:1515–1518

Joyard J, Douce R (1977) Site of synthesis of phosphatidic acid and diacylglycerol in spinach chloroplasts. Biochim Biophys Acta 486:273–286

Julienne M, Vergnolle C, Kader JC (1981) Activity of phosphatidylcholine-transfer pro-tein from spinach (Spinacia oleracea) leaves with mitochondria and chloroplasts. Bio-chem J 197:763–766

Kader JC (1975) Stimulation of phospholipid exchange between mitochondria and micro-somal fractions by protein isolated from potato tuber. Biochim Biophys Acta 380:31–44

Kamp op den JAF (1981) The asymmetric architecture of membranes. In: Finean JB, Michell RH (eds) Membrane strucutre. Elsevier, Amsterdam, pp 83–126

Kinney AJ, Clarkson DT, Loughman BC (1982) The effect of temperature on phospholip-id biosynthesis in rye roots. In: Wintermans JFGM, Kuiper PJC (eds) Biochemistry and metabolism of plant lipids. Elsevier, Amsterdam, pp 437–440

Koenig F (1971) Konzentration einiger Lipide in den Chloroplasten von Zea mays und Antirrhinum majus. Z Naturforsch T B 26:1180–1187

Kuhn D, Stumpf PK (1980) unpublished observations. Quoted in: Stumpf PK, Kuhn DN, Murphy DJ, Pollard MR, McKeon T, MacCarthy J (1980) Oleic acid, the central substrate. In: Mazliak P, Benveniste P, Costes C, Douce R (eds) Biogenesis and function of plant lipids. Elsevier, Amsterdam, pp 3–10

Kuhn DN, Knauf M, Stumpf PK (1981) Subcellular localisation of acetyl-CoA synthetase in leaf protoplasts of Spinacia oleracea. Arch Biochem Biophys 209:441–451

Loewus FA, Loewus MW (1980) Myo-inositol: biosynthesis and metabolism. In: Preiss J (ed) The biochemistry of plants, vol III. Academic Press, London New York, pp 43–76

Lord JM (1975) Evidence that phosphatidylcholine and phosphatidylethanolamine are synthesised by a single enzyme present in the endoplasmic reticulum of castor bean endosperm. Biochem J 151:451–453

Lord JM (1976) Phospholipid synthesis and exchange in castor bean endosperm homogenates. Plant Physiol 57:218–223

Lord JM, Kagawa T, Beevers H (1972) Intracellular distribution of enzymes of the cytidine diphosphate choline pathway in castor bean endosperm. Proc Natl Acad Sci USA 69:2429–2432

Lyons JM (1973) Chilling injury in plants. Annu Rev Plant Physiol 24:445–466

MacCarthy JJ, Stumpf PK (1980a) Fatty acid composition and biosynthesis in cell suspension cultures of *Glycine max* (L.) Merr, *Catharanthus roseus* G. Don and *Nicotiana tabacum* L. Planta 147:384–388

MacCarthy JJ, Stumpf PK (1980b) The effect of different temperatures on fatty acid synthesis and polyunsaturation in cell suspension cultures. Planta 147:389–395

Macher BA, Mudd JB (1974) Biosynthesis of phosphatidylethanolamine by enzyme preparations from plant tissues. Plant Physiol 53:171–175

Mackender RO, Leech RM (1975) The galactolipid, phospholipid and fatty acid composition of the chloroplast envelope membranes of *Vicia faba* L. Plant Physiol 53:496–502

Marshall MO, Kates M (1972) Biosynthesis of phosphatidylglycerol by cell-free preparations from spinach leaves. Biochim Biophys Acta 260:558–570

Marshall MO, Kates M (1974) Biosynthesis of nitrogenous phospholipids in spinach leaves. Can J Biochem 52:469–482

Mazliak P (1977) Glyco- and phospholipids of biomembranes in higher plants. In: Tevini M, Lichtenthaler HK (eds) Lipids and lipid polymers in higher plants. Springer, Berlin Heidelberg New York, pp 48–74

Mazliak P, Kader JC (1980) Phospholipid-exchange systems. In: Stumpf PK, Conn EE (eds) The biochemistry of plants, vol IV. Academic Press, London New York, pp 283–300

Mazliak P, Kane O (1978) Changes in the mitochondrial lipids of mango fruits during low temperature storage. In: Ducet G, Lance C (eds) Plant mitochondria. Elsevier, Amsterdam, pp 389–394

McCarty RE, Douce R, Benson AA (1973) The acyl lipids of highly purified plant mitochondria. Biochim Biophys Acta 316:266–270

McNamee MG, McConnell HM (1973) Transmembrane potentials and phospholipid flip-flop in excitable membrane vesicles. Biochemistry 12:2951–2958

McNeil PH, Thomas DR (1976) The effect of carnitine on palmitate oxidation by pea cotyledon mitochondria. J Exp Bot 27:1163–1180

Meunier D, Mazliak P (1972) Différence de composition lipidique entre les deux membranes des mitochondries de pomme de terre. CR Acad Sci 275:213–216

Miflin BJ, Lea PJ (1977) Amino acid metabolism. Annu Rev Plant Physiol 28:299–329

Moore TS (1974) Phosphatidylglycerol synthesis in castor bean endosperm. Kinetics, requirements and intracellular localisation. Plant Physiol 54:164–168

Moore TS (1975) Phosphatidylserine synthesis in castor bean endosperm. Plant Physiol 56:177–180

Moore TS (1976) Phosphatidylcholine synthesis in castor bean endosperm. Plant Physiol 57:383–386

Moore TS (1982) Phospholipid biosynthesis. Annu Rev Plant Physiol 33:235–259

Moore TS, Sexton JC (1978) Phosphatidate phosphatase of castor bean endosperm. Plant Physiol 61S:69

Moore TS, Lord JM, Kagawa T, Beevers H (1973) Enzymes of phospholipid metabolism in the endoplasmic reticulum of castor bean endosperm. Plant Physiol 52:50–53

Moreau F, Dupont J, Lance C (1974) Phospholipid and fatty acid composition of outer and inner membranes of plant mitochondria. Biochim Biophys Acta 345:295–304

Morgan MC, Moore TS (1978) Phospholipid exchange between castor bean endosperm organelles. Plant Physiol 62 (Suppl) 80

Morre DJ, Nyquist S, Rivera E (1970) Lecithin biosynthetic enzymes of onion stem and the distribution of phosphorylcholine cytidyl transferase amongst cell fractions. Plant Physiol 45:800–804

Mudd JB (1980) Phospholipid biosynthesis. In: Stumpf PK, Conn EE (eds) The biochemistry of plants, vol IV. Academic Press, London New York, pp 249–282

Murphy DJ, Kuhn DN (1981) The lack of a phospholipid-exchange-protein activity in soluble fractions of *Spinacia oleracea* leaves. Biochem J 194:257–264

Murphy DJ, Leech RM (1981) Photosynthesis of lipids from $^{14}CO_2$ in *Spinacia oleracea*. Plant Physiol 68:762–765

Murphy DJ, Stumpf PK (1981) The origin of chloroplastic acetyl-CoA. Arch Biochem Biophys 212:730–739

Neuburger M, Journet E-P, Bligny R, Carde J-P, Douce R (1982) Purification of plant mitochondria by isopycnic centrifugation in density gradients of Percoll. Arch Biochem Biophys 217:312–323

Nobel PS (1974) Temperature dependence of the permeability of chloroplasts from chilling-sensitive and chilling-resistant plants. Planta 115:369–372

Ohmori M, Yamada M (1974) Composition of phosphoglycerides and glycosylglycerides in castor bean mitochondria. Plant Cell Physiol 15:1129–1132

Ohnishi J, Yamada M (1982) Glycolipid synthesis in *Avena* leaves during greening of etiolated seedlings. 3. Synthesis of α-linolenoyl-monogalactosyl diacylglycerol from liposomal linoleoyl-phosphatidylcholine by *Avena* plastids in the presence of phosphatidylcholine-exchange protein. Plant Cell Physiol 23:767–774

Oursel A (1979) Calcium inhibition of phospholipid biosynthesis in a calcicolous plant (*Vicia faba*); comparison with a calcifuge one (*Lupinus luteus*). In: Appelqvist L-A, Liljenberg C (eds) Advances in the biochemistry and physiology of plant lipids. Elsevier, Amsterdam, pp 421–426

Oursel A, Lamant A, Salsac L, Mazliak P (1973) Étude comparée des lipides et de la fixation passive du calcium dans les racines et les fractions subcellulaires du *Lupinus luteus* et de la *Vicia faba*. Phytochemistry 12:1865–1874

Parrish DJ, Leopold AC (1978) Confounding of alternate respiration by lipoxygenase activity. Plant Physiol 62:470–472

Parsons DF, Williams GR, Thomson W, Wilson DF, Chance B (1967) Improvements in the procedure for purification of mitochondrial outer and inner membranes. Comparison of the outer membrane with smooth endoplasmic reticulum. In: Qualiariello E, Papa S, Slater EC, Tager JM (eds) Mitochondrial structure and compartmentation. Adriatica Editrice, Bari, pp 29–70

Patterson GW (1970) Effect of culture temperature on fatty acid composition of *Chlorella sorokiniana*. Lipids 5:597–600

Peterman TK, Siedow JN (1981) Lipoxygenase, the alternate oxidase? Plant Physiol 67:S-151

Pham Thi AT, Flood C, Silva de JV (1982) Effects of water stress on lipid and fatty acid composition of cotton leaves. In: Wintermans JFGM, Kuiper PJC (eds) Biochemistry and metabolism of plant lipids. Elsevier, Amsterdam, pp 451–454

Raison JK (1980) Membrane lipids. In: Stumpf PK, Conn EE (eds) The biochemistry of plants, vol IV. Academic Press, London New York, pp 57–83

Raison JK, Chapman EA (1976) Membrane phase changes in chilling-sensitive *Vigna radiata* and their significance to growth. Aust J Plant Physiol 3:291–293

Raison JK, Chapman EA, White PY (1977) Wheat mitochondria. Oxidative activity and membrane lipid structure as a function of temperature. Plant Physiol 59:623–627

Rebeille F, Bligny R, Douce R (1980) Role de l'oxygène et de la température sur la composition en acides gras des cellules isolées d'érable (*Acer pseudoplatanus* L.). Biochim Biophys Acta 620:1–9

Roche de la IA, Pomeroy MK, Andrews CJ (1975) Changes in fatty acid composition in wheat cultivars of contrasting hardiness. Cryobiology 12:506–512

Roughan PG, Slack CR (1982) Cellular organisation of glycolipid metabolism. Annu Rev Physiol 33:97–132

Sanchez J, Khor HT, Harwood JL (1983) Studies on the incorporation of fatty acids into lipids and acylthioesters by a particulate fraction from germinating peas. Phytochemistry 22:849–854

Sastry PS, Kates M (1966) Biosynthesis of lipids in plants II. Incorporation of glycerophosphate-^{32}P into phosphatides by cell-free preparations from Spinach leaves. Can J Biochem 44:459–467

Schwertener HA, Biale JB (1973) Lipid composition of plant mitochondria and chloroplasts. J Lipid Res 14:235–241

Sen A, Williams WP, Brain APR, Quinn PJ (1982) Bilayer and nonbilayer transformations in aqueous dispersions of mixed sn-3-galactosyldiacylglycerols isolated from chloroplasts. Biochim Biophys Acta 685:297–306

Sexton JC, Moore TS (1978) Phosphatidylinositol synthesis in castor bean endosperm. Cytidine diphosphate diglyceride: inositol transferase. Plant Physiol 62:978–980

Sexton JC, Moore TS (1981) Phosphatidylinositol synthesis by a Mn^{2+}-dependent exchange enzyme in castor bean endosperm. Plant Physiol 68:18–22

Shingles R, Hill RD (1981) An alternative to the alternate pathway. Plant Physiol 67:S-65

Siminovitch D, Singh J, Roche de la IA (1975) Studies on membranes in plant cells resistant to extreme freezing. 1. Augmentation of phospholipids and membrane substance without changes in unsaturation of fatty acids during hardening of black locust bark. Cryobiology 12:144–153

Singer SJ, Nicolson GL (1972) The fluid mosaic model of the structure of cell membranes. Science 175:720–731

Singh J, Roche de la IA, Siminovitch D (1977) Differential scanning calorimeter analysis of membrane lipids isolated from hardened and unhardened black locust bark and from winter rye seedlings. Cryobiology 14:620–624

Sklar LA, Hudson BS, Simoni RD (1975) Conjugated polyene fatty acids as membrane probes: preliminary characterization. Proc Natl Acad Sci USA 72:1649–1653

Smith MM, Hodson MJ, Optik H, Wainwright SJ (1982) Salt-induced ultrastructural damage to mitochondria in root tips of a salt-sensitive ecotype of *Agrostis stolonifera*. J Exp Bot 33:886–895

Sparace SA, Moore TS (1979) Phospholipid metabolism in plant mitochondria. Submitochondrial sites of synthesis. Plant Physiol 63:963–972

Sparace SA, Moore TS (1981) Phospholipid metabolism in plant mitochondria. II. Submitochondrial sites of synthesis of phosphatidylcholine and phosphatidylethanolamine. Plant Physiol 67:261–265

Stuiver CEE, Kok de LJ, Hendriks AE, Kuiper PJC (1982) The effect of salinity on phospholipid content and composition of two *Plantago* species, differing in salt tolerance. In: Wintermans JFGM, Kuiper PJC (eds) Biochemistry and metabolism of plant lipids. Elsevier, Amsterdam, pp 455–458

Stumpf PK (1980) Biosynthesis of saturated and unsaturated fatty acids. In: Stumpf PK, Conn EE (eds) The biochemistry of plants, vol IV. Academic Press, London New York, pp 177–204

Stumpf PK (1981) Plants, fatty acids, compartments. Trends Biochem Sci 6:173–176

Stumpf PK, Barber GA (1956) Fat metabolism in higher plants. VII. β-Oxidation of fatty acids by peanut mitochondria. Plant Physiol 31:304–308

Sumida S, Mudd JB (1968) Biosynthesis of cytidine diphosphate diglyceride by cauliflower mitochondria. Plant Physiol 43:1162–1164

Sumida S, Mudd JB (1970) Biosynthesis of cytidine diphosphate diglyceride by enzyme preparations from cauliflower. Plant Physiol 45:719–722

Tanaka T, Yamada M (1979) A phosphatidylcholine exchange protein isolated from germinated castor bean endosperms. Plant Cell Physiol 20:533–542

Tanaka T, Yamada M (1982) Properties of phospholipid exchange proteins from germinated castor bean endosperms. In: Wintermans JFGM, Kuiper PJC (eds) Biochemistry and metabolism of plant lipids. Elsevier, Amsterdam, pp 96–104

Tanaka T, Ohnishi J-I, Yamada M (1980) The occurrence of phosphatidylcholine exchange protein in leaves. Biochem Biophys Res Commun 96:394–399

Thomas DR, McNeil PH (1976) The effect of carnitine on the oxidation of saturated fatty acids by pea cotyledon mitochondria. Planta 132:61–63

Thomas DR, Jalil MNH, Cooke RJ, Yong BCS, Ariffin A, McNeil PH, Wood C (1982) The synthesis of palmitoylcarnitine by etiochloroplasts of greening barley leaves. Planta 154:60–65

Threlfall DR, Griffiths WT (1966) Biosynthesis of terpenoid quinones. In: Goodwin TW (ed) Biochemistry of chloroplasts, vol II. Academic Press, London New York, pp 257–282

Tyler NJ, Gusta LV, Fowler DB (1981) The effect of a water stress on the cold hardiness of winter wheat. Can J Bot 59:1717–1725

Vick B, Beevers H (1977) Phosphatidic acid synthesis by castor bean endosperm. Plant Physiol 59:459–463

Vick B, Beevers H (1978) Fatty acid synthesis in endosperm of young castor bean seedlings. Plant Physiol 62:173–178

Warren G (1981) Membrane proteins: structure and assembly. In: Finean JB, Michell RH (eds) Membrane structure. Elsevier, Amsterdam, pp 251–277

Weaire PJ, Kekwick RGO (1975) The synthesis of fatty acids in avocado mesocarp and cauliflower seed tissue. Biochem J 146:425–437

Wharfe JM, Harwood JL (1979) Lipid metabolism in germinating seeds. Purification of ethanolamine kinase from soya bean. Biochim Biophys Acta 575:102–111

Williams M, Randall DD (1979) Pyruvate dehydrogenase complex from chloroplasts of *Pisum sativum* L. Plant Physiol 64:1099–1103

Wirtz WA, Zilversmit DB (1961) Participation of soluble liver proteins in the exchange of membrane phospholipids. Biochim Biophys Acta 193:552–557

Yamada M, Stumpf PK (1965) Fat metabolism in higher plants XXIV. A soluble β-oxidation system from germinating seeds of *Ricinus communis*. Plant Physiol 40:653–658

Yoshida S, Sakai A (1973) Phospholipid changes associated with cold hardiness of cortical cells from popular stem. Plant Cell Physiol 14:353–359

Zilkey BF, Canvin DT (1972) Localisation of oleic acid biosynthesis in the proplastids of developing castor endosperm. Can J Bot 50:323–326

4 Plant Mitochondrial Cytochromes

G. Ducet

1 Introduction

This chapter on plant mitochondrial cytochromes will report mainly on higher plants. Cytochromes from fungus, especially from yeast and from *Neurospora*, have been more extensively studied, as they are easier to prepare than those from higher plants. The starting material is homogenous, can be grown in large quantities and genetically manipulated, and thus provides a good tool for studies on biosynthesis. However, mitochondria from higher plants are now relatively easy to prepare in a high state of purity, by density gradient centrifugation in sucrose or Percoll, and they remain functional in oxidative and phosphorylative activities (see Chap. 1, this Vol.).

Higher plants contain cytochromes not only in mitochondria but also in chloroplasts and other membrane systems. In chloroplasts cytochromes are involved in electron transfer, even if the exact details of their functioning is not completely known. In other membranous systems, studies are scarce, but indicate the presence of b type cytochromes in microsomes and also in cellular fractions enriched in plasmalemma and perhaps in tonoplast membranes. Whether these b type cytochromes are involved in particular electron transport is not known, but they appear to be linked to special metabolism at certain steps of plant development: for example, these systems are inducible, much like those found in animal tissues, for detoxification. They are probably linked to the NAD or NADP cytochrome c reductases in these membranes, but their roles in metabolism are unknown; cytochrome c is certainly not the only physiological acceptor.

The presence of cytochromes in higher plants was reported in the first paper published by Keilin (1925) relating the presence in animal tissues and yeast of a new kind of pigment, the cytochrome. These discoveries were made possible by the use of the ocular microspectroscope and the Hartridge reversion spectroscope owing to their low dispersion. If MacMunn was in fact the first to have observed spectroscopically the cytochromes, which he named histohematin, his observations were only done on animal tissues and Keilin was the first to show their presence in higher plants.

Among the Japanese workers who, in the 1930's, developed numerous studies on cytochromes, Yakushiji (1934) did spectroscopic investigations on the presence of cytochromes in higher plants and algea. In soybean cotyledons he observed four bands characteristic of cytochromes, one of which, at 630–640 nm, he attributed to cytochrome a_2 (at that time found in some bacteria), in addition to the a, b, c components; moreover, he found that seed respiration was inhib-

ited by carbon monoxide. Other observations were made on either nonchloro-
phyllous tissues or acetone powder from green leaves (where he could see a
sharp band at 550 nm, which eventually shifted to between 550 and 565 nm).

In the same period, LUNDEGÅRDH and BURSTRÖM (1933, 1935) associated
anion uptake in roots with the part of respiration sensitive to cyanide or carbon
monoxide: they postulated an anionic respiration giving rise to an active transport of anions bound to a flux of electrons through an organized respiratory
system, including cytochromes (see LUNDEGÅRDH 1960b, for details). This concept was retained by MITCHELL (1961) when formulating the chemi-osmotic
theory and the proton pumping redox loops.

The basic ideas of anionic respiration bound to the cytochrome system were
initially in favor of a localization of the system near the outer surface of the
cytoplasm. When it was shown that the cytochrome system is localized in mitochondria, LUNDEGÅRDH proposed that the organelles must manage at times
to come into close contact with the outer plasma membrane, and then move
to the tonoplast (LUNDEGÅRDH 1960b).

However, the views of LUNDEGÅRDH were critized by SMITH and CHANCE
(1958) on the basis of slow kinetic responses of the cytochromes in relation
to anion uptake. They suggested that the main response in LUNDEGÅRDH's experiments was the one of a broad "cytochrome b" band, and that perhaps this
band could be related to a nonrespiratory hemoprotein system involved in anion
transport.

Knowledge of mitochondrial cytochromes rests on mitochondrial purity.
Higher plant mitochondria were isolated earlier by differential centrifugation
(MILLERD et al. 1951, GODDARD and BONNER 1960, BONNER 1967) and later
by density gradient centrifugation. DOUCE et al. (1972) developed a method
which is generally used now for the preparation of purified mitochondria which
have retained their outer membrane. LOOMIS (1974) has reviewed the many
pitfalls which must be bypassed for establishing grinding and assay media
adapted to the preparation of mitochondria from various plant materials. Mitochondria from leaves were more difficult to prepare than from nonchlorophyl-
lous tissues, due to chloroplast brittleness: chloroplast fragments heavily
contaminate mitochondrial fractions. However, in density gradient centrifugation of pea and tobacco leaf homogenates, DUCET (1960) could show a band
specifically stained by Janus green, which spectroscopically exhibited cytochrome bands and had a high cytochrome oxidase activity.

A number of reviews have been published on plant cytochromes including
a chapter by LUNDEGÅRDH in the first Encyclopedia of Plant Physiology (1960a).
The book, *Cytochromes*, by LEMBERG and BARRETT (1973) contains many references devoted to plant cytochromes. This chapter will focus mainly on more
recent developments rather than seeking to present a complete survey of the
subject.

2 Cytochrome Estimation

Keilin observed cytochromes and estimated their concentrations in tissues with the help of the microspectroscope, coupled to a double wedge trough to match absorbance in the specimen to that of a reference solution whose optical path length can vary continuously (see Hartree (1955) and Keilin (1966) for more details on this apparatus). The Hartridge reversion spectroscope has been used for precise measurements of the band location (accuracy of 0.1 nm) in the visible part of the spectrum. These spectroscopes are useful even now, as they allow a direct observation of living tissues or cellular preparations when illuminated with high intensity light. Moreover, the visual spectra show finer details than those obtained with recording spectrometers: the relatively large bandwidth of the latter minimizes the fine splitting of the absorption bands. Hill (1975) wrote a vivid chapter on the days of visual spectroscopy developed mainly in Great Britain.

Elliot and Keilin (1934), Keilin and Mann (1937) and Mann (1937) used the microspectroscope to study the distribution of hematin in plant tissues, by dithionite reduction followed by pyridine injection. The intensity of the pyridine-hemochromogen band was matched with an external standard (from crystallized hemin).

Low temperature (liquid air or liquid nitrogen, 77 K) intensifies absorption bands, and this was first applied by Keilin and Hartree (1949, 1950) to the study of cytochromes. They showed that intensification is maximal after establishment of a microcrystalline state. The absorption bands are sharpened, sometimes split and generally blue-shifted, but as freezing can induce aggregation resulting in a red shift, some components may have the same absorbance maximum at room and at low temperature. However, it was not really possible to quantify the cytochrome content of tissues and cell fractions until the advent of spectrometers adapted to scattering objects, pioneered by Chance (1954, 1957) and Lundegårdh (1951, 1960a).

Scattering of light transmitted through turbid samples gives rise to "sieve effects" and mutual shading, as discussed by Lundegårdh (1960a). Difference spectra help to eliminate these effects, if the changes result only from chemical differences. It is generally accepted that the light scattering is the same for the two samples in a difference spectrum, in the case of narrower wavelength regions. However, if physical state differences affect the chemical differences, distortion of the spectra is to be expected.

Estimation of the cytochrome content in tissues and organelles is generally achieved with the absorption coefficients established by Chance and Williams (1955a) for difference spectra. The molar extinction coefficients (MEC) are related to the absorption spectra of the pure compounds, and thus can apply only to cytochrome c (with different values according to the various origins). For the other cytochromes, MEC depends on the purity attained: for this reason the MEC vary for cytochrome oxidase, depending on the enzyme sources and authors. Two kinds of MEC are used in analysis of difference spectra: one reports the absorbance changes, at the wavelength of the maxima, between completely reduced or oxidized forms, whereas the other relates absorption changes at the wavelengths of the maxima to an isobestic point whose wavelength is derived from spectra of the pure compounds.

Simply applying MEC to the determination of cytochromes can lead to various errors. For example, there are cross-interferences due to band overlap, even in low-temperature spectra; there are also changes in the base line level and slope even for difference spectra, due to the dependence of wavelength of light scattering on redox states (as will be seen later). Various methods have been proposed to minimize these effects, either by computation of molecular extinction coefficients at different wavelengths for purified cytochromes (WILLIAMS 1964, VANNESTE 1966) or by differential reductions in difference spectra (MEUNIER 1973). Assuming that Beer's law is followed, as each cytochrome contributes to some extent (predictably) to the absorption maximum of the others, one derives a set of linear equations which can be solved. However, the precision is not good, since the exact contributions depend on the slope of the absorption bands and therefore on the bandwidth of the spectrometers.

CLAISSE et al. (1970) tried to compute the cytochrome contents of yeast cells from absolute spectra recorded on thick paste compressed between two lucite plates. As the light scattering was high in these conditions, they used two linear base lines (drawn empirically between 500 nm and either 560 or 630 nm) to measure the absorbances at the band maxima, and then took the mean values. Trying different paste thicknesses, they obtained linear responses between thickness and cytochrome c and a absorbances. By comparing cytochrome c absorbance to the real concentration (determined subsequently by extraction from the yeast), they could estimate the cytochrome c and a concentrations in normal and mutant yeasts. The method was extended to the determination of cytochrome c_1 in mutants of *Saccharomyces cerevisiae,* from low temperature spectra. As cytochromes c, c_1 and b have been purified from yeast, they computed the interference coefficient of each cytochrome to the absorption of the others, using spectra of the pure compounds. When applied to a complex spectrum, this leads to an easily solved 3*3 matrix, and the results obtained independently with different base lines were in good accordance. However, these interference coefficients cannot be used for spectra from other materials because the shapes of the low temperature spectra of the pure c and c_1 cytochromes are peculiar to yeast: there is no shoulder on the blue side of the alpha band of cytochrome c in contrast to mammalian and higher plant cytochromes c, but on the other hand there is a shoulder on the blue side of the alpha band of cytochrome c_1, which is absent in c_1 from higher plants (LANCE and BONNER 1968, DUCET and DIANO 1978 a).

Theoretically, knowing the shape of the spectra of each pure individual component, it is possible to analyze a complex spectrum and separate each individual contribution. Because some components have not been purified, are not stable, or exist only in complexes (as is the case for the different forms of chlorophylls), it has sometimes been assumed that the shape of absorption bands can be represented mathematically (i.e., Lorentzian, Gaussian functions; orthogonal polynomials). However, this is not applicable to the alpha bands of cytochromes. Derivatization of computed spectra has also been tried, using

1st, 2nd, and 4th derivatives, to estimate the number of constituents. Experimental errors become very important and interpretation of results difficult, as seen in the case of the multiple forms of chlorophylls. Denis and Deyrieux (1977) concluded that when neither the composite spectral forms nor their relative contributions are known, the problem of resolving the components is basically impossible to solve and only the number of components can be estimated (with some caution) from a few spectra. Additional information is necessary for the resolution of individual components. In principle, this is possible by simultaneously analyzing spectrometric and potentiometric data.

The method introduced by Dutton et al. (1970) has been theoretically refined especially by Denis et al. (1980a) and it is experimentally possible to distinguish the c and c_1 cytochromes in yeast mitochondria because of their different redox potentials in situ (Denis et al. 1980b). Recordings have been made either by the dual wavelength method of Chance (1954) or with a rapid scan spectrometer (Chantrel et al. 1967, Denis and Ducet 1975). During the course of the titration, a more or less reversible change of the opacity of the suspension was observed, indicative of light scattering changes. For correction of spectra, the absorbance changes were estimated in relation to a straight base line connecting points at 540 and 575 nm. From this line the relative contribution of the cytochromes c and c_1 was computed at various wavelengths, as well as their respective redox potentials. The results showed that the midpoint potential values of in situ cytochromes c (285 ± 5 mV) and c_1 (220 ± 10 mV) are the same as their values when isolated (290 ± 1.2 mV and 229 ± 2 mV). Their relative contributions were 2 to 1 at the maximum amplitude compared to values of 1.3–1.5 derived from the low temperature spectra by Claisse and Pajot (1974). Similar variations in light scattering during redox titrations were reported by Hendler and Schrager (1979) for *E. coli* membrane preparations, and were corrected by an analogous *three points method*. However, when analyses for yeast were done from the dual wavelength records, the midpoint potentials computed were 257 ± 2 mV for the c and 197 ± 22 mV for the c_1 cytochromes, and the relative contribution were 4–6 to 1. This could have led to the conclusion that different in situ and in vitro values corresponded to interactions of cytochromes with the membrane, but when the results were corrected for light scattering this was found not to be true.

Potentiometric titration is another way of computing contributions of various cytochromes to absorption at a given wavelength. This was applied by Dutton and Storey (1971) to the resolution of the cytochromes of mung bean mitochondria. However, the spectra computed may sometimes have more anomalous shapes than expected from experimental errors.

More information can be found on mitochondrial cytochromes, for isolation, purification and spectrometric measurements, in *Methods in Enzymology*, Volumes 52 and 53, edited by S. Fleischer and L. Packer, on biological oxidations.

3 Mitochondrial Cytochromes in Higher Plants

As observed by KEILIN (1925) in etiolated plant tissues, three types of cyto-chromes, a, b and c cytochromes, are localized in mitochondria, but some b's are also found in chloroplasts and cellular membranes (microsomes from the endoplasmic reticulum, and perhaps in plasmalemma and tonoplast), as constit-uents of peculiar electron transfer systems. A c type cytochrome, cytochrome f, is strictly localized in chloroplasts.

3.1 The c Cytochromes

Higher plant mitochondria contain two c cytochromes: c and c_1 as shown first by BONNER (1965) and LANCE and BONNER (1968)[1].

3.1.1 Cytochrome c

Plant cytochrome c was extracted from pollen by OKUNUKI (1939) and purified from wheat germ for the first time by GODDARD (1944), either by direct hot acetic acid extraction of the germs or by phosphate extraction of an acetone powder, with a yield of 4.3 mg kg^{-1} of wheat germ (about 0.35 nmol g^{-1}). The cytochrome c from wheat germ was crystallized by HAGIHARA et al. (1958). The amino acid sequence of this cytochrome was established by STEVENS and GANZER (1967), but the higher plant cytochromes c were more extensively stud-ied by BOULTER and coworkers and their sequences published after 1970. Extrac-tions were done from dark-grown seedlings, at acidic pH, in the presence of ascorbic acid and EDTA. The cytochrome c was absorbed first on Amberlite and purified subsequently on molecular sieves and cationic resins (RICHARDSON et al. 1970). Some improvements were introduced (RICHARDSON et al. 1971), by starting from 50 to 150 kg of seeds, germinated in the dark. DIANO and MARTINEZ (1971) started from gradient purified mitochondria (white potato), extracting the cytochrome c by salts according to JACOBS and SANADI (1960): after absorption on Amberlite this gives a starting product of higher quality.

Higher plant cytochromes c are characterized by a longer amino acid se-quence (mainly at the N terminal but also at the C terminal) than those from the mammalian tissues and the presence of two residues of a peculiar amino acid: trimethyl-lysine (DELANGE et al. 1969).

The phylogenetic implications of the sequences determined on more than 20 cytochromes c have been discussed (BOULTER et al. 1970, 1972, BOULTER 1973). The results indicate that in plants of the same family, sequences can differ in up to eight positions; however, they are more closely related to each other than to those from other families. [They are always found on the same line of descent on affinity trees derived from sequence data (BOULTER 1973)]. This means that considerable evolutionary diversity may exist. MARTINEZ et al.

---

1 Further data on plant cytochromes are presented by RAMSHAW (1982) in Vol. 14A of this encyclopedia (D. BOULTER and B. PARTHIER eds.) pp. 241–246. For photo-synthetic cytochromes see W.A. CRAMER's Chap. 12 in Vol. 5 of this encyclopedia (A. TREBST and M. AVRON eds.) pp. 237–246 (1977)

(1974) determined the potato cytochrome c sequence and found five differences in amino acid positions with respect to tomato, also a Solanaceae. This was criticized by BROWN and BOULTER (1975), who suggested that the assumption that peptides of the same electrophoretic mobility and the same amino acid composition are identical is not valid. There was, however, a strong homology between the sequences of potato and tomato cytochromes c confirming the observations of BOULTER (1973).

MORI and MORITA (1980) found that cytochrome c from rice had the highest content of proline (nine residues) of all the known cytochromes c and differs also by the presence of a serine residue at 96, instead of alanine as generally observed. However, this cytochrome shows the same degree of homology as the other plant cytochromes c.

3.1.2 Cytochrome c_1

Cytochrome c_1 was discovered by YAKUSHIJI and OKUNUKI (1940) in ox heart muscle as a band at 552 nm after reduction with succinate in the presence of cyanide. OKUNUKI and YAKUSHIJI (1941) reported some properties of this new cytochrome. It was rediscovered by KEILIN and HARTREE (1949), when they introduced the technique of low-temperature spectroscopy. They presented it as a new cytochrome: cytochrome e. However, it soon became evident that cytochrome e was identical with cytochrome c_1 (KEILIN and HARTREE 1955).

Cytochrome c_1 was postulated to be present in plant mitochondria (LUNDE-GÅRDH 1958, MARTIN and MORTON 1957, WISKICH et al. 1960), but the appearance of a band at 554 nm in the presence of HOQNO or amytal raised some doubts: BONNER (1961) rejected the presence of this cytochrome in plants because the component with an alpha band at 552 nm (at -190 °C) was not reduced by ascorbate. However, the published absorption spectra of mitochondria (CHANCE and HACKETT 1959, BONNER 1961) showed that the alpha band, at room temperature, was located around 551 nm instead of 550 nm as it should have been with cytochrome c as a single c type cytochrome in the respiratory chain: this indicated that there was another component present. The definitive evidence for cytochrome c_1 being a component of plant mitochondria was presented by BONNER (1965): when cytochrome c was extracted by extensive washing with NaCl, an ascorbate-reducible cytochrome remained strongly bound to the mitochondrial membrane, showing a symetrical band at 549 nm at 77 K. This was confirmed by LANCE and BONNER (1968) on *Helianthus tuberosus* mitochondria: when cytochrome c was extracted by phosphate washing, the residue showed at 77 K a symmetrical band centered at 549 nm. As mitochondria at 77 K show bands of equal magnitude only at 547, 552, and 557 nm, they concluded that the band at 552 nm did not correspond to the c_1 component, in contrast to the one found in animal or yeast mitochondria. MEUNIER et al. (1969) and DUCET et al. (1970b) extracted the cytochrome c from gradient purified white potato mitochondria. By comparing the mitochondria spectra before and after extraction, they concluded that cytochrome c_1 was present in equal quantities to cytochrome c. In difference spectra (ascorbate-TMPD reduced minus oxidized) bands were pinpointed at 552–553, 522, and 422–423 nm (room temperature).

Using methods developed with animal or yeast mitochondria, DUCET and DIANO (1971, 1978a) succeeded in extracting the b–c_1 complex from potato tuber mitochondria and then partially purified the c_1 component, contaminated with iron-sulfur compounds (Rieske factor, also present in plant mitochondria, as demonstrated later by PRINCE et al. 1981), which distort the Soret but not the alpha bands. The maxima were at 552–553 nm at room temperature, and 551 nm at 77 K. A characteristic feature of cytochrome c_1 is the low-temperature splitting of the beta band, narrower than the one of cytochrome c, when observed with a hand spectroscope. The position of the alpha band (551 nm at 77 K) is 2 nm farther to the red than that described by LANCE and BONNER (1968), and the same is true also for the c cytochrome (549 instead of 547 nm). The higher values are similar to the ones found for animal or yeast mitochondria, and the earlier low values may have just resulted from wavelength calibration problems.

By means of O_2 pulses, STOREY (1969) obtained a kinetic spectrum for the c cytochromes which could be differentiated by their different half-times of oxidation. Curiously enough, the half-time for oxidation of cytochrome c_1 (2.2 ms) is shorter than for cytochrome c (3.0 ms), in contrast with what can be expected if cytochrome c reacted directly with cytochrome a. However, MEUNIER et al. (1969), using white potato mitochondria depleted of cytochrome c also observed that ascorbate-TMPD reduced the cytochromes c_1 and a as if there were direct electron transfer between c_1 and a.

The midpoint potentials for the cytochromes c and c_1 were found to be indistinguishable by DUTTON and STOREY (1971), with a common value of 235 mV. This has been discussed previously in the section on cytochrome determination, with regard to the work of DENIS et al. (1980a, b).

3.2 The b Cytochromes

3.2.1 The Various Cytochromes b

The early reviews on plant cytochromes (HILL and HARTREE 1953, HARTREE 1955, SMITH and CHANCE 1958, LUNDEGÅRDH 1960a, HACKETT (1959, 1964) described numerous b cytochromes with absorption bands around 560 nm, following KEILIN'S (1925) definition. Their cellular location was undefined except for the mitochondrial b cytochrome and cytochrome b_6 confined to chloroplasts. The others, b_3, b_5, b_7 and some soluble cytochromes were attributed to various cellular structures including mitochondria. HILL and SCARISBRICK (1951) purified a soluble cytochrome b_3 from broad bean leaves and from etiolated tissues. SHICHI and HACKETT (1957a, b) and SHICHI et al. (1963a) isolated soluble cytochromes b from large batches of mung bean and subsequently purified two components: cytochrome b-555 (maxima in the reduced state 423, 527, 555–559 (double band) nm, at room temperature, and 426, 552, 559 nm at 77 K) and cytochrome b-561 [maxima 427, 531, 561 (single band) nm, at room temperature and 427, 560 nm at 77 K]. These preparations showed no peroxidase activity, and their midpoint redox potentials were −30 mV (b-555) and 0 mV (b-561) at pH 7.0. SHICHI et al. (1963b) also purified a cytochrome b-559 (maxima at

425, 529, 559 nm) from the same material, and suggested that it was identical to the b_3 of Hill and Scarisbrick (1951). Shichi and Hackett (1966), from the 77 K spectrum, which shows bands at 598, 564, 559, and 555 nm, suggested that cytochrome b-555 is present in mung bean microscomes as well as in mitochondria. Digestion of mung bean mitochondria with steapsin led to a soluble b component, spectrally identical to b-555. Raw and Mahler (1959) also obtained a soluble b_5 component in pig liver mitochondria digested with pancreatine. Shichi and Hackett (1966) proposed that b-555 could be a mediator in a "b shunt" in cyanide-insensitive respiration. However, considering the preparation procedure used by these authors (long duration, water extraction, precipitation of impurities at pH 5.0, in particular), it is probable that the soluble cytochromes b resulted from endogenous protease action on insoluble membranous cytochromes.

Some assays of solubilization of plant mitochondria were done with cholate used for extracting mammalian mitochondria cytochromes, as well as with other reagents (proteases, detergents). No clear results were obtained by Sisler and Evans (1959) (tomato roots), Bonner and Smith (1958) (skunk cabbage), Crane (1957) (cauliflower buds). Miller et al. (1958) obtained a cholate extract from soybean roots with cytochrome oxidase activity and some solubilization of cytochromes. As aroids often show a cyanide-insensitive respiration (see Chap. 8, this Vol.), their cytochrome content has been particularly investigated, as well as the oxidative activities of their mitochondria (Bendall and Hill 1956, Bendall 1958).

An important spectrometric study of plant mitochondria was carried out by Chance and Hackett (1959) on skunk cabbage organelles. These authors concluded, from difference spectra during which the a and c cytochromes were kept either completely reduced or oxidized, that three b type cytochromes were present: one autoxidizable in presence of cyanide (band maxima at 427, 558.5 nm), one reducible in presence of HOQNO (band maxima at 429.5, 560.5 nm), and one reduced by dithionite after anaerobiosis (band maxima at 425, 557.5 nm). At 77 K, the b components had a major peak at 556 nm with shoulder at 558, 567 nm. Absorbance changes (not clearly defined) were said to indicate a high content of b and c relatively to a_3 and more a_3 than a. It was concluded that the b component (b_7) is different in skunk cabbage from the one seen in *Arum* (Bendall and Hill 1956). Kinetic measurements performed with the regenerative flow apparatus of Chance and Williams (1955b), showed that the b components (followed at 558–540 nm), were active in electron transfer to the cytochromes c, a, a_3, oxidation of c preceding oxidation of b_7, which itself preceeded oxidation of b-563, as interpreted later by Bonner (In: Chance et al. 1968, p 315).

Yocum and Bonner (1958) also suggested that cytochrome b was probably a multicomponent, and this was developed at length at the Canberra symposium on hematin enzymes (Bonner 1961), along with the peculiar methodology used to spectrally separate the various components of mitochondria. At the same meeting Estabrook (1961) gave a general survey of spectrometric studies of cytochromes frozen in liquid nitrogen, among which results on wheat germ mitochondria reduced by succinate or dithionite were included.

BONNER's paper (1961) gave numerous examples of low temperature spectra under various conditions for mitochondria from different origins, and also for plant microsomes and chloroplasts. The recorded spectra showed clearly three distinct absorption bands at 554, 557, 562 nm in antimycin- or HOQNO-treated mitochondria, the observations being confined to the alpha and beta bands. Studies of the Soret bands by BONNER and VOSS (1961) have shown complexities which were resolved by BENDALL and BONNER (1966) upon demonstration of differences between the a and a_3 cytochromes in this spectral region: a has two bands (445 and 438 nm) and a_3 has one (445 nm).

These results were extended by LANCE and BONNER (1968) to tightly coupled mitochondria prepared from a variety of plant sources. In the presence of anti-mycin, the low-temperature difference spectra showed three b cytochromes with alpha bands at 552, 557, 561 nm, a beta band at 525 nm and another band at 534 nm, which is correlated with the one at 561 nm. Using the absorbance coefficients of CHANCE and WILLIAMS (1955a), they computed b cytochrome concentrations and found that b-552 and b-557 had similar concentrations, which were about twice that of b-561. However, errors may have occurred due to strong band overlapping in this study.

MEUNIER et al. (1969) and DUCET et al. (1970b) studied the b cytochromes of white potato mitochondria at room temperature: they found that antimycin further reduced the cytochromes in the presence of succinate. The spectra corresponding to the increased reduction in the presence of antimycin showed two absorption bands, 557–558 and 565 nm, while those only in the presence of succinate showed just one band at 560 nm. A b cytochrome reducible only by dithionite (band at 558 nm) was also present. It was also observed that with limiting quantities of antimycin, which retarded but did not completely inhibit transfer, a reoxidation of b-558 and b-565 occurred when cytochromes c and a, a_3, oxidized after addition of antimycin, were again reduced upon anaerobiosis. Subsequently the b cytochromes slowly became reduced. No explanation was given at that time, but these results were later found again and interpreted by LAMBOWITZ and BONNER (1973) as showing the anaerobic oxidation of the b cytochromes.

STOREY did a thorough optical and kinetic study of the respiratory chain of plant mitochondria, especially those from skunk cabbage and mung bean. STOREY (1969) and STOREY and BAHR (1969) showed that ascorbate-TMPD could reduce cytochromes b-560 (b-557 77 K) and b-556 (b-553 77 K) and that cytochrome b-566 (b-562 77 K) appeared free of interference from the others in a difference spectrum between energized and deenergized succinate-reduced, anaerobic mitochondria. Taking advantage of the different rates of reduction, STOREY (1969) succeeded in recording relatively pure difference spectra of b-557 (b-553, shoulder at 548, 77 K), and b-566 (b-562, ancillary band 555 nm, beta band 534, 77 K). Kinetic spectra of the cytochromes of mung bean mitochondria were obtained by means of O_2 pulses. It was shown that b-560 and b-566 reacted with a half-time of 8 ms and 16 ms respectively. However, when the spectra were recorded on a larger time scale, a complex spectrum was obtained with a maximum at 552 nm and a large shoulder on the red side. The maximum was attributed to c-552 (STOREY 1969), but it was later shown (DUTTON and

Table 1. Percent absorbance contribution of different cytochromes at selected wavelengths

Cytochrome component	Wave length in nm				
	549	552	556	560	565
	% Absorbance contribution				
c-549			26		
b-556	25	30	67		
b-560			7	78	43
b-566				22	57

Storey 1971) that the component had a mid-point potential of 75 mV and so could not be a c cytochrome. Therefore it seems that spectra calculated from kinetic measurements cannot be taken as equivalent to optical spectra.

Dutton and Storey (1971) applied the potentiometric method developed by Wilson and Dutton (1970) to the determination of cytochrome E_m values. The results for the b cytochromes were interpreted as showing the presence of three b entities with midpoint potentials of 75 ± 5 mV (b-556); 42 ± 5 mV (b-560); -77 ± 5 mV (b-566). It was subsequently revealed that 30% of the absorbance changes at 552 nm were due to b-560. Table 1 shows the relative contributions of the cytochromes at various wavelengths, calculated from the results of Dutton and Storey (1971):

Taking into account the real absorbance values at each wavelength, the resulting spectrum for cytochrome b-556 is evidently very unusual and cannot be taken as valid.

Lambowitz and Bonner (1974a, b) studied the b cytochromes, using spectrometric and potentiometric data. They followed the recommendation of the IUB nomenclature committee in naming according to the position of the alpha peak in difference spectra recorded at room temperature. They obtained partial spectral resolution of high and low potential b cytochromes: b-556 (b-553 at 77 K) and b-560 (b-557) were largely reduced by ascorbate (E_m from 60 to 80 mV); b-566 (b-563) was only partially reduced by succinate (E_m -75 mV) but completely by dithionite. Taking advantage of the increased reduction of the low potential b cytochromes in energized, compared to uncoupled, succinate reduced mitochondria, which results from reversed electron transport, they observed a component at 557–558 nm (b-554–555) along with b-566 (b-563). This compound was distinct from the high potential b-555. This is not a double band component for b-566, as sometimes proposed (but also dismissed by Wikström 1973) for animal mitochondria. Higuti et al. (1975) have shown that b-562, b-566, and b-558 can be reduced separately in rat liver mitochondria. Lambowitz and Bonner (1974b) found that antimycin induced an increased reduction of b-558 and b-566, as well as a shift of 1 to 2 nm to the red for b-560. The enhanced reduction was dependent upon the presence of oxygen or another suitable oxidant, and slowly relaxed after anaerobiosis, as shown earlier (Lambowitz and Bonner 1973). As previously reported by Lance and

Table 2. Spectral and potentiometric characteristics of cytochromes b from higher plants

Cytochrome component	Absorbance peaks		Mid point potential (mV)
	Room temp.	77 K	
b-556	556	553–554	75–100
b-560	560	557	40– 80
b-558	557–558	553–555	− 70 to − 105
b-566	566	561–563	− 75
b dith. red.	557–561	554, 560 dbl. band.	− 100

BONNER (1968), a dithionite-reducible b component could be resolved, in both room- and low-temperature spectra. This was present in relatively large amounts and did not seem to be an artifact due to either incomplete reduction of the known cytochromes b, or to contamination by microsomes or b cytochromes localized in the external mitochondrial membrane.

Neither NADH nor NAD-linked substrates reduced the b cytochromes further than did succinate; in fact b-566 stayed even more oxidized with these substrates than with succinate.

Potentiometric titrations were recorded according to DUTTON et al. (1970) and DUTTON and STOREY (1971). Table 2 summarizes the results of LAMBOWITZ and BONNER (1974b).

The uncertainty of the potential determinations can be explained by the relatively rough analysis method used. As discussed previously, the dual wavelength method prevents observation of light scattering and the results cannot be accepted as accurate. Experiments were also done to follow the response of b-566 to changes in energy state, since it had been proposed that this cytochrome plays a direct role in energy transduction by changing its midpoint potential between energized and nonenergized states (WILSON and DUTTON 1970). The results suggested that b-566 (and b-558) had an energy-linked function: in mung bean, after ATP addition, these cytochromes showed an increased reduction, reversed by uncouplers. In potato mitochondria, where the respiratory chain is inaccessible to ATP, there was an increased reduction of b-566 (and b-558) at anaerobiosis (suppressed by uncouplers) followed by a slow reoxidation after anaerobiosis. These results seemed to show a special role for cytochromes b-566 (and b-558) in energy transduction as they apparently changed their midpoint potential. To demonstrate directly such a change, one has to use mediators for measuring the potential. However, the b-566 reduction caused by ATP was collapsed by mediators and at the same time the rate of ATP hydrolysis doubled. This provided evidence that the increased reduction of b-566 and (b-558) upon addition of ADP resulted from reversed electron transport, the mediator bypassing a coupling site (LAMBOWITZ et al. 1974). WIKSTRÖM and LAMBOWITZ (1974) obtained the same result with rat liver mitochondria, when the mediator concentration was large enough to overcome the effect of reversed electron transfer. Thus, the ATP-induced E_m shift is likely to be due to reversed electron transfer and not to an energized form of cytochrome b-566.

Storey (1974a) observed that cytochrome b-566 can be reduced by succinate through an energy-linked process, but that this reaction was slow (as compared to reduction of endogenous pyridine nucleotides) and required constant input of energy, so that reoxidation was rapid when the energy source was exhausted. These results and others (Storey 1973) led him to conclude that cytochrome b-566 does not belong to the main sequence of the plant mitochondria respiratory chain. He tentatively suggested that cytochrome b-566 was part of the enzyme complex which carried out hydroxylations, producing ubiquinone from a precursor in the plant mitochondrial membrane. However, Storey (1978) upon comparing interactions of antimycin A or 8-hydroxyquinoleine (HOQNO) with the b cytochromes of mung bean mitochondria, later concluded, following the proposals of Von Jagow (1975), that antimycin A promoted formation of a complex between cytochromes b-560 and b-566. This is accompanied by a shift of the mid-point potential from -77 mV to $+5$ mV. The last value was calculated from the intramitochondrial redox potential, IMPh, (redox state of cytochrome b-560 whose E_m 7.2 is $+42$ mV) and the redox state of cytochrome b-566 at this IMPh (Storey 1973). HOQNO also induced a shift of the E_m 7.2, from -77 mV to -29 mV, but acted as an uncoupler.

These studies done by Storey, well summarized elsewhere (Storey 1980), are all based on the postulated existence of high energy chemical intermediates for oxidative phosphorylation. This argument was further developed later (Storey et al. 1981). However, the anomalous behavior of cytochrome b-566 can be analyzed in another way, as will be discussed below.

Douce et al. (1978) studied the effect of a terpene, beta pinene, on the oxidative properties of purified intact mung bean and white potato mitochondria. This compound acted as an uncoupler at low concentration and as an inhibitor of electron transfer at higher concentrations with malate and succinate, but not with ascorbate-TMPD, tetramethyl-p-phenylenediamine, as substrates. Difference spectra at 77 K showed that the b cytochromes were much less reduced by beta pinene than they were with antimycin and Douce et al. (1978) concluded that beta pinene inhibits the electron flow at the level of the quinone pool. As it did not increase the activity of antimycine-sensitive NADH-cytochrome c reductase, it was concluded that beta pinene was not a detergent. However, the difference spectrum presents anomalous bands between 450 and 520 nm and one is led to conclude that beta pinene is a chaotropic agent which causes the destruction of carotenoid pigments apparently present in gradient purified mung bean mitochondria in the same way as thiocyanate acts on skunk cabbage mitochondria (Bendall and Bonner 1971). This can also lead to cytochrome alterations, as shown by Dupont et al. (1982).

Apart from yeast and *Neurospora*, cytochromes in fungi have not been systematically studied. As stated in the introduction, it is not the purpose of this review to cover fungus cytochromes. However, it may be significant to report on the works of Gallinet (Gallinet 1978, Denis and Gallinet 1981) on mitochondria from the common mushroom *Agaricus bisporus*. Gallinet found only two b cytochromes (b-560 and b-566) in this mushroom by low-temperature

spectral studies, b-556 being lacking. Potentiometric studies (DENIS and GAL-
LINET 1981) using the method of DENIS et al. (1980a), also detected only two
b cytochromes with relatively high E_m values of 177 (or 192.9 mV) and 51.7 mV
respectively, but only one absorption maximum at 561 nm. Moreover, there
was no component detected with a potential of -70 mV, as expected for cytochrome b-566. Nonetheless spectral studies with and without antimycin showed
a component with the behavior and absorption band expected for cytochrome b-
566.

According to results from genetic investigations (WEISS 1976, CLAISSE et al.
1978, SPYRIDAKIS and CLAISSE 1978), yeast mitochondria cytochromes b derive
from one single structural gene. The different spectral and potentiometric properties related to the b cytochromes could result from differences in the environment of a single protein, not only in the mitochondrial membrane but also
in the "solubilized" b–c_1 complex. The results of DENIS and GALLINET can
also be interpreted as showing an effect of the redox potential on the light
scattering of mitochondria, leading to apparent peak displacement of absorption
bands.

No purification of the mitochondrial b cytochromes from higher plants has
been done yet, and therefore no molecular data are available to characterize
them relative to their mammalian or fungus counterparts.

3.2.2 The Cytochrome b–c_1 Complex

Despite the numerous studies done on the cytochrome b–c_1 complex from animal, yeast mitochondria (see WIKSTRÖM 1973, RIESKE 1976, for reviews) there
is only one report on the extraction of the same complex from higher plant
mitochondria. DUCET and DIANO (1978a) purified the b–c_1 complex from white
potato mitochondria by deoxycholate extraction, in the presence of KCl, followed by ammonium sulfate fractionation, in the presence of cholate, according
to the methods developed by RIESKE (1967) and ODA (1978). The complex was
dissociated according to YU et al. (1972), using beta mercaptoethanol to solubilize cytochrome c_1 contaminated with the Rieske iron-sulfur protein. The resulting insoluble cytochrome b was solubilized in phosphate buffer with the help
of guanidium hydrochloride. It was spectrally devoid of other cytochromes
and showed only one alpha band at room (560 nm) and at low (77 K, 557 nm)
temperature, while in the complex, three b cytochromes (553, 557, and 563 nm
at 77 K) were present. During the course of the dissociation, cytochrome b-563
disappeared first, followed by cytochrome b-553; cytochrome b-557 remained
insoluble. The spectrum of the isolated cytochrome b was very like the one
obtained by WEISS (1972) from *Neurospora*. However, as no protein assays
by SDS gel chromatography were done, the subunit composition could not
be ascertained, to verify if there was only one cytochrome b protein.

In *Neurospora* mitochondria, WEISS and KOLB (1979) separated succinate-
ubiquinone reductase and ubiquinone-cytochrome c reductase, each of them
containing cytochromes b genetically distinct, as deduced from the differential
effect of chloramphenicol on their biosynthesis. This is reminiscent of results
of YU et al. (1975), who found multiple cytochrome b proteins in succinate-

cytochrome c reductase from heart-muscle mitochondria. This would mean that only the b cytochrome of the bc_1 complex depends on the mitochondrial genome, and, as there is only one mitochondrial gene coding for it, the multiple cytochrome b forms reflect different topological arrangements of this protein in the complex.

3.2.3 Cytochromes b of the External Mitochondrial Membrane

There is only one report on the cytochromes present in the external membrane of higher plant mitochondria, by Moreau (1978). Using the method for preparation of external membrane developed by Moreau and Lance (1972), Moreau studied the electron transfer systems of outer membrane isolated from cauliflower mitochondria and compared them to those present in microsomes from the same tissue. Several b type cytochromes could be detected by difference spectrometry. Two cytochromes b, and even a third one, are present in the outer membrane: one reducible by ascorbate and NADH, cytochrome b-559, one reducible only by dithionite, cytochrome b-558, and perhaps a cytochrome b-554, reducible by NADH and not by ascorbate. This is similar to the b cytochromes studied in higher plant microsomes by Rich and Bendall (1975). They showed that cauliflower microsomes have a cytochrome analogous to the cytochrome b_5 of animal microsomes, with peak at 555 nm and E_m around 0 mV, a b cytochrome reducible by ascorbate (peak at 559.5 nm, $E_m + 135$ mV) and possibly components reducible only by dithionite. Moreover, cauliflower microsomes contain cytochrome P-450 and its degradation product, P-420. Moreau (1978) could not find cytochrome P-450 in the outer membrane of cauliflower mitochondria, which was correlated to the lack of NADPH-cytochrome c reductase activity in this membrane.

3.3 The a Cytochromes

As with mitochondria from other eukaryotes, higher plant mitochondria have two a cytochromes, a and a_3, which can be distinguished by their spectral properties but have not yet been separated. In fact only one heme a is known. Cytochrome oxidase is a complex of two hemes a, plus two copper compounds not yet well defined. It must be remembered that Keilin considered the oxidase to be the cytochrome a_3, only, because it was thought to be the one which reacted directly with oxygen; now cytochrome oxidase means the complex of a and a_3.

Apart from spectroscopic observations (Keilin 1925, Okunuki 1939), the first demonstration of the presence of cytochrome oxidase in etiolated plant tissues was by Hill and Bhagvat (1939). Rosenberg and Ducet (1949) demonstrated the presence of cytochrome oxidase in a chloroplastic fraction of leaves. McClendon (1953) showed that this enzyme was not localized in chloroplasts but in a lighter fraction, presumably representing mitochondria contaminated by chloroplast fragments. Bhagvat and Hill (1951) gave a survey of the presence of cytochrome oxidase in higher plants. The results were based on spectroscopic observations and measurements of oxidative activities, increased in the presence of cytochrome c. In higher plants the first spectrometric observations of cytochrome oxidase, among other cytochromes, were made by LUNDEGÅRDH (1951) on wheat roots, but the results were critized for their level of low resolution by SMITH and CHANCE (1958). BONNER and YOCUM (1956), YOCUM and HACKETT (1957), HACKETT and HAAS (1958), YOCUM and BONNER (1958), with more sensitive spectrometers, were able to measure the height of absorption bands and to calculate the concentrations and turnover numbers of the plant source cytochromes, including cytochrome oxidase, in tissue slices from various plant amongst which was the spadix of Araceae.

BONNER (1961) reported numerous low temperature spectra in which the a cytochromes were clearly seen with the alpha band at 597–598 nm. At room temperature, the alpha band is located at 603 nm instead of at 605 nm as in animal mitochondria. No beta bands for cytochrome oxidase are noticeable in mitochondrial spectra. The Soret bands have been shown by BENDALL and BONNER (1966) to be distinct for cytochrome a and a_3: the maxima are respectively at 438 nm and 445 nm (double band) for a, and at 445 nm (single band) for a_3. This was demonstrated by using cyanide, which inhibits electron transfer between a and a_3 by combining with the oxidized form of a_3. However, with skunk cabbage mitochondria, which have a cyanide-insensitive respiration, the cyanide complex of a_3 becomes relatively rapidly reduced after anaerobiosis, and BENDALL and BONNER (1966) could record a difference spectrum between reduced a_3 and a_3-CN, which was very similar to the one which YONETANI (1960) had described for purified cytochrome oxidase preparations (trough at 588 nm, peak at 447 nm).

LANCE and BONNER (1968) obtained similar results with mitochondria from different sources. They used azide to spectrally separate the a and a_3 components, because this also complexes only with cytochrome a_3. Another thorough study of the optical properties of cytochrome oxidase and its compounds with cyanide and azide was done by BENDALL and BONNER (1971) on mitochondria from skunk cabbage. The effects of azide were complex and difficult to interpret: azide seemed to induce a small (2 nm) shift to the red for cytochrome a. Cytochrome a_3 seemed to undergo oxidation and reduction in the presence of both cyanide and azide, which may indicate a slow leak through the inhibited cytochrome oxidase. Another interesting result from this study on skunk cabbage mitochondria was the presence in seemingly large quantities of carotenoid pigments. This could be seen in a difference spectrum when one sample was treated with thiocyanate (used to inhibit the alternate pathway but also a chaotropic agent). The presence of carotenoid pigments could result from contamination of the mitochondria by the microsomal fraction, since they were not gradient-purified. Aroid spadix contain a large amount of starch enclosed in amyloplasts which have an external membrane easily broken as shown by FISHWICK and WRIGHT (1980) for potato amyloplasts. These membranes contain, like other plastids, carotenoid pigments. However, mitochondria may have endogenous carotenoids, as shown by COSTES et al. (1976). This can also be derived from the results of DOUCE et al. (1978). In this last case, the presence of carotenoids in gradient-purified mung bean mitochondria is deduced from the difference spectrum in presence of beta pinene, the shape of which is very similar to the one recorded by Bendall and Bonner (1971) in the presence of thiocyanate. However, carefully prepared mitochondria show very low carotenoid peaks (Neuberger et al. 1982).

Analysis of the spectra obtained by Bendall and Bonner (1966) led Storey (1980) to suggest that absorbance of the band at 445 nm was 60% due to cytochrome a and 40% to cytochrome a_3, and that the band at 603 nm (599 nm at 77 K) could be attributed mostly to cytochrome a. Storey (1970b) obtained similar results, and applied them to measurements of oxidation rates of the cytochromes a by oxygen pulses in mung bean mitochondria. Using 438 nm (with 455 nm as reference wavelength) and 603 nm (620 nm reference), the recorded half-time was 2.0 ms. With 445 nm (455 nm reference) the half-time was 1.4 ms, giving the calculated value of 0.9 ms for oxidation of cytochrome a_3, close to that reported for animal mitochondria.

Bonner and Plesnicar (1967), with mitochondria from black valentine beans, *Phaseolus vulgaris,* observed a double band in the red (589 and 597 nm) in a difference spectrum between mitochondria having reached anaerobiosis in State 4 (energized state) and those under aerobic conditions. They suggested that this corresponded to a cross-over point between cytochrome a and cytochrome a_3, also seen by the different oxidation-reduction response of the two cytochromes to the addition of ATP. However, they followed the redox state of cytochrome a at 438 nm where a large interference from the b cytochromes (peak 425–430 nm) may occur: that the redox changes attributed to cytochrome a were in fact those of b cytochromes was shown later by Ducet (1978).

Dutton and Storey (1971) performed redox titrations of mung bean mitochondria in order to compute the midpoint potential of the a cytochromes: they reported values of $+380$ mV for cytochrome a_3 and $+190$ mV for cytochrome a. The gap of nearly 200 mV between the midpoint potentials seemed to confirm the suggestion of Bonner and Plesnicar (1967) that a site of energy conservation could be located between cytochromes a and a_3. This will be discussed later.

Inhibitors of cytochrome oxidase have been used from the beginning to demonstrate the enzyme activity in intact tissues and extracts. In isolated mitochondria, cyanide has been shown sometimes to inhibit oxygen uptake only partially, giving a basis to the long-known cyanide insensitivity of the respiration of various tissues from some higher plants. This is discussed at length in Lance et al. Chapter 8, this Volume. Ikuma and Bonner (1967) computed a K_i around 2 µM for inhibition of the cyanide-sensitive State 3 rate of oxygen consumption in plant mitochondria. Storey (1970b), in mung bean mitochondria partially sensitive to cyanide, observed a lengthening of the cycle time of oxidation and re-reduction following oxygen pulses when cyanide was present, compared to the cycle time in its absence. The concentration of cyanide which gave half maximum lengthening of the cycle time was 2 µM, and that which inhibited 50% cytochrome a oxidation was 30 µM, the latter representing the dissociation constant of the reduced cytochrome a_3 cyanide complex.

Carbon monoxide has been used to characterize cytochrome oxidase, due to the light reversibility of its inhibition of oxygen uptake: it is therefore possible to compute an action spectrum. This has been used by PLESNICAR et al. (1967) to differentiate between cytochrome oxidase and a peroxidase contamination of mitochondria from mung bean, since the carbon monoxide-peroxidase complex is not dissociated by light. The carbon monoxide complex of cytochrome oxidase exhibited peaks at 430 nm and 590 nm, which appeared in the uncompensated action spectrum for the release of oxygen uptake by carbon monoxide-inhibited mung bean mitochondria.

Cytochrome oxidase from animal mitochondria contains copper as shown by MORRISON et al. (1963). In plant mitochondria, extraction and purification of cytochrome oxidase has not reached a sufficient state of completion for copper analysis to be done and a stoichiometry between a cytochromes and copper proposed. However, an indirect demonstration of the need for copper has been given by BLIGNY and DOUCE (1977, 1978). These authors grew sycamore (*Acer pseudoplatanus* L.) cells on copper-deficient nutrient medium, and subsequently isolated mitochondria. Those from copper-deficient cells had a normal content of cytochromes, except for cytochrome oxidase, which was present at a level around 1/20th of that of the mitochondria from copper-sufficient cells. Mitochondria from copper-deficient cells had oxidative and phosphorylative activities identical to those from normal cells, but the turnover number for cytochrome oxidase in copper-deficient mitochondria was more than 20 times higher than that in copper-sufficient mitochondria. However, the cyanide sensitivity was changed: the half-maximal inhibition of oxygen uptake was obtained at greater concentration (20 μM) in the normal cells compared to copper-deficient cells (2 μM). No alternative oxidase pathway developed during copper deficiency. All these results showed that cytochromes a, a_3 are present in very large excess in higher plant mitochondria, as had been proposed earlier (DUCET and ROSENBERG 1962).

Few attempts have been made to purify cytochrome oxidase from higher plants. The enzyme was partially purified from potato mitochondria by DUCET et al. (1970a), using deoxycholate extraction in the presence of salts. The enzyme was spectroscopically pure, as judged from the absence of contaminating cytochromes b and c_1. Peaks of absorption bands were different from those in animal mitochondria. From the absorbance change induced by carbon monoxide, it was concluded that about equal amounts of cytochromes a and a_3 were present.

MILLER et al. (1958) used cholate and ammonium sulfate fractionation to extract cytochrome oxidase from soybean roots. Some activity was recovered, but spectral studies indicated contamination by b and c cytochromes. SRIVASTAVA and SARKISSIAN (1971) used also deoxycholate to solubilize cytochrome oxidase from mitochondria of several higher plants, but they measured only oxidative activities and associated K_m's to study the interspecific differences in cytochrome oxidase activity. BOMHOFF and SPENCER (1977) isolated cytochrome oxidase from etiolated pea seedlings using Triton but they, too, measured only oxidative activities after optimizing the assay conditions (pH and ionic strength).

MAESHIMA and ASAHI (1978, 1979) purified cytochrome oxidase from sweet potato roots (*Ipomoea batatas*) by deoxycholate extraction followed by DEAE cellulose column chromatography. The heme a content was 12.4 nmol mg^{-1}

protein, with a phospholid content of 2.5% from phosphorus determination. Absorption spectra were given for the oxidized and reduced forms, as well as a difference spectrum between these. The results were similar to those of Ducet et al. (1970a) for the wavelengths of the absorption peaks. They found five subunits, from SDS-urea polyacrylamide gel electrophoresis, with molecular weights of 39,000; 33,500; 26,000; 20,000 and 5,700, the last one probably being multiple. Maeshima and Asahi (1981a, b) also studied the biogenesis of cytochrome oxidase during aging of sweet potato slices: cytochrome oxidase activity increased during aging, but not if chloramphenicol (an inhibitor of mitochondrial protein synthesis) was present. Other results indicated that the two smaller subunits were cytoplasmically synthesized, while the three larger ones depended on mitochondrial protein synthesis. Matsuoka et al. (1981) studied the subunit composition of pea cytochrome oxidase and also found five subunits, but with higher molecular weights than for sweet potato: 39,000; 33,000; 28,500; 16,500; and 8,000–6,000. Moreover, two additional polypeptides (M.W. 13,000, 10,000) coprecipitated in an immunoprecipitate from partially purified pea cytochrome oxidase. As the cytochrome oxidase with five subunits was enzymatically active, they concluded that the two additional subunits were not involved in the oxidation of reduced cytochrome c, and they proposed that they participated in either the proton pump or the topological arrangement of the enzyme in the membrane. These results show that higher plant cytochrome oxidase differs from the cytochrome oxidase isolated from animal and fungal cells which apparently contain six to eight subunits (Malmström 1979).

The functioning of cytochrome oxidase is complex and not completely understood. In higher plants, only a few assays have been tried by Denis and co-workers, using potato mitochondria. Denis and Bonner (1978) used the low temperature flash photolysis technique of Chance et al. (1975) to investigate the cytochrome oxidase-oxygen reaction in mitochondria treated with ferricyanide. This treatment induces what is called the mixed valence state of cytochrome oxidase, in which heme a and its associated copper, Cu_A, are oxidized while heme a_3 and its associated copper, Cu_B, are reduced. The complex was initially stabilized by carbon monoxide and low temperature ($-120\,°C$), after which it was cautiously mixed with dissolved oxygen. A flash of strong light dissociated the carbon monoxide complex and then, by increasing the temperature to $-100\,°C$, oxygenation began at a slow rate which could be followed optically by repetitive scanning between appropriate wavelengths (sweep rate of 45 s per scan). The first complex formed (A_2) is supposed to correspond to oxygen binding, with no electron transfer. Then follows the formation of compound C in which Cu_B is oxidized and the oxygen reduced. The sequence was suggested to be as follows:

$$a^{3+}Cu^{2+}a_3^{2+}Cu^{1+}+O_2 \rightarrow a^{3+}Cu^{2+}a_3^{2+}Cu^{1+}O_2 \rightarrow a^{3+}Cu^{2+}a_3^{2+}Cu^{2+}O_2^-$$

The results of Denis and Bonner (1978) suggested a lower affinity for oxygen of potato cytochrome oxidase than the one of beef heart. They investigated the effect of cyanide and found that it prevented compound A_2 formation, but not that of compound C. Moreover, compound C formation showed two

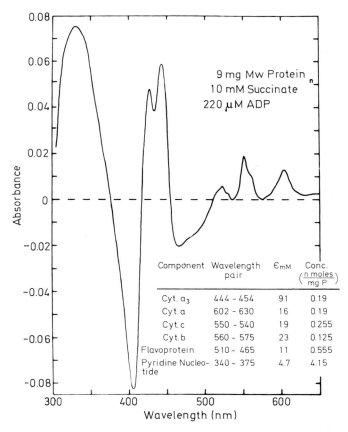

Fig. 1. Difference spectrum of mung bean hypocotyl mitochondria between anaerobic (succinate reduced) and aerobic states at room temperature (BOWMAN et al. 1976). *Table in inset* shows the wavelength pairs, the mM absorbance coefficients (ε) used and the calculated content (in nmol mg^{-1} protein) of each electron carrier

phases when observed at 609 nm and only one when followed at 597 nm: this anomalous behavior suggested that heme a_3 may have special environments in potato mitochondria.

DENIS and CLORE (1981) used rapid wavelength scanning (DENIS and DUCET 1975) to follow the first steps after flash photolysis. As the spectra were numerically recorded, computation by sophisticated methods was performed to obtain best fit computed kinetic curves and spectra. The previous results of DENIS and BONNER (1978) were confirmed: only one compound C was found instead of two as in beef heart mitochondria (DENIS 1981); the reaction was much slower at 173 K in potato mitochondria than in the beef heart system, by a factor ranging from 3 to 6; the C compound exhibited a split alpha band with a prominent peak at 607 nm and a side peak at 594 nm; the A_2 compound had a shoulder at 584 nm and a peak at 591 nm. The split alpha band of

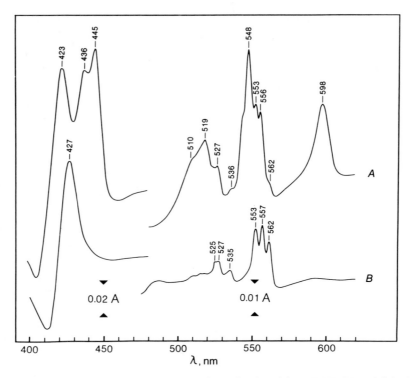

Fig. 2. Difference spectra of potato tuber mitochondria at 77 K (Unpublished experiments of Bligny and Douce). Protein concentration: 3.3 mg ml^{-1}. Optical path: 2 mm. *A* Difference spectrum between anaerobic (succinate reduced) and aerobic preparation. *B* Difference spectrum between succinate reduced mitochondria, in presence of antimycin A (5 µg ml^{-1}) and aerobic preparation

compound C indicated the presence of two electronic events (charge transfer transitions from O_2^-).

That potato mitochondria cytochrome oxidase reacts differently from the one of animal mitochondria, was further shown by Denis and Richaud (1982) when they studied the dynamics of carbon monoxide recombination to fully reduced cytochrome oxidase of potato mitochondria after flash photolysis. A four-step mechanism was found, even when the CO concentration was decreased to 0.2%, lower than the cytochrome oxidase concentration in the assay. Dual wavelength scanning experiments failed to detect optical forms correlated with the resolved phases. These results, which excluded a diffusion effect for the carbon monoxide recombination, are in good accord with the existence of a heme pocket in cytochrome oxidase. Richaud (1983) extended their results and found that in potato mitochondria Cu_A and Cu_B are spectrally visible, contrary to mammalian mitochondria in which the Cu_B is silent. This quite new Cu_B difference spectrum (red − ox) is characterized by a broad positive band centered at 812 nm with an extinction coefficient of 4.3 mM^{-1} cm^{-1}, while the Cu_A difference spectrum (red − ox) has a trough at 812 nm and an extinction coefficient of $-2.0\ mM^{-1}\ cm^{-1}$, analogous to its mammalian counterpart.

Figures 1 and 2 summarize the spectral observations on plant cytochromes. Figure 1 is a typical room-temperature difference spectrum of mung bean mitochondria, showing the cytochrome bands located at 427, 530, 560 nm (cytochromes b), 522, 552 nm (cytochromes c) and 443, 603 nm (cytochromes a). The trough at 466 nm and the band at 332 nm are indicative of respectively reduced flavoproteins and pyridine nucleotides. Figure 2 shows low temperature difference spectra of purified potato tuber mitochondria when reduced by succinate with and without antimycin A. The characteristic bands of the b cytochromes are clearly seen: b-562 (with beta band at 535 nm), and b-553 and b-557. The a cytochrome has its split band at 436 and 445 nm, while cytochrome a_3 has only one band at 445 nm.

4 Influence of the Membrane Potential on the Redox States of the Mitochondrial Cytochromes

BONNER and PLESNICAR (1967) have shown that ADP addition to mitochondria oxidizing substrates (induction of State 3) induced an oxidation of the b cytochromes, followed spectrally at 560 nm. When State 4 became established (ADP exhausted), the b cytochromes became re-reduced. DUCET (1978) made similar observations when following the b cytochromes at 427 nm versus 410 nm, and found that these were the only ones to show a redox response when ADP was added to the mitochondria: the c cytochromes did not exhibit redox changes, when followed at 552 nm (referred to 540 nm) and if the a cytochromes displayed oxidation and reduction, when followed at 440 nm (refered to 455 nm), this was due to interference from the b cytochromes at this wavelength. When bovine serum albumine (BSA) was added in the assay medium the amplitude of the redox changes, induced by ADP, was increased, as well as the respiratory coupling (RC). DUCET and DIANO (1978 b) repeated these experiments, recording the spectra by rapid scan spectrometry between 370 and 625 nm: it was then possible to compute difference spectra between states 3 and 4, and the results indicated that only the b cytochromes exhibited a redox response, the amplitude of which was increased in presence of BSA. In this case, at anaerobiosis, cytochrome a, a_3 reached a level of reduction which remained steady, at least for 2 min, while a biphasic reduction of these cytochromes was observed when BSA was missing, the second phase being slow. More than 1 min was needed before the final level of reduction was reached, and the final extent of reduction was about double that in presence of BSA, as if, in this last case, only part of cytochrome aa_3 was available to reduction.

BSA, which is used as chelator for free fatty acids, was shown to decrease the proton conductance of potato mitochondria membranes (DUCET 1979a). It was therefore logical to try the effects of uncouplers (which increase the proton conductance) on the behavior of the cytochromes during oxidation of substrates and the accompanying phosphorylation. Ducet (1979b) investigated by rapid scan spectrometry the oxidation of succinate by potato mitochondria, recording spectra every 2 s from before addition of succinate until extended anaerobiosis, followed by a final reduction with dithionite. Phosphorylation steps were included during aerobiosis. Difference spectra could then be computed at will. Apart from the already known redox changes of the b cytochromes between States 3 and 4, a fast change in the base line level was observed during the transition from State 4 to State 3 (addition of ADP), in the form of absorbance increase in the whole spectrum. The change was even faster than the redox change of the b cytochromes and was completed in less than 0.1 s, which is about the mixing time of ADP to the cuvette content. On return to State 4, the base line returned to the initial level concomitantly upon reduction of the b cytochromes with a half-time of about 4 s. Difference spectra exhibited peaks at 603, 551, 446–440, and 421 nm, with no shoulder on the red side of the 551 nm peak: in energized mitochondria (having reached anaerobiosis in State 4), no more reduction of b cytochromes had occurred, and some oxidation was even detected. Difference spectra between dithionite reduced mitochondria and anaerobic ones, exhibited a weak band at 605 nm, a large one centered at 560 nm and large Soret bands with two distinct peaks at 444 and 430 nm. When an uncoupler was added during aerobiosis, the b cytochromes were partially oxidized at once and at anaerobiosis the Soret bands exhibited two distinct peaks around 444 and 430 nm; further reduction by dithionite showed only a b cytochrome spectrum. When the uncoupler was added after anaerobiosis, a distinct peak at 445–442 nm was seen instead of a shoulder on the red side of the b cytochromes which were partially re-oxidized. These results indicate that cytochrome a_3 remained oxidized at anaerobiosis, while cytochrome a (and the other components of the chain) were reduced, as if electron transfer was inhibited between the two cytochromes at anaerobiosis.

These results were obtained with coupled mitochondria oxidizing succinate, but were extended to the oxidation of other substrates with similar results (Ducet 1981), with the exception that for alpha-ketoglutarate oxidation the reduction of the b cytochromes in aerobiosis was about half that exhibited for succinate or ascorbate oxidation. In anaerobiosis, after addition of uncoupler, oxidation of the b cytochromes was shown to precede the extra reduction of the a cytochromes.

The behavior of the b cytochromes was interpreted by Bonner and Plesnicar (1967) as suggesting a cross-over point (meaning a site of energy conservation in the sense of Chance and Williams 1956) between the b and c cytochromes. The same suggestion can account for the different behavior of cytochromes a and a_3 between States 3 and 4, but we raised some doubts on this interpretation (see above).

The interpretation is more straightforward when it is assumed that electron transfer induces a proton motive force, consisting of a combination of a proton gradient and a membrane potential according to Mitchell (1969). A membrane potential influences directly the apparent redox potential of any component of the electron transfer chain which senses the potential, as explained by Mit-

CHELL (1969), WIKSTRÖM (1973) and more thoroughly by WALZ (1979):

$$E''_h = E'_h - \psi \quad \text{and} \quad E''_m = E'_m + \psi$$

where E_h is the redox potential, E_m is the midpoint potential and $''$ and $'$ are the regions of potential respectively ψ and 0.

For mitochondria, numerous results indicate that oxidation of external substrates induces a membrane potential, negative inside. As discussed by MITCHELL (1969), the spatial distribution of the potential is unknown, but the equipotential surfaces are probably complex, because of the heterogeneous membrane. However, one can predict a lowering of the apparent E_m for a compound included in the membrane and sensing the potential: it would appear more oxidized than it really is. If the membrane potential decreases, the compound will seem to become more reduced.

As ADP decreases the membrane potential in plant mitochondria (MOORE et al. 1978, DUCET et al. 1983) the b cytochromes in the mitochondrial membrane will appear more oxidized in State 3 than in State 4. The redox behavior of the b cytochromes can be considered therefore as a probe for membrane potential changes during the transitions between States 3 and 4. It was computed (DUCET 1980) that the 35 mV difference in the membrane potential between State 4 and State 3, measured by MOORE et al. (1978) for potato mitochondria, was sufficient to account for the observed changes in cytochrome b oxidation.

That cytochrome a_3 is not reduced at anaerobiosis has been observed previously by OSHINO et al. (1974): in this study it was slowly reduced after anaerobiosis. This was suggested to be due to a high phosphate potential, opposing the reduction, which was destroyed after anaerobiosis by an active ATPase present in the mammalian mitochondria. This explanation does not apply to the results of DUCET (1980, 1981), because potato mitochondria have very low ATPase activity, which is not increased by uncouplers. In this case the anomalous behavior of cytochrome a_3 at anaerobiosis can be explained by assuming that this compound senses the membrane potential, while cytochrome a is not influenced by it. It thus seems that the two hemes are located in two different subunits and the one which possesses the binding site for cytochrome c (and heme a) is accessible from the cytoplasmic side of the inner membrane (DEPIERRE and ERNSTER 1977).

It therefore appears that cytochromes b (or part of them) and a_3 can sense the membrane potential, whereas cytochromes c and a do not sense it. However, changes in the base line level during the transition State 4 to State 3, which are faster than the redox changes of the b cytochromes, can be interpreted as a primary event, leading subsequently to the redox changes. As these redox changes correspond to electronic reequilibration in an electric field, we suggest that the primary event is a change in the membrane potential. This change (charge separation?) is very fast, because the estimated time of 0.1 s probably reflects the time taken to mix ADP with the mitochondrial suspension.

In photosynthesis, the primary event is thought to lead to electrochromism of some of the photosynthetic pigments (JUNGE and JACKSON 1982) and this causes a shift of the absorption bands of some carotenoids, and consequently gives rise to characteristic difference spectra. This has been quantitatively analyzed and the response (of absorbance of the bands in difference spectra) is approximately proportional to the intensity of the electric field.

However, in mitochondrial difference spectra, anomalous bands which could be interpreted as electrochromism effects are not observed. We therefore suggested another interpretation for the fast changes of the base line level, considering it as an electrostriction effect, apparent as variations of mitochondria membrane thickness induced by changes of the membrane potential. Intuitively, one can sense that these thickness changes are translated in refractive index alterations, leading to scattering modulation. The fast change in base line level is effectively due to a variation in light scattering in the same way as swelling and shrinking of mitochondria cause volume changes which also lead to light scattering alterations.

5 Miscellany

If something is now known about mitochondrial cytochrome biosynthesis in yeast or *Neurospora,* the only reports concerning higher plant mitochondria were established from studies on cytoplasmic male sterility, among which work by Forde and Leaver (Forde and Leaver 1979, 1980, Forde et al. 1978). Fox and Leaver (1981) could identify a mitochondrial gene, coding for cytochrome oxidase subunit II, in *Zea mays*. The review by Leaver and Forde (1980) can also be consulted.

6 Cytochromes in the Respiratory Chain of Higher Plant Mitochondria

All studies on plant as well as fungal and mammalian mitochondria lead to eliminate the early concept of a linear electron transport chain in favor of respiratory components assembled in complexes, electron transfer occurring through collision controlled processes (Schneider et al. 1980). The isolated complexes II (succinate ubiquinone reductase), III (b–c_1 complex) and IV (cytochrome oxidase) contain cytochromes but, until now, their spatial arrangement and functioning are under investigation without giving a definitive answer, in particular as concerns their proton pumping activities. This is discussed at length by Moore and Rich (Chap. 6, this Vol.) on the organisation of the respiratory chain. The situation is analogous to that in photosynthesis, with the existence of complexes, including a bf one, and the separate location of the two photosynthetic systems.

References

Bendall DS (1958) Cytochromes and some respiratory enzymes in mitochondria from the spadix of *Arum maculatum*. Biochem J 70:381–390

Bendall DS, Bonner WD (1966) Optical properties of plant cytochrome oxidase. In: Chance B, Estabrook RW, Yonetani T (eds) Hemes and hemoproteins. Academic Press, London New York, pp 485–488

Bendall DS, Bonner WD (1971) Cyanide-insensitive respiration in plant mitochondria. Plant Physiol 47:236–245

Bendall DS, Hill R (1956) Cytochrome components in the spadix of *Arum maculatum*. New Phytol 55:206–212

Bhagvat K, Hill R (1951) Cytochrome oxidase in higher plants. New Phytol 50:112–120

Bligny R, Douce R (1977) Mitochondria of isolated plant cells (*Acer pseudoplatanus*). II Copper deficiency effects on cytochrome c oxidase and oxygen uptake. Plant Physiol 60:675–679

Bligny R, Douce R (1978) The cytochrome c oxidase of higher plant cells (*Acer pseudoplatanus* L): Copper deficiency effects and turnover number. In: Ducet G, Lance C (eds) Plant mitochondria. Elsevier/North Holland Biomedical Press, Amsterdam New York, pp 43–50

Bomhoff G, Spencer M (1977) Optimum pH and ionic strength for the assay of cytochrome c oxidase from pea cotyledon mitochondria. Can J Biochem 55:1114–1117

Bonner WD (1961) The cytochromes of plant tissues. In: Falk JE, Lemberg R, Morton RK (eds) Haematin enzymes. Pergamon Press, Oxford New York, pp 479–497

Bonner WD (1963) Higher plant cytochromes. Proc 5th Int Congr Biochem, vol II. Pergamon Press, Oxford New York, pp 50–62

Bonner WD (1965) Mitochondria and electron transport. In: Bonner J, Varner JE (eds) Plant biochemistry. Academic Press, London New York, pp 89–123

Bonner WD (1967) A general method for the preparation of plant mitochondria. In: Eastbrook RW, Pullman ME (eds) Methods in enzymology, vol X. Academic Press, London New York, pp 126–133

Bonner WD, Plesnicar M (1967) Electron transport carriers in plant mitochondria. Nature (London) 214:616–617

Bonner WD, Smith S (1958) Properties of cytochrome isolated from plant tissues. Plant Physiol 33:vii

Bonner WD, Voss DO (1961) Some characteristics of mitochondria extracted from higher plants. Nature (London) 1961:682–684

Bonner WD, Yocum CS (1956) Spectroscopic and enzymatic observations on the spadix of skunk cabbage. Plant Physiol 31:xli

Bowman EJ, Ikuma H, Stein HJ (1976) Citric acid cycle activity in mitochondria isolated from mung bean hypocotyls. Plant Physiol 58:426–432

Boulter D (1973) In: Bendz G, Santesson J (eds) Chemistry in botanical classification. Nobel Symp 25. Academic Press, London New York, pp 211–216

Boulter D, Thompson EW, Ramshaw JAM, Richardson M (1970) Higher plant cytochrome c. Nature (London) 228:552–554

Boulter D, Ramshaw JAM, Thompson EW, Richardson M, Brown RH (1972) A phylogeny of higher plants based on the aminoacid sequences of cytochrome c and its biological implications. Proc R Soc London Ser B 181:441–455

Brown R, Boulter D (1975) A re-examination of the amino acid sequence data of cytochrome c from *Solanum tuberosum* (potato) and *Lycopersicum esculentum* (tomato). FEBS Lett 51:66–67

Chance B (1954) Spectrophotometry of intracellular respiratory pigments. Science 120:767–775

Chance B (1957) Techniques for the assay of the respiratory enzymes. In: Colowick SP, Kaplan NO (eds) Methods in enzymology, vol IV. Academic Press, London New York, pp 273–329

G. Ducet:98 G. Ducet:

Chance B (1961) Energy transfer and conservation in the respiratory chain. In: Falk JE, Lemberg R, Morton RK (eds) Haematin enzymes. Pergamon Press, Oxford New York, pp 597–622

Chance B, Hackett DP (1959) The electron transfer system of skunk cabbage mitochondria. Plant Physiol 34:33–49

Chance B, Williams GR (1955a) Respiratory enzymes in oxidative phosphorylation. II Difference spectra. J Biol Chem 217:395–407

Chance B, Williams GR (1955b) Respiratory enzymes in oxidative phosphorylation. IV. The respiratory chain. J Biol Chem 217:429–438

Chance B, Williams GR (1956) The respiratory chain and oxidative phosphorylation. Adv Enzymol 17:65–134

Chance B, Bonner WD, Storey BT (1968) Electron transport in respiration. Annu Rev Plant Physiol 19:295–320

Chance B, Saronio C, Leigh JS (1975) Functional intermediates in the reaction of cytochrome oxidase with oxygen. J Biol Chem 250:9226–9237

Chantrel H, Denis M, Baldy A (1967) Spectromètre à réseau à exploration rapide. Rev Phys Appl 2:245–248

Claisse ML, Pajot PF (1974) Presence of cytochrome c_1 in cytoplasmic "petite" mutants of *Saccharomyces cerevisiae*. Eur J Biochem 49:49–59

Claisse ML, Péré-Aubert GA, Clavilier LP, Slonimski PP (1970) Méthode d'estimation de la concentration des cytochromes dans les cellules entières de levure. Eur J Biochem 16:430–438

Claisse ML, Spyridakis A, Wambier-Kluppel ML, Pajot P, Slonimski FP (1978) Mosaic organization and expression of the mitochondrial DNA region controlling cytochrome c reductase and oxidase. II Analysis of proteins translated from the box region. In: Bacila M, Horecker BL, Stoppani AWN (eds) Biochemistry and genetics of yeast. Academic Press, London New York, pp 369–390

Costes C, Burghoffer C, Carrayol E, Ducet G, Diano M (1976) Occurrence of carotenoids in nonplastidial materials from potato tuber cells. Plant Sci Lett 6:253–259

Crane FL (1957) Electron transport and cytochromes of subcellular particles from cauliflower buds. Plant Physiol 32:619–625

DeLange RJ, Glazer AN, Smith EL (1969) Presence and location of an unusual aminoacid, N-Trimethyllysine, in cytochrome c, from wheat germ and *Neurospora*. J Biol Chem 244:1385–1388

Denis M (1981) Resolution of two compound C-type intermediates in the reaction with oxygen of mixed-valence state membrane-bound cytochrome oxidase. Biochim Biophys Acta 634:30–40

Denis M, Bonner WD (1978) Visible spectral properties of compound C observed from cytochrome oxidase-oxygen reaction in ferricyanide pretreated potato mitochondria. In: Ducet G, Lance C (eds) Plant mitochondria. Elsevier/North Holland Biomedical Press, Amsterdam New York, pp 35–42

Denis M, Clore GM (1981) Reaction of mixed valence state cytochrome oxidase with oxygen in plant mitochondria. A study by low temperature flash photolysis and rapid wavelength scanning optical spectrometry. Plant Physiol 68:229–235

Denis M, Deyrieux R (1977) Is multicomponent spectra analysis coming to a deadlock? J Theor Biol 69:301–309

Denis M, Ducet G (1975) Spectrométrie rapide. Application à l'étude de phénomènes biologiques. Cas des milieux structurés. Physiol Vég 13:621–632

Denis M, Gallinet JP (1981) Optical and potentiometric study of the b and c type cytochromes in mushroom *Agaricus bisporus* Lge mitochondria. Plant Physiol 68:658–663

Denis M, Richaud P (1982) Dynamics of carbon monoxide recombination to fully reduced cytochrome c oxidase in plant mitochondria after low temperature flash photolysis. Biochem J 206:379–385

Denis M, Neau E, Blein JP (1980a) Multicomponent redox systems: analytical representation $A(U_h)$, method of analysis and theoretical resolution limits. Bioelectrochem Bioenerg 7:757–773

Denis M, Neau E, Agalidis I, Pajot P (1980b) Multicomponent redox systems: potentiometric resolution of cytochromes c and c_1 in yeast mitochondria. Bioelectrochem Bioenerg 7:775–785

DePierre J, Ernster L (1977) Enzyme topology of intracellular membranes. Annu Rev Biochem 46:201–262

Diano M, Ducet G (1971) Isolement d'un cytochrome b et du cytochrome c_1 des mitochondries du tubercule de pomme de terre. CR 273:943–945

Diano M, Martinez G (1971) Le cytochrome c des mitochondries du tubercule de pomme de terre. CR Acad Sci 272:2692–2694

Douce R, Christensen EL, Bonner WD (1972) Preparation of intact plant mitochondria. Biochim Biophys Acta 275:148–160

Douce R, Neuburger M, Bligny R, Pauly G (1978) Effects of beta pinene on the oxidative properties of purified intact plant mitochondria. In: Ducet G, Lance C (eds) Plant mitochondria. Elsevier/North Holland Biomedical Press, Amsterdam New York, pp 207–214

Ducet G (1960) Fractionnement cellulaire de feuilles par centrifugation en gradient de densité. Ann Physiol Vég 2:19–28

Ducet G (1978) Action de la sérum albumine sur les mitochondries du tubercule de pomme de terre. Physiol Vég 16:753–772

Ducet G (1979a) An investigation by rapid scan spectrometry of succinate oxidation by coupled potato tuber mitochondria. Plant Physiol 63:(suppl) 147

Ducet G (1979b) Influence of bovine serum albumin on the proton conductance of potato mitochondria membranes. Planta 147:122–126

Ducet G (1980) Succinate oxidation by coupled potato tuber mitochondria followed by rapid scan spectrometry. Physiol Plant 50:241–250

Ducet G (1981) Rapid scan spectrometry: oxidation of substrates by coupled potato mitochondria. Physiol Plant 52:161–166

Ducet G (1982) Respiratory coupling in potato tuber mitochondria. Effect of mitochondrial aging. Physiol Vég 20:187–199

Ducet G, Diano M (1978a) On the dissociation of the cytochrome bc_1 of potato mitochondria. Plant Sci Lett 11:217–226

Ducet G, Diano M (1978b) Spectral changes followed by rapid scan spectrometry during succinate oxidation by potato tuber mitochondria. In: Ducet G, Lance C (eds) Plant mitochondria. Elsevier/North Holland Biomedical Press, Amsterdam New York, pp 51–58

Ducet G, Rosenberg AJ (1962) Leaf respiration. Annu Rev Plant Physiol 13:171–200

Ducet G, Diano M, Denis M (1970a) Purification de la cytochrome oxydase des mitochondries du tubercule de pomme de terre. CR Acad Sci 270:2288–2291

Ducet G, Cotte-Martinon M, Coulomb P, Diano M, Meunier D (1970b) Les mitochondries du tubercule de pomme de terre. Physiol Vég 8:35–54

Ducet G, Gidrol X, Richaud P (1983) Membrane potential changes in coupled potato mitochondria. Physiol Vég 21:385–394

Dupont J, Rustin P, Lance C (1982) Interaction between mitochondrial cytochromes and linoleic acid hydroperoxide. Possible confusion with lipoxygenase and alternative pathway. Plant Physiol 69:1308–1314

Dutton PL, Storey BT (1971) The respiratory chain of plant mitochondria. IX. Oxidation-reduction potentials of the cytochromes of mung bean mitochondria. Plant Physiol 47:282–288

Dutton PL, Wilson DF, Lee CP (1970) Oxidation reduction potentials of cytochromes in mitochondria. Biochemistry 9:5077–5082

Elliot KAC, Keilin D (1934) The haematin content of horse-radish peroxidase. Proc R Soc London Ser B 114:210–222

Estabrook RW (1961) Spectrophotometric studies of cytochromes cooled in liquid nitrogen. In: Falk JE, Lemberg R, Morton RK (eds) Haematin enzymes. Pergamon Press, Oxford New York, pp 436–457

Fishwick MJ, Wright AJ (1980) Isolation and characterization of amyloplast envelope membranes from *Solanum tuberosum*. Phytochemistry 19:55–59

Forde BG, Leaver CJ (1979) Mitochondrial genome expression in maize: possible involvement of variant mitochondrial polypeptides in cytoplasmic male sterility. In: Davies DR, Hopwood DA (eds) The plant genome. Innes Charity, Norwich, pp 131–146

Forde BG, Leaver CJ (1980) Nuclear and cytoplasmic genes controlling synthesis of variant mitochondrial polypeptides in male-sterile maize. Proc Natl Acad Sci USA 77:418–422

Forde BG, Oliver RJC, Leaver CJ (1978) Variation in mitochondrial translation products associated with male-sterile cytoplasm in maize. Proc Natl Acad Sci USA 75:2841–3845

Fox TD, Leaver CJ (1981) The zea mays mitochondrial gene coding cytochrome oxidase subunit II has an intervening sequence and does not contain TGA codons. Cell 26:315–323

Gallinet JP (1978) Redox properties of two cytochromes b in *Agaricus campestris* (F.) var bisporus mitochondria. In: Ducet G, Lance C (eds) Plant mitochondria. Elsevier/North Holland Biomedical Press, Amsterdam New Yok, pp 27–34

Goddard D (1944) Cytochrome c and cytochrome oxidase from wheat germ. Plant Physiol 31:270–276

Goddard DR, Bonner WD (1960) Cellular respiration. In: Steward FC (ed) Plant physiology, vol Ia. Academic Press, London New York, pp 209–312

Hackett DP (1959) Respiratory mechanisms in higher plants. Annu Rev Plant Physiol 10:111–146

Hackett DP (1964) Enzymes of terminal respiration. In: Linskens F, Sanwal BD, Tracey MV (eds) Modern methods of plant analysis, vol VII. Springer, Berlin Göttingen Heidelberg, pp 647–694

Hackett DP, Haas DW (1958) Oxidative phosphorylation and functional cytochromes in skunk cabbage mitochondria. Plant Physiol 33:27–32

Hagihara B, Tagawa K, Morikawa I, Shin M, Okunuki K (1958) Crystalline cytochrome c. Crystallization of cytochrome c from wheat germ. Nature (London) 181:1588–1590

Hartree EF (1955) Haematin compounds. In: Paech K, Tracey MC (eds) Modern methods of plant analysis, vol IV. Springer, Berlin Göttigen Heidelberg, pp 197–245

Hendler RW, Shrager RI (1979) Potentiometric analysis of *Escherichia coli* cytochromes in the absorbance optical range of 500 nm to 700 nm. J Biol Chem 254:11288–11299

Higuti T, Mizuno S, Muraoka S (1975) Stepwise reduction of cytochromes b-562, b-566 and b-558 in rat liver mitochondria. Biochim Biophys Acta 396:36–47

Hill R (1975) Days of visual spectroscopy. Annu Rev Plant Physiol 26:1–11

Hill R, Bhagvat K (1939) Cytochrome oxidase in flowering plants. Nature (London) 143:726–728

Hill R, Hartree EF (1953) Hematin compounds in plants. Annu Rev Plant Physiol 4:115–150

Hill R, Scarisbrick R (1951) The haematin compounds of leaves. New Phytol 50:98–111

Ikuma H, Bonner WD (1967) Properties of higher plant mitochondria. III. Effects of respiratory inhibitors. Plant Physiol 42:1535–1544

Jagow G von (1975) The thermodynamics of the antimycin dependent reduction of cytochrome b-566 under exclusion of electron flow. In: Quagliariello E, Papa S, Palmieri F, Slater EC, Silipandri N (eds) Electron transfer chains and oxidative phosporylation. North Holland Publ Comp, Amsterdam Oxford, pp 23–30

Jacobs EE, Sanadi DR (1960) The reversible removal of cytochrome c from mitochondria. J Biol Chem 235:531–534

Junge W, Jackson JB (1982) The development of electrochemical potential gradient across photosynthetic membranes. In: Govindjee (ed) Photosynthesis vol I. Academic Press, London New York, pp 589–646

Keilin D (1925) On cytochrome, a respiratory pigment, common to animals, yeast, and higher plants. Proc R Soc London Ser B 98:312–339

Keilin D (1966) The history of cell respiration and cytochrome. Cambridge Univ Press, Cambridge

Keilin D, Hartree EF (1949) Effect of low temperature on the absorption spectra of haemoproteins; with observations on the absorption spectrum of oxygen. Nature (London) 164:254–259

Keilin D, Hartree EF (1950) Further observations on absorption spectra at low temperature. Nature (London) 165:504–505

Keilin D, Hartree EJ (1955) Relationship between certain components of the cytochrome system. Nature (London) 176:200–206

Keilin D, Mann T (1937) On the haematin compound of peroxidase. Proc R Soc London Ser B 122:119–133

Lambowitz AM, Bonner WD (1973) The effect of antimycin on the b-cytochromes of plant mitochondria. Oxidation-reduction behavior of cytochrome b-566. Biochem Biophys Res Commun 52:703–711

Lambowitz AM, Bonner WD (1974a) The b cytochromes of plant mitochondria. In: Ernster L, Estabrook RW, Slater EC (eds) Dynamics of energy-transducing membranes. Elsevier, Amsterdam, pp 77–92

Lambowitz AM, Bonner WD (1974b) The b-cytochromes of plant mitochondria. A spectrophotometric and potentiometric study. J Biol Chem 249:2428–2440

Lambowitz AM, Bonner WD, Wikström MKF (1974) On the lack of ATP-induced midpoint potential shift for cytochrome b-566 in plant mitochondria. Proc Natl Acad Sci USA 71:1183–1187

Lance C, Bonner WD (1968) The respiratory chain components of higher plant mitochondria. Plant Physiol 43:756–766

Leaver CJ, Forde BG (1980) Mitochondrial genome expression in higher plants. In: Leaver CJ (ed) Genome expression in plants. Plenum Press, New York London, pp 407–425

Lemberg R, Barrett J (1973) Cytochromes. Academic Press, London New York

Loomis WD (1974) Overcoming problems of phenolics and quinones in the isolation of plant enzymes and organelles. In: Fleischer S, Packer L (eds) Methods in enzymology, vol XXXI: Biomembranes, pt A. Academic Press, London New York, pp 528–544

Lundegårdh H (1951) Spectroscopic evidence of the participation of the cytochrome–cytochrome oxidase system in the active transport of salts. Ark Kem 3:69–79

Lundegårdh H (1958) Spectrophotometric investigations on enzyme systems in living objects. III Respiratory enzymes in homogenates of wheat roots. Biochim Biophys Acta 27:355–365

Lundegårdh H (1960a) The cytochrome-cytochrome oxidase system. In: Ruhland W (ed) Encyclopedia of plant physiology, vol XII/1. Springer, Berlin Göttingen Heidelberg, pp 311–364

Lundegårdh H (1960b) Anion respiration. In: Ruhland W (ed) Encyclopedia of plant physiology, vol XII/2. Springer, Berlin Göttigen Heidelberg, pp 185–233

Lundegårdh H, Burström H (1933) Untersuchungen über die Salzaufnahme der Pflanzen. III Quantitative Beziehungen zwischen Atmung und Anionenaufnahme. Biochem Z 261:235–251

Lundegårdh H, Burström H (1935) Untersuchungen über die Atmungsvorgänge in Pflanzenwurzeln. Biochem Z 277:223–251

Maeshima M, Asahi T (1978) Purification and characterization of sweet potato cytochrome c oxidase. Arch Biochem Biophys 187:423–430

Maeshima M, Asahi T (1979) Sweet potato cytochrome c oxidase: its properties and biosynthesis in wounded root tissue. In: King TE, Orii Y, Chance B, Okunuki K (eds) Cytochrome oxidase. Elsevier/North Holland Biomedical Press, Amsterdam New York, pp 375–382

Maeshima M, Asahi T (1981a) Mechanism of increase in cytochrome c oxidase activity in sweet potato root tissue during aging of slices. J Biochem (Tokyo) 90:391–397

Maeshima M, Asahi T (1981b) Presence of an inactive protein immunologically analogous to cytochrome c oxidase in the inner membrane of sweet potato root mitochondria. J Biochem (Tokyo) 90:399–406

Malmström BG (1979) Cytochrome oxidase. Structure and catalytic activity. Biochim Biophys Acta 549:281–303

Mann T (1937) Haematin compounds in plants and their relation to peroxidase. Doct Thes, Cambridge Univ

Martin EM, Morton RK (1957) Haem pigments of cytoplasmic particles from non-photosynthetic plant tissues. Biochem J 65:404–413

Martinez G, Rochat H, Ducet G (1974) The amino acid sequence of cytochrome c from *Solanum tuberosum* (potato). FEBS Lett 47:212–217

Matsuoka M, Maeshima M, Asahi T (1981) The subunit composition of pea cytochrome c oxidase. J Biochem (Tokyo) 90:649–655

McClendon JH (1953) The intracellular localization of enzymes in tobacco leaves. II Cytochrome oxidase, catalase, polyphenol oxidase. Am J Bot 40:260–266

Meunier D (1973) Les cytochromes des mitochondries du tubercule de pomme de terre. Thes Doct Etat, Univ Marseille-Luminy

Meunier D, Diano M, Ducet G (1969) Caractérisation des cytochromes de type c et de type b dans les mitochondries du tubercule de pomme de terre (*Solanum tuberosum*). CR Acad Sci 269:1070–1073

Miller GW, Evans HJ, Sisler E (1958) The properties of cytochrome oxidase in cholate extracts from soybean roots. Plant Physiol 33:124–131

Millerd A, Bonner J, Axelrod B, Bandurski R (1951) Oxidative and phosphorylative activity of plant mitochondria. Proc Natl Acad Sci USA 37:855–862

Mitchell P (1961) Coupling of phosphorylation to electron and hydrogen transfer by a chemi-osmotic type of mechanism. Nature (London) 191:144–148

Mitchell P (1969) Chemiosmotic coupling and energy transduction. In: Cole A (ed) Theoretical and experimental biophysics. Dekker, New York London, pp 160–216

Moore A, Bonner WD, Rich PR (1978) The determination of the proton motive force during cyanide insensitive respiration in plant mitochondria. Arch Biochem Biophys 186:298–306

Moreau F (1978) The electron transport system of outer membranes of plant mitochondria. In: Ducet G, Lance C (eds) Plant mitochondria. Elsevier/North Holland Biomedical Press, Amsterdam New York, pp 77–84

Moreau F, Lance C (1972) Isolement et propriétés des membranes externes et internes de mitochondries végétales. Biochimie 54:1335–1348

Mori E, Morita Y (1980) Amino acid sequence of cytochrome c from rice. J Biochem (Tokyo) 87:249–266

Morrison M, Horie S, Mason HS (1963) Cytochrome c oxidase components. II A study of the copper in cytochrome c oxidase. J Biol Chem 238:2220–2224

Neuberger M, Journet EP, Bligny R, Carde JP, Douce R (1982) Purification of plant mitochondria by isopycnic centrifugation in density gradients of Percoll. Arch Biochem Biophys 217:312–323

Oda T (1968) Macromolecular structure and properties of mitochondrial cytochrome $(b+c_1)$ complex, cytochrome oxidase, and ATPase. In: Okunuki K, Kamen MD, Sekuzu I (eds) Structure and function of cytochromes. Univ Tokyo Press, Tokyo, pp 500–515

Okunuki K (1939) Über den Gaswechsel der Pollen II. Acta Phytochim 11:27–64

Okunuki K, Yakushiji E (1941) Über die Eigenschaften und die Verbreitung der c1-Komponente des Cytochroms. Proc Imp Acad Jpn 17:263–265

Oshino N, Sugano T, Oshino R, Chance B (1974) The redox behavior of mitochondrial cytochrome oxidase during aerobic-anaerobic transition and its relation to ATP hydrolysis. In: Ernster L, Estabrook RW, Slater EC (eds) Dynamics of energy transducing membrane. Elsevier, Amsterdam New York, pp 201–214

Plesnicar M, Bonner WD, Storey BT (1967) Peroxidase associated with higher plant mitochondria. Plant Physiol 42:366–370

Prince RC, Bonner WD, Bershak PA (1981) On the occurence of the Rieske iron-sulfur cluster in plant mitochondria. Fed Proc 40:1667 (735)

Raw R, Mahler HR (1959) Studies on electron transport enzymes. III. Cytochrome b_5 of pig liver mitochondria. J Biol Chem 234:1867–1869

Rich R, Bendall DS (1975) Cytochrome components of plant microsomes. Eur J Biochem 55:333–341

Richardson M, Laycock MV, Ramshaw JAM, Thompson EWS, Boulter D (1970) Isolation and purification of cytochrome c from some species of higher plants. Phytochemistry 9:2271–2280

Richardson M, Richardson D, Ramshaw JAM, Thompson EW, Boulter D (1971) An inproved method for the purification of cytochrome c from higher plants. J Biochem (Tokyo) 69:811–813

Richaud P (1983) Contribution à l'étude de la cytochrome c oxydase dans les mitochondries de tubercule de pomme de terre (*Solanum tuberosum* L.). Thes Doct 3. Cycle, Univ Marseille-Luminy

Rieske JS (1967) Preparation and properties of reduced coenzyme Q-cytochrome c reductase (Complex III of the respiratory chain). Meth Enzymol 10:239–245

Rieske JS (1976) Composition, structure, and function of complex III of the respiratory chain. Biochim Biophys Acta 456:195–247

Rosenberg AJ, Ducet G (1949) Activité cytochromoxydasique chez l'épinard. CR Acad Sci 229:391–393

Schneider H, Lemasters JJ, Höchli M, Hackenbrock CR (1980) Fusion of liposomes with mitochondrial inner membranes. Proc Natl Acad Sci USA 77:442–446

Shichi H, Hackett DP (1957a) Studies of the b type cytochromes from mung bean seedlings. I. Purification of cytochromes b-555 and b-561. J Biol Chem 237:2955–2958

Shichi H, Hackett DP (1957b) Studies of the b type cytochromes from mung bean seedlings. II. Some properties of cytochromes b-555 and b-561. J Biol Chem 237:2959–2964

Shichi H, Hackett DP (1966) A possible role for cytochrome b-555 in the mung bean mitochondrial electron transport system. J Biochem (Tokyo) 59:84–88

Shichi H, Hackett DP, Funatsu G (1963a) Studies on the b type cytochromes from mung bean seedlings. III. The structure of cytochrome b-555. J Biol Chem 238:1156–1161

Shichi H, Kasinsky HE, Hackett DP (1963b) Studies on the b cytochromes from mung bean seedlings. IV. Purification and properties of cytochrome b_3 (b-559). J Biol Chem 238:1162–1166

Sisler EC, Evans HJ (1959) Electron transport mechanisms in tobacco roots. I. Studies on the cytochrome and related systems. Plant Physiol 34:81–90

Smith L, Chance B (1958) Cytochromes in plants. Annu Rev Plant Physiol 9:449–482

Spyridakis A, Claisse ML (1978) Yeast cytochrome b: Structural modifications in yeast mitochondrial mutants. In: Ducet G, Lance C (eds) Plant mitochondria. Elsevier/North Holland Biomedical Press, Amsterdam New York, pp 11–18

Srivastava HK, Sarkissian IV (1971) Purification of cytochrome oxidase from mitochondria of higher plants. Phytochemistry 10:977–980

Stevens FC, Ganzer AN (1967) The amino acid sequence of wheat germ cytochrome c from mitochondria. J Biol Chem 242:2764–2779

Storey BT (1969) The respiratory chain of plant mitochondria. III. Oxidation rates of the cytochromes c and b in mung bean mitochondria reduced with succinate. Plant Physiol 44:413–421

Storey BT (1970a) The respiratory chain of plant mitochondria. IV. Oxidation rates of the respiratory carriers of mung bean mitochondria in the presence of cyanide. Plant Physiol 45:447–454

Storey BT (1970b) The respiratory chain of plant mitochondria. V. Reaction of reduced cytochromes a and a_3 in mung bean mitochondria with oxygen in the presence of cyanide. Plant Physiol 45:455–460

Storey BT (1970c) The respiratory chain of plant mitochondria. VIII. Reduction kinetics of the respiratory chain carriers of mung bean mitochondria with NADH as substrate. Plant Physiol 46:625–630

Storey BT (1971) The respiratory chain of plant mitochondria. XI. Electron transport from succinate to endogenous pyridine nucleotide in mung bean mitochondria. Plant Physiol 48:694–701

Storey BT (1972a) The respiratory chain of plant mitochondria. XII. Some aspects of the energy-linked reverse electron transport from the cytochromes c to the cytochromes b in mung bean mitochondria. Plant Physiol 49:314–322

Storey BT (1972b) The respiratory chain of plant mitochondria. XIII. Redox state changes of cytochrome b-562 in mung bean seedling mitochondria treated with antimycin A. Biochim Biophys Acta 267:48–64

Storey BT (1973) The respiratory chain of plant mitochondria. XV. Equilibration of cytochromes c-549, b-553, b-557 and ubiquinone in mung bean mitochondria: placement of cytochrome b-557 and estimation of the mid-point potential of ubiquinone. Biochim Biophys Acta 292:592–603

Storey BT (1974a) The respiratory chain of plant mitochondria. XVI. Interaction of cytochrome b-562 with the respiratory chain of coupled and uncoupled mung bean mitochondria: evidence for its exclusion from the main sequence of the chain. Plant Physiol 53:840–845

Storey BT (1974b) The respiratory chain of plant mitochondria. XVII. Flavoprotein-cytochrome b-562 interaction in antimycin treated skunk cabbage mitochondria. Plant Physiol 53:846–850

Storey BT (1978) Some interactions of antimycin A, 2-heptyl-4-hydroxyquinoline-oxide (HOQNO), ATP, and fluorooxalacetate (FOAA) with the b cytochromes of mung bean mitochondria. In: Ducet G, Lance C (eds) Plant mitochondria. Elsevier/North Holland Biomedical Press, Amsterdam New York, pp 19–26

Storey BT (1980) Electron transport and energy coupling in plant mitochondria. In: Davies DD (ed) The biochemistry of plants, a comprehensive treatise, vol II: Metabolism and respiration. Academic Press, London New York, pp 125–195

Storey BT, Bahr JT (1969) The respiratory chain of plant mitochondria. I. Electron transport between succinate and oxygen in skunk cabbage mitochondria. Plant Physiol 44:115–125

Storey BT, Lee CP, Wikström M (1981) Discussion forum: is transmembrane proton electrochemical potential essential to mitochondrial energy coupling? Trends Biochem Sci 6:166–170

Vanneste WM (1966) Molecular proportion of the fixed cytochrome components of the respiratory chain of Keilin-Hartree particles and beef heart mitochondria. Biochim Biophys Acta 113:175–178

Walz D (1979) Thermodynamics of oxidation-reduction reaction and its application to bioenergetics. Biochem Biophys Acta 505:279–353

Weiss H (1972) Cytochrome b in Neurospora crassa mitochondria. A membrane protein containing subunits of cytoplasmic and mitochondrial origin. Eur J Biochem 30:469–478

Weiss H (1976) Subunit composition and biogenesis of mitochondrial cytochrome b. Biochim Biophys Acta 456:291–313

Weiss H, Kolb HJ (1979) Isolation of mitochondrial succinate ubiquinone reductase, cytochrome c reductase and cytochrome c oxidase from Neurospora crassa using non-ionic detergent. Eur J Biochem 99:139–149

Wikström MKF (1973) The different cytochrome b components in the respiratory chain of animal mitochondria and their role in electron transport and energy conservation. Biochim Biophys Acta 301:155–193

Wikström MKF, Lambowitz AM (1974) The interaction of redox mediators with the "second phosphorylation site"; significance for the cytochrome b_T hypothesis. FEBS Lett 40:149–153

Williams JN (1964) A method for the simultaneous quantitative estimation of cytochromes a, b, c_1 and c in mitochondria. Arch Biochem Biophys 107:537–543

Wilson DF, Dutton PL (1970) Energy dependent changes in the oxidation-reduction potentials of cytochrome b. Biochem Biophys Res Commun 39:59–64

Wiskich JT, Morton RK, Robertson RN (1960) The respiratory chain of beet root mitochondria. Aust J Biol Sc 13:109–122

Yakushiji E (1934) Über das Vorkommen des Cytochroms in höheren Pflanzen und in Algen. Acta Phytochim 8:325–329

Yakushiji E, Okunuki K (1940) Über eine neue Cytochromkomponente und ihre Funktion. Proc Imp Acad Jpn 16:299–302

Yocum CS, Bonner WD (1958) Electron transport in *Symplocarpus foetidus*. Plant Physiol 33: vi

Yocum CS, Hackett DP (1957) Participation of cytochromes in the respiration of the aroid spadix. Plant Physiol 32:187–191

Yonetani T (1960) Studies on cytochrome oxidase. I. Absolute and difference absorption spectra. J Biol Chem 235:845–852

Yu CA, Yu L, King TE (1972) Preparation and properties of cardiac cytochrome c_1. J Biol Chem 247:1012–1019

Yu CA, Yu L, King TE (1975) The presence of multiple cytochrome b proteins in succinate-cytochrome reductase. Biochem Biophys Res Commun 66:1194–1200

5 The Outer Membrane of Plant Mitochondria

C.A. MANNELLA

1 Perspective

The current definition of mitochondrial compartments derives from a liason established in the 1950's between biochemistry and electron microscopy. The double membrane nature of this organelle was initially recognized by electron microscopists around 1953 (see SJØSTRAND 1977). A few years later, biochemists divided the mitochondrion into sucrose-penetrable and -nonpenetrable volumes (e.g., WERKHEISER and BARTLEY 1957) which were assigned in collaborative biochemical/ultrastructural studies to the inter-membrane and matrix spaces respectively (PFAFF et al. 1968). (However, these assignments were not unambiguous; see TEDESCHI 1965).

Once it was established (after considerable early confusion) that all the components of oxidative phosphorylation are contained on or in the space bounded by the inner mitochondrial membrane, interest in the outer membrane waned. Its one notable characteristic is the apparent ease by which it is penetrated by low molecular-weight molecules and ions (WERKHEISER and BARTLEY 1957, O'BRIEN and BRIERLEY 1965). But as recently as 1977, reviewers were undecided whether this extreme permeability was due to damage incurred by the outer membrane during mitochondrial isolation or, in stead to some "remarkable" architectural feature of this membrane (DEPIERRE and ERNSTER 1977).

While it is true that some (especially early) cell fractionation procedures can be damaging to organellar membranes, this is not necessarily the case. In the particular instance of higher plants, application of density gradient centrifugation procedures has led to isolation of mitochondrial fractions with predominantly intact outer as well as inner membranes (DOUCE et al. 1972a). Outer membrane integrity can be assayed in terms of the accessibility of exogenous cytochrome c ($M_r = 12,000$) to inner membrane redox sites. The general finding is that metabolite diffusion through intact outer mitochondrial membranes does not appear to be rate-controlling for inner membrane oxidation reactions in vitro. However, there is increasing evidence that the outer membrane of liver mitochondria detectably limits the accessibility of substrates to intermembrane-space enzymes (BRDICZKA 1978, ULVÍK 1983). Thus, while the outer mitochondrial membrane may be very permeable to small molecules, it is not necessarily "freely" permeable, as sometimes claimed.

Since the available biochemical data indicate that all specific mitochondrial carrier systems are confined to the inner membrane, what accounts for the large permeability of the intact outer mitochondrial membrane? The answer, at the time of writing, appears to be numerous, large, nonspecific passive diffusion channels in the outer mitochondrial membranes of all eukaryotes. These as yet incompletely described outer membrane components are at the forefront of a recent renewal of interest in this membrane system. The first evidence for the occurrence of pores in the mitochondrial outer membrane was provided by electron micrographs of higher plant mitochondria, a full decade prior to their functional detection (in vitro) and subsequent isolation (PARSONS et al. 1965; SCHEIN et al. 1976).

What follows (except for Sect. 5.5) is a summary of the current state of understanding of the composition, structure, and (to a lesser extent) function of the outer membrane of plant mitochondria. The issue of the biological role of this membrane cannot be adequately addressed since, to a large degree, the interesting questions are just beginning to be asked. In addition to being the first barrier to the passage of small molecules and ions into the mitochondrion, the outer membrane probably mediates several other interactions between the mitochondrion and the rest of the cell. There is evidence that the outer membrane influences mitochondrial shape, for example in sperm (HRUDKA 1978). Defects in mitochondrial outer membrane conformation may be directly responsible for sperm tail disorganization in selenium-deficient mice (WALLACE et al. 1981). Involvement of the outer membrane in mitochondrial movement is suggested by numerous electron microscopic observations of direct mitochondrial contacts with microtubules (e.g., SMITH et al. 1977) and microfilaments (e.g., BRADLEY and SATIR 1979). Similarly there are several observations of viruses directly attached to the outer mitochondrial membranes of infected plant cells (see MUNN 1974). Are there specific receptors for these entities on the outer mitochondrial membrane? There is growing evidence for specific outer-membrane receptors for cytosolic enzymes (e.g., hexokinase in mammalian mitochondria, FELGNER et al. 1979) and for hormones which may directly effect mitochondrial activity (triiodothyronine in mammals, HASHIZUME and ICHIKAWA 1982; phytochrome in plants, ROUX et al. 1981). Whether hormonal signals might be transmitted between outer and inner membranes is undetermined. There is evidence of close coordination between outer and inner mitochondrial membranes in the process of import of cytosolically synthesized protein precursors (see review by ADES 1982). Identification and characterization of the outer-mitochondrial membrane components involved in this mechanism and in the others mentioned above is just underway and so will not be addressed further in this review.

2 Isolation of Mitochondrial Membranes

The state of confusion brought about by early attempts at mitochondrial membrane fractionation was due to a large extent to the use of overly drastic, essentially uncontrolled disruption procedures. Two protocols which eventually gave true separation of the outer and inner membranes of rodent liver mitochondria involved lysis of the outer membrane (1) by osmotic swelling of the inner membrane (PARSONS et al. 1966, 1967) or (2) by the action of the cardiac glycoside, digitonin (SCHNAITMAN et al. 1967, SCHNAITMAN and GREENAWALT 1968, LEVY et al. 1967) followed by density gradient centrifugation. In both cases mitochondrial membrane integrity was monitored either ultrastructurally or biochemically so that outer membrane lysis could be maximized while minimizing inner membrane disruption. Successful separations of outer and inner membranes from mitochondria of several plant tissues were subsequently reported which employed either osmotic swelling techniques (mung bean, DOUCE et al. 1972b, 1973; cauliflower, MOREAU and LANCE 1972; potato, MANNELLA and BONNER 1975a; castor bean, SPARACE and MOORE 1979) or digitonin treatment (potato, MEUNIER et al. 1971; turnip, DAY and WISKICH 1975).

A critical experimental evaluation of the two mitochondrial fractionation procedures (hypoosmotic swelling versus digitonin action) applied to one kind of plant mitochondria has not yet been undertaken. There are several general conclusions that might be drawn from available data on the fractionation of both plant and animal mitochondria (cf. above-mentioned reports and a comparative study of rat liver mitochondrial fractionation by COLBEAU et al. 1971). For example, it appears that digitonin procedures give more complete separation of outer from inner mitochondrial membranes, resulting in larger outer membrane yields and inner membrane fractions more free of outer membrane activities. This advantage must be balanced against observations that the glycoside inhibits mitochondrial membrane enzyme activities (e.g., cytochrome oxidase and monoamine oxidase), although with some degree of reversibility (see also WOJTCZAK and ZALUSKA 1969). There are also numerous indications of relatively higher levels of inner-membrane marker activities in digitonin-prepared outer mitochondrial membrane fractions. Largely because digitonin is a membrane-active agent (and therefore potentially denaturing), systematic structural studies on the outer membranes of plant, fungal, or animal mitochondria have used membrane fractions isolated by hypoosmotic swelling (THOMPSON et al. 1968, PARSONS et al. 1966, 1967, MANNELLA and BONNER 1975b, MANNELLA 1982). On the other hand, digitonin treatment has become the method of choice for preparing so-called mitoplast fractions, i.e., outer-membrane-free mitochondria that retain some degree of respiratory coupling, which is not the case with osmotically swollen inner membrane "ghosts".

A problem encountered in the use of hypoosmotic lysis for preparing outer membranes from plant mitochondria is that these membranes appear more resistant to osmotic shock than their animal counterparts. Nonetheless final osmolarities can usually be defined at which outer membrane lysis is maximal (based on accessibility of exogenous cytochrome c to the inner membrane) while maintaining inner membrane integrity (based on release of soluble enzymes from the matrix). Inclusion of a few tenths mM EDTA or EGTA in the lysis medium may be needed to achieve maximal lysis of plant mitochondrial outer membranes, due to the protective effects of divalent metal ions on outer membrane integrity (MANNELLA 1974, MANNELLA and BONNER 1975a). Figure 1A shows the relative distributions of an outer and an inner membrane marker enzyme activity after step-gradient centrifugation of mung bean mitochondria with hypo-osmotically lysed outer membranes. Fractions at the 0.6/0.9 M sucrose interface routinely contain 30% of the total outer mitochondrial mem-

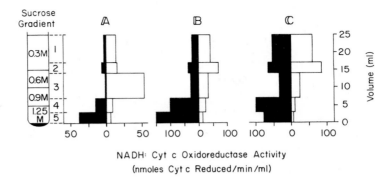

Fig. 1. Distribution of NADH:cytochrome c oxidoreductase activities after step-gradient centrifugation of mung bean mitochondria disrupted by various techniques. *White blocks* outer membrane antimycin-insensitive activity; *black blocks* inner membrane antimycin-sensitive activity. *A* Hypoosmotic treatment followed by contraction in 0.3 M sucrose; *B* Swelling-contraction followed by passing of mitochondrial suspension through Yeda press (Flow rate 0.2 ml min⁻¹); *C* Swelling-contraction followed by sonication of mitochondrial suspension (15 s Branson sonifier with microtip, 3 W output, 4 °C). Note use of different scales on the abscissae. (DOUCE et al. 1973)

brane marker activity and can be shown by spectrophotometric, enzymatic, and ultrastructural criteria to be essentially free of inner mitochondrial membranes and microsomes (DOUCE et al. 1973, MANNELLA 1974). (The latter characteristic will, of course, depend critically on the purity of the initial mitochondrial fractions, see DOUCE et al. 1972a). As indicated by the results of Fig. 1B and C, attempts to increase outer membrane yields by shearing the swollen mitochondria result mainly in increased contamination of the light outer membrane fractions with inner membrane fragments.

3 Lipid Composition of the Mitochondrial Outer Membrane

The apparent buoyant densities of plant outer mitochondrial membranes (based on near-equilibrium sedimentation in sucrose gradients) are in the range 1.08–1.12 g cc⁻¹, essentially the same as those of the corresponding fungal and liver membranes (Table 1). This is despite the wide variation in lipid-protein ratios reported for the outer mitochondrial membranes of the different organisms.

Table 1 summarizes the available details of lipid composition of outer mitochondrial membranes isolated from several sources by procedures involving only osmotic swelling (plus, in the case of fungi, mechanical shearing after contraction of the lysed mitochondria). As shown in this table, the total lipid content of outer mitochondrial membranes can vary from 23 to 69%, although leaving out the fungal membranes narrows the range considerably (27–48%).

Table 1. Lipid composition of outer mitochondrial membranes

Source	Apparent buoyant density (g cc^{-1})	Total lipid (Mass ratio)	Phospholipid: protein (Mass ratio)	Major phospholipids[a] PC PE PI (Molar ratios)			Sterol: phospholipid (Molar ratio)
Plants							
Potato[b,c]	1.09–1.12	0.35	0.50	0.36	0.64		1:17
Mung bean[b]	1.08–1.11	0.36	0.49				1:5
Cauliflower[d]	1.09–1.14		0.63	0.42	0.24	0.21	
Fungi							
N. crassa[e]	1.08–1.09	0.69	1.44	0.35	0.32	0.07	1:3
S. cerevisiae[f]	1.04–1.08	0.23	0.29				1:29
Animal, liver							
Guinea pig[g]	1.09–1.12	0.48	0.88	0.55	0.25	0.14	1:17
Rat[h]	1.12–1.14	0.27	0.35	0.39	0.35	0.09	1:8

[a] Abbreviations: PC, phosphatidyl choline; PE, phosphatidyl ethanolamine; PI, phosphatidyl inositol.
[b] MANNELLA 1974, MANNELLA and BONNER 1975a.
[c] McCARTY et al. 1973.
[d] MOREAU et al. 1974, MOREAU and LANCE 1972.
[e] HALLERMAYER and NEUPERT 1974, MANNELLA 1982.
[f] BOTTEMA and PARKS 1980.
[g] PARSONS et al. 1967, PARSONS and YANO 1967.
[h] PARSONS et al. 1966, COLBEAU et al. 1971.

3.1 Lipid Classes

The major lipids of plant, fungal, and animal outer mitochondrial membranes are phospholipids, in particular the zwitterions, phosphatidyl choline and phosphatidyl ethanolamine. Minor phospholipid components often detected in outer mitochondrial membrane fractions include the lysoforms of these two phospholipids, phosphatidyl glycerol, and the acidic phospholipids phosphatidyl inositol, phosphatidyl serine, and phosphatidic acid (listed in order of their usual detection levels).

Cardiolipin is sometimes detected in outer mitochondrial membrane fractions, but at levels attributable to inner membrane contamination. Sterols, on the other hand, are consistently detected in outer mitochondrial membrane fractions, comprising 2 to 15% of the total lipid mass. The question is sometimes raised whether these sterols are a true component of this membrane or the result of contamination of the fractions with sterol-rich plasma membranes. In the case of liver, quantitative comparisons between plasma membrane marker enzyme activities and cholesterol levels in different outer mitochondrial membrane fractions (COLBEAU et al. 1971) suggest strongly that the cholesterol in these fractions originates predominantly from the outer mitochondrial membranes themselves. Ultrastructural experiments with the drug, filipin, might help settle this issue. This polyene antibiotic interacts specifically with 3β-hydroxy sterols in membranes, the complexes forming 25-nm protuberances in freeze-fracture membrane faces (VERKLEIJ et al. 1973). Detectability of the distinctive filipin–sterol complexes on fracture faces of outer mitochondrial membranes which are supposed to have a high sterol content (e.g., those of mung beans and *N. crassa*, Table 1) should indicate whether sterols are, in fact, an important integral component of these membranes.

Glycolipids are not detected in either the outer or inner membranes of gradient-purified mitochondria from beans, potatoes, or cauliflower (McCARTY et al. 1973, MOREAU et al. 1974). This indicates that earlier reports of galactolipids in plant mitochondria reflected plastid contamination of mitochondrial fractions prepared by differential centrifugation alone. There is a recent report (GATEAU et al. 1980) that the outer membranes of liver mitochondria contain small amounts of glucolipid (dolichylglucosyl phosphate and glucosylceramide) associated with endogenous glucosyl transferase activity. There have been no reports of similar findings with plant or fungal mitochondria.

Finally there are spectrophotometric indications that carotenoids are present in the outer membranes of potato (MANNELLA 1974) and fungal (HALLERMAYER and NEUPERT 1974) mitochondria (see, however, NEUBURGER, Chap. 1, this Vol.).

3.2 Fatty Acid Composition: Temperature Modulation

The percent composition of fatty acids of outer membranes of mung bean hypocotyl mitochondria are summarized in Table 2, along with the comparable figures for the inner mitochondrial and light microsomal membranes of the same tissue. The observed trend, that the inner mitochondrial membrane is

Table 2. Percent composition of fatty acids in bean membranes as a function of growth temperature. (MANNELLA and BONNER 1975a)

Fatty acid[a]	Mitochondrial membranes						Light microsomal membranes
	Outer			Inner			
	14 °C[b]	28 °C	40 °C	15 °C	28 °C	40 °C	28 °C
C_{16}	25	25	25	14.5	15.5	19.5	34
C_{18}	3.5	5	4	1.5	2	3	5
$C_{18:1}$	5.5	7	7.5	4	5	7.5	13
$C_{18:2}$	28.5	30	36	35	36	41	27
$C_{18:3}$	37	35	27.5	45	41	29	20.5
db/fa[c]	1.74	1.72	1.62	2.09	2.00	1.77	1.29

[a] Symbols stand for palmitic acid (C_{16}), stearic acid (C_{18}), oleic acid. ($C_{18:1}$), lineoleic acid ($C_{18:2}$), and linolenic acid ($C_{18:3}$).
[b] Temperatures are those at which beans were grown.
[c] db/fa is the average number of double bonds per fatty acid chain for each type of membrane.

the most desaturated of these membranes, also holds true for other plants (Moreau et al. 1974) and liver (Levy et al. 1968, Colbeau et al. 1971). This appears
to reflect in large part the high degree of desaturation of the fatty acyl chains
of the cardiolipin of the inner membrane (Douce and Lance 1972, Colbeau
et al. 1971).

In the experiments of Table 2, the effect of varying growth temperature
on the fatty acid composition of bean mitochondrial membranes was also determined. The observed increase in acyl chain desaturation with decreasing environmental temperature is consistent with the well-documented trend in bulk lipids
of poikilotherms. In the case at hand, the greatest differences in fatty acid
composition occur between the two higher temperatures (40 °C and the normal
growth temperature 28 °C) and specifically in the relative amounts of linoleic
and linolenic acids.

3.3 Lipid Phase Transitions

It is clear from the data presented in the preceding sections that the outer
mitochondrial membrane (like most other biomembranes) contains a complex
mixture of lipids. It is not surprising, then, that differential thermal analysis
(DTA) of these membranes over the temperature range −10° to 70 °C (Mannella 1974) does not display sharp maxima, which are diagnostic for the order–
disorder phase transitions of simpler lipid mixtures. Instead, for the outer mitochondrial membranes of mung beans grown at 28 °C, there is a broad DTA
maximum from (about) 20° to 45 °C. Similar, wide DTA or DSC (differential
scanning calorimetry) maxima are commonly observed with other natural membranes and have the following interpretation according to modern models of
lipid domain structure (e.g., Linden and Fox 1975). Above the upper temperature limit of the broad thermotropic transition, all the lipids in the membrane
bilayer are in a random or fluid state. As the temperature is decreased, certain
lipids form "solid" patches as their disorder–order phase transition temperatures are reached. The number and extent of these lateral phase separations
increase with decreasing temperature as more and more lipid classes "freeze
out" of the fluid phase until, at the lower temperature limit of the DTA maximum, all lipids are in the ordered state. Freeze-fracture electron microscopy
indicates that, during this processes of lateral phase separation, membrane proteins segregate into the fluid lipid phases, which consequently appear as condensed patches of intramembrane particles at or below the lower temperature
limit (as illustrated for liver mitochondrial membranes by Hackenbrock et al.
1976).

The percentage of membrane lipids involved in the broad thermotropic transition observed with plant outer mitochondrial membranes at 20° to 45 °C is
unknown. Wide-angle X-ray diffraction patterns from these membranes at temperatures ranging from 5° to 50 °C display a single, broad peak at $(0.45 \text{ nm})^{-1}$,
(Mannella 1974), indicating that the bulk of the membrane lipids are in the
disordered state over this temperature range (e.g., see Engelman 1971). Perhaps
analogously, in the case of liver mitochondrial membranes, there is a weak

Fig. 2. Temperature dependence of the antimycin-insensitive NADH:cytochrome c oxidoreductase of outer membranes of mung bean mitochondria. *Closed circles* hypocotyls grown 5 days at 28 °C (referred to in text). *Open circles* hypocotyls grown 3 days at 28 °C followed by 4 days at 18 °C. (MANNELLA 1974)

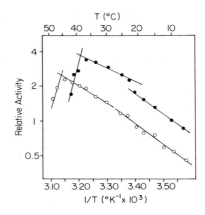

thermotropic transition detected by DSC over the temperature range 15° to 40 °C (BACH et al. 1978), while the bulk of the lateral phase separations of the lipids in these membranes occur at much lower temperatures, 0° to −30 °C (BLAZYK and STEIM 1972, HACKENBROCK et al. 1976). It is interesting that the DTA evidence indicates that lipid phase separations occur in bean outer mitochondrial membranes at the plant's growth temperature (28 °C). This suggests that the presence of a certain, probably small fraction of ordered lipids is important to the physiological functioning of this membrane. This is somewhat unexpected from the point of view of the current model of the fluid-mosaic biological membrane (SINGER and NICOLSON 1972). However, this would not be the first example of thermotropic phase transitions occurring in the membranes of an organism at or near its growth temperature (e.g., review by QUINN 1981).

That the lipid phase separations in the plant outer mitochondrial membranes may be functionally important is suggested by the temperature dependence of the NADH: cytochrome c oxidoreductase activity associated with these membranes. Figure 2 is a semi-log plot of the steady-state velocity of this electron transfer system in bean outer mitochondrial membranes versus (temperature)$^{-1}$. There are two breaks or discontinuities in such "pseudo-Arrhenius" plots. As temperature increases, there is an upward inflection and associated decrease in slope at 22 °C, near the lower temperature limit of the DTA maximum for these membranes. As the temperature increases further, there is a sharp downward deflection at 39 °C, just before the upper temperature limit of the DTA maximum. Both effects are totally reversible with temperature, indicating that they are associated with changes in lipid/protein organization and not with protein denaturation.

It would be premature to attempt to explain the reaction-velocity curves of Fig. 2 on a molecular basis. The theoretical framework of the thermodynamics of membrane-bound enzymes is still being formulated (e.g., OVERATH et al. 1976, WYNN-WILLIAMS 1976, JAHNIG and BRAMHALL 1982) and the plant outer mitochondrial membrane enzyme system is incompletely characterized. However, it is worth noting that the reversible decrease in enzyme activity observed as the plant outer mitochondrial membrane becomes totally fluid is difficult to reconcile with models in which simple lateral diffusion of electron carriers is rate-controlling (as appears to be the case with the analogous rat liver microsomal electron transfer system; STRITTMATTER and ROGERS 1975). The actual situation might instead be like that proposed by WYNN-WILLIAMS (1976). Assume that the enzyme system is confined to one of several coexisting lipid phases in the membrane and that the conformation of the rate-controlling component is critically dependent on the concentration of particular lipids within this phase. The temperature-dependent changes in apparent activation energy observed in experiments like those of Fig. 2 might then be due to exchange of lipids among the separate phases as different lipid phase-transition temperatures are reached, thus altering the concentration of the critical lipids in the enzyme-containing phase.

Clearly the question of the number and types of lipids involved in phase separations in the outer membranes of plant mitochondria at physiological temperatures needs further investigation.

4 Enzymes of the Mitochondrial Outer Membrane

Compared with the plethora of enzymes and electron carriers on the inner mitochondrial membrane, relatively few enzymatic activities have been assigned with certainty to the outer membrane of this organelle.

MOREAU and LANCE (1972) have reported that plant outer mitochondrial membranes lack both monoamine oxidase and kynurenine hydroxylase. These two enzyme activities have proven to be useful markers for liver mitochondrial outer membranes since they are absent from microsomal fractions of the same animal tissue (SCHNAITMAN et al. 1967, OKAMOTO et al. 1967).

Several enzymes of lipid metabolism have been assigned to the outer membranes of animal and plant mitochondria, including long-chain fatty acyl-CoA synthetase (heart, WHEREAT et al. 1969), glycerophosphate acyltransferase (liver, NIMMO 1979) and choline- and ethanolaminephosphotransferases (castor bean, SPARACE and MOORE 1981). Most of the phospholipase A activity of liver mitochondria appears to be associated with outer membrane fractions (VIGNAIS et al. 1969). While plant mitochondria also have considerable phospholipase A activity (DOUCE et al. 1968), it has not yet been localized within the plant organelle. Finally there are no reports on the occurrence in plant mitochondrial outer membranes of two interesting enzyme activities recently detected in the animal membrane, namely cyclic AMP phosphodiesterase (CERCEK and HOUSLAY 1982) and glucosyl transferase (see Sect. 3.1).

4.1 The NADH: Cytochrome c Oxidoreductase System

The one enzyme system that may be common to the outer mitochondrial membranes of all organisms is an electron transfer system composed of a flavoprotein NADH dehydrogenase and a b-type cytochrome. This electron carrier system

Fig. 3. Difference spectrum of an outer membrane fraction from mung bean mitochondria. Dithionite-reduced *minus* aerobic; liquid-N$_2$ temperature. (MANNELLA 1974)

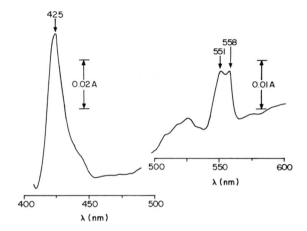

was first detected in outer membranes of rodent liver mitochondria (SOTTOCASA et al. 1967) and has subsequently been found in each type of plant mitochondrial outer membrane thus far isolated (e.g., MOREAU and LANCE 1972, DOUCE et al. 1972b).

Externally added oxidized cytochrome c can serve as electron acceptor for this redox system, probably via interaction with the b cytochrome. (For example, mild protease treatment of isolated plant mitochondrial outer membranes results in parallel decrease in the b cytochrome content and steady-state NADH: cytochrome c oxidoreductase activity of these membranes, but does not decrease the activity of the dehydrogenase; MANNELLA 1974.) The NADH-reducible b cytochrome associated with the outer mitochondrial membrane redox system is spectrophotometrically similar to the NADH-reducible b cytochrome of microsomal fractions in liver as well as in plants (Figs. 3 and 4). However, in the case of liver, it has recently been shown (ITO 1980, cf. also FUKUSHIMA and SATO 1973) that the outer mitochondrial membrane cytochrome is distinct in primary structure from the microsomal cytochrome (called cytochrome b-5).

Since the NADH-reducible b cytochrome of outer mitochondrial membranes is the only cytochrome component of this membrane, spectrophotometric criteria can sometimes provide rapid assessment of the purity of outer membrane fractions from either inner membranes (which contain multiple a, b and c type cytochromes) or microsomes (which can contain an ascorbate-reducible b cytochrome, illustrated in Fig. 4, and cytochrome P-450 in addition to cytochrome b-5). However, one or both of the additional microsomal pigments can be absent from particular plant tissues, so that enzymatic as well as spectral assays for microsomal contamination generally need to be applied to plant mitochondrial outer membrane fractions (e.g., presence of NADPH dehydrogenase activity).

Unlike the NADH: cytochrome c oxidoreductase activities associated with inner-membrane-bound dehydrogenases, the outer membrane (and microsomal) activities are insensitive to antimycin A and rotenone. That an iron-sulfur center is involved in the plant mitochondrial outer membrane dehydrogenase reaction

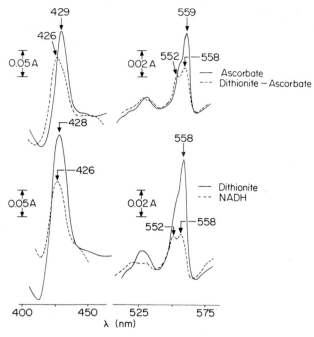

Fig. 4. Difference spectra of light microsomal membranes isolated from mung bean hypo-cotyls. Liquid-N_2 temperature. *Top, solid line* 4 mM ascorbate-reduced *minus* aerobic; *dashed line* dithionite-reduced *minus* ascorbate-reduced. *Bottom, solid line* dithionite-re-duced *minus* aerobic; *dashed line* 2 mM NADH-reduced *minus* aerobic. Note (1) difference between dithionite *minus* aerobic spectra above and of Fig. 3 and (2) similarity between NADH-reducible b cytochrome above and the outer membrane b cytochrome of Fig. 3. (MANNELLA 1974)

is suggested by its sensitivity both to SH-reagents (DAY and WISKICH 1975, MOREAU 1978) and to thenoyltrifluoroacetone, an iron chelator (MANNELLA 1974, MOREAU 1978). However, the inhibition by thenoyltrifluoroacetone is in-complete and other nonheme iron inhibitors, such as hydroxamic acids (MAN-NELLA 1974, MOREAU 1978) and o-phenanthroline (MANNELLA 1974), exert no inhibition on the system. Thus definite assignment of an iron-sulfur center to this enzyme system in plant outer mitochondrial membranes must await electron spin resonance data. In the case of kidney mitochondria, a 2Fe-2S protein has been isolated which appears functionally related to the outer membrane NADH: cytochrome c oxidoreductase (BACKSTROM et al. 1978). Finally, the outer membrane NADH dehydrogenase system of plant mitochondria has been shown to be inhibited by di- and trichlorophenoxyacetic acids, the common herbicides 2,4-D and 2,4,5-T (MANNELLA and BONNER 1978).

4.1.1 Possible Functions of the Outer Membrane Redox Chain

The physiological role of the outer mitochondrial membrane NADH dehydroge-nase–cytochrome b system is uncertain. There is evidence that it participates in NADH-semidehydroascorbate reductase activity, which supports cholesterol side-chain cleavage in outer membranes of adrenal cortex mitochondria (ITO et al. 1981, NATARAJAN and HARDING 1983). It may also be involved in the NADH-dependent hydroxylation of aryl hydrocarbons reported to occur in liver mitochondrial outer membranes (UEMURA and CHIESARA 1976). It does not appear to take part in the desaturation of activated fatty acids in either liver (MUGNAI and BODDI 1977, cf. also FUKUSHIMA and SATO 1973) or plant mitochondrial outer membranes (MANNELLA 1974, DIZENGREMEL et al. 1978), as does the similar antimycin-insensitive NADH:cytochrome c oxidoreductase activity of liver and plant microsomes (OSHINO et al. 1971, DIZENGREMEL et al. 1978).

There have been numerous suggestions in the literature that the outer mitochondrial membrane electron transfer system might serve to provide reducing equivalents from cytosolic NADH to the inner membrane respiratory chain, via a cytochrome c shuttle in the intermembrane space. While this is a reasonable hypothesis, in vitro demonstrations of this shuttle mechanism are generally unconvincing. Antimycin-insensitive oxidation of externally added NADH by intact animal or plant mitochondria using endogenous cytochrome c is very slow, even under conditions expected to release endogenous cytochrome c from the inner membrane (MATLIB and O'BRIEN 1976, BERNARDI and AZZONE 1981, MOREAU 1976). There are other factors which argue against a physiological role for the intermembrane shuttle system. For example there are indications that most, if not all, of the cytochrome c-reducing sites on the outer membrane of mitochondria are on the cytoplasmic side of this membrane, not on the side facing the inner membrane, as would be required for operation of the cytochrome c shuttle (FUKUSHIMA et al. 1971, MANNELLA 1974). Also, in the case of plant and fungal mitochondria, such an intermembrane electron transfer system for NADH oxidation appears to be redundant. These mitochondria have a NADH dehydrogenase which is directly linked to the respiratory chain and which is positioned on the outer surface of their inner membranes, so that it is accessible to cytosolic NADH (provided, of course, that the outer membrane is as permeable to NADH in situ as it is in vitro; VON JAGOW and KLINGENBERG 1970, DOUCE et al. 1973).

Increases observed in the activity of antimycin-insensitive NADH oxidation of potato mitochondria upon aging of tuber slices have been attributed to increased participation of the outer membrane electron transfer system in inner membrane respiration via intermembrane cytochrome c (VAN DER PLAS et al. 1976, DIZENGREMEL 1977). Interpretation of these observations is complicated by the inner membrane NADH dehydrogenase of potato mitochondria, which detaches rather easily from the inner membrane (e.g., during osmotic swelling) and can then react directly with cytochrome c, giving rise to an increased antimycin-insensitive NADH: cytochrome c oxidoreductase activity (MANNELLA 1974). In general, whether the antimycin-insensitive NADH oxidase activity of plant mitochondria actually involves the outer membrane NADH dehydrogenase should be assessed by determining the specificity of NADH dehydrogenation. (The outer membrane dehydrogenase acts on the $4\alpha H$ atom of NADH while the inner membrane dehydrogenase is specific for the $4\beta H$; DOUCE et al. 1973.)

5 Channel-Formers of the Outer Mitochondrial Membrane

5.1 Structural Evidence for the Existence of Pores

Electron microscopy provided the first clues to the physical basis of the large permeability of the mitochondrial outer membrane. PARSONS and co-workers (1965, 1966) found that negatively stained outer membranes of plant mitochondria were covered with random, close-packed arrays of stain-accumulating sites (Fig. 5A). These discrete stain centers, each 2 to 3 nm in diameter, were referred to as "pits", and it was suggested that they might represent projections of the stain-filled interiors of aqueous pores in these membranes. Outer membrane vesicles isolated from rodent liver mitochondria did not display this pitted appearance. However, in negatively stained specimens of freshly lysed liver mitochondria, narrow extrusions of the outer membrane were sometimes seen which bore projecting subunits shaped like hollow cylinders (PARSONS 1963). The cylindrical subunits on these outer mitochondrial membranes (shown in Fig. 5B) had an outer diameter of about 6 nm and an inner diameter of about 2 nm, the latter dimension being similar to the size of the pits in the plant membrane.

The maximum pore size inferred from electron microscopic images of outer mitochondrial membranes was generally consistent with the known sieving characteristics of these membranes. For example, the largest crystallographic dimensions of molecules known to rapidly penetrate intact mitochondrial outer membranes, such as sucrose and ATP, are on the order of 1 nm (BROWN and LEVY 1963, KENNARD et al. 1970). Molecules of this size would be expected to passively diffuse through a pore 2 to 3 nm wide. On the other hand, crystallographic data indicates holocytochrome c to be an ellipsoid with axes $3.0 \times 3.4 \times 3.4$ nm^3 (DICKERSON et al. 1971). A rigid molecule of this size should not pass through a 3-nm pore and this protein is, in fact, excluded by intact outer membranes of both animal and plant mitochondria (WOJTCZAK and ZALUSKA 1969, DOUCE et al. 1973, MANNELLA 1974).

Subsequent to the electron microscopic observations described above, outer membranes isolated from liver and plant mitochondria were subjected to X-ray diffraction analysis (THOMPSON et al. 1968, MANNELLA and BONNER 1975b). Specimens of the isolated membranes were prepared by a combination of centrifugation and slow, partial dehydration. With X-ray beams incident normal to the sedimentation axis, evenly spaced arcs were observed in the small-angle region of the diffraction patterns on the meridion (i.e., the axis parallel to the direction of centrifugation). This indicated that the membranes within the specimens were flattened into multilayered stacks with the membrane planes predominantly normal to the sedimentation axis. With this specimen geometry, membranes containing prominent in-plane subunits should display maxima in their X-ray diffraction patterns at right angles to the meridional reflections. The diffraction patterns from liver outer mitochondrial membranes showed no such equatorial maxima (THOMPSON et al. 1968), consistent with the general absence of substructure in the negative-stain electron images of the isolated membranes. (Recall that the hollow-cylinder subunits were seen only rarely and never after isolation of the outer membranes from the liver mitochondria.)

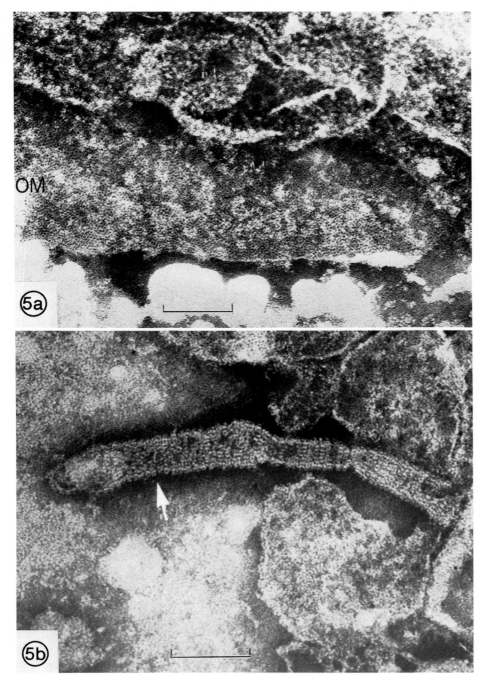

Fig. 5. a Electron microscopic image of negatively stained potato mitochondrion showing pitted appearance of outer membrane (*OM*). *Bar* 100 nm. (PARSONS et al. 1965). **b** Negative-stain image of outer-membrane protrusion of mammalian mitochondrion showing projecting pore-like subunits. *Bar* 100 nm. (D.F. PARSONS, unpublished micrograph)

The oriented plant outer mitochondrial membranes, on the other hand, gave intense equatorial diffraction, appearing as four or five diffuse arcs at low angle (Mannella and Bonner 1975b). Radial Patterson function analysis of the X-ray diffraction data indicated that the structures responsible for the equatorial diffraction from the plant membrane have a diameter in the membrane plane of 4 to 5 nm, which equals the closest center-to-center spacing observed for the "pits" in negative-stain images of the same membranes. Thus the X-ray diffraction experiments appeared to confirm the existence of the in-plane, pore-like subunits detected by electron microscopy in the outer membranes of plant mitochondria.

5.2 Trypsin-Insensitive Polypeptides of the Plant Membrane

In addition, the X-ray diffraction experiments of Mannella and Bonner (1975b) identified the probable polypeptide components of the outer mitochondrial membrane subunits. In SDS-PAGE patterns of outer membranes isolated from plant mitochondria (Fig. 6), more than half of the total Coomassie stain is contained in a polypeptide doublet near M_r 30,000 (actually 28,000 and 29,000) (Mannella and Bonner 1975a). These outer membrane polypeptides (called band I proteins) are insensitive to high salt or EDTA treatment (Mannella 1974), sonication (Fig. 6, lane d) and trypsinization (lanes b and c), indicating that they are integral membrane proteins that probably do not extend far from the surface of the membranes.

While treatment of the isolated outer mitochondrial membranes with trypsin does not effect the band I polypeptides, it does cause release of almost all other polypeptides from these membranes (Fig. 6, lane c). The same degree of trypsinization was found not to diminish the equatorial X-ray diffraction from these membranes (Mannella and Bonner 1975b), suggesting that the in-plane subunits are composed of the M_r 30,000 band I polypeptides.

5.3 The Pore-Forming Polypeptides

That the band I polypeptides of plant mitochondrial outer membranes have pore-forming activity was directly demonstrated several years later by Zalman and co-workers (1980). They were able to isolate these polypeptides in sucrose density gradients after Triton/NaCl solubilization of outer membranes isolated from mung bean mitochondria. Incorporation of the 30,000 M_r proteins into liposomes rendered the vesicles permeable to sucrose but not to dextrans of molecular weight 30,000 to 80,000. By measuring the retention of oligo- and polysaccharides by band I-containing liposomes, Zalman et al. defined the sieving characteristics of the putative reconstituted outer mitochondrial membrane channels. There was rapid diffusion of molecules smaller than M_r 1,000, and increasingly limited diffusion as molecular size increased from 1,000 to an exclusion limit of M_r 4,000 to 6,000. This behavior was in good agreement with the known in vitro permeability of the outer membranes of isolated mitochondria (see Sect. 1 and 5.1).

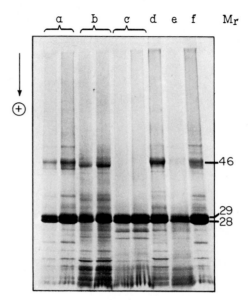

Fig. 6. SDS-polyacrylamide gradient (7.5–15%) gel electrophoresis of outer membranes isolated from mung bean mitochondria by hypoosmotic lysis. Specimens run in *right-hand lanes* of *a*, *b* and *c* and in lanes *d*, *e* and *f* correspond to approximately the same amount of outer membrane protein (20 µg) prior to the respective treatments. *Left-hand lanes* of *a*, *b* and *c* were loaded with half as much membrane protein. *a*, *f* controls; *b* outer membranes of mitochondria that were treated with trypsin (0.7 mg mg^{-1} outer membrane protein, 30 min, 25 °C) prior to osmotic lysis; *c* outer membranes that were treated with trypsin after isolation (same conditions as *b*); *d* outer membranes sonicated after isolation (15 s, Branson W185 sonifier with microtip, 12 W output, 4 °C); *e* outer membranes that were sonicated (as in *d*) and then trypsin-treated (as in *c*). The major polypeptides at M$_r$ 28,000 and 29,000 are unaffected by either sonication or trypsinization although there are indications of increased sensitivity of these polypeptides to trypsin following sonication (suggesting that the latter process alters the fundamental organization of these membranes). The polypeptide at M$_r$ 46,000 is more sensitive to trypsin in isolated outer membranes than in intact mitochondria, suggesting that it is located on the side of the outer membrane facing the inner membrane. (C.A. MANNELLA and N.-H. CHUA, unpublished data)

In fact, the first published report of pore-forming activity associated with outer mitochondrial membranes was by COLOMBINI (1979). His study was a follow-up to the discovery by SCHEIN and co-workers (1976) that *Paramaecium* mitochondrial material induced voltage-dependent, anion-selective channels (VDAC) in planar phospholipid bilayers. COLOMBINI (1979) demonstrated that similar ionic channels could be induced in bilayers by Triton extracts from a variety of animal and fungal mitochondria (Fig. 7) and localized this VDAC activity (in the case of liver mitochondria) to outer membrane fractions.

For outer membranes of the fungus, *Neurospora crassa*, upward of 90% of the total protein is concentrated in a single band on SDS-polyacrylamide gels at M$_r$ 31,000 (MANNELLA 1982, cf. also NEUPERT and LUDWIG 1971). Like the outer membranes of plant mitochondria, those of *Neurospora* mitochondria

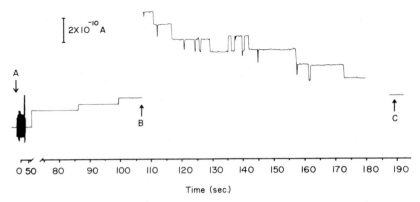

Fig. 7. Insertion of ion channels from *Neurospora* outer mitochondrial membranes into a phospholipid membrane. A Triton X-100 extract of the isolated membranes was added to the aqueous phase on one side of a planar bilayer membrane at time *A*. There is an insertion (upward deflection) of a triplet channel at t = 60 s, followed by two single-channel insertions. (The *Neurospora* channels tend to form aggregates of three and multiples of three; Mannella et al. 1983.) At *B*, the transmembrane voltage is increased from 10 mV to 40 mV and the recorder sensitivity is increased, causing the offset observed. Immediately following the voltage increase, the five channels can be seen to shift one at a time to their less conductive state. At *C* the transmembrane potential is returned to 10 mV and the recorder is returned to its initial sensitivity, showing that the membrane conductivity returns to the value displayed just prior to *B*. (Mannella et al. 1983)

may have a pitted appearance. Unlike the pore-like subunits of the plant membrane, however, the fungal subunits form extended ordered arrays. Examples of electron microscopic images of negatively stained *Neurospora* mitochondrial outer membranes are presented in Fig. 8. As might be expected from the preceding discussion, the fungal mitochondrial outer membrane fractions show a high specific activity of VDAC insertion in in vitro bilayer experiments (Mannella et al. 1983). Antibodies against the M_r 31,000 *Neurospora* protein (1) inhibit the in vitro insertion of *Neurospora* VDAC into planar lipid membranes and (2) bind specifically to membranes containing crystalline arrays of the stain-accumulating subunits in *Neurospora* outer mitochondrial membrane fractions (Mannella et al. 1982). Thus the link between the pore-like subunits detected by electron microscopy and the polypeptides responsible for the pore-forming activity of outer mitochondrial membranes has been established immunologically.

A summary of the characteristics of the channels which have been isolated by several laboratories from outer mitochondrial membranes is contained in Table 3. In each case the polypeptides display a molecular weight of about 30,000, as determined by SDS-PAGE. The possibility that these mitochondrial pore-formers may be evolutionarily related to the so-called "porins" of bacterial outer envelopes has often been raised. (Characteristics of some bacterial pore-forming proteins are also summarized in Table 3). Whether there are, in fact, regions of homology between the polypeptides that form passive diffusion channels in the outer membranes of bacteria and of mitochondria will not be known until the mitochondrial polypeptides are sequenced.

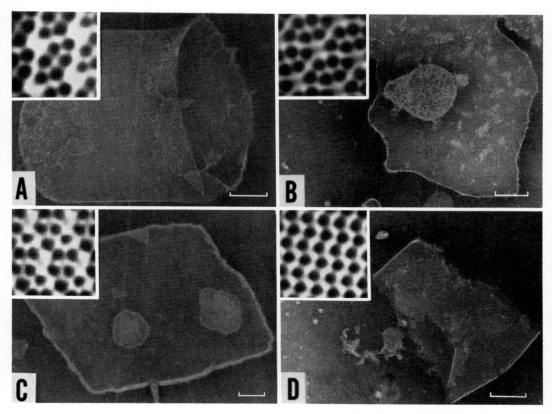

Fig. 8 A–D. Electron microscopic images of crystalline outer membranes from *Neurospora crassa* mitochondria. *Inserts* are computer-filtered images of single periodic layers, 27×27 nm^2, near the center of each membrane. The crystallographic unit cell in **A–C** contains six stain-accumulating sites (each a transmembrane channel, see Sect. 5.4.2) arranged in a hexagon (best visualized in **A**). The membranes are negatively stained with phosphotungstate, an anionic stain. **A** Control. **B** 1 mM MgCl$_2$ included in stain solution. **C** EDTA-pretreated membrane, no divalent cations in stain. Note the apparent increase in anionic stain accumulation at and between the putative pores with Mg^{2+} and the decrease in same associated with chelator treatment. **D** Phospholipase A$_2$-treated membrane. The channels in **D** are arranged in offset zigzag rows with no large stain-free patches as in **A–C**, implying that the unstained regions of the native arrays are composed primarily of lipid. *Bars* 100 nm. (MANNELLA and FRANK 1982 b)

5.4 Structure and Function of the Channels

5.4.1 Model from X-Ray Diffraction

A prediction for the channel structure in plant mitochondrial outer membranes (based on available data and several simplifying assumptions) is presented in Fig. 9. The channel is defined by two concentric cylinders aligned normal to

Table 3. Outer membrane channels of mitochondria and bacteria

Source	Polypeptide M_r	Ion step conductance in 1 M KCl (nS)	Exclusion limit M_r	Reference
Mitochondria				
Mung bean	30,000	–	4,000–6,000	Zalman et al. (1980)
N. crassa	–	4.5	3,400–6,800	Colombini (1980a)
	31,000	5.0	–	Freitag et al. (1982)
Rat liver	32,000	4.0	–	Colombini (1980b, c)
	32,000	4.5	(>426)	Fiek et al. (1982)
	30,000	–	(>342)	Linden et al. (1982a)
Guinea pig heart	–	4.5	–	Schreibmayer et al. (1983)
Bacteria				
E. coli	36,500	2.1	600	Schindler and Rosenbusch (1978)
S. typhimurium	38–40,000	2.3	600	Benz et al. (1980)
P. aeruginosa	35,500	–	3,000–9,000	Hancock et al. (1979)

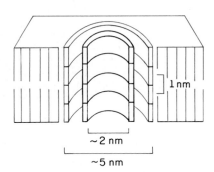

Fig. 9. Predicted structure for the channels in the outer membranes of plant mitochondria

the membrane plane. This is the simplest radially symmetric, transversely isotropic shape consistent with the normal-incidence X-ray diffraction from these membranes (Mannella 1981). The cylinders do not extend past the membrane surface, in agreement with (1) the insensitivity of the M_r 30,000 polypeptides to trypsin and (2) the close spacing between the membranes in the X-ray diffraction specimens. (The fundamental spacing of the lamellar diffraction from the plant outer mitochondrial membranes approaches that of simple phospholipid-water phases, i.e., about 5 nm; see Mannella 1981.) The horizontal delineation within the cylinders of the model of Fig. 9 is meant to represent the carbon backbones of polypeptides running approximately parallel to the membrane plane. This is inferred from the presence of arcs of diffracted intensity weakly aligned on the meridion at $(1.1 \text{ nm})^{-1}$ in the grazing-incidence X-ray patterns from these membranes. That the same patterns did not show maxima at

$(0.52 \text{ nm})^{-1}$ suggests that α-helical structure is not prominent. It may be relevant that recent model studies indicate that tilted β-sheets can form cylinders the size of those in the channel model of Fig. 9 (SALEMME and WEATHERFORD 1981, SALEMME 1981).

5.4.2 Structure from Electron Microscopy

The actual shape and size of the channels formed by M_r 30,000 polypeptides in fungal outer mitochondrial membranes is being determined by three dimensional electron microscopic techniques. The planar crystals of channels found in outer membrane fractions from *Neurospora* mitochondria are suited to reconstruction schemes like those recently applied to stained gap junctions (UNWIN and ZAMPIGHI 1980) and unstained bacteriorhodopsin arrays (HENDERSON and UNWIN 1975). The shape of the *Neurospora* channels has been reconstructed in three dimensions from tilt-series electron images recorded from a negatively stained crystalline outer mitochondrial membrane (MANNELLA et al. 1984). Each of the stain-accumulating sites seen in Fourier-filtered projection images like those of Fig. 8 is a single, stain-filled channel through the 5-nm-thick membrane. There is no lateral fusion of the channels in the bilayer, as is the case with one type of *E. coli* porin channels (ENGEL et al. 1983). Higher resolution reconstructions of the *Neurospora* arrays in the absence of stain are in progress, which should allow direct visualization of the protein defining the channels. Similar studies are underway to define the complexes which the pores form with other proteins. For example, there is evidence that at least one class of channels in the outer membrane of liver mitochondria interacts with cytosolic hexokinase (LINDEN et al. 1972b, FIEK et al. 1982), which raises the possibility of directed diffusion of the products of this enzyme reaction into the mitochondrion. (Might specificity be imposed on the channels in situ by interaction with cytoplasmic components such as enzymes, which may be lost in the process of isolating the mitochondria?) Hexokinase activity has been detected in mitochondrial fractions from plant tissue (DRY et al. 1983, TANNER et al. 1983) but a specific outer membrane receptor for this enzyme has not as yet been identified.

5.4.3 Mechanism of Ion Selectivity

In in vitro bilayer experiments, the mitochondrial outer membrane channels display a selectivity for anions over cations that is unexpected for pores of such large diameter. This implies (COLOMBINI 1980a, ROOS et al. 1982) that fixed charges in and around the channel entrance might control its permeability toward ions. Experiments with succinic anhydride further implicate fixed charges in the mechanism of ion selectivity of these channels. Succinylation, which attaches negatively charged carboxyl groups to basic amino acid residues, changes the mitochondrial channel selectivity from anionic to cationic, and also abolishes the voltage dependence of the associated step conductance (DORING and COLOMBINI 1984). Electron microscopic images of succinylated channel arrays (MANNELLA and FRANK 1984) have the same lattice parameters as those of the untreated arrays, indicating no large-scale changes in protein conformation or organization. Two-dimensional difference maps of averaged, Fourier-filtered array images indicate the occurrence of one locus on each channel where anionic stain (phosphotungstate) accumulates less in succinylated than in control arrays. Since phosphotungstate is known to stain basic amino acids in proteins (TZAPH-LIDOU et al. 1982), the stain minima in the difference maps could represent clusters of basic amino acids involved in confering ion selectivity (and/or voltage dependence) on the channels. Repeating this experiment with three-dimensional data should determine whether these basic amino acids are located on the lip of the channel or along its interior surface.

The accumulation of anionic stain molecules around the pores increases with addition of divalent cations and decreases with EDTA pretreatment of the membranes (Fig. 8), suggesting that bound metal ions also play a role in defining the surface charge on the channel array (MANNELLA and FRANK 1982b). There are several observations that the (antimycin-sensitive) NADH oxidation of intact plant and fungal mitochondria is stimulated by polyvalent cations (JOHNSTON et al. 1979, MOLLER et al. 1982). The ion dependence of this effect is consistent with electrical screening, and the stimulation has been interpreted in terms of increased accessibility of anionic NADH to the inner membrane dehydrogenase (JOHNSTON et al. 1979). The possibility that this increased substrate accessibility reflects increased permeability of the outer mitochondrial membrane toward anions when fixed negative charges at or near the channels are screened should be investigated.

5.5 Speculation on a Regulatory Role for Outer Membrane Channels

As noted several times above, the in vitro voltage gating of channels isolated from the outer mitochondrial membrane raises the possibility that the permeability of this membrane might be regulated in situ. Mechanisms already proposed to establish the small transmembrane potential which might close the pores (20 mV) include intrinsic charge assymmetry or assymetry induced by macromolecule binding or enzyme activity (COLOMBINI 1979, FREITAG et al. 1982). It has also been suggested that the outer membrane channels might interact directly with inner membrane transport systems (ROOS et al. 1982).

We propose another mechanism for channel regulation, one based on electrostatic interaction between the outer and inner mitochondrial membranes. That adjacent phospholipid membrane surfaces exert long-range electrostatic forces on each other has been brought out by the theoretical and experimental studies of RAND, PARSEGIAN and others (reviewed in RAND 1981). The distance over which these forces act is determined both by the surface charge density on the membranes and the dielectric constant of the medium between them. Extrapolating these model studies to the case of the outer and inner mitochondrial membranes is difficult because there is no consensus on either the magnitude of the surface potential on the membranes or on the composition of the fluid in the intermembrane space. If one assumes a Debye length in the intermembrane space of 1 nm (corresponding to physiological saline) and a surface potential

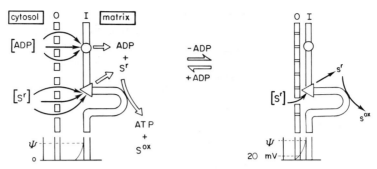

Fig. 10. Hypothetical interactions between outer membrane (*O*) channels and inner membrane (*I*) potential. *Large* and *small letters* indicate relative levels of ADP or of substrate (oxidized, S^{ox}; reduced, S^r). Inner membrane transporters of substrate and ADP are indicated by *triangle* and *circle*. When mitochondria oxidize substrate in the presence of excess ADP, the inner membrane is very convoluted and the matrix space is condensed (cf. HACKENBROCK 1968). The inner membrane field, Ψ, is essentially zero at the outer membrane (because of the distance between the two membranes) and so the outer membrane channels are open (*left*). If ADP is exhausted (*right*), the mitochondria convert to the so-called orthodox conformation, in which the outer and inner membranes are on average much closer together. The channels on outer membrane surfaces which are closely apposed to the inner membrane close down in response to the latter's close electric field. If enough channels close, diffusion of substrate through the outer membrane becomes the rate-controlling step for respiration

of 60 mV on the outer face of the inner membrane, a transmembrane potential sufficient to close the outer membrane channels (20 mV) would be induced in the outer membrane whenever the two membranes were within about 1 nm of each other. There is in fact considerable ultrastructural evidence of usually close proximity between the two mitochondrial membranes, the extent of which can vary with metabolic state (e.g., SJØSTRAND 1978, HACKENBROCK 1968). Thus the channels in the outer mitochondrial membrane might open and close in response to factors which modulate the intermembrane distance. A hypothetical scheme whereby outer membrane channels might function in a feedback loop to regulate substrate oxidation according to the availability of phosphate acceptor is presented in Fig. 10.

There is no direct experimental evidence for or against physiological regulation of the permeability of the outer mitochondrial membrane, either by interaction with inner-membrane potentials or by any other mechanism. There is indirect support for the former mechanism, namely recent reports of correlations between inner-membrane energy state and two other outer membrane activities, monoamine oxidation (SMITH and REID 1978, SMITH 1980) and apocytochrome c import (SCHLEYER et al. 1982).

The absence of detailed information about mitochondrial outer membrane permeability is due in no small measure to the general presumption that this membrane is "freely" permeable (or uncontrollably leaky) toward small molecules. Thus current studies of the ion or metabolite permeability of mitochondria often employ mitoplasts or submitochondrial particles, neither of which contain outer membranes. The ion selectivity and voltage sensitivity of the channels isolated from the outer mitochondrial membrane argue for experimental re-investigation of the permeability of the intact outer membrane and its role in the regulation of mitochondrial metabolism.

References

Ades IZ (1982) Transport of newly synthesized proteins into mitochondria – a review. Mol Cell Biochem 43:113–127

Bach D, Bursuker I, Miller IR (1978) Differential scanning calorimetry of rat liver mito-chondria. Experientia 34:717–718

Backstrom D, Lorusso M, Anderson K, Ehrenberg A (1978) Characterization of the iron-sulfur protein of the mitochondrial outer membrane partially purified from beef kidney cortex. Biochim Biophys Acta 502:276–288

Benz R, Ishii J, Nakae T (1980) Determination of ion permeability through the channels made of porins from the outer membrane of *Salmonella typhimurium* in lipid bilayer membranes. J Mem Biol 56:19–29

Bernardi P, Azzone GF (1982) Cytochrome c as an electron shuttle between the outer and inner mitochondrial membranes. J Biol Chem 256:7187–7192

Blazyk JF, Steim JM (1972) Phase transitions in mammalian membranes. Biochim Bio-phys Acta 266:737–741

Bottema CK, Parks LW (1980) Sterol analysis of the inner and outer mitochondrial membranes in yeast. Lipids 15:987–992

Bradley TJ, Satir P (1979) Evidence of microfilament-associated mitochondrial move-ment. J Supramol Struct 12:165–175

Brdiczka D (1978) Impermeability of the outer mitochondrial membrane for small polar molecules. Hoppe-Seylers Z Physiol Chem 359:1063

Brown GM, Levy HA (1963) Sucrose: precise determination of crystal and molecular structure by neutron diffraction. Science 141:921–923

Cercek B, Houslay MD (1982) Submitochondrial localization and assymetric disposition of two peripheral cyclic nucleotide phosphodiesterases. Biochem J 207:123–132

Colbeau A, Nachbaur J, Vignais PM (1971) Enzymic characterization and lipid composi-tion of rat liver subcellular membranes. Biochim Biophys Acta 249:462–492

Colombini M (1979) A candidate for the permeability pathway of the outer mitochondrial membrane. Nature (London) 279:643–645

Colombini M (1980a) Pore size and properties of channels from mitochondria isolated from *Neurospora crassa*. J Memb Biol 53:79–84

Colombini M (1980b) Structure and mode of action of a voltage dependent anion-selec-tive channel (VDAC) located in the outer mitochondrial membrane. Ann NY Acad Sci 341:552–562

Colombini M (1980c) On the structure of a channel-forming protein functionally purified from mitochondria. Fed Proc 39:1812

Day DA, Wiskich JT (1975) Isolation and properties of the outer membrane of plant mitochondria. Arch Biochem Biophys 171:117–123

DePierre JW, Ernster L (1977) Enzyme topology of intracellular membranes. Annu Rev Biochem 46:201–262

Dickerson RE, Takano T, Eisenberg D, Kallai OB, Samson L, Cooper A, Margoliash E (1971) Ferricytochrome c: general features of the horse and bonito proteins at 2.8 A resolution. J Biol Chem 246:1511–1535

Dizengremel P (1977) Increased participation of the outer mitochondrial membrane in the oxidation of exogenous NADH during aging of potato slices. Plant Sci Lett 8:283–289

Dizengremel P, Kader J-C, Mazliak P, Lance C (1978) Electron transport and fatty acid synthesis in microsomes and mitochondrial membranes of plant tissues. Plant Sci Lett 11:151–157

Doring C, Colombini M (1984) On the nature of the molecular mechanism underlying the voltage dependence of the channel-forming protein, VDAC. Biophys J 45:44–46

Douce R, Lance C (1972) Altération des activités oxydatives et phosphorylantes des mitochondries de chou-fleur sous l'action de phospholipases et du vieillissement. Physiol Veg 10:181–198

Douce R, Guillot-Salomon T, Lance C, Signol M (1968) Étude compasée de la composition en phospholipides de mitochondries et de chloroplastes isolés de quelques tissue végétaux. Bull Soc Fr Physiol Veg 14:351–373

Douce R, Christensen EL, Bonner WD Jr (1972a) Preparation of intact plant mitochondria. Biochim Biophys Acta 275:148–160

Douce R, Manella CA, Bonner WD Jr (1972b) Site of the biosynthesis of CDP-diglyceride in plant mitochondria. Biochem Biophys Res Comm 49:1504–1509

Douce R, Mannella CA, Bonner WD Jr (1973) The external NADH dehydrogenases of intact plant mitochondria. Biochim Biophys Acta 292:105–116

Dry IB, Nash D, Wiskich JT (1983) The mitochondrial localization of hexokinase in pea leaves. Planta 158:152–156

Engel A, Dorset DL, Massalski A, Rosenbusch JP (1983) The low-resolution 3D-structure of porin, a channel forming protein in *E. coli* outer membranes. In: Bailey GW (ed) Proc 41st annu meet Electron Microsc Soc Am. San Francisco Press, San Francisco, pp 440–441

Engelman DL (1971) Lipid bilayer structure in the membrane of *Mycoplasma laidlawii*. J Mol Biol 58:153–165

Felgner PL, Messer JL, Wilson JE (1979) Purification of a hexokinase-binding protein from the outer mitochondrial membrane. J Biol Chem 254:4946–4949

Fiek C, Benz R, Roos N, Brdiczka D (1982) Evidence for identity between the hexokinase-binding protein and the mitochondrial porin in the outer membrane of rat liver mitochondria. Biochim Biophys Acta 688:429–440

Freitag H, Neupert W, Benz R (1982) Purification and characterization of a pore protein of the outer mitochondrial membrane from *Neurospora crassa*. Eur J Biochem 123:629–639

Fukushima K, Sato R (1973) Purification and characterization of cytochrome b_5-like hemoprotein associated with outer mitochondrial membrane of rat liver. J Biochem 74:161–173

Fukushima K, Ito A, Omura T, Sato R (1972) Occurrence of different types of cytochrome b_5-like hemoprotein in liver mitochondria and their intramitochondrial localization. J Biochem 71:447–461

Gateau O, Morelis R, Louisot P (1980) Glucosyltransferase activities in liver mitochondria. Eur J Biochem 112:193–201

Hackenbrock CR (1968) Chemical and physical fixation of isolated mitochondria in low-energy and high-energy states. Proc Natl Acad Sci USA 61:598–605

Hackenbrock CR, Hochli M, Chau RM (1976) Calorimetric and freeze-fracture analysis of lipid phase transitions and lateral translational motion of intramembrane particles in mitochondrial membranes. Biochim Biophys Acta 455:466–484

Hallermayer G, Neupert W (1974) Lipid composition of mitochondrial outer and inner membranes of *Neurospora crassa*. Hoppe-Seylers Z Physiol Chem 355:279–288

Hancock REW, Decad GM, Nikaido H (1979) Identification of the protein producing transmembrane diffusion pores in the outer membrane of *Pseudomonas aeruginosa* PA01. Biochim Biophys Acta 554:323–331

Hashizume K, Ichikawa K (1982) Localization of 3,5,3′-L-triiodothyronine receptor in rat kidney mitochondrial membranes. Biochem Biophys Res Commun 106:920–926

Henderson R, Unwin PNT (1975) Three-dimensional model of purple membrane obtained by electron microscopy. Nature (London) 257:28–32

Hrudka F (1978) A morphological and cytochemical study on isolated sperm mitochondria. J Ultrastruct Res 63:1–19

Ito A (1980) Cytochrome b$_5$-like hemoprotein of outer mitochondrial membrane; OM cytochrome b. J Biochem 87:63–71

Ito A, Hayashi S, Yoshida T (1981) Participation of a cytochrome b$_5$-like hemoprotein of outer mitochondrial membrane (OM cytochrome b) in NADH-semidehydroascorbic acid reductase activity of rat liver. Biochem Biophys Res Commun 101:591–598

Jagow von G, Klingenberg M (1970) Pathways of hydrogen in mitochondria of *Saccharomyces carlsbergensis*. Eur J Biochem 12:583–592

Jahnig F, Bramhall J (1982) The origin of a break in Arrhenius plots of membrane processes. Biochim Biophys Acta 690:310–313

Johnston SP, Moller IM, Palmer JM (1979) The stimulation of exogenous NADH oxidation in Jerusalem artichoke mitochondria by screening of charges on the membranes. FEBS Lett 108:28–32

Kennard O, Isaacs NW, Coppola JC, Kirby AJ, Warren S, Motherwell WDS, Watson DG, Wampler DL, Chenery DH, Larson AC, Kerr KA, Riva di Sanseverino L (1970) Three-dimensional structure of adenosine triphosphate. Nature (London) 225:333–336

Levy M, Toury R, Andre J (1967) Séparation des membranes mitochondriales purification et caractérisation enzymatique de la membrane externe. Biochim Biophys Acta 135:599–613

Levy M, Toury R, Sauner M-T, Andre J (1968) Recent findings on the biochemical and enzymatic composition of the two isolated mitochondrial membranes in relation to their structure. In: Ernster L, Drahota Z (eds) Mitochondria: structure and function. Academic Press, London NewYork, pp 33–42

Linden CD, Fox CF (1975) Membrane physical state and function. Acc Chem Res 8:321–327

Linden M, Gellerfors P, Nelson BD (1982a) Purification of a protein having pore-forming activity from the rat liver mitochondrial outer membrane. Biochem J 208:77–82

Linden M, Gellerfors P, Nelson BD (1982b) Pore protein and the hexokinase-binding protein from the outer membrane of rat liver mitochondria are identical. FEBS Lett 141:189–192

Mannella CA (1974) Composition and structure of the outer membranes of plant mitochondria. PhD Thesis, Univ Pennsylvania

Mannella CA (1981) Structure of the outer mitochondrial membrane. Analysis of X-ray diffraction from the plant membrane. Biochim Biophys Acta 645:33–40

Mannella CA (1982) Structure of the outer mitochondrial membrane: Ordered arrays of pore-like subunits in outer-membrane fractions from *Neurospora crassa* mitochondria. J Cell Biol 94:680–687

Mannella CA, Bonner WD Jr (1975a) Biochemical characteristics of the outer membranes of plant mitochondria. Biochim Biophys Acta 413:213–225

Mannella CA, Bonner WD Jr (1975b) X-ray diffraction from oriented outer mitochondrial membranes. Detection of in-plane subunit structure. Biochim Biophys Acta 413:226–233

Mannella CA, Bonner WD Jr (1978) 2,4-dichlorophenoxyacetic acid inhibits the outer membrane NADH dehydrogenase of plant mitochondria. Plant Physiol 62:468–469

Mannella CA, Frank J (1982a) Effects of divalent metal ions and chelators on the structure of outer mitochondrial membranes from *Neurospora crassa*. Biophys J 37:3–4

Mannella CA, Frank J (1982b) Negative-staining characteristics of pore-like subunits of the mitochondrial outer membrane: Implications for surface-charge distribution. In: LePoole JB, Zeitler E, Thomas G, Schimmel G, Weichan C, Bassewitz von K (eds) Electron microscopy 1982. Dtsch Ges Elektronenmikrosk, Frankfurt, pp 41–42

Mannella CA, Frank J (1984) Electron microscopic stains as probes of the surface charge of mitochondrial outer membrane channels. Biophys J 45:139–141

Mannella CA, Cognon B, Colombini M (1982) Immunological evidence for identity between the subunits of the crystalline arrays in *Neurospora* outer mitochondrial membranes and the pore-forming components of the membranes. J Cell Biol 95:256a

Mannella CA, Colombini M, Frank J (1983) Structural and functional evidence for multiple pore complexes in the outer membrane of *Neurospora crassa* mitochondria. Proc Natl Acad Sci USA 80:2243–2247

Mannella CA, Radermacher M, Frank J (1984) Three-dimensional reconstruction of negatively stained crystalline arrays of mitochondrial pore protein. In: Proc 8th Eur Congr Electron Microsc. 8th Eur Congr Electron Microsc Found, Budapest, vol. 3, pp 1491–1492

Matlib MA, O'Brien PJ (1976) Properties of rat liver mitochondria with intermembrane cytochrome c. Arch Biochem Biophys 173:27–33

McCarty RE, Douce R, Benson AA (1973) The acyl lipids of highly purified plant mitochondria. Biochim Biophys Acta 316:266–270

Meunier D, Pianeta C, Coulomb P (1971) Obtention de membranes externes et de membranes internes à partir de mitochondries issues du tubercule de pomme de terre. CR Acad Sci 272:1376–1379

Moller IM, Schwitzguebel J-P, Palmer JM (1982) Binding and screening by cations and the effect on exogenous NAD(P)H oxidation in *Neurospora crassa* mitochondria. Eur J Biochem 123:81–88

Moreau F (1976) Electron transfer between outer and inner membranes in plant mitochondria. Plant Sci Lett 6:215–221

Moreau F (1978) The electron transport system of outer membranes of plant mitochondria. In: Ducet G, Lance C (eds) Plant mitochondria. Elsevier/North-Holland Biomedical Press, Amsterdam New York, pp 77–84

Moreau F, Lance C (1972) Isolement et propriétés des membranes externes et internes de mitochondries végétales. Biochimie 54:1335–1348

Moreau F, DuPont J, Lance C (1974) Phospholipid and fatty acid composition of outer and inner membranes of plant mitochondria. Biochim Biophys Acta 345:294–304

Mugnai G, Boddi V (1977) Stearoyl-CoA desaturase in mitochondrial membrane fractions. Ital J Biochem 26:245–253

Munn EA (1974) The structure of mitochondria. Academic Press, London New York

Natarajan R, Harding B (1983) Support of cholesterol side chain cleavage activity by adrenal cortex outer mitochondrial membrane NADH-semidehydroascorbate reductase. Fed Proc 42:2065

Neupert W, Ludwig GD (1971) Site of biosynthesis of outer and inner membrane proteins of *Neurospora crassa* mitochondria. Eur J Biochem 19:523–532

Nimmo HG (1979) The location of glycerol phosphate acyltransferase and fatty acyl CoA synthase on the inner surface of the mitochondrial outer membrane. FEBS Lett 101:262–264

O'Brien RL, Brierely G (1965) Compartmentation of heart mitochondria. J Biol Chem 240:4527–4531

Okamoto H, Yamamoto S, Nozaki M, Hayaishi O (1967) On the submitochondrial localization of L-Kynurenine-3-hydroxylase. Biochem Biophys Res Commun 26:309–314

Oshino N, Imai Y, Sato R (1971) A function of cytochrome b_5 in fatty acid desaturation by rat liver microsomes. J Biochem 69:155–167

Overath P, Thilo L, Trauble H (1976) Lipid phase transitions and membrane function. Trends Biochem Sci 1:186–189

Parsons DF (1963) Mitochondrial structure: two types of subunits on negatively stained mitochondrial membranes. Science 140:985–987

Parsons DF, Yano Y (1967) The cholesterol content of the outer and inner membranes of guinea-pig liver mitochondria. Biochim Biophys Acta 135:362–364

Parsons DF, Bonner WD Jr, Verboon JG (1965) Electron microscopy of isolated plant mitochondria and plastids using both the thin-section and negative-staining techniques. Can J Bot 43:647–655

Parsons DF, Williams GR, Chance B (1966) Characteristics of isolated and purified preparations of the outer and inner membranes of mitochondria. Ann NY Acad Sci 137:643–666

Parsons DF, Williams GR, Thompson W, Wilson D, Chance B (1967) Improvements in the procedure for purification of mitochondrial outer and inner membrane. Comparison of the outer membrane with smooth endoplasmic reticulum. In: Quagliariello E, Papa S, Slater EC, Tager JM (eds) Mitochondrial structure and compartmentation. Adriatica Editrice, Bari, pp 29–70

Pfaff E, Klingenberg M, Ritt E, Vogell W (1968) Korrelation des unspezifisch permeablen mitochondrialen Raumes mit dem "Intermembran-Raum." Eur J Biochem 5:222–232

Plas van der LHW, Jobse PA, Verleur JD (1976) Cytochrome-c-dependent, antimycin-A-resistant respiration in mitochondria from potato tuber. Biochim Biophys Acta 430:1–12

Quinn PJ (1981) The fluidity of cell membranes and its regulation. Prog Biophys Mol Biol 38:1–104

Rand RP (1981) Interacting phospholipid bilayers: Measured forces and induced structural changes. Annu Rev Biophys Bioeng 10:277–314

Roos N, Benz R, Brdiczka D (1982) Identification and characterization of the pore forming protein in the outer membrane of rat liver mitochondria. Biochim Biophys Acta 686:204–214

Roux SJ, McEntire K, Slocum RD, Cedel RE, Hale CC (1981) Phytochrome induces photoreversible calcium fluxes in a purified mitochondrial fraction from oats. Proc Natl Acad Sci USA 78:283–287

Salemme FR (1981) Conformational and geometrical properties of β-sheets in proteins. III. Isotropically stressed configurations. J Mol Biol 146:143–156

Salemme FR, Weatherford DW (1981) Conformational and geometrical properties of β-sheets in proteins. I. Parallel β-sheets. J Mol Biol 146:101–117

Schein SJ, Colombini M, Finkelstein A (1976) Reconstitution in planar lipid bilayers of a voltage-dependent anion selective channel obtained from paramecium mitochondria. J Membr Biol 30:99–120

Schindler H, Rosenbusch JP (1978) Matrix protein from *Escherichia coli* outer membranes forms voltage-controlled channels in lipid bilayers. Proc Natl Acad Sci USA 75:3751–3755

Schleyer M, Schmidt B, Neupert W (1982) Requirement of a membrane potential for the posttranslational transfer of proteins into mitochondria. Eur J Biochem 125:109–116

Schnaitman CA, Greenawalt JW (1968) Enzymatic properties of the inner and outer membranes of rat liver mitochondria. J Cell Biol 38:158–175

Schnaitman CA, Erwin VG, Greenawalt JW (1967) The submitochondrial localization of monoamine oxidase. J Cell Biol 32:719–735

Schreibmayer W, Hagauer H, Tritthart HA (1983) Incorporation of a voltage sensitive pore from guinea pig heart mitochondria into black lipid membranes and characterization of electrical properties. Z Naturforsch 38C:664–667

Singer SJ, Nicolson GL (1972) The fluid mosaic model of the structure of cell membranes. Science 175:720–731

Sjøstrand FS (1977) The arrangement of mitochondrial membranes a new structural feature of the inner mitochondrial membranes. J Ultrastruct Res 59:292–319

Sjøstrand FS (1978) The structure of mitochondrial membranes: a new concept. J Ultrastruct Res 64:217–245

Smith DS, Jarlfors U, Cayer ML (1977) Structural cross-bridges between microtubules and mitochondria in central axons of an insect (*Periplaneta americana*). J Cell Sci 27:255–272

Smith GS (1980) Changes in monoamine oxidase activity associated with the uncoupling of rat liver mitochondria. FEBS Lett 121:303–305

Smith GS, Reid RA (1978) The influence of respiratory state on monoamine oxidase activity in rat liver mitochondria. Biochem J 176:1011–1014

Sottocasa GL, Kuylenstierna B, Ernster L, Bergstrand A (1967) An electron transport system associated with the outer membrane of liver mitochondria. J Cell Biol 32:415–438

Sparace SA, Moore TS Jr (1979) Phospholipid metabolism in plant mitochondria. Submitochondrial sites of synthesis. Plant Physiol 63:963–972

Sparace SA, Moore TS Jr (1981) Phospholipid metabolism in plant mitochondria. II. Submitochondrial sites of synthesis of phosphatidylcholine and phosphatidylethanolamine. Plant Physiol 67:261–265

Strittmatter P, Rogers MJ (1975) Apparent depencence of interactions between cyto-chrome b_5 and cytochrome b_5 reductase upon translational diffusion in dimyristoyl lecithin liposomes. Proc Natl Acad Sci USA 72:2658–2661

Tanner GT, Copeland L, Turner JF (1983) Subcellular localization of hexose kinases in pea stems: mitochondrial hexokinase. Plant Physiol 72:659–663

Tedeschi H (1965) Some observations on the permeability of mitochondria to sucrose. J Cell Biol 25:229–242

Thompson JE, Coleman R, Finean JB (1968) Comparative X-ray diffraction and electron microscope studies of isolated mitochondrial membranes. Biochim Biophys Acta 150:405–415

Tzaphlidou M, Chapman JA, Meek KM (1982) A study of positive staining for electron microscopy using collagen as a model system. I. Staining by phosphotungstate and tungstate ions. Micron 13:119–131

Uemura T, Chiesara E (1976) NADH-dependent aryl hydrocarbon hydroxylase in rat liver mitochondrial outer membrane. Eur J Biochem 66:293–307

Ulvik RJ (1983) Reduction of exogenous flavins and mobilization of iron from ferritin by isolated mitochondria. J Bioenerg Biomembr 15:151–160

Unwin PNT, Zampighi G (1980) Structure of the junction between communicating cells. Nature (London) 283:545–549

Verkleij AJ, Kruyff De B, Gerritsen WF, Demel RA, Deenen van LLM, Ververgaert PHJT (1973) Freeze-etch electron microscopy of erythrocytes, *Acholesplasma laidlawii* cells and liposomal membranes after the action of filipin and amphotericin B. Biochim Biophys Acta 291:577–581

Vignais PM, Nachbauer J, Vignais PV (1969) A critical approach to the study of the localization of phospholipase-A in mitochondria. In: Ernster L, Drahota Z (eds) Mitochondria: structure and function. Academic Press, London New York, pp 43–58

Wallace E, Cooper GW, Calvin HI (1981) Abnormal shape and arrangement of mitochon-dria in selenium-deficient mouse sperm. J Cell Biol 91:190a

Werkheiser WC, Bartley W (1957) The study of steady-state concentrations of internal solutes of mitochondria by rapid centrifugal transfer to a fixation medium. Biochem J 66:79–91

Whereat AF, Orishimo MW, Nelson J, Phillips SJ (1969) The location of different synthet-ic systems for fatty acids in inner and outer mitochondrial membranes from rabbit heart. J Biol Chem 244:6498–6506

Wojtczak L, Zaluska H (1969) On the impermeability of the outer mitochondrial mem-brane to cytochrome c. I. Studies on whole mitochondria. Biochim Biophys Acta 193:64–72

Wynn-Williams AT (1976) An explanation of apparent sudden changes in the activation energy of membrane enzymes. Biochem J 157:279–281

Zalman LS, Nikaido H, Kagawa Y (1980) Mitochondrial outer membrane contains a protein producing nonspecific diffusion channels. J Biol Chem 255:1771–1774

6 Organization of the Respiratory Chain and Oxidative Phosphorylation

A.L. MOORE and P.R. RICH

1 Introduction

The fundamental structure, function and organization of the respiratory chain in plant mitochondria is similar to that found in mammalian systems. However, in addition to this basic structure, plant mitochondria are unique in so far as they can possess routes of substrate oxidation and a terminal oxidase not normally encountered in mammalian systems. Such a complex respiratory system may enable plant cells to function under a wide variety of metabolic conditions which, when necessary, can by-pass all of the energy conservation sites.

It is the intention of this chapter to provide a comprehensive view of the organization of the plant respiratory chain complexes and their linkage to ATP synthesis. The respiratory chain components, such as cytochromes, are only treated in terms of their organization into discrete multiprotein complexes. For a more detailed account of their structure the reader is referred to Chap. 4, this Volume. The phenomenon of cyanide insensitive respiration is dealt with only as an electron transport activity of the respiratory chain; its structure, function and operation in intact tissues is described in Chapt. 8, this Volume. Similarly, the NADH dehydrogenases are considered only with respect to their position within the respiratory chain and how they can be utilized as a means of studying the organization of the respiratory chain; for a detailed analysis of their structure and properties the reader is referred to Chap. 7, this Volume. It is not within the scope of this review to give a complete account either of the structure of the respiratory chain complexes or the mechanism of oxidative phosphorylation. This chapter is therefore restricted to a survey of the most pertinent and recent research from plant studies, and for completion we have included relevant data from mammalian studies. In the final section oxidative phosphorylation has been considered in terms of the role of the H^+ electrochemical gradient. Recent data on H^+/O stoichiometry of the plant respiratory chain have been included and their relevance to the organization of the respiratory chain is considered. The reader is also referred to a number of relevant general plant reviews (PALMER 1979, STOREY 1980, HANSON and DAY 1980, DAY et al. 1980, MOORE and RICH 1980, MOORE and PROUDLOVE 1983, MOORE and COTTINGHAM 1983).

2 Organization of Respiratory Components

2.1 The Basic Functional Units

Since the fractionation work of GREEN, HATEFI and colleagues (HATEFI 1978), many lines of evidence have produced a picture in which the respiratory components are arranged into discrete multiprotein units, in which the individual proteins do not associate/dissociate within the time scale of the electron transfer events. In mammalian mitochondria, five major complexes which are associated with electron transfer and oxidative phosphorylation may be identified – four of these catalyze component reactions of the electron transfer pathway, whilst the fifth is the ATPase. All have been purified and characterized in some detail from mammalian sources. Together they account for the complete sequence of reactions which comprise respiration-driven ATP synthesis.

The four electron transfer complexes are complex I (internal NADH → ubiquinone), complex II (succinate → ubiquinone), complex III (ubiquinol → cytochrome c and complex IV (cytochrome c → oxygen). In higher plant mitochondria, we must add two further components. One of these is a second NADH dehydrogenase located in the inner membrane but with its binding site for NADH facing the cytosolic medium (DOUCE et al. 1973). The second is a cyanide-insensitive alternative oxidase, which may be naturally present or inducible in many types of plant mitochondria (BONNER and RICH 1978). Neither of these has counterparts in mammalian systems and it is only recently that progress has been made in the characterization of their basic structures.

Two further small redox molecules must be added to the list of electron transferring multiprotein complexes in order to define the full electron transfer sequence: ubiquinone-10, a hydrophobic redox agent which is necessary for redox connection of complexes I and II to complex III, and the small hydrophilic cytochrome c, which allows electron transfer from complex III to complex IV.

These electron transfer components and their interrelations are summarized in Fig. 1. It is our intention to briefly review the information available on

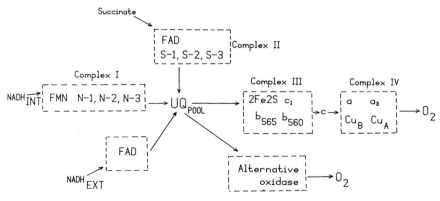

Fig. 1. The higher plant respiratory chain. The known components of the plant system are arranged into their presumed complexes. *2Fe2S* the Rieske iron-sulfur protein

the structure of these complexes and to discuss how the dehydrogenases and oxidases may interact via the ubiquinone.

2.1.1 Complex I, NADH Dehydrogenase

Complex I is able to catalyze the reduction of ubiquinone-10, or a range of other ubiquinone analogues, by NADH (Crane 1977). In the intact membrane, the NADH-oxidizing site faces internally into the matrix space, so that exogenously added NADH is not oxidized. The mammalian enzyme has been purified (Hatefi and Rieske 1967) and reconstituted into lipid vesicles in such a way that proton pumping associated with electron transfer may be observed (Ragan and Hinkle 1975). However, it is the least well-characterized of the mitochondrial complexes, and probably also the least well understood mechanistically. Its molecular weight is around 850,000 and it has at least 26 distinct polypeptides (Heron et al. 1979), only some of which contain redox active centers. Each complex contains FMN and 16–28 acid labile sulfurs and iron atoms per FMN. These are contained in a number of distinct iron-sulfur proteins, which to some extent may be identified separately by EPR (Ohnishi and Salerno 1982).

The analogous complex from higher plants has not yet been isolated, and therefore no specific comparison with the mammalian counterpart may be made. However, EPR signals closely resembling the iron-sulfur components have been detected in a number of plant mitochondrial preparations (Rich and Bonner 1978a, Cammack and Palmer 1977) and from this we may provisionally conclude that site I of plant mitochondria is roughly analogous to its mammalian counterpart (see Table 1). Such a conclusion is partially corroborated by the sensitivities of site I from both sources to the specific site I inhibitors, rotenone

Table 1. Iron-sulfur centers attributed to complex I and succinate dehydrogenase

	Beef heart submitochondrial[a]	Higher plants[b]
	$E_{m_{7.2}}$ (mV)	
Site I		
N-1a	−370	n.d
N-1b	−220	−240, −260
N-2	− 80	−110, 60
N-3	−240	−275
N-4	−240	n.d.
N-5	−275	n.d.
Succinate dehydrogenase		
S-1	0	−7, 60
S-2	−260[c]	−240, −225
S-3	70 to 120	85, 65

n.d. Not detected.

[a] Ohnishi (1979)
[b] Rich and Bonner (1978a) and Cammack and Palmer (1977)
[c] In succinate cytochrome c reductase

and piericidin A (PALMER 1976). However, plant mitochondrial preparations complicate the analogy by being only partially sensitive to these inhibitors, caused by a by-pass around these sites of inhibition which is not coupled to proton translocation (BRUNTON and PALMER 1973, DAY and WISKICH 1974a, b). There are several possible explanations for this inhibitor-insensitive pathway and its relation to oxidation of NADH produced by malate dehydrogenase and malic enzyme and the reader is referred to PALMER and WARD (Chap. 7, this Vol.) for detailed discussion.

2.1.2 Complex II, Succinate Dehydrogenase

Mammalian complex II has been purified and characterized in some detail. Although the early preparations showed little activity when reconstituted (SINGER et al. 1956), later procedures produced material of high purity and full activity (KING 1963, HANSTEIN et al. 1971, ACKRELL et al. 1977). Such preparations contain one covalently bound FAD, eight nonheme irons and eight acid labile sulfurs per molecule. Each active complex consists of two large nonidentical and several smaller subunits. The largest subunit, the flavo-iron-sulfur protein (the Fp subunit), has a molecular weight of 70,000 and contains the FAD and two spin-coupled binuclear 2 Fe-2 S iron-sulfur clusters, termed centers S-1 and S-2 (OHNISHI and SALERNO 1982). The other large subunit of molecular weight 27,000, the Fe-S protein (the Ip subunit), contains one tetranuclear 4 Fe-4 S iron-sulfur cluster which has been termed center S-3 (in an earlier nomenclature this was termed a HiPIP – a high potential iron-sulfur protein).

Surface labeling and immunoprecipitation studies with mitochondria and Submitochondrial particles (SMP's) (BELL et al. 1979) and more recently, studies of the interaction of complex II subunits with phospholipids having a reactive nitrene group in the head group region and on the methyl terminus of one of the fatty acid side chains (GIRDLESTONE et al. 1981) have indicated how the subunits are arranged in the mitochondrial membrane (OHNISHI and SALERNO 1982, CAPALDI 1972). In such a model, the Fp subunit is located in the matrix space, whereas the Ip subunit and the two smaller proteins which are found in reconstitutively active preparations are mostly buried, but with the whole complex spanning the membrane. A further series of investigations have produced estimations of the distance between redox centers based upon the effects of redox state of one component on the relaxation rate of an interacting species (SALERNO et al. 1979, ALBRACHT 1980). Such studies have been combined with measurements of distances of components from membrane surfaces by dysprosium relaxation of EPR signals (OHNISHI and SALERNO 1982), and also with the orientations of the redox centers with respect to the membrane surface as elucidated by the dependence of EPR spectral features on the angle between the membranes and the applied magnetic field in samples of dried multilayers of mitochondrial membranes (SALERNO et al. 1979). The type of picture which emerges is summarized in Fig. 2.

The two major subunits of succinate dehydrogenase have been characterized in sweet potato (HATTORI and ASAHI 1982a, b, HATTORI et al. 1983) and mung bean (BURKE et al. 1982) with molecular weights of 65,000–67,000 and

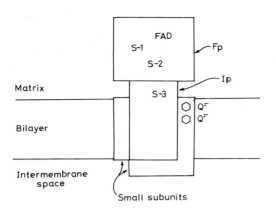

Fig. 2. Topological arrangement of succinate dehydrogenase in the inner mitochondrial membrane. (After OHNISHI 1979 and OHNISHI and SALERNO 1982)

26,000–30,000 for the Fp and Ip subunits respectively. In addition, there have been several reports of EPR signals in plant mitochondria which may be attributable to centers S-1, S-2 and S-3, (CAMMACK and PALMER 1977, MOORE et al. 1976, RICH et al. 1977, RICH and BONNER 1978a, b, BURKE et al. 1982) and their EPR spectral features and redox potentials are remarkably similar to those of their mammalian counterparts (Table 1). Furthermore, a signal which has been reported in mammalian mitochondria which probably arises from an interaction between center S-3 and two bound ubisemiquinone species (RUZICKA et al. 1975, INGLEDEW et al. 1976) [which are probably bound to the 13,500 M.W. protein (CAPALDI 1982)] may also be observed in a variety of plant mitochondria (MOORE et al. 1976, RICH et al. 1977). Unfortunately, the signal in the higher plant system was rather more labile and this lability precluded the possibility of comparison of redox properties in the two systems. Nevertheless, such similarity of EPR spectra, taken together with the rather similar sensitivity to inhibitors such as 2-tenoyltrifluoroacetone (TTFA) (WILSON 1971), points to a strong analogy of the plant succinate dehydrogenase to mammalian complex II.

2.1.3 Complex III, the Cytochrome bc_1 Complex

Mammalian complex III has been isolated in a purified form and in its isolated state is still capable of catalyzing the reduction of cytochrome c by ubiquinol (HATEFI and RIESKE 1967). The purified enzyme reveals eight major polypeptides on an SDS polyacrylamide gel and a specific function has been ascribed to most of these bands. In the presence of urea, a similar gel reveals up to ten bands (CAPALDI 1982). Four redox centres are detectable – a 2 Fe-2 S iron-sulfur center which is generally termed the Rieske center (TRUMPOWER 1981), a c-type heme and two b-type hemes. The cytochromes b may be distinguished since they have different spectral properties and midpoint potentials – cytochrome b-562 has an α-band maximum at 562 nm and an E_{m_7} of around 90 mV whereas cytochrome b-566 has an α-band at 566 nm with a shoulder at 558 nm and an E_{m_7} of around -30 mV (NELSON and GELLERFORS 1976). It is likely that both heme groups are contained in the same polypeptide (HAUSKA et al. 1983).

Evidence is accumulating to suggest that the complex may be dimeric in its natural state (CAPALDI 1982). Indeed, some workers have suggested that there may be four distinct cytochromes b and two distinct Rieske centers in the intact operative dimer and a complex mode of operation based upon this possibility has been made (DE VRIES et al. 1982), although without substantial experimental evidence at present.

The mechanism of operation of the complex has received much attention. One reason for this is that a multiprotein of the cytochrome bc complex type forms a central feature of the energy transducing apparatus not only in mitochondrial but also in chloroplast and various photosynthetic bacterial electron transfer chains. A second reason is that, besides the $2H^+/2e^-$ which are deposited on the C-side of the membrane when a quinol reduces the complex, a further $2H^+/2e^-$ are "pumped" across the membrane as the electrons are transferred through the complex to cytochrome c. Complex III therefore forms an ideal system with which to study the fundamental mechanisms of proton motive function. This was made even more useful by the demonstration that the isolated, purified complex could be reincorporated into lipid vesicles in such a way that this proton motive function was retained and could be easily demonstrated (LEUNG and HINKLE 1975, GUERRIERI and NELSON 1975).

Two highly specific and stoichiometric inhibitors of the bc_1 complexes have been of particular use in the elucidation of the electron transfer route. The first of these was antimycin A, an inhibitor which classically inhibits between cytochromes b and c_1 in the complex. The second and much more recent is myxothiazol, an inhibitor which has been shown to inhibit electron donation into the complex at the site where a quinol would normally react (THIERBACH and REICHENBACH 1981). By using a combination of these two inhibitors, it has been possible to show that there are two sites with which a quinol/quinone couple may interact with the complex. Such a finding is very much in line with the "Q-cycle" formulation of MITCHELL (1976) for electron transfer through the complex. This Q-cycle was originally formulated in order to explain the proton motive stoichiometry of the complex and also to explain several rather unusual kinetic responses of the intact and isolated complex. The most novel of these is the "oxidant-induced reduction" of cytochrome b, a phenomenon most clearly observed when antimycin A is also present to block the reoxidation of cytochrome b (RIESKE 1976). The observation is that, in the presence of a donor such as quinol, the cytochrome c_1 is reduced and the cytochrome b is oxidized in the resting state. When, however, a pulse of oxidant for cytochrome c_1 is added (for example, ferricyanide or cytochrome c), the cytochrome c_1 becomes oxidized but the cytochrome b becomes reduced and only slowly reoxidizes when the oxidant has been consumed. WIKSTRÖM and BERDEN (1972) were the first to offer an explanation for this effect in terms of a concerted reaction where one electron from quinol reduced a cytochrome b, whereas the second reduced the Rieske center cytochrome c_1 region. A similar concerted reaction was incorporated into the Q-cycle formation by MITCHELL (1976). This reaction was suggested to occur at the outer surface of the inner membrane (the P or positively charged phase), so that the protons associated with the quinol oxidation were liberated into this phase. The electron on cytochrome c_1 then ultimately went to cytochrome oxidase, whereas the electron donated to cytochrome b traversed the membrane from b-566 to b-562. The b-562 then reduced the quinone pool while in contact with the inner phase (the N or negatively charged phase), so that one H^+ was then taken up from this phase:

$$^1/_2 Q + H^+_{in} + \text{b-562}_{red} \rightarrow \,^1/_2 QH_2 + \text{b-562}_{ox}$$

The details by which this cytochrome b might reduce quinone in two $n=1$ stages to quinol are complex and beyond the scope of this article. A number of recent reviews may be consulted for further discussion of this point (Crofts and Wraight 1983, Hauska et al. 1983, Rich 1984) and for the possible role of "protein-bound" quinone species in such schemes.

An alternative scheme for the mechanism of proton translocation by the complex was presented by Wikström et al. (1981a). The mechanism of electron transfer through the complex is envisaged to be essentially the same as that described above, but the cytochromes b were proposed to be on the same side of the membrane as the quinol oxidation site. Proton movement was then achieved by "conformational proton pumps" (Papa 1982) driven by the redox cycling (and concurrent conformational changes) of the cytochromes b. With one pump associated with each cytochrome b, the observed stoichiometry of $4H^+$ appearing on the outside for each two electrons passing through the complex to cytochrome c could be achieved. Figure 3 illustrates such a "b-cycle" together with a Mitchellian "Q-cycle" for comparison.

A cytochrome bc_1 complex has been isolated from higher plants by Ducet and Diano (1978) which displays optical characteristics very similar to those of the mammalian complex. Unfortunately no SDS-PAGE gels or enzymic activities of the isolated materials were given. Although there are more b type cytochromes present in plant mitochondria than in mammalian mitochondria, (Storey 1980) Dutton and Storey (1971) and Lambowitz and Bonner (1974) have identified two of the complement which would appear to be good candidates for bc_1 complex constituents. These are b-560 ($E_{m_7} + 42$ to $+79$ mV) and b-565 ($E_{m_7} - 75$ mV). Also, a membrane-bound c-type cytochrome with an α-band and at 549 nm, indicative of cytochrome c_1, has been described (Bonner 1965, Lance and Bonner 1968). Until recently, one rather disturbing anomaly in plant mitochondria was the apparent lack of an EPR signal attributable to a Rieske-type iron-sulfur center (Rich and Bonner 1978a). However, Prince et al. (1981) have indicated that although the signal may be somewhat weaker than the mammalian counterpart, it may be detected, particularly when the quinone analogue 5-n-undecyl-6-hydroxy-4,7-dioxobenzothiazole, UHDBT, (Trumpower 1981) is added so that it binds to and alters the EPR spectrum of the Rieske center. All the above findings point to there being a distinct cytochrome bc_1 complex, especially in view of the evidence for the operation of a Q-cycle in plant mitochondria (Rich and Moore 1976). It is highly likely that it is structurally analogous to the mammalian and photosynthetic bacterial counterparts, since it too is stoichiometrically inhibited by antimycin A and myxothiazol. Elucidation of the polypeptide composition of the plant complex would be particularly interesting, since although the mammalian and bacterial enzymes are mechanistically extremely similar, their polypeptide compositions are quite different (Hauska et al. 1983).

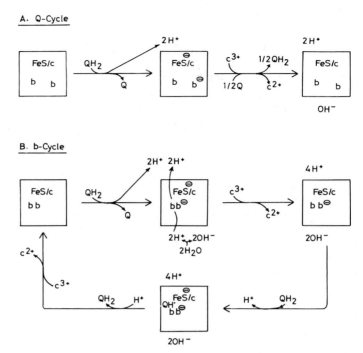

A. Q-Cycle

B. b-Cycle

Fig. 3A, B. Possible representations of the Q-cycle and the b-cycle. In both models the Rieske center and c_1 are assumed to be in rapid equilibrium and both b cytochromes are involved. In the Q-cycle, the cytochrome b is reoxidized by quinone on the inside of the membrane. In the b-cycle, the cytochrome b is reoxidized by a quinone species on the outside of the membrane and the required additional proton movements are brought about by pumps associated with the redox cycling of the cytochromes b

2.1.4 Complex IV, Cytochrome Oxidase

Mammalian cytochrome oxidase is generally described as having seven major polypeptide components, although up to 12 proteins may actually be present in purified preparations (CAPALDI 1982). Each complex contains two a-hemes and two copper atoms, all of which are located on subunits I and II. Two-dimensional crystals of the enzyme in lipid bilayers have allowed a three-dimensional model to be produced. This information, together with data from a variety of additional techniques, has produced a detailed model of cytochrome oxidase structure. The complex spans the membrane and it is likely that all four redox centers are close to the C-side of the membrane. The reader is referred to several recent reviews for detailed description (WIKSTRÖM et al. 1981a, CAPALDI 1982). Functionally, it is agreed that the reaction catalyzed is that of cytochrome c reoxidation from the C-side of the membrane coupled to reduction of molecular oxygen to water with the uptake of protons from the matrix. What is more controversial is whether, in addition to this proton-consuming reaction, a proton pump also exists which translocates an additonal proton from internal to external phases for each equivalent transferred. For recent reviews of the conflicting views see (Wikström et al. 1981 a, b, Mitchell and Moyle 1983).

Several preparations of cytochrome oxidase from higher plant sources have been reported (Srivasta and Sarkissian 1971, Bomhoff and Spencer 1977). In the most recent, an active, highly purified, cytochrome oxidase preparation has been isolated from sweet potato mitochondria by Maeshima and Asahi (1978). At least five major polypeptides were detected, but the molecular weights did not obviously correspond to the mammalian values. Perhaps of interest is the observation that two of the subunits do correspond reasonably closely with the *Paracoccus* enzyme, which has been found to contain only two subunits at 45,000 and 28,000 M.W. (Ludwig and Schatz 1980), and these presumably contain all the redox components of the larger mammalian complex. The spectral properties are similar, although not identical, to the mammalian enzyme (Bendall and Bonner 1971) and some differences are also observed in intermediates observed after flash photolysis of carbon monoxide from the enzyme at low temperatures (Denis and Richaud 1982). However the redox potentials of $+380$ mV and $+190$ mV for cytochromes a_3 and a at pH 7.2 are close to mammalian values (Dutton and Storey 1971). Furthermore, Mitchell and Moore (1984) have recently found evidence for a proton pumping function associated with the plant oxidase (see Sect. 4.3).

2.1.5 The Alternative Oxidase

Many higher plant mitochondria contain an oxidase activity which is capable of oxidizing components which are able to donate to the ubiquinone pool (Bonner and Rich 1978, Siedow 1982). The most dramatic demonstration of this oxidase activity is found in the spadices of many aroid species, where the capacity of this "alternative oxidase" may exceed the capacity of the cytochrome oxidase by a factor of 10 or more. As has been repeatedly reviewed, the nature of this oxidase has remained remarkably elusive and many hypotheses of its nature have appeared. The difficulty has arisen because it apparently does not possess any clear optical or EPR characteristics. Despite this, it became clear from many angles of work that it was at the level of ubiquinone that the alternative oxidase derived its reducing equivalents and that the final product of the redox process was water (Bonner and Rich 1978, Moore and Rich 1980, Storey 1980, Siedow 1982, Laties 1982, Moore and Cottingham 1983):

$$2UQH_2 + O_2 \rightarrow 2UQ + 2H_2O.$$

That this reaction was catalyzed by the oxidase was later confirmed when two laboratories (Huq and Palmer 1978b, Rich 1978) independently produced a quinol oxidase assay for the activity which allowed partial purification of this respiratory function. The solubilized activity oxidized a number of quinol substrates with the concomitant reduction of molecular oxygen to water in a cyanide-insensitive and hydroxamate-sensitive manner.

Recently, some speculations were made to suggest that the alternative oxidase activity might be completely artifactual (Kelly 1982) or might be caused by a cycling between fatty acid and fatty acid peroxy radicals, and so might not be enzymic in origin (RUSTIN et al. 1983). The suggestions of KELLY were well dealt with by PALMER (1982), who drew attention to a number of convincing observations that such an alternative oxidase activity, which is distinct from lipoxygenase activity, does in fact exist. The hypothesis of RUSTIN et al. (1983) was unfortunately devoid of any experimental evidence which could be critically examined. However, of those conditions which the authors listed as promoting such a nonenzymic fatty acid radical oxidation, i.e., visible light, UV light, heat, high O_2 partial pressure, none is in fact found to increase the oxidase activity and in fact the very opposite may even be found, i.e., loss of activity even on mild warming. The lack of such activity in most mitochondria is also rather difficult to reconcile with this hypothesis. Recently, BONNER, CLARKE and RICH (unpublished) have succeeded in taking the oxidase purification considerably further and have produced a preparation from *Arum maculatum* of increased specific activity. The preparation was almost colorless and mostly separated from a copper-containing protein and carotenoid which had copurified with the oxidase in the previous cruder preparations (RICH 1978). Metal analyses of the most active fractions revealed a significant content of iron which correlated with the quinol oxidase activity. SDS-PAGE of the best preparations produced only one (or perhaps two closely spaced) strongly-staining bands of protein in the molecular weight range 30,000–35,000. The quinol oxidase activity of this preparations was around five times greater than the starting material and retained normal inhibitor sensitivities. Such observations strongly suggest an enzymic origin of the alternative oxidase and are in conflict with the suggestions of RUSTIN et al. (1983).

2.1.6 The External NADH Dehydrogenase

It is well established that plant mitochondria can oxidize exogenous NADH by an NADH dehydrogenase located on the external face of the inner membrane (PALMER and PASSAM 1970, COLEMAN, PALMER 1971, DAY and WISKICH 1974a). The oxidation of NADH by this system is insensitive to inhibitors of complex I, such as rotenone and piericidin A, but sensitive to antimycin A. It is specific for the 4 β hydrogen atom of NADH and is coupled to two sites of phosphorylation, suggesting that it donates reducing equivalents into the respiratory chain at the level of ubiquinone. This notion is confirmed by the finding that in intact mitochondria, NADH:ferricyanide oxidoreductase activity is antimycin A-sensitive (DOUCE et al. 1973). To date, relatively little information is available on either the subunit or polypeptide composition of this dehydrogenase. Recently, however, there have been two reports on its solubilization and partial purification (KLEIN et al. 1983, COOK and CAMMACK 1983) from cauliflower and *Arum maculatum* mitochondria. Both groups obtained a solubilized NADH dehydrogenase that was insensitive to rotenone and interestingly unaffected by either EGTA or Ca^{2+}. The membrane-bound dehydrogenase is generally stimulated by calcium and inhibited by EGTA (COLEMAN and PALMER 1971, MØLLER et al. 1981, MOORE and ÅKERMAN 1982). Our laboratory has also solubilized the external NADH dehydrogenase from phosphate-washed *Arum macula-*

tum mitochondria (Cottingham and Moore, unpublished). The dehydrogenase was purified by chromatography, and an SDS-PAGE gel when silver-stained revealed only three polypeptides (the major band of which had a molecular weight of 78,000) noncovalently bound FAD, and undetectable amounts of cytochrome or nonheme iron. Purified preparations oxidize NADH but not NADPH in a rotenone insensitive fashion with a K_m of 28 μM and either UQ-1 or DCPIP can act as electron acceptors. As with other preparations, very little sensitivity to either Ca^{2+} or EGTA was observed. The reason for this loss of Ca^{2+} sensitivity is at present unclear, although similar observations have been made with solubilized preparations of glycerophosphate dehydrogenase (Wernette et al. 1981).

3 Connection Between the Functional Units

3.1 The Sidedness of the Reactions of the Alternative Oxidase, Succinate Dehydrogenase and the External NADH Dehydrogenase

In some plant mitochondria which possess an alternative oxidase, equivalents from both external NADH and from succinate, may be used by the alternative oxidase (although the former pathway may be less favored than the latter: see Sect. 3.2.3). Both pathways have been shown to be nonproton motive (see Sect. 4.3). This is of particular interest, since the succinate dehydrogenase donates reducing equivalents into the respiratory chain from the inside, whereas the external NADH dehydrogenase donates reducing equivalents from the outside (Douce et al. 1973).

From such elementary observations, several interesting deductions may be made concerning the interconnection of these respiratory components. In the first place, it becomes impossible for both the succinate dehydrogenase and NADH dehydrogenase to directly donate to the oxidase as electron (rather than H-atom) donors. If such were the case, then oxidation of one of the two substrates would have to be electrogenic, depending upon which side of the membrane the oxidase reduced the oxygen and consumed protons. This is not the case. Furthermore, as it is extremely unlikely that the oxidase could act as an electron acceptor for one substrate and a hydrogen atom acceptor for another, we may deduce that the oxidase is reduced by an H-atom donor in both cases. However, succinate dehydrogenase acts an electron donor, since it is center S-3 (a one-electron redox agent) which is the species in the complex which is able to interact with and reduce an added acceptor. The conclusion is that there is an H-atom redox agent which acts as an intermediate in the donation of equivalents from these two dehydrogenases to the alternative oxidase and the most probable candidate for such an intermediate is the ubiquinone. The experimental evidence for the involvement of ubiquinone comes from several types of observation and will be discussed in more detail in Sect. 3.2.

Figure 4 illustrates the envisaged sidedness of these reaction sites and the following points may be noted:

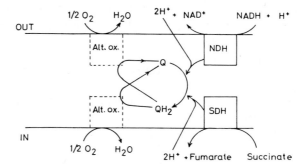

Fig. 4. Inter-relations of succinate and external NADH oxidations through the alternative oxidase. Proton release on the outside occurs when NADH is oxidized, but overall a net $1\,H^+/2e^-$ is taken up as quinone is reduced to quinol. Protons are transiently released inside during succinate oxidation but are subsequently taken up again as quinone is reduced to quinol. The alternative oxidase reaction may occur on either side of the membrane, but its location is presently unknown: *SDH* succinate dehydrogenase; *NDH*, *NADH* dehydrogenase; *Alt. ox.* alternative oxidase

1. protons released into the inside by succinate dehydrogenase as it oxidises succinate to fumarate are retaken up as quinone is reduced to quinol;
2. Protons associated with oxidation of external NADH are consumed as quinone is reduced to quinol (whether there is transient proton release, or whether hydrogen atom donation occurs is not known). Overall, a net alkalinization of the outside which is nonelectrogenic will occur (this has been observed experimentally, MITCHELL and MOORE 1984);
3. since the quinol carries both protons and electrons the reaction site of the oxidase with molecular oxygen cannot be deduced. It may be in contact with either side of the membrane since the net reaction catalyzed is electroneutral and not proton-utilizing.

3.2 The Role of Ubiquinone as a Mobile Redox "Pool"

3.2.1 Mobility Between Components

In general, although the cytochrome bc_1 and cytochrome oxidase complexes are present in mitochondria in roughly stoichiometric quantities, the dehydrogenases tend to be substoichiometric in quantity (KRÖGER and KLINGENBERG 1967). Despite this, it is quite possible for a single substrate, such as succinate, to reduce all of the cytochrome, bc_1 or cytochrome oxidase complexes rapidly. It would appear that there is mobility of some sort in the way in which the complexes donate to each other, rather than their existing in tight $1:1$ stoichiometric multicomplexes. The notion of mobility of some sort is also indicated by the well-documented observation that a number of stoichiometric inhibitors, such as antimycin A, actually inhibit the respiratory chain "sigmoidally". An explanation for such effects with the cytochrome bc_1 complex inhibitor, antimycin A, was provided by Kröger and Klingenberg (1973 a, b). If, for example, in the case of antimycin A, an excess of cytochrome bc_1 activity is present in the untreated mitochondria, then inhibition of a significant fraction of these bc_1 complexes will not affect respiratory rate, since electrons from a given dehydrogenase are not confined to a single bc_1 complex. The idea that there is mobility of some sort between the complexes is further strengthened by an extension of the approach of Kröger and Klingenberg to cover loosely bound inhibitors (Moreira et al. 1980, Cottingham and Moore 1983), from studies which showed that when several substrates are oxidized simultaneously, they behaved as if they competed for a common reaction step rather than being independent (Gutman 1980), and from the experiments of Schneider et al. (1980), which demonstrated by lipid dilution that diffusion was necessary for interaction to occur.

3.2.2 The Role of Ubiquinone in Providing Mobility

In 1962, Green had suggested that ubiquinone acts as a lipid soluble cofactor between dehydrogenases and bc_1 complexes and many of the subsequent investigations which indicated mobility were interpreted in terms of mobility of this quinone, rather than of the protein complexes. The quinone acted as a "pool" capable of transferring reducing equivalents from a dehydrogenase to any bc_1 complex in the system.

Ubiquinone is a substituted 1,4 benzoquinone and all biological forms have a long, extremely hydrophobic side chain of isoprenyl units. It is usually present in a five- to tenfold molar excess over the cytochrome bc_1 complex (Kröger and Klingenberg 1967). In mammalian (Crane 1977) and plant mitochondria (Huq and Palmer 1978 a) this side chain is composed of 10 isoprenyl units, although ubiquinone extracted from other sources may have different chain lengths (Crane 1977). The quinone has often been pictured as having its side chain parallel to the lipid molecules in the membrane with a diffusion coefficient close to the phospholipid value of $D = 10^{-8}$ cm^2 s^{-1} (Cherry 1979). Transmembrane movement of reducing equivalents would then occur by "flipping" of the quinone, so that the head group moved to the opposite surface of the membrane. More recently, a rather different picture of quinone structure in the membrane has emerged. Quinn and colleagues (Quinn 1980, Quinn and Esfahani 1980, Katsikas and Quinn 1982) used surface pressure studies of monolayers and calorimetric studies of bilayers, and produced a model in which the quinone molecules reside at the center of the bilayer. Recent NMR work of Kingsley and Feigenson (1981) has complemented such studies, as has the observation of Chance (1972) that only fluorescent probes which can penetrate 14 Å into the membrane can be fluorescence-quenched, presumably by the quinone ring. A picture in which the quinone is in the middle of the bilayer, with the head group only occasionally reaching the membrane surfaces, is favored by the above data. In such a location, mobility could be rather more rapid than if the side chain were arranged across the membrane. However, a recent attempt was made by Crofts and Wraight (1983) to estimate the diffusion coefficient for ubiquinone in bacterial chromatophores. By observing the lag in cytochrome b reduction after a flash, and assuming random component distribution, a diffusion coefficient closer to 10^{-9} cm^2 s^{-1} was estimated. RICH (1983) has calculated the number of cytochrome bc$_1$ dimers which are within range of a quinol, based upon these diffusion coefficients. This number decreases as the turnover time of the respiratory chain increases and hence is lower in "state 3" than in "state 4" mitochondria[1]. By using a Q/bc$_1$ ratio of 5 and values of 200 s^{-1} and 20 s^{-1} for the turnover number of bc$_1$ monomers in state 3 and state 4 respectively, $D = 10^{-8}$ cm^2 s^{-1} indicated that around 80 (state 3) and 800 (state 4) bc$_1$ dimers were potentially within range of a quinol. A value of $D = 10^{-9}$ cm^2 s^{-1} decreased these numbers by a factor of 10.

From the above, it appears that ubiquinone does indeed act as an electron-redistributing pool between the multiprotein complexes. Some doubt was, however, cast on such a notion with the finding of quinone-binding proteins of stabilized semi-quinone species and with the finding in photochemical systems of quinone species such as Q$_z$ (DUTTON and PRINCE 1978), which possess properties distinct from the quinone pool. Suggestions were even made that all ubiquinone might be protein-bound (for a review, see YU and YU 1981). However, a possible means of reconciliation of these opposing viewpoints was offered (RICH 1981), when it was suggested that the observations of "bound quinones" actually represented enzyme intermediates. In such a view, the quinone does act as a mobile pool but enzymic intermediates may be observed under appropriate conditions. Because of binding equilibria, these enzyme intermediates will have different properties from the bulk quinone, although, at least as fully oxidized or fully reduced Q species, they are able to exchange rapidly with pool quinone. More detailed discussion is beyond the scope of this article and the reader is referred to two recent reviews (RICH 1984, CROFTS and WRAIGHT 1983).

3.2.3 The Relation of the Quinone Pool to Control of Electron Flow Through the Cytochrome and Alternative Oxidases

In many types of plant mitochondria which possess an alternative oxidase, electrons from several different substrates have access to both the cytochrome oxidase and the alternative oxidase. The notion of a freely diffusing quinone pool which interconnects the randomly arranged dehydrogenases and oxidases is of relevance to the way in which relative flux through the two oxidases is "controlled". This question of relative flux has received much attention, since it has a bearing on the in vivo efficiency and physiological significance of the electron transfer routes.

The most generally used model for control of flux through the pathways is that derived by BAHR and BONNER (1973a, b). They described a method of determining the electron flux rate through the cytochrome chain (V$_{cyt}$) and the alternative pathway (V$_{alt}$) by titration with the alternative oxidase inhibitors, the benzhydroxamic acids (SCHONBAUM et al. 1971). They found that in both state 3 and in state 4 mitochondria, the cytochrome pathway was always maximally active whether the alternative pathway was operating or not, and con-

---

1 Definition of the different states see Chap. 10, this Vol., p. 283

cluded on the basis of this that both pathways could not be competing for a common intermediate, since alternative oxidase activity was not affecting the activity of the cytochrome pathway.

The fraction of alternative oxidase activity actually operative varied between 0 and 1, depending upon a proposed flavoprotein which was in equilibrium with, but at lower potential than, a component A, which donated to the cytochrome chain. Only when A was very reduced, i.e., when dehydrogenase activities more than saturated the cytochrome chain activity, would the flavoprotein become reduced and allow the alternative pathway to operate.

However, the model of BAHR and BONNER is based on several assumptions which may not be justified. The most important of these concerns those which led them to rule out a Q-pool model as a control mechanism. In particular, the model was ruled out because modulation of alternative oxidase activity did not affect electron flow through the cytochrome chain. This is not necessarily a valid deduction, since it is totally dependent on whether the cytochrome chain is actually limited by the collisional reaction rate between QH_2 and cytochrome bc_1 complex, i.e.:

$$V_{cyt} = k_1 [QH_2][bc_1].$$

If this is limiting, then it is true that altering alternative oxidase activity would alter cytochrome chain activity because of changes in the level of QH_2. However, if it is not the limiting reaction, then cytochrome chain activity will be independent of alternative oxidase activity and a simple Q-pool model becomes perfectly valid as a means of regulation through the different pathways. In state 4 mitochondria, it is clear that the cytochrome chain will be rate-limited by processes within the complexes because of the proton motive force, and it is probably also the case that in state 3 rate limitation at a step other than that required by BAHR and BONNER is the case.

Our general conclusion, then, is that control of the relative flux through the two oxidase pathways is accommodated within the Q-pool model and at present does not require the postulation of additional components to explain the majority of the reported literature.

3.3 Some Instances Where Ideal Q-Pool Behaviour Is Not Observed

Although the Q-pool model in its general form is able to account for much of the behavior of branched respiratory pathways, there are some notable experimental observations which indicate that reservations must be exercised in using such an ideal model in all cases.

It should be emphasized here that the "pool" equations, such as those of KRÖGER and KLINGENBERG (1973a, b), are dependent upon reactants behaving as if in ideal homogenous solution. This is a necessary requirement of any model which incorporates rate constants. Secondly, it requires that respiratory chain reactions must always be limited by donation into and out of the ubiquinone pool via collisions between a protein complex and a quinone or quinol, such that the velocity of the overall process is given by (KRÖGER and KLINGEN-BERG 1973a, b):

$$v = \frac{V_1 \cdot V_2}{V_1 + V_2},$$

where v is observed respiratory activity and V_1 and V_2 are the total capacities for donation into and out of the pool respectively.

Such ideal behavior may often not be the case (for detailed discussion see RICH 1984). For example: respiratory rate may become so high that a quinol only has time to travel between nearest-neighbor complexes (RICH 1984); an internal rate constant of one of the complexes may become rate-controlling such that the rate becomes zero order with Q or QH_2 – in such a case the complex will have a measurable K_m for Q or QH_2; components may not be randomly arranged in the membrane – they may bind to other complexes or may be concentrated in particular regions of the membrane because of, for example, surface charges or membrane curvature; RAGAN and HERON (1978) have provided evidence that stoichiometic complexes between dehydrogenases and bc_1 complex may form in membranes at sufficiently high protein/lipid ratios. Furthermore, they suggest that donation from dehydrogenase to bc_1 complex via quinone will only occur during the formation of such a transient dimer and have pointed out that such a model can remain consistent with "Q-pool" behaviour. In such a model, selectivity of reaction route from a donor to a specific acceptor becomes possible.

With these possibilities in mind, two recent examples of deviation from classical Q-pool behavior will be discussed. The first of these is more difficult to fit into the Q-pool model than the second. The first example concerns the now well-documented observation that many plant mitochondria are able to oxidize succinate quite readily via the alternative oxidase and yet reducing equivalents from the external NADH dehydrogenase have only extremely limited access to the alternative oxidase. A particularly clear demonstration of this was given by HUQ and PALMER (1978a) with mitochondria isolated from cassava tissue. They demonstrated that although antimycin A drastically inhibited external NADH oxidation, SHAM-sensitive respiration could be restored with succinate. Such a result is incompatible with an ideally behaving quinone pool. To the present authors, the most likely explanation lies in a nonrandom distribution of membrane complexes such that the external NADH dehydrogenases are on average further from alternative oxidase complexes than are the succinate dehydrogenase units. The long diffusion pathway of the connecting quinone severely limits (but does not prevent) NADH oxidation by the alternative oxidase. Until further experimental data are available, however, we cannot rule out the other possibility; that the behavior is caused by a more specific association between succinate dehydrogenase and the alternative oxidase (however see COTTINGHAM and MOORE 1983).

A second example of possible contradiction of ideal Q-pool behavior may be found in a recent report by DRY et al. (1983), who found that glycine was preferentially oxidized by pea leaf mitochondria at maximal rates even when

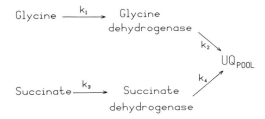

Fig. 5. A simplified kinetic model of succinate and glycine oxidation in pea leaf mitochondria. Rate constants are proposed to be such that glycine oxidation is rate-limited by a term in k_1, whereas succinate oxidation is limited by the second order k_4 process; see text for details. Glycine dehydrogenase refers to the re-oxidation of NADH via NADH dehydrogenase

several other substrates were present in the glycine. However, consideration of a more detailed kinetic model provides a plausible explanation which is still entirely consistent with the mobile pool model. Referring to Fig. 5, we may postulate that glycine oxidation is limited by the dehydrogenase (a term included in the K_1 rate constant) and that the rate constant k_2 is very large. Succinate dehydrogenase, on the other hand, is k_4 limited with a rate given by

$$\text{rate} = K_4[\text{SDH}][\text{Q}].$$

The outcome of such a model is that succinate oxidation can be inhibited by glycine, and yet glycine oxidation will be unaffected by the addition of other respiratory substrates.

In summary, then, we have adopted the Q-pool model as the most reasonable explanation for the behavior of branched electron transfer chains. To this we have added the proviso that the model is only an approximation, since it requires homogeneous solution kinetics to be operative. Although aberrant behavior can be observed in some instances, we believe that kinetic refinement of the general Q-pool model, together with possible consideration of nonhomogeneous component organization, represents a useful way in which to proceed at the present time.

4 Oxidative Phosphorylation

4.1 Background

Oxidative phosphorylation is the process whereby the energy released from the oxidation-reduction reactions of the electron transfer chain is used for the synthesis of ATP. Although a number of hypotheses have been postulated to describe the mechanism of oxidative phosphorylation such as the chemical (SLATER 1953) and the conformational (BOYER 1965) hypotheses, none has stood the test of time and it is now widely agreed (see BOYER et al. 1977) that the coupling of oxidoreductions to phosphorylation is via protonic currents as independently proposed by MITCHELL and WILLIAMS in 1961. In its original form, the essential tenets of the chemiosmotic hypothesis as proposed by MITCHELL (1961) are: (a) an outwardly directed electrogenic proton translocating respiratory chain (b) an inner membrane which is impermeable to H^+, OH^- and other ions except by specific exchange-diffusion systems and (c) a reversible anisotropic proton translocating ATP-synthase. Thus protons ejected by the respiratory chain generate a proton electrochemical gradient which is used to drive H^+ re-entry via the ATP synthase. Although there is general consensus that the coupling of electron transfer to phosphorylation is mediated via proton currents, controversy still exists on whether the nature of these currents is purely osmotic (MITCHELL 1961, 1966, 1976), more localized to the coupling membrane (WILLIAMS 1978, WESTERHOFF et al. 1981, KELL 1979) or both (PADAN and ROTTENBERG 1973, ZORATTI et al. 1983). Furthermore, the mechanism by which the proton current is generated, its coupling to phosphorylation, and even whether it is the direct and sole driving force remains a subject of intense research and debate.

4.2 Proton Electrochemical Gradient

The proton electrochemical gradient $\Delta\bar{\mu}_{H^+}$ (or the protonmotive force, Δp, when expressed in mV) is comprised of two components, a membrane potential $\Delta\psi$, and a pH gradient, ΔpH. Since both of these components are interconvertible, it is the magnitude of Δp, and not its individual components, that is important in a consideration of its adequacy as a driving force for ATP synthesis.

Considerable attention has been directed towards the quantification of Δp, since a single demonstration of ATP synthesis in the absence of a sufficient Δp would be enough to disprove the chemiosmotic hypothesis.

All of the techniques used to estimate Δp involve separate determination of both $\Delta\psi$ and ΔpH. The most commonly used techniques are based either upon the steady-state distribution of a permeant ion between the internal and external aqueous phases, altered spectral properties of an optical probe, or ion-specific electrodes.

4.2.1 Steady-State Ion Distribution

The most common method for determination of $\Delta\psi$ and ΔpH is to measure the steady-state distribution of a permeant ion and, assuming electrochemical equilibrium, to calculate $\Delta\psi$ or ΔpH from its internal and external concentrations using the Nernst ($\Delta\psi$) and Henderson-Hasselbalch (ΔpH) equations.

In order to determine $\Delta\psi$ or ΔpH using this technique, a number of assumptions are generally made such as (a) the indicator should pass freely and rapidly across the membrane and not be transported by an energy-linked system, (b) it should not precipitate or bind to the membrane, or if it does its activity coefficient should be readily calculated, (c) it should be detectable at low concentrations so that the measured system is not perturbed and (d) it should not be metabolized. If the indicator fulfills these basic criteria, then it may be used as either a $\Delta\psi$ or pH probe. K^+ in the presence of valinomycin is usually employed to measure $\Delta\psi$, whereas weak acids or bases are used to measure ΔpH. More recently a number of synthetic cations, originally developed by

Table 2. Typical values for the protonmotive force across energy-transducing membranes

Material	Conditions	$\Delta\Psi$ (mV)	-60ΔpH (mV)	Δp (mV)	Reference
Mung bean	State 4	126	36	162	Moore et al. (1978a)
Potato	State 4	126	27	153	Moore et al. (1978a)
Arum spadix	State 4	132	26	158	Moore et al. (1978a)
Rat liver	State 4	150	78	228	Nicholls (1974)
Brown fat	Proton channel open	79	−25	54	Nicholls (1979)
Mung bean SMP's	ATP	97	127	224	Moore and Bonner (1981)
Beef heart SMP's	NADH	145	0	145	Sorgato et al. (1978)

Skulachev, such as TPMP$^+$ or TPP$^+$, have been successfully used to calculate $\Delta\psi$. Several techniques can be used to measure the transmembrane distribution of these probes such as filtration (Nicholls 1974), sedimentation (Padan and Rottenberg 1973), silicone-oil centrifugation (Wiechmann et al. 1975), or flow dialysis (Sorgato et al. 1978). Each has its own advantages and disadvantages, and these have been extensively reviewed (Rottenberg 1979). Table 2 compares the values for Δp obtained with plant mitochondria with those determined using mammalian tissues. Under state 4 conditions respiration generates a Δp of approximately 160 mV which is comprised of a $\Delta\psi$ of −126 to −132 mV and a pH gradient of about 0.5 pH unit. Since incubation conditions are not identical, some variability is expected, nevertheless the values with plants are very comparable to those obtained by Padan and Rottenberg (1973), Wiechman et al. (1975) and Rottenberg (1979). They are somewhat lower than those reported by Mitchell and Moyle (1969) and Nicholls (1974), but this may be a reflection of the lower matrix space (0.4 µl mg^{-1} protein compared with 1.0 µl mg^{-1}) estimated by these workers. The 228 mV reported by Nicholls (1974) may also be an overestimate in view of the large contribution by the pH gradient. Ferguson and Sorgato (1982) have suggested that binding of acetate (used to estimate ΔpH) and methylamine (a marker for the extramitochondrial space), either to the filters employed or to mitochondria, may have also led to an overestimate of this component. In the light of these suggestions, the values observed for plant mitochondria are very comparable to published data from other tissues. Interestingly, *Arum spadix* mitochondria were able to maintain a protonmotive force (Moore et al. 1978a), suggesting they are not de-energized due to the presence of a proton channel as is the case with brown fat mitochondria (Nicholls 1974) (both of these tissues are thermogenic).

The lipophilic cations, TPP$^+$ and TPMP$^+$, are now widely used as indicators of $\Delta\psi$ in mitochondria and cells. Although at high $\Delta\psi$ values TPMP$^+$ and ^{86}Rb$^+$ plus valinomycin give similar distributions, at low values the differences between the accumulation of these two probes are quite significant (Rottenberg 1979). This is generally presumed to be due to extensive binding of the probe TPMP$^+$ to proteins and phospholipids. It is, of course, possible to substract the amount of probe bound under de-energized conditions from the total amount accumulated under energized conditions, but recently LOLKEMA et al. (1982) have suggested that this procedure can lead to significant errors, especially when the site of binding is unknown, and that in the absence of an energy supply the membrane potential is assumed to be zero. Both NICHOLLS (1974) and MOORE et al. (1978a, b) have reported considerable potentials under non-energized conditions. It is perhaps wiser to compare the distribution of $TPMP^+$ with that of K^+ plus valinomycin, although some binding of K^+ to biological material has been also reported (see TEDESCHI 1981). If, however, corrections are made for this binding, the results do suggest that both of these probes are able to equilibrate with $\Delta\psi$ in accordance with the Nernst equation. Perhaps a more serious objection, certainly with respect to $\Delta\psi$ measurements in plant mitochondria, has been raised by KIMPEL and HANSON (1978), who presented data to suggest that not only does valinomycin increase K^+ permeability, but that it also increases the activity of the endogenous K^+/H^+ antiporter, thus increasing the influx of K^+ via $\Delta\psi$ and efflux by ΔpH. Hence determination of ΔpH and $\Delta\psi$ after valinomycin addition will not reveal the condition existing previously. Since it is the total value of Δp that is important, however, and as long as conditions are chosen so as not to uncouple oxidative phosphorylation, such measurements are still valid.

Δp can also be measured in submitochondrial particles under energised conditions if $^{14}SCN^-$ and ^3H-methylamine are used as $\Delta\psi$ and ΔpH probes respectively (MOORE and BONNER 1981). The values obtained are similar to those observed with intact mitochondria, although the extent of the pH gradient is debatable (see SORGATO et al. 1978, 1982a, b).

4.2.2 Spectroscopic Probes

Changes in the fluorescence or light absorption of extrinsic probes offer an alternative technique of monitoring transmembrane electrical potentials and pH gradients. Extrinsic probes have the advantage over steady-state ion distributions in so far as they do not significantly perturb the membrane to which they are applied and also allow a continuous monitor of changes in $\Delta\psi$ and ΔpH. In most cases fluorescent or light absorption changes are due to binding to the membrane, penetration into the matrix, or stacking on the membrane surface (see ROTTENBERG 1979, BASHFORD and SMITH 1979). Most of the difficulties encountered in their use as $\Delta\psi$ or ΔpH probes are those of calibration. The best procedure is to calibrate by comparison with steady-state ion distributions under comparable conditions (ROTTENBERG 1979), or by imposition of a valinomycin-mediated K^+ diffusion potential (ÅKERMAN and WIKSTRÖM 1976).

In plant mitochondria, fluorescence or absorption changes of the cationic dye safranine 0 have been used as a measure of $\Delta\psi$ (WILSON 1980, MOORE et al. 1980, MOORE and BONNER 1981, 1982, MOREAU and ROMANI 1982). WILSON (1980) and MOREAU and ROMANI (1982) employed safranine to determine whether energy conservation is associated with electron transport via the alternative oxidase. In the presence of oligomycin, the addition of SHAM was found to decrease succinate-dependent $\Delta\psi$ in cyanide-inhibited mitochondria (WILSON

1980), from which it was concluded that alternative oxidase activity generates a proton gradient. Similar conclusions were reached by Moreau and Romani (1982), although it must be pointed out that malate was used in these experiments and consequently the observed $\Delta\psi$ in the presence of KCN was probably due to proton pumping at site 1. This is contrary to the conclusion reached in our laboratory (Moore et al. 1978a, Moore 1978a, Moore and Bonner 1982), which was based on steady-state ion distribution, oxygen pulse, and spectrophotometric techniques. With all of these techniques we were unable to demonstrate a $\Delta\psi$ generated during cyanide-insensitive respiration. It is conceivable that the potentials observed in the presence of cyanide by other workers are due to an energy-independent Donnan potential, low membrane proton conductance preventing total dissipation of the membrane potential. Other suitable probes that have been used to estimate $\Delta\psi$ are anilinonaphthalene-1-sulfonate (ANS$^-$) and the merocyanine dyes (Waggoner 1976, Bashford and Smith 1979). Again, each of these probes must be calibrated in order to relate spectral changes to the magnitude of $\Delta\psi$. It is important to recognize, however, that both of these probes have considerable side effects. Walsh Kinnally et al. (1978) examined several carbo- and merocyanine dyes and concluded that some acted as electron transport inhibitors. A similar conclusion was reached by Howard and Wilson (1979) who reported that even at concentrations of $< 1\ \mu M$ the cyanine dye [dis-C_2-(5)] acted as both an inhibitor of electron transport and an uncoupler of oxidative phosphorylation. Similarly, ANS$^-$ has been used by some workers to monitor $\Delta\psi$ (Wilson 1980), even though the fluorescence of this dye is known to be altered by several factors such as surface potential or protein conformation (Ferguson et al. 1976). Obviously considerable caution must be exercised in the use of these dyes in the light of the above observations.

The oxonol dyes have proved useful as extrinsic probes of membrane potential in submitochondrial particles both in animal (Bashford and Thayer 1977) bacterial (Bashford et al. 1979) and plant systems (Moore and Bonner 1981). Either respiration or ATP hydrolysis causes a red shift in the absorption of oxonol-VI, indicative of a membrane potential. Although, as with other extrinsic probes, calibration is a problem, this dye has been successfully used to investigate the effect of the phosphorylation potential on the magnitude of probe responses to respiration. It was found that the phosphorylation potential imposed by the addition of adenine nucleotides altered the membrane potential of respiring submitochondrial particles with a null point ΔGp of approx. 10.5 kcal mol^{-1} (Bashford and Thayer 1977) and 11.6 kcal mol^{-1} with plant tissues (Moore and Bonner 1981).

4.2.3 Ion-Specific Electrodes

In recent years mitochondrial membrane potentials have been estimated using electrodes that are sensitive to lipid-soluble ions such as dibenzyldemethylammonium (DDA$^+$) and tetraphenyl phosphonium (TPP$^+$) (Kamo et al. 1979). Ion-specific electrodes have advantages in that the technique allows the monitoring of changes in membrane potentials both continuously and easily in circumstances where microelectrodes cannot be used, and where high absorption may interfere with extrinsic probe measurements (i.e., with green leaf mitochondria). Values obtained using this technique are similar to those obtained with other methods (namely around -180 mV to -200 mV) (MOORE and ÅKERMAN 1982, DUCET 1983, MANDOLINO et al. 1983).

4.2.4 Magnitude of Δp and its Response to the Metabolic State

The agreement between the values of $\Delta \psi$ using different techniques suggests that all of the methods give reasonable estimates for the sizes of the components of Δp. A reasonable value for Δp under state 4 conditions is between 150 and 180 mV (Table 2). The addition of ADP results in a rapid fall of approx. 10–30 mV which subsequently is restored to its initial value upon transition to state 4. Its rapid collapse by uncouplers of oxidative phosphorylation and response to respiratory inhibitors is in general agreement with the notion that the protonmotive force is the obligatory link between respiration and ATP synthesis as originally envisaged by the chemiosmotic hypothesis (however see Sect. 4.5). However, whether the redox-generated Δp is thermodynamically sufficient when compared to the driving force for proton translocation (ΔE) and the energy necessary to synthesize ATP (ΔGp) is still controversial. Prior to a discussion of this important topic, however, a consideration of the number of protons translocated per pair of electrons passing from a substrate to oxygen, and the number required to synthesize ATP, is necessary.

4.3 Mechanism for Generating Δp

4.3.1 H$^+$/O Ratios

In its simplest form, the chemiosmotic hypothesis (MITCHELL 1966, 1979) states that the respiratory chain is arranged in loops of alternating hydrogen and electron carriers and requires a fixed stoichiometry of $2H^+$ ejected for every two electrons traversing a redox loop or energy-conserving site. According to the conformational type of mechanism (PAPA 1976, BOYER et al. 1977) which can be envisaged as a co-operative linkage between the transfer of reducing equivalents by the respiratory chain and translocation of protons by ionizable groups in the apoproteins, the H^+ stoichiometry ($H^+/2\bar{e}$) may be different from 2.

Experimental determinations of H^+/O stoichiometry have produced contrasting results. A H^+/O ratio of 4 for succinate respiration, as suggested by the chemiosmotic mechanism, is supported by considerable experimental evidence (MITCHELL and MOYLE 1967, HINKLE and HORSTMAN 1971, PAPA et al. 1980a, b, MITCHELL et al. 1978, MITCHELL 1979) although the finding by BRAND et al. (1976a, b) that N-ethylmaleimide, which inhibits the H/Pi symporter, increases the stoichiometry to 6 started a lively debate. Such higher stoichiometries have been confirmed by other laboratories (WIKSTRÖM and KRAB 1979, 1980, AL-SHAWI and BRAND 1981), although some have reported values of 8 (REYNA-

A.L. Moore and P.R. Rich:

Table 3. H$^+$/O Ratios

	Substrate	Treatment	H$^+$/O	$t_{1/2_{(s)}}$
Mung bean[a]	Malate	–	6.96±0.14 [76]	4.32±0.26
	Malate	+NEM	5.87±0.17 [37]	2.70±0.11
	Succinate	–	4.58±0.13 [62]	2.66±0.11
	Succinate	+NEM	6.27±0.21 [42]	1.71±0.06
	NADH	–	4.64±0.34 [41]	3.06±0.19
	NADH	+NEM	5.43±0.34 [35]	2.34±0.18
Rat liver[a]	Asc/TMPD	–	2.58±0.13 [15]	3.16±0.3
Rat liver[a]	Succinate	–	4.84±0.15 [8]	17.3 ±1.9
	Succinate	+NEM	6.3 ±0.14 [18]	13.3 ±0.46
Yeast[b]	Succinate	Fe(CN)$_6^{2-}$	2.71±0.4 [57]	5.66±1.50

The results are the means±S.E.M. for the numbers of observations shown in parentheses. The $t_{1/2}$ values are obtained from back-extrapolation of the initial fast decay phase. The control with ascorbate/TMPD includes 0.17 nmol mg^{-1} protein. [NEM] is 1 µmol mg^{-1} protein (mung bean) or 40 nmol mg^{-1} protein (rat liver). Results taken from [a] Mitchell and Moore (1984); [b] Villalobo et al. (1981). (Note that ferricyanide was used as electron acceptor in these experiments).

Abbreviations: Asc, ascorbate; TMPD, tetramethyl-p-phenylene diamine; NEM, N-ethyl-maleimide

Farje et al. 1976, Alexandre et al. 1978, 1980, Vercesi et al. 1978, Azzone et al. 1978c, Alexandre and Lehninger 1979, Pozzan et al. 1979a, b).

Table 3 presents a comparison of the H$^+$/O ratios obtained with mung bean mitochondria with published data from other systems. H$^+$/O ratios were obtained using the conventional O$_2$ pulse technique (Mitchell and Moyle 1967). Although the current estimates are consistent with a stoichiometry of 6 for succinate several points are worthy of note. Firstly the $t_{1/2}$ of H$^+$ backflow into plant mitochondria is considerably faster than the initial decay times for rat liver. The $t_{1/2}$ values remained relatively constant and could not be related to either respiratory control indices or membrane H$^+$ conductance. They are, however, quite comparable to those observed in yeast mitochondria (Villalobo et al. 1981) and may be related to the presence of an endogenous electroneutral K$^+$/H$^+$ antiporter. Indeed there is considerable evidence in the literature in support of such an antiporter in plants (Hensley and Hanson 1975, Moore and Wilson 1977, Kimpel and Hanson 1978, Jung and Brierley 1979, Fluegel and Hanson 1981), although controversy surrounds the question of whether it functions as a unidirectional cation-extruding system (Jung and Brierley 1979) or merely facilitates K$^+$ uptake (Fluegel and Hanson 1981). Turnip (Moore and Wilson 1977) and mung bean mitochondria (Huber and Moreland 1979) certainly display passive swelling in potassium acetate, suggesting the operation of such a carrier during salt influx. Presumably, under energized conditions the antiporter operates in the reverse direction, being driven by opposing H$^+$ and K$^+$ gradients. Thus under the conditions of the oxygen pulse experiments in which K$^+$ (+val) is included to act as a permeant cation-compensating ion, rapid H$^+$ back decay is seen as a result of internal K$^+$ exchanging for H$^+$ and its cycling via valinomycin. The existence of such an antiporter

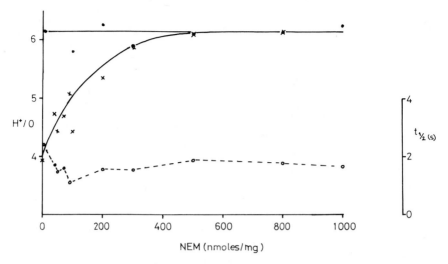

Fig. 6. The effect of NEM on H^+/O ratios. Mung bean mitochondria ($\simeq 12$ mg) were pre-incubated with 1 mM NADH, 200 ng mg^{-1} valinomycin in a de-oxygenated medium containing 0.25 M mannitol, 49 mM Li Cl, 1 mM Tris-HEPES pH 7.2 and the appropriate concentration of NEM. Electron flow was initiated by the addition of 10 ng atoms O (air-saturated distilled water). H^+/O ratios determined by back-extrapolation of the fast decay to a time to which half of the added oxygen was calculated have been used. (MITCHELL and MOORE 1984)

obviously has some implications with respect to oxidative phosphorylation in plant mitochondria. Since it is electroneutral, the membrane potential component is not dissipated and uncoupling is not observed. Incidentally, it is also apparent that its turnover is not sufficiently rapid to dissipate ΔpH, since such gradients have been readily measured under state 4 conditions (MOORE et al. 1978b). However, it may contribute significantly to state 4 respiration rates (FLUEGEL and HANSON 1981) if a mechanism exists for the ready penetration of K^+ (JUNG et al. 1977). Obviously the simultaneous functioning of both carriers will result in an uncoupling of oxidative phosphorylation if some sort of control mechanism such as that suggested for mammalian mitochondria (GARLID 1980) does not exist.

The addition of N-ethylmaleimide to inhibit the phosphate carrier increases the observed stoichiometry, consistent with the original suggestion of BRAND et al. (1976a, b). Figure 6 shows that high concentrations of NEM are required to increase the observed stoichiometry in plant mitochondria and that furthermore, with some mitochondria, high stoichiometries are seen in the absence of NEM. This effect must be presumably related to the level of endogenous phosphate which has been reported to vary in the range 19–35 nmol mg^{-1} protein (JUNG and HANSON 1975, DAY and HANSON 1977). The reason why such high concentrations of NEM are required to inhibit phosphate movement (0.5–1.0 μmol mg^{-1} protein; DESANTIS et al. 1975) when compared to the 20–40 nmol mg^{-1} quoted for rat liver (BRAND et al. 1976a) is unclear. The necessity for such high concentrations, however, presents a further problem, since under these conditions NEM induces K^+ permeability (JUNG et al. 1977), and this may be responsible for the increase in H^+ backflow rates shown in Table 3.

4.3.2 H^+/Site Ratios

The possession of an alternative oxidase, a rotenone-insensitive pathway around site I, and an external NADH dehydrogenase by plant mitochondria, enables the respiratory chain to be dissected on a H^+/site basis. For instance, in the presence of antimycin, site I alone can be investigated, whereas H^+/O ratios for site I can also be obtained by the difference in the overall stoichiometry in the presence and absence of rotenone. Such a systematic dissection is not possible using intact mammalian mitochondria. Some typical values are presented in Table 4 (MITCHELL and MOORE 1984). Note that when malate was used as substrate the stoichiometries were obtained in the absence of NEM since this compound was found to inhibit NAD^+-linked substrates (MITCHELL and MOORE 1984). Results from both mung bean and *S. guttatum* mitochondria suggest that either by difference ($\pm$ rotenone) or in the presence of antimycin, the H^+/O ratio for site I is close to 2. The H^+/O ratio with succinate (sites 2 and 3) is close to 6 whereas for sites 2 and 3 it is 4 and 2 respectively. With either succinate or NADH as substrate, in the presence of antimycin, no net acidification was observed, consistent with the notion that the alternative oxidase is non-electrogenic (MOORE 1978a). The values in Table 4 agree with other determinations (BRAND et al. 1976a, b) and are similar to published steady-state measurements (BRAND et al. 1978, WIKSTRÖM and KRAB 1980, WIKSTRÖM et al. 1981a), but differ from those obtained using rapid kinetic methods (ALEXANDRE et al. 1978, PAPA et al. 1980a, b, 1983a, b). They are not consistent in the strictest sense with the redox loop formulations described by MITCHELL (1979) in so much as they support the controversial notion that cytochrome oxidase possesses a proton pumping function (WIKSTRÖM and KRAB 1980, REYNARFARJE et al. 1982, CASEY and AZZI 1983, WIKSTRÖM and PENTTILÄ 1982, however see MITCHELL and MOYLE 1983, PAPA et al. 1983a, b). Since the same H^+/O stoichi-

Table 4. H^+/Site ratios

Tissue	Substrate	Treatment	Sites	H^+/O
Mung bean	Malate	Control	1+2+3	6.96±0.14 [76]
		+rotenone	2+3	4.88±0.34 [29]
		(by difference)	1	2.08
	Succinate	Control	2+3	6.27±0.21 [42]
		+AA	alt. ox.	0
	Ascorbate+TMPD	+AA	3	2.58±0.13 [15]
		(by difference)	2	3.69
Sauromatum guttatum	Malate	+AA	1	2.57±0.47 [5]

The results are the mean ± S.E.M. for the numbers of observations shown in parentheses. Additions made at the following concentrations, rotenone (1 μM); antimycin A (0.17 nmol mg^{-1} protein), substrates (5 mM).

Fig. 7. A vectorial model of the plant mitochondrial respiratory chain. Abbreviations: *I–IV* the respiratory chain complexes; *a* the external NADH dehydrogenase; *b* the alternative oxidase; *c* cytochrome c; *Q* oxidized ubiquinone; QH_2 reduced ubiquinol

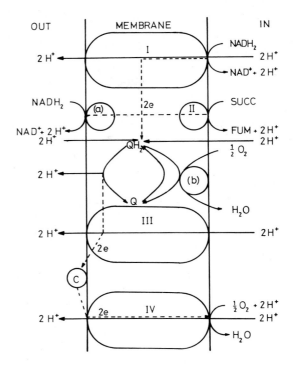

ometry was obtained with two different site 2 substrates, namely succinate and NADH (Table 3) and taking into account the deductions made concerning the sidedness of these reactions (see Sect. 3.1), it may be concluded that both of these substrates feed reducing equivalents into a common pathway prior to the bc_1 complex. A simple model to accommodate such findings is shown in Fig. 7. This shows complexes I, III and IV each pumping $2H^+/2\bar{e}$ by mechanisms that could be a direct redox loop mechanism or a conformational pump for complex I, a 'b' cycle (WIKSTRÖM and KRAB 1980, WIKSTRÖM et al. 1981a), proton pump (PAPA 1976, 1982, PAPA et al. 1982) or Q cycle (MITCHELL 1976, MITCHELL and MOYLE 1982) for complex III and a proton pump for complex IV (WIKSTRÖM and KRAB 1979). In the case of NAD^+-linked substrates and succinate, Q is reduced by matrix H^+s whereas protons are derived from the medium when external NADH is the substrate. This would seem to be the only way of rationalizing the observations made earlier (see Sect. 3.1 – however see ALEXANDRE et al. 1980, ALEXANDRE and LEHNINGER 1982).

4.3.3 H^+/ATP Ratios

The number of protons translocated by the ATP synthetase to synthesize one ATP molecule may be calculated either by measuring the transient proton extrusion during the hydrolysis of a small pulse of ATP or by thermodynamic analysis of Δp and ΔGp for the $ATP \rightleftharpoons ADP + P_i$ reaction under equilibrium conditions.

According to the chemiosmotic hypothesis, the ATPase has a strict stoichiometry of 2 (Mitchell and Moyle 1968). This has been supported by experimental
evidence using both mitochondria and submitochondrial particles (Mitchell
and Moyle 1968, Moyle and Mitchell 1973, Thayer and Hinkle 1973, Brand
and Lehninger 1977) in which the number of H^+ ejected during the hydrolysis
of a small pulse of ATP was measured. Using rate measurements Lehninger
et al. (1977), however, presented evidence for ratios close to 3, which was later
supported by determinations using a new steady-state technique (Alexandre
et al. 1978) in which the rate of H^+ ejection was measured (as opposed to
the extent) alongside the rate of ATP hydrolysis. However, both pulse and
rate experiments are technically difficult, since corrections due to scalar protons
both from the ATPase itself and the entry of ATP have to be accounted for.
Perhaps a more quantitative approach is by thermodynamic analysis of both
ΔGp and Δp under equilibrium conditions. In mitochondria, the external ΔGp
is related to Δp not only by the H^+/ATP ratio but also by an extra term
which accounts for the energy of transport of ADP and P_i into the matrix
in exchange for ATP. The internal ΔGp is related to Δp by H^+/ATP ratio
alone. Current estimates of Δp and external ΔGp in mammalian tissues are
consistent with a value of 3 (see Ferguson and Sorgato 1982). A similar
result has been obtained for mung bean mitochondria (Moore and Bonner
1981), whereas with submitochondrial particles, the H^+/ATP ratio alone was
close to two (Moore and Bonner 1981). This result although consistent with
kinetic methods (Thayer and Hinkle 1973) and current estimates of the H^+/O
stoichiometry, is not readily reconciled with recent measurements using mammalian submitochondrial particles (Sorgato et al. 1978, 1982b, Branca et al. 1981,
Berry and Hinkle 1983), which point to a minimum H^+/ATP value of three.
The reason for this discrepancy is unclear but may be due, in part, to the
system used by Moore and Bonner (1981) not achieving a full equilibration
of ΔGp with Δp. Kinetic limitations, imposed by the adenine nucleotide translocator, may have prevented ΔGp from reaching its maximum value or alternatively an overestimation of the value of Δp cannot be excluded. The values for
ΔGp and Δp in plant mitochondria are, however, in agreement with a recent
report by Berry and Hinkle (1983) and the ratios obtained, namely 3.4 and
2.2 for mitochondria and submitochondrial particles respectively, may merely
be minimal values. Whatever the explanation, a final judgement on the H^+/ATP
ratio cannot be made until further work has been performed in resolving the
discrepancy between the thermodynamic and kinetic methods.

4.4 Thermodynamic Competence of Δp

The question of whether the redox generated Δp is thermodynamically sufficient
when compared to the driving force for proton translocation (ΔE) and the
energy necessary to synthesise ATP (ΔGp) is still controversial. An answer
to this question not only requires a knowledge of the redox span under steady-
state conditions but also the H^+ stoichiometry of the respiratory chain and

Table 5. Effect of various stoichiometries on steady-state Δp values

	H$^+$/O			H$^+$/ATP		
	8	6	4	4	3	2
Δp^a (mV)	190	253	380	165	220	330

^{a} Δp calculated on the basis of:
1. $2\Delta E = n\Delta p$ where ΔE for succinate$\rightarrow O_2 = 760$ mV (WIKSTRÖM et al. 1981a) n = H$^+$/O at values of 8, 6 and 4.
2. $\Delta Gp = n.\Delta p$ where ΔGp for ATP hydrolysis = 659 mV (MOORE and BONNER 1981) and n = H$^+$/ATP at values of 4, 3 and 2.

ATPase. It can be seen from the previous sections that there are problems inherent in the measurement of each of these variables. For instance, if the H$^+$/O stoichiometry for succinate is considered to be n = 6 (Table 5) and the redox span (ΔE) from succinate to O_2, 760 mV (WIKSTRÖM et al. 1981a), then Δp would have to have a value of 253 mV since

$$2\,\Delta E = n\Delta p.$$

This is certainly far larger than the current estimates for Δp presented in Table 2. Furthermore, if the extramitochondrial ΔGp is taken as 659 mV (MOORE and BONNER 1981) and a value of 3 for the H$^+$/ATP ratio, then the Δp in equilibrium with ΔGp would have a value of 220 mV since

$$\Delta Gp = H^+/ATP \cdot F \cdot \Delta p.$$

Table 5 summarizes the minimum Δp required at various H$^+$/O and H$^+$/ATP ratios. These values are obviously at variance with the chemiosmotic hypothesis in its earliest form, which is restricted to a H$^+$/O stoichiometry of 4 (for succinate) and a H$^+$/ATP ratio of 2 at a Δp of 250 mV.

The stoichiometries for proton extrusion by the respiratory chain and for H$^+$ re-entry during ATP synthesis also have to be consistent with the overall stoichiometry for ATP synthesis (P/O ratio) which is generally considered to be 3 for NAD$^+$-linked substrates and 2 for FAD-linked substrates (however see HINKLE and YU 1979). On the basis of these values, combined with a H$^+$/ATP ratio of 3, the H$^+$/O ratios would be 9 and 6 for NADH$_{int}$ and succinate respectively. It is, of course, also conceivable that ATP synthesis stoichiometries are not necessarily equal. This notion is supported by some experimental evidence (WIKSTRÖM and KRAB 1979, and BRAND et al. 1978), and is consistent with a smaller redox span for succinate-cytochrome c than for cytochrome c-O_2. Obviously H$^+$/O stoichiometries higher than 6 and 4 for NAD$^+$- and FAD-linked substrates, respectively, cannot be accommodated by direct loop mechanisms, and other possibilities such as conformational proton pumps or more complex electron transfer pathways at sites I and III would be required.

4.5 Is Δp an Obligate Intermediate?

While it is widely accepted that the operation of the redox and ATP-driven proton pumps creates a H^+ electrochemical gradient, the mechanism by which these pumps are coupled is still debatable. In the chemiosmotic mechanism (Mitchell 1966), the coupling between the two pumps is achieved through a delocalized Δp between the bulk aqueous phases and involves a H^+-catalyzed reaction at the active site of the ATPase. Other models differ principally with respect to the molecular mechanisms and the role of the bulk phase Δp. In the model advocated by Williams (1961, 1978), protons are generated in membrane-controlled spaces, and not in equilibrium with one another. It is envisaged that these energized protons drive ATP synthesis; Δp thus represents an energy store rather than an intermediate in coupling (Williams 1978, Kell 1979). In localized chemiosmosis (van Dam et al. 1978), domains of respiratory chain complexes and ATPases are postulated where H^+ do not freely equilibrate with the bulk phase and where a Δp higher than in the bulk phase is established (however see Westerhoff et al. 1981). Such an idea was further refined in the electrodic view formulated by Kell and colleagues (Kell 1979, Kell and Morris 1981).

If the bulk phase Δp is the obligate intermediate between electron transport and ATP synthesis, it requires that changes in its value should be paralleled both by a proportional change in the respiratory rate, ΔGp and the rate of ATP synthesis. In the literature there are some significant observations which argue against this view. For instance, the addition of an uncoupler was found to reduce Δp considerably more than ΔGp (Azzone et al. 1978a, Westerhoff et al. 1981) and the stimulation of respiration upon addition of ADP was accompanied by a smaller decrease in Δp than was seen when respiration was increased to the same extent with an uncoupler (Padan and Rottenberg 1973, Nicholls 1974, Azzone et al. 1978b, Moore 1978b, Zoratti et al. 1981). This observation is only compatible with the chemiosmotic mechanism if it is assumed that the activity of the respiratory chain is allosterically regulated by adenine nucleotides (Westerhoff et al. 1981 – however see Zoratti et al. 1983). The general insensitivity of Δp to inhibition of electron transfer (Pietrobon et al. 1981, 1983, Mandolino et al. 1983), which has been interpreted either as evidence for variable proton leaks (Nicholls 1974) or redox slips (Pietrobon et al. 1981, 1983), is also at odds with the views of delocalized coupling. Similarly, when the rate of respiratory flow is decreased, a parallel inhibition of the rate of phosphorylation is observed, while very limited effects can be detected on the extent of Δp (Sorgato et al. 1980, Zoratti et al. 1981, 1982, 1983, Mandolino et al. 1983). Such observations are difficult to reconcile with Δp as being the sole and obligate intermediate between electron transfer and ATP synthesis, suggesting that there may be some degree of direct interaction with the respiratory chain and ATPase (Williams 1961, 1978, Padan and Rottenberg 1973, Zoratti et al. 1981, 1983, Westerhoff et al. 1981, Kell and Morris 1981).

Acknowledgements. Experimental work described in this review was supported by the Science Research Council and the Agricultural Research Council (ALM), and the Venture Research Unit of British Petroleum Company p.l.c. (PRR).

References

Ackrell BAC, Kearney EB, Coles CJ (1977) Isolation of reconsitutively active succinate dehydrogenase in highly purified state. J Biol Chem 252:6963–6965

Åkerman KEO, Wikström MKF (1976) Safranine as a probe of the mitochondrial membrane potential. FEBS Lett 68:191–196

Albracht SPJ (1980) Prosthetic groups in succinate dehydrogenase. Biochim Biophys Acta 612:11–28

Alexandre A, Lehninger AL (1979) Stoichiometry of H^+ translocation coupled to electron flow from succinate to cytochrome c in mitochondria. J Biol Chem 254:11555–11560

Alexandre A, Lehninger AL (1982) Pathways of respiration-coupled H^+ extrusion via ubiquinone in rat liver mitochondria. In: Trumpower BL (ed) Functions of quinones in energy conserving systems. Academic Press, London New York, pp 541–552

Alexandre A, Reynafarje B, Lehninger AL (1978) Stoichiometry of vectorial H^+ movements coupled to electron transport and to ATP synthesis in mitochondria. Proc Natl Acad Sci USA 75:5296–5300

Alexandre A, Galiazzo F, Lehninger AL (1980) On the location of the H^+ extruding steps in site 2 of the mitochondrial electron transport chain. J Biol Chem 255:10721–10730

Al-Shawi MK, Brand MD (1981) Steady state H^+/O stoichiometry of liver mitochondria. Biochem J 200:539–546

Azzone GF, Pozzan T, Massari S (1978a) Proton electrochemical gradient and phosphate potential in mitochondria. Biochim Biophys Acta 501:307–316

Azzone GF, Pozzan T, Masaari S, Bragadin M (1978b) Proton electrochemical gradient and rate of controlled respiration in mitochondria. Biochim Biophys Acta 501:296–306

Azzone GF, Pozzan T, DiVirgilio F, Miconi V (1978c) The three proton pumps of mitochondrial respiratory chain. In: Dutton PL, Leigh JS, Scarp A (eds) Frontiers of biological energetics, vol I. Academic Press, London New York, pp 375–383

Bahr JT, Bonner WD Jr (1973a) Cyanide-insensitive respiration I. The steady states of skunk cabbage spadix and bean hypocotyl mitochondria. J Biol Chem 248:3441–3445

Bahr JT, Bonner WD Jr (1973b) Cyanide-insensitive respiration II. Control of the alternative pathway. J Biol Chem 248:3446–3450

Bashford CL, Smith JC (1979) The use of optical probes to monitor membrane potential. In: Fleischer S, Packer L (eds) Methods in enzymology, vol 55. Academic Press, London New York, pp 569–586

Bashford CL, Thayer WS (1977) Thermodynamics of the electrochemical proton gradient in bovine heart submitochondrial particles. J Biol Chem 252:3459–3463

Bashford CL, Chance B, Prince RC (1979) Oxonol dyes as monitors of membrane potential. Their behaviour in photosynthetic bacteria. Biochim Biophys Acta 545:46–57

Bell RL, Sweetland J, Ludwid B, Capaldi RA (1979) Labelling of complex III with [^{34}S]diazobenzene-sulphonate: Orientation of the electron transfer segment in the mitochondrial inner membrane. Proc Natl Acad Sci USA 76:741–745

Bendall DS, Bonner WD Jr (1971) Cyanide-insensitive respiration in plant mitochondria. Plant Physiol 47:236–245

Berry EA, Hinkle PC (1983) Measurement of the electrochemical proton gradient in submitochondrial particles. J Biol Chem 258:1474–1486

Bomhoff GH, Spencer M (1977) Optimum pH and ionic strength for the assay of cytochrome c oxidase from pea cotyledon mitochondria. Can J Biochem 55:1114–1117

Bonner WD Jr (1965) Mitochondria and electron transport. In: Bonner J, Varner JE (eds) Plant biochemistry. Academic Press, London New York, pp 89–123

Bonner WD Jr, Rich PR (1978) Molecular aspects of cyanide/antimycin-resistant respiration. In: Ducet G, Lance C (eds) Plant mitochondria. Elsevier, Amsterdam, pp 241–247

Boyer PD (1965) Carboxyl activation as possible common reaction in substrate level and oxidative phosphorylation and in muscle contraction. In: King TE, Mason HS, Morrison M (eds) Oxidases and related redox systems. Wiley, New York, pp 994–1008

Boyer PD, Chance B, Ernster L, Mitchell P, Racker E, Slater EC (1977) Oxidative phosphorylation and photophosphorylation. In: Snell EE, Boyer PD, Meister A, Richardson CC (eds) Annu Rev Biochem 46. Palo Alto, pp 955–1026

Branca D, Ferguson SJ, Sorgato MC (1981) Clarification of factors influencing the nature and magnitude of the protonmotive force in bovine heart submitochondrial particles. Eur J Biochem 116:341–346

Brand MD, Lehninger AL (1977) H^+/ATP ratio during ATP hydrolysis by mitochondria: modification of the chemiosmotic theory. Proc Natl Acad Sci USA 74:1955–1959

Brand MD, Reynafarje B, Lehninger AL (1976a) Re-evaluation of the H^+/site ratio of mitochondrial electron transport with the oxygen pulse technique. J Biol Chem 251:5670–5679

Brand MD, Reynafarje B, Lehninger AL (1976b) Stoichiometric relationship between energy-dependent proton ejection and electron transport in mitochondria. Proc Natl Acad Sci USA 73:437–441

Brand MD, Harper WG, Nicholls DG, Ingledew WJ (1978) Unequal charge separation by different coupling spans of the mitochondrial electron transport chain. FEBS Lett 95:125–129

Brunton CJ, Palmer JM (1973) Pathways for the oxidation of malate and reduced pyridine nucleotide by wheat mitochondria. Eur J Biochem 39:283–291

Burke JJ, Siedow JN, Moreland DE (1982) Succinate dehydrogenase. A partial purification from mung bean hypocotyls and soybean cotyledons. Plant Physiol 70:1577–1581

Cammack R, Palmer JM (1977) EPR studies of iron-sulfur proteins of plant mitochondria. Ann NY Acad Sci 222:816–822

Capaldi RA (1982) Arrangement of proteins in the mitochondrial inner membrane. Biochim Biophys Acta 694:291–306

Casey RP, Azzi A (1983) An evaluation of the evidence for H^+ pumping by reconstituted cytochrome oxidase in the light of recent criticism. FEBS Lett 154:237–242

Chance B (1972) On probe and ubiquinone interactions in mitochondrial membranes. In: Azzone GF, Carafoli E, Lehninger AL, Quagliariello E, Siliprandi N (eds) Biochemistry and biophysics of mitochondrial membranes. Academic Press, London New York, pp 85–99

Cherry RJ (1979) Rotational and lateral diffusion of membrane proteins. Biochim Biophys Acta 559:289–327

Coleman JOD, Palmer JM (1971) The role of Ca^{2+} in the oxidation of exogenous NADH by plant mitochondria. FEBS Lett 17:203–208

Cook ND, Cammack R (1983) Partial purification and characterisation of a rotenone-insensitive NADH dehydrogenase from *Arum maculatum* mitochondria using affinity chromatography. Abst 15th FEBS Meeting, p 257

Cottingham IR, Moore AL (1983) Ubiquinone pool behaviour in plant mitochondria. Biochim Biophys Acta 724:191–200

Crane FL (1977) Hydroquinone dehydrogenases. Annu Rev Biochem 46:439–469

Crofts AR, Wraight CA (1983) The electrochemical domain of photosynthesis. Biochim Biophys Acta 726:149–186

Dam van K, Casey RP, van der Meer R, Groen AK, Westerhoff HV (1978) The energy balance of oxidative phosphorylation. In: Dutton PL, Leigh JS, Scarpa A (eds) Frontiers of biological energetics, vol I. Academic Press, London New York, pp 430–438

Day DA, Hanson JB (1977) Effect of phosphate and uncouplers on substrate transport and oxidation by isolated corn mitochondria. Plant Physiol 59:139–144

Day DA, Wiskich JT (1974a) The oxidation of malate and exogenous reduced nicotinamide adenine dinucleotide by isolated plant mitochondria. Plant Physiol 53:104–109

Day DA, Wiskich JT (1974b) The effects of exogenous nicotinamide adenine dinucleotide on the oxidation of NAD-linked substrates by isolated plant mitochondria. Plant Physiol 54:360–363

Day DA, Arron GP, Laties GG (1980) Nature and control of respiratory pathways in plants: the interaction of cyanide-resistant respiration with the cyanide-sensitive pathway. In: Stumpf PK, Conn EE (eds) The biochemistry of plants, vol II. Academic Press, London New York, pp 198–241

Denis M, Richaud P (1982) Dynamics of carbon monoxide recombination to fully reduced cytochrome c oxidase in plant mitochondria after low temperature flash photolysis. Biochem J 206:379–385

Douce R, Mannella CA, Bonner WD Jr (1973) The external NADH dehydrogenases of intact plant mitochondria. Biochim Biophys Acta 292:105–116

Dry IB, Day DA, Wiskich JT (1983) Preferential oxidation of glycine by the respiratory chain of pea leaf mitochondria. FEBS Lett 158:154–158

Ducet G (1983) Membrane potential in potato tuber mitochondria. Plant Physiol 72S: 172

Ducet G, Diano M (1978) On the dissociation of the cytochrome bc_1 of potato mitochondria. Plant Sci Lett 11:217–226

Dutton PL, Prince RC (1978) Reaction-centre-driven cytochrome interactions in electron and proton translocation and energy coupling. In: Clayton RK, Sistrom WR (eds) The photosynthetic bacteria. Plenum Press, New York London, pp 525–570

Dutton PL, Storey BT (1971) The respiratory chain of plant mitochondria IX. Oxidation-reduction potentials of the cytochromes of mung bean mitochondria. Plant Physiol 47:282–288

Ferguson SJ, Sorgato MC (1982) Proton electrochemical gradients in energy-transduction processes. In: Snell EE, Boyer PD, Meister A, Richardson CC (eds) Annu Rev Biochem 51. Palo Alto, pp 185–217

Ferguson SJ, Lloyd WJ, Radda GK (1976) On the nature of the energised state of sub-mitochondrial particles: investigations with N-aryl naphthalene sulphonate probes. Biochim Biophys Acta 423:174–188

Fluegel MJ, Hanson JB (1981) Mechanisms of passive potassium influx in corn mitochondria. Plant Physiol 68:267–271

Garlid KD (1980) On the mechanism of regulation of the mitochondrial K^+/H^+ exchanger. J Biol Chem 255:11273–11279

Girdlestone J, Bisson R, Capaldi RA (1981) Interaction of succinate-ubiquinone reductase (complex II) with (arylazido) phospholipids. Biochemistry 20:152–156

Green DE (1962) Structure and function of subcellular particles. Comp Biochem Physiol 4:81–122

Guerrieri F, Nelson BD (1975) Studies on the characteristics of a proton pump in phospholipid vesicles inlayed with purified complex III from beef heart mitochondria. FEBS Lett 54:339–342

Gutman M (1980) Electron flux through the mitochondrial ubiquinone. Biochim Biophys Acta 594:53–84

Hanson JB, Day DA (1980) Plant mitochondria. In: Stumpf PK, Conn EE (eds) The biochemistry of plants, vol I. Academic Press, London New York, pp 315–358

Hanstein WG, Davis KA, Ghalamber MA, Hatefi Y (1971) Succinate dehydrogenase II. Enzymatic properties. Biochemistry 10:2517–2524

Hatefi Y (1978) Introduction-preparation and properties of the enzymes and enzyme complexes of the mitochondrial oxidative phosphorylation system. In: Fleischer S, Packer L (eds) Methods in enzymology, vol 153. Academic Press, London New York, pp 3–47

Hatefi Y, Rieske JS (1967) Preparation and properties of DPNH-coenzyme Q reductase (complex I of the respiratory chain). In: Fleischer S, Packer L (eds) Methods in enzymology, vol 10. Academic Press, London, New York, pp 235–239

Hattori T, Asahi T (1982a) The presence of two forms of succinate dehydrogenase in sweet potato root mitochondria. Plant Cell Physiol 23:515–523

Hattori T, Asahi T (1982b) The mechanism of the increase in succinate dehydrogenase activity in wounded sweet potato root tissue. Plant Cell Physiol 23:525–532

Hattori T, Iwasaki V, Sakajo S, Asahi T (1983) Cell-free synthesis of succinate dehydrogenase and mitochondrial adenosine triphosphatase of sweet potato. Biochem Biophys Res Commun 113:235–240

Hauska G, Hurt E, Babellini N, Lockau W (1983) Comparative aspects of quinol-cyto-chrome c/plastocyanin oxidoreductases. Biochim Biophys Acta 726:97–134

Hensley JR, Hanson JB (1975) Action of valinomycin in uncoupling corn mitochondria. Plant Physiol 56:13–18

Heron C, Smith S, Ragan CI (1979) An analysis of the polypeptide composition of bovine heart mitochondrial NADH-ubiquinone oxidoreductase by two-dimensional polyacrylamide gel electrophoresis. Biochem J 181:435–443

Hinkle PC, Horstman LL (1971) Respiration-driven proton transport in submitochon-drial particles. J Biol Chem 246:6024–6028

Hinkle PC, Yu ML (1979) The phosphorus/oxygen ratio of mitochondrial oxidative phosphorylation. J Biol Chem 254:2450–2455

Howard PH Jr, Wilson SB (1979) Effects of the cyanine dye 3,3'-dipropylthiocarbocyan-ine on mitochondrial energy conservation. Biochem J 180:699–672

Huber SC, Moreland DE (1979) Permeability properties of the inner membrane of mung bean mitochondria and changes during energisation. Plant Physiol 64:115–119

Huq S, Palmer JM (1978a) The involvement and possible role of quinone in cyanide-resistant respiration. In: Ducet G, Lance C (eds) Plant mitochondria. Elsevier, Am-sterdam, pp 225–232

Huq S, Palmer JM (1978b) Oxidation of durohydroquinone via the cyanide-insensitive respiratory pathway in higher plant mitochondria. FEBS Lett 92:317–320

Ingledew WJ, Salero JC, Ohnishi T (1976) Studies on electron paramagnetic resonance spectra manifested by a respiratory chain hydrogen carrier. Arch Biochem Biophys 177:176–184

Jung DW, Brierley GP (1979) Swelling and contraction of potato mitochondria. Plant Physiol 64:948–953

Jung DW, Hanson JB (1975) Activation of 2,4-DNP-stimulated ATPase activity in cauli-flower and corn mitochondria. Arch Biochem Biophys 168:358–365

Jung DW, Chavez E, Brierley GP (1977) Energy-dependent exchange of K^+ in heart mitochondria: K^+ influx. Arch Biochem Biophys 183:452–459

Kamo N, Muratsugu M, Hongoh R, Kobatake Y (1979) Membrane potential of mito-chondria measured with an electrode sensitive to tetraphenyl phosphonium and rela-tionship between proton electrochemical potential and phosphorylation potential in the steady state. J Membr Biol 49:105–121

Katsikas H, Quinn PJ (1982) The distribution of ubiquinone-10 in phospholipid bi-layers. (A study using differential scanning calorimetry). Eur J Biochem 124:165–169

Kearney EB, Singer TP (1956) Studies on succinic dehydrogenase 1) Preparation and assay of the soluble dehydrogenase. J Biol Chem 219:963–975

Kell DB (1979) On the functional proton current pathway of electron transport phosphor-ylation: An electrodic view. Biochim Biophys Acta 549:55–99

Kell DB, Morris JG (1981) Proton-coupled membrane energy transduction: pathways, mechanisms and control. In: Palmieri F, Quagliariello E, Siliprandi N, Slater EC (eds) Vectorial reactions in electron and ion transport in mitochondria and bacteria. Elsevier/North Holland Biomedical Press, Amsterdam New York, pp 339–347

Kelly GJ (1982) How widespread are cyanide-resistant mitochondria in plants? Trends Biochem Sci 7:233

Kimpel JA, Hanson JB (1978) Efflux and influx of potassium salts elicited by valinomycin in plant mitochondria. Plant Sci Lett 11:329–335

King TE (1963) Reconstitution of respiratory chain enzyme systems XII. Some observa-tions on the reconstitution of the succinate oxidase system from heart muscle. J Biol Chem 238:4037–4051

Kingsley DB, Feigenson GW (1981) ^{1}H-NMR study of the location and motion of ubiquinones in perdeuterated phosphatidyl-choline bilayers. Biochim Biophys Acta 635:602–618

Klein RP, Maxwell JT, Burke JJ (1983) NAD(P)H dehydrogenases: a partial purification from cauliflower mitochondria. Plant Physiol 72S:171

Kröger A, Klingenberg M (1967) On the role of ubiquinone. Curr Top Bioenerg 1:151–193

Kröger A, Klingenberg M (1973a) The kinetics of the redox reactions of ubiquinone related to the electron-transport activity in the respiratory chain. Eur J Biochem 24:358–368

Kröger A, Klingenberg M (1973b) Further evidence for the pool function of ubiquinone as derived from the inhibitions of the electron transport by antimcycin. Eur J Biochem 39:313–323

Lambowitz AW, Bonner WD Jr (1974) The b-cytochromes of plant mitochondria. A spectrophotometric and potentiometric study. J Biol Chem 249:2428–2440

Lance C, Bonner WD Jr (1968) The respiratory chain components of higher plant mitochondria. Plant Physiol 43:756–766

Laties GG (1982) The cyanide-resistant alternative path in higher plant respiration. Annu Rev Plant Physiol 33:519–555

Lehninger AL, Reynafarje B, Alexandre A (1971) Stoichiometry of proton movements coupled to mitochondrial electron transport and ATP hydrolysis. In: Dam van K, Gelder van BF (eds) Structure and function of energy transducing membranes. Elsevier/North Holland Biomedical Press, Amsterdam New York, pp 95–106

Leung KH, Hinkle PC (1975) Reconstitution of ion transport and respiratory control in vesicles formed from reduced coenzyme Q-cytochrome c reductase and phospholipids. J Biol Chem 250:8467–8471

Lolkema JS, Hellingwerf KJ, Könings WN (1982) The effect of probe binding on the quantitative determination of the proton motive force in bacteria. Biochim Biophys Acta 681:85–94

Ludwig B, Schatz G (1980) A two-subunit cytochrome c oxidase (cytochrome aa$_3$) from *Paracoccus denitrificans*. Proc Natl Acad Sci USA 77:196–200

Maeshima M, Asahi T (1978) Purification and characterisation of sweet potato cytochrome c oxidase. Arch Biochem Biophys 187:423–430

Mandolino G, Santis De A, Melandri BA (1983) Localised coupling in oxidative phosphorylation by mitochondria from Jerusalem artichoke (*Helianthus tuberosus*). Biochim Biophys Acta 723:428–439

Mitchell JA, Moore AL (1984) Proton stoichiometry of plant mitochondria. Biochem Soc Trans 12:849–850

Mitchell P (1961) Coupling of phosphorylation to electron and hydrogen transfer by a chemi-osmotic type of mechanism. Nature (London) 191:144–150

Mitchell P (1966) Chemiosmotic coupling in oxidative and photosynthetic phosphorylation. Biol Rev 41:445–502

Mitchell P (1976) Possible molecular mechanisms of the protonmotive function of cytochrome systems. J Theor Biol 62:327–367

Mitchell P (1977) A commentary on alternative hypotheses of protonic coupling in the membrane systems catalysing oxidative and photosynthetic phosphorylation. FEBS Lett 78:1–20

Mitchell P (1979) Compartmentation and communication in living systems. Ligand conduction: a general catalytic principle in chemical osmotic and chemiosmotic reaction systems. Eur J Biochem 95:1–20

Mitchell P, Moyle J (1967) Respiration-driven proton translocation in rat liver mitochondria. Biochem J 105:1147–1162

Mitchell P, Moyle J (1968) Proton translocation coupled to ATP hydrolysis rat liver mitochondria. Eur J Biochem 4:530–539

Mitchell P, Moyle J (1969) Estimation of membrane potential and pH differences across the cristae membrane of rat liver mitochondria. Eur J Biochem 7:471–484

Mitchell P, Moyle J (1982) Protonmotive mechanisms of quinone function. In: Trumpower BL (ed) Functions of quinones in energy conserving systems. Academic Press, London New York, pp 553–575

Mitchell P, Moyle J (1983) Alternative hypotheses of proton ejection in cytochrome oxidase vesicles. Transmembrane proton pumping or redox-linked deprotonation of phospholipid-cytochrome c complexes. FEBS Lett 151:167–178

Mitchell P, Moyle J, Mitchell R (1978) Proton stoichiometry of redox and ATPase systems. In: Dutton PL, Leigh JS, Scarpa A (eds) Frontiers of biological energetics vol 1. Academic Press, London New York, pp 394–402

Møller IM, Johnston SP, Palmer JM (1981) A specific role for Ca^{2+} in the oxidation of exogenous NADH by Jerusalem artichoke (*Helianthus tuberosus*) mitochondria. Biochem J 194:487–495

Moore AL (1978a) An evaluation of H^+ translocation via the alternative pathway in mung bean mitochondria. In: Degn H, Lloyd D, Hill GC (eds) Functions of alternative terminal oxidases, vol 49. Pergamon Press, Oxford New York, pp 141–148

Moore AL (1978b) The electrochemical gradient of protons as an intermediate in energy transduction in plant mitochondria. In: Ducet G, Lance C (eds) Plant mitochondria. Elsevier/North Holland Biomedical Press, Amsterdam New York, pp 85–92

Moore AL, Åkerman KEO (1982) Ca^{2+} stimulation of the external NADH dehydrogenase in Jerusalem artichoke (*Helianthus tuberosus*) mitochondria. Biochem Biophys Res Commun 109:513–517

Moore AL, Bonner WD Jr (1981) A comparison of the phosphorylation potential and electrochemical proton gradient in mung bean mitochondria and phosphorylating sub-mitochondrial particles. Biochim Biophys Acta 634:117–128

Moore AL, Bonner WD Jr (1982) Measurements of membrane potentials in plant mitochondria with the safranine method. Plant Physiol 70:1271–1276

Moore AL, Cottingham IR (1983) Characteristics of the higher plant respiratory chain. In: Robb DA, Pierpoint WS (eds) Metals and micronutrients: uptake and utilisation by plants. Academic Press, London New York, pp 170–204

Moore AL, Proudlove MO (1983) Mitochondrial and sub-mitochondrial particles. In: Hall JL, Moore AL (eds) Isolation of membranes and organelles from plant cells. Academic Press, London New York, pp 153–184

Moore AL, Rich PR (1980) The bioenergetics of plant mitochondria. Trends Biochem Sci 5:284–288

Moore AL, Wilson SB (1977) Translocation of some anions, cations and acids in turnip (*Brassica napus* L.) mitochondria. J Exp Bot 28:607–618

Moore AL, Rich PR, Ingledew WJ, Bonner WD Jr (1976) A complex EPR signal in mung bean mitochondria and its possible relation to the alternative pathway. Biochem Biophys Res Commun 72:1099–1107

Moore AL, Bonner WD Jr, Rich PR (1978a) The determination of the protonmotive force during cyanide-insensitive respiration in plant mitochondria. Arch Biochem Biophys 186:298–306

Moore AL, Rich PR, Bonner WD Jr (1978b) Factors influencing the components of the total protonmotive force in mung bean mitochondria. J Exp Bot 29:1–12

Moore AL, Linnett PE, Beechey RB (1980) Dibutylchloromethyltin chloride, a potent inhibitor of electron transport in plant mitochondria. J Bioenerget Biomembr 12:309–323

Moreau F, Romani R (1982) Malate oxidation and cyanide-insensitive respiration in avacado mitochondria during the climacteric cycle. Plant Physiol 70:1385–1390

Moreira MTF, Rich PR, Bendall DS (1980) A method for demonstrating the involvement of a mobile quinone pool in an electron transport chain by the use of loosely bound inhibitors. 1st EBEC Short Rep, pp 61–62, Patron editore

Moyle J, Mitchell P (1973) Proton translocation quotient for the adenosine triphosphatase of rat liver mitochondria. FEBS Lett 30:317–320

Nelson BD, Gellerfors P (1976) The redox properties of the cytochromes of purified complex III. Biochim Biophys Acta 357:358–364

Nicholls DG (1974) The influence of respiration and ATP hydrolysis on the proton electro chemical gradient across the inner membrane of rat liver mitochondria as determined by ion distribution. Eur J Biochem 50:305–315

Nicholls DG (1979) Brown adipose tissue mitochondria. Biochim Biophys Acta 549:1–29

Ohnishi T (1979) Mitochondrial iron-sulfur flavodehydrogenases. In: Capaldi RA (ed) Membrane proteins in energy transduction. Marcel Dekker, New York, pp 1–87

Ohnishi T, Salerno JC (1982) Iron-sulfur clusters in the mitochondrial electron-transport chain. In: Spiro TG (ed) Iron-sulfur proteins, vol IV. John Wiley, New York, pp 285–326

Padan E, Rottenberg (1973) Respiratory control and the proton electrochemical gradient in mitochondria. Eur J Biochem 40:431–437

Palmer JM (1976) The organisation and regulation of electron transport in plant mitochondria. Annu Rev Plant Physiol 27:133–157

Palmer JM (1979) The uniqueness of plant mitochondria. Biochem Soc Trans 7:246–252

Palmer JM (1982) Ample evidence for cyanide-resistant oxidases in plants. Trends Biochem Sci 7:357

Palmer JM, Passam HC (1970) The oxidation of exogenous NADH by plant mitochondria. Biochem J 122:16p

Papa S (1976) Proton translocation reactions in the respiratory chains. Biochim Biophys Acta 456:39–84

Papa S (1982) Molecular mechanism of proton translocation by the cytochrome system and ATPase of mitochondria. Role of proteins. J Bioenerget Biomembr 14:69–86

Papa S, Capuano F, Market M, Altamura N (1980a) The H^+/O stoichiometry of mitochondrial respiration. FEBS Lett 111:243–248

Papa S, Guerrieri F, Lorusso M, Izzo G, Boffoli D, Capuano F, Capitanio N, Altamura N (1980b) The H^+/O stoichiometry of respiration-liked proton translocation in the cytochrome system of mitochondria. Biochem J 192:203–218

Papa S, Guerrieri F, Lorusso M, Izzo G, Capuano F (1982) The proton translocation function of the ubiquinone-cytochrome c oxidoreductase of mitochondria. In: Trumpower BL (ed) Functions of quinones in energy conserving systems. Academic Press, London New York, pp 527–537

Papa S, Lorusso M, Capitanio N, Nitto de E (1983a) Characteristics of redox-linked proton ejection in cytochrome c oxidase reconstituted in phospholipid vesicles: New observations in support mechanisms different from proton pumping. FEBS Lett 157:7–14

Papa S, Guerrieri F, Izzo G, Boffoli D (1983b) Mechanism of proton translocation associated to oxidation of N,N,N',N'-tetramethyl-p-phenylenediamine in rat liver mitochondria. FEBS Lett 157:15–20

Pietrobon D, Azzone GF, Walz D (1981) Effect of funicoulosin and antimycin A on the redox-driven H^+-pumps in mitochondria: on the nature of "leaks". Eur J Biochem 177:389–394

Pietrobon D, Zoratti M, Azzone GF (1983) Molecular slipping in redox and ATPase H^+ pumps. Biochim Biophys Acta 723:317–321

Pozzan T, Virgilio Di F, Bragadin M, Ciconi V, Azzone GF (1979a) H^+/site, charge/site and ATP/site ratios in mitochondrial electron transport. Proc Natl Acad Sci USA 76:2123–2127

Pozzan T, Miconi V, Virgilio Di F, Azzone GF (1979b) H^+/site, charge/site and ATP/site ratios at coupling sites I and II in mitochondrial electron transport. J Biol Chem 254:10200–10205

Prince RC, Bonner WD Jr, Bershak PA (1981) On the occurrence of the Rieske iron-sulfur cluster in plant mitochondria. Fed Proc 40:1667

Quinn PJ (1980) The interaction of coenzyme Q with phospholipid monolayers. Biochem Int 1:77–88

Quinn PJ, Esfahani MA (1980) Ubiquinones have surface-active properties switched to transport electrons and protons across membranes. Biochem J 185:715–722

Ragan CI (1976) NADH-ubiquinone oxidoreductase. Biochim Biophys Acta 456:249–290

Ragan CI, Heron C (1978) The interaction between mitochondrial NADH-ubiquinone oxidoreductase and ubiquinol-cytochrome c oxidoreductase. Biochem J 174:783–790

Ragan CI, Hinkle PC (1975) Ion transport and respiratory control in vesicles formed from reduced nicotinamide adenine dinucleotide coenzyme Q reductase and phospholipids. J Biol Chem 250:8472–8476

Reynafarje B, Brand MD, Lehninger AL (1976) Evaluation of the H^+/site ratio of mitochondria electron transport from rate measurements. J Biol Chem 251:744–751

Reynafarje B, Alexandre A, Davies P, Lehninger AL (1982) Proton translocation stoichiometry of cytochrome oxidase: use of a fast-responding oxygen electrode. Proc Natl Acad Sci USA 79:7218–7222

Rich PR (1978) Quinol oxidation in *Arum maculatum* mitochondria and its application to the assay, solubilisation and partial purification of the alternative oxidase. FEBS Lett 96:252–256

Rich PR (1981) A generalised model for the equilibration of quinone pools with their biological donors and acceptors in membrane-bound electron transfer chains. FEBS Lett 130:173–178

Rich PR (1984) Electron and proton transfers through quinones and cytochrome bc complexes. Biochim Biophys Acta 168:53–79

Rich PR, Bonner WD Jr (1978a) The nature and location of cyanide and antimycin resistant respiration in higher plants. Proc 11th FEBS Meeting 49 B6:149–158

Rich PR, Bonner WD Jr (1978b) EPR studies of higher plant mitochondria II. Centre S-3 succinate dehydrogenase and its relation to alternative respiratory oxidations. Biochim Biophys Acta 501:381–395

Rich PR, Moore AL (1976) The involvement of the protonmotive ubiquinone cycle in the respiratory chain of higher plants and its relation to the branchpoint of the alternate pathway. FEBS Lett 65:339–344

Rich PR, Moore AL, Ingledew WJ, Bonner WD Jr (1977) EPR studies of higher plant mitochondria. I. Ubisemiquinone and its relation to alternative respiratory oxidations. Biochim Biophys Acta 462:501–514

Rieske JS (1976) Composition, structure and function of Complex III of the respiratory chain. Biochim Biophys Acta 456:195–247

Rottenberg H (1979) The measurement of membrane potential and pH in cells, organelles and vesicles. In: Fleischer S, Packer L (eds) Methods in enzymology, vol 55. Academic Press, London New York, pp 547–569

Rustin P, Dupont J, Lance C (1983) A role for fatty acid peroxy radicals in the cyanide-insensitive pathway of plant mitochondria. Trends Biochem Sci 8:155–157

Ruzicka FJ, Beinert M, Schepler KL, Dunham WR, Sands RH (1975) Interaction of ubisemiquinone with a paramagnetic component in heart tissue. Proc Natl Acad Sci USA 72:2886–2890

Salerno JC, Blum H, Ohnishi T (1979) The orientation of iron-sulfur clusters and a spin-coupled ubiquinone pair in the mitochondrial membrane. Biochim Biophys Acta 547:270–281

Santis de A, Borraccio G, Arrigoni O, Palmieri F (1975) The mechanism of phosphate permeation in purified bean mitochondria. Plant Cell Physiol 16:911–923

Schneider H, Lemasters JJ, Hochli M, Hackenbrock CR (1980) Fusion of liposomes with mitochondrial inner membranes. Proc Natl Acad Sci USA 77:442–446

Schonbaum GR, Bonner WD Jr, Storey BT, Bahr JT (1971) Specific inhibition of the cyanide-insensitive respiratory pathway in plant mitochondria by hydroxamic acids. Plant Physiol 47:124–128

Siedow JN (1982) The nature of the cyanide-resistant pathway in plant mitochondria. In: Creasy LL, Hrazdina G (eds) Recent advances in phytochemistry, vol XVI. Plenum Press, New York London, pp 47–83

Singer TP, Kearney EB, Bernath P (1956) Studies on succinate dehydrogenase II. Isolation and properties of the dehydrogenase from beef heart. J Biol Chem 223:599–613

Slater EC (1953) Mechanism of phosphorylation in the respiratory chain. Nature (London) 172:975–978

Sorgato MC, Ferguson SJ, Kell DB, John P (1978) The proton motive force in submitochondrial particles. Biochem J 174:237–256

Sorgato MC, Branca D, Ferguson SJ (1980) The rate of ATP synthesis by submitochondrial particles can be independent of the magnitude of the proton motive force. Biochim Biophys Acta 188:945–948

Sorgato MC, Galiazzo F, Valente M, Cavallini L, Ferguson SJ (1982a) Hydrolysis of ITP generates a membrane potential in submitochondrial particles. Biochim Biophys Acta 681:319–322

Sorgato MC, Galiazzo F, Panato L, Ferguson SJ (1982b) Estimation of H^+ stoichiometry of mitochondrial ATPase by comparison of proton motive forces with clamped phosphorylation potentials in sub-mitochondrial particles. Biochim Biophys Acta 682:184–188

Srivasta HK, Sarkissian IV (1971) Purification of cytochrome oxidase from mitochondria of higher plants. Phytochemistry 10:977–998

Storey BT (1980) Electron transport and energy coupling in plant mitochondria. In: Stumpf PK, Conn EE (eds) The biochemistry of plants, vol II. Academic Press, London New York, pp 125–195

Tedeschi H (1981) The transport of cations in mitochondria. Biochim Biophys Acta 639:157–196

Thayer WS, Hinkle PC (1973) Stoichiometry of adenosine triphosphate-driven proton translocation in bovine heart submitochondrial particles. J Biol Chem 248:5395–5402

Thierbach G, Reichenbach H (1981) Myxothiazol, a new inhibitor of the cytochrome b-c_1 segment of the respiratory chain. Biochim Biophys Acta 638:282–289

Trumpower BL (1981) Function of the iron-sulfur protein of the cytochrome b-c_1 segment in electron transfer and energy-conserving reactions of the mitochondrial respiratory chain. Biochim Biophys Acta 639:129–155

Vercesi A, Reynafarje B, Lehninger AL (1978) Stoichiometry of H^+ ejection and Ca^{2+} uptake coupled to electron transport in rat heart mitochondria. J Biol Chem 253:6379–6385

Villalobo A, Briquet M, Goffeau A (1981) Electrogenic proton ejection coupled to electron transport through the energy-conserving site 2 and K^+/H^+ exchange in yeast mitochondria. Biochim Biophys Acta 637:124–129

Vries de S, Albracht SPJ, Berden JA, Slater EC (1982) The pathway of electronics through QH_2: cytochrome c oxidoreductase studied by pre-steady state kinetics. Biochim Biophys Acta 681:41–53

Waggoner AS (1976) Optical probes of membrane potential. J Membr Biol 27:317–334

Walsh Kinally K, Tedeschi H, Maloff BL (1978) Use of dyes to estimate the electrical potential of the mitochondrial membrane 17:3419–3428

Wernette ME, Ochs RS, Lardy HA (1981) Ca^{2+} stimulation of rat liver mitochondrial glycerophosphate dehydrogenase. J Biol Chem 256:12767–12771

Westerhoff HV, Simonetti ALM, Dam van K (1981) The hypothesis of localised chemiosmosis is unsatisfactory. Biochem J 200:193–202

Wiechmann AHCA, Beem EP, Dam van K (1975) The relation between H^+ translocation and ATP synthesis in mitochondria. In: Quagliariello E (ed) Electron transfer chains and oxidative phosphorylation. Elsevier/North Holland Biomedical Press, Amsterdam New York, pp 335–342

Wikström MKF, Berden JA (1972) Oxidoreduction of cytochrome b in the presence of antimycin. Biochim Biophys Acta 283:403–420

Wikström MKF, Krab K (1979) Proton-pumping cytochrome c oxidase. Biochim Biophys Acta 549:177–222

Wikström MKF, Krab M (1980) Respiration-linked H^+ translocation in mitochondria: stoichiometry and mechanism. In: Sanadi DR (ed) Current topics in bioenergetics, vol X. Academic Press, London New York, pp 51–101

Wikström MKF, Penttilä T (1982) Critical evaluation of the proton-translocating property of cytochrome oxidase in rat liver mitochondria. FEBS Lett 144:183–189

Wikström MKF, Krab K, Saraste M (1981a) Proton-translocating cytochrome complexes. In: Snell EE, Boyer PD, Meister A, Richardson CC (eds) Annu Rev Biochem 50. Palo Alto, pp 623–655

Wikström MKF, Krab K, Saraste M (1981b) Cytochrome oxidase-A synthesis. Academic Press, London New York

Williams RJP (1961) Possible functions of chains of catalysts. J Theor Biol 1:1–13

Williams RJP (1978) The multifarious couplings of energy transduction. Biochim Biophys Acta 505:1–44

Wilson SB (1971) Studies on the cyanide-insensitive oxidase of plant mitochondria. FEBS Lett 15:49–52

Wilson SB (1980) Energy conservation by the plant mitochondrial cyanide-insensitive
   oxidase. Biochem J 190:349–360
Yu C-A, Yu L (1981) Ubiquinone-binding proteins. Biochim Biophys Acta 639:99–128
Zoratti M, Pietrobon D, Conover T, Azzone GF (1981) On the role of $\Delta\bar{\mu}H^+$ as an
   intermediate in ATP synthesis. In: Palmieri F, Quagliariello E, Siliprandi N, Slater
   EC (eds) Vectorial reactions in electron and ion transport in mitochondria and bacteria. Elsevier/North Holland Biomedical Press, Amsterdam New York, pp 331–338
Zoratti M, Pietrobon D, Azzone GF (1982) On the relationship between rate of ATP
   synthesis and H^+ electro-chemical gradient in rat liver mitochondria. Eur J Biochem
   126:443–451
Zoratti M, Pietrobon D, Azzone GF (1983) Studies on the relationship between ATP
   synthesis and transport and the proton electrochemical gradient in rat liver mitochondria. Biochim Biophys Acta 723:59–70

7 The Oxidation of NADH by Plant Mitochondria

J.M. PALMER and J.A. WARD

1 Introduction

Intensive investigation of molecular mechanisms involved in plant respiration began in the late 1930's. Before this period interest centered mainly on cytological observations using the uptake of Janus Green B to detect the mitochondria. These early studies were reviewed by NEWCOMER (1940), who was not convinced that mitochondria were the centers of respiratory metabolism.

Research during the period of 1937 until 1950 concentrated on the isolation of soluble proteins from plant material or on spectral analysis of intact tissue or tissue homogenates. BERGER and AVERY (1943) investigated the dehydrogenases present in cell-free extracts of *Avena* coleoptiles and found evidence for most of the NAD^+-linked dehydrogenases associated with the tricarboxylic acid cycle. However, they found no evidence for succinic dehydrogenase, and concluded that the usual 4-carbon dicarboxylic acid cycle might not function in *Avena* tissue. Other studies showed that plant cells contained considerable quantities of $NADP^+$ (WHATLEY 1951) and a variety of soluble $NADP^+$-linked dehydrogenases were isolated from plant tissue (CONN et al. 1949). At this time there was little evidence concerning the mechanism by which reduced pyridine nucleotides were reoxidized by molecular oxygen. LOCKHART (1939) isolated an unstable membrane-bound diaphorase similar to the diaphorase obtained from yeast by WARBURG and CHRISTIAN (1932), as a suspension from pea and bean seedlings which was specific for NADH, however the natural electron acceptor was not identified.

Studies on the soluble oxidase systems revealed that polyphenol oxidase enzymes were widespread in plant tissue. KUBOWITZ (1938) isolated and purified potato polyphenol oxidase and showed it to be a copper-containing enzyme capable of oxidizing ortho-diphenols to the corresponding quinones with the formation of water. Data showing that quinones and the polyphenol oxidase could bring about the complete oxidation of NADH or NADPH (JAMES 1953) led to the suggestion that this enzyme system could be a major route for the oxidation of reduced pyridine nucleotide coenzymes. A second pathway involving the action of glutathione reductase and ascorbic acid oxidase was reported to be responsible for the oxidation of reduced pyridine nucleotides (BOSWELL and WHITING 1940 and BEEVERS 1954).

HILL and BHAGVAT (1939) were the first to investigate the role of cytochromes in oxidizing respiratory substrates in plant tissues. They isolated a particulate fraction from germinating seeds containing cytochrome oxidase, which was capable of mediating electron flow between succinate and oxygen in the presence of soluble cytochrome c; no evidence was available to suggest that the pyridine nucleotides could be reoxidized by the cytochrome pathway. Experiments by Levy and Schade (1948) cast doubt on the role of the soluble oxidases as the main system for oxidation of reduced pyridine nucleotides in potato tissue and suggested that the cytochrome system was likely to be the principal route by which all respiratory coenzymes were re-oxidized. Dubuy et al. (1950) were amongst the first to use isotonic sucrose to prepare intact mitochondria from plant tissue, demonstrating that they contained most of the cytochrome oxidase. Bhagvat and Hill (1951) extended their study of the cytochrome system and showed the presence of cytochrome a, b and c in plant tissue and provided evidence that they were all active in the oxidation of succinate. By demonstrating that isolated mung bean mitochondria could oxidize NAD^+-linked Krebs cycle acids using the cytochrome chain coupled to ATP synthesis, Millerd et al. (1951) finally established that the NADH produced during respiratory metabolism was reoxidized by the cytochrome system. Brummond and Burris (1953) used ^{14}C-labeled pyruvate to demonstrate that isolated plant mitochondria were capable of catalyzing all the reactions associated with the Krebs cycle. Smillie (1955) showed that active mitochondria could be obtained from all types of plant tissue, both green and etiolated, and concluded that plant mitochondria were comparable with animal mitochondria. Hackett (1955a), reviewing progress since 1950, concluded that mitochondria were the main centers of respiration carrying out the Krebs cycle reactions, terminal stages of oxidation and energy-trapping to form ATP.

Evidence soon accumulated which suggested that certain differences exist between animal and plant mitochondria. Both Davies (1953) and Laties (1953) reported that the addition of NAD^+ resulted in a marked stimulation of malate oxidation, which led to the conclusion that added NAD^+ was necessary to obtain maximal rates of oxidation of all NAD^+-linked acids. These observations were in marked contrast to the results obtained with mammalian mitochondria, which were found to be impermeable to both NADH and NAD^+ (Lehninger 1951). The observations that plant mitochondria were apparently permeable to NAD^+ were quickly followed by observations that they readily oxidized added NADH. Hackett (1955b) first reported that potato mitochondria oxidized exogenous NADH and that the mitochondria preparation contained both a particulate and soluble diaphorase that could react with 2,6 dichlorophenolindophenol. Humphreys and Conn (1956) carried out the first systematic measurements of NADH oxidation using lupin mitochondria. Their results showed that mitochondria were able to oxidize exogenous NADH and NAD^+-linked Krebs cycle acids via an antimycin A-sensitive mechanism. The addition of soluble cytochrome c was observed to stimulate only the oxidation of external NADH, via an antimycin A-insensitive pathway. The cytochrome c-dependent component of exogenous NADH oxidation was compared with a similar phenomenon reported in mammalian mitochondria (Lehninger 1951). Humphreys and Conn (1956) recognized that the component of exogenous NADH oxidase not dependent on cytochrome c and inhibited by antimycin A had no parallel in mammalian mitochondria. They proposed that external NADH may gain access to the matrix space via a transmembrane transhydrogenase. However, they observed that the oxidation of internal NADH was stimulated by AMP and Mg^{2+}, whilst the oxidation of external NADH was not, and concluded that the oxidation of external and internal NADH was not achieved at a common site.

CRANE (1957) found an antimycin A-resistant NADH-cytochrome c oxidoreductase in mitochondrial preparations which was associated with contaminating microsomes. This raised the possibility that the cytochrome c-dependent component of NADH oxidation by isolated mitochondria might be an artifact caused by the contamination of the preparation with microsomes. MARTIN and MORTON (1956) compared the NADH-cytochrome c reductases in microsomes and mitochondria obtained from the petioles of silver beet (*Beta vulgaris*). They found that antimycin A caused a 35% reduction of the velocity in the mitochondrial fraction, but no inhibition in the microsomal fraction, and concluded that the velocity of the antimycin A-resistant NADH-cytochrome c reductase in the mitochondria could not be reasonably attributed to contaminating microsomes. Their extensive investigations into the distribution of the NADH dehydrogenase system found that the NADH-cytochrome c reductases were associated with membrane systems, whilst the NADH-diaphorase (reducing 2:6 dichlorophenolindophenol) was easily solubilized.

HACKETT (1957) was the first to study the oxidation of exogenous NADH in skunk cabbage (*Symplocarpus foetidus*) mitochondria, which contain an active cyanide-resistant oxidase in addition to the normal cyanide-sensitive cytochrome oxidase. His results indicated a very complex interaction between the dehydrogenases and the two terminal oxidases. External NADH oxidation was almost completely inhibited by both antimycin A and cyanide, whilst the oxidation of internal NADH and succinate were only partially inhibited by cyanide. In contrast, antimycin A caused a 60% inhibition of succinic oxidase and only a 20% inhibition of internal NADH oxidase, although the control rates with each donor were comparable. HACKETT suggested that the oxidation of internal NADH could occur by a pathway that by-passed the antimycin A-sensitive step. This pathway was not apparently available for electrons coming from either succinate or external NADH.

ZELITCH and BARBER (1960) reported that the oxidation of both internal NADH and external NADH appeared to be coupled to three sites of ATP synthesis. BONNER and VOSS (1961) showed that respiratory control could be demonstrated using either malate or external NADH as electron donors. This clear demonstration that the oxidation of external NADH was coupled to ATP formation finally dispelled claims that the oxidation was associated with contaminating microsomes. BONNER and VOSS (1961) concluded that the oxidation of external NADH was a phenomenon common to all plant tissues and that it represented one of the marked differences between plant and animal tissues. They also drew attention to the rapid oxidation of malate, in the absence of an oxaloacetate trap, as another significant difference. Spectral analysis carried out by BONNER and VOSS (1961) confirmed that the cytochrome system was the principal pathway by which external NADH was oxidized.

HACKETT (1961) observed that salts markedly stimulated the oxidation of exogenous NADH, the evidence showing that the rate of oxidation responded in a general way to the ionic strength. He also suggested that there was an additional requirement for a divalent cation in order to obtain maximal rates of NADH oxidation. Further, there was no effect of ions on the NADH oxidase in digitonin fragments, or on the activity of diaphorase in intact mitochondria. HACKETT (1961) therefore concluded that the external NADH had free access to the external flavoprotein responsible for diaphorase activity, but needed ions to gain access to the enzyme responsible for donating electrons to the cytochrome chain. He suggested that the surface properties of the membrane were important in allowing the oxidation of external NADH to proceed, and that the role of ions was important in allowing external NADH to gain access to the respiratory chain.

BAKER and LIEBERMAN (1962) reported that the oxidation of exogenous NADH appeared to be coupled to only two sites of ATP synthesis, rather than three sites previously reported by ZELITCH and BARBER (1960), while the oxidation of NAD^+-linked Krebs cycle acids yielded 3 mol of ATP per two electrons transferred to molecular oxygen. Such stoichiometry was consistent with the view that external NADH may be oxidized by a pathway distinct from that used for internal NADH. CUNNINGHAM (1964) coupled the study of NADH oxidation with structural observations using the electron microscope. He observed that antimycin A-resistant oxidation of NADH, obtained in the presence of added cytochrome c, increased in velocity when the mitochondria were structurally damaged. However, the antimycin A-sensitive oxidation of exogenous NADH, catalyzed by the endogenous cytochromes, required highly intact organelles. He also found that P/O associated with the oxidation of exogenous NADH, in intact mitochondria, was only 1.2. CUNNINGHAM (1964) suggested that the antimycin A-sensitive pathway for the oxidation of external NADH differed from the pathway for the oxidation of internal NADH in by-passing the flavoprotein responsible for the coupling of electron flux to the first site of ATP synthesis; he proposed that electrons from external NADH then joined the main respiratory chain just before cytochrome b in order to account for the antimycin A sensitivity.

Therefore by 1964 there was good evidence to support the view that there were three pathways by which NADH could be oxidized by most types of plant mitochondria. One pathway was responsible for the oxidation of endogenous NADH, produced by the activity of the NAD^+-linked dehydrogenases of the Krebs cycle, this pathway being coupled to three sites of ATP synthesis and inhibited by antimycin A. The other two pathways appeared to be responsible for the oxidation of external NADH. One of these pathways was clearly distinct from the internal pathway, being dependent on exogenous soluble cytochrome c and resistant to antimycin A. The enzymes catalyzing this pathway were associated with the mitochondrion, but had similarities with a system located in the microsomes; MARTIN and MORTON (1956) had already suggested that this enzyme system might not play a role in cellular respiration. The second pathway for the oxidation of external NADH has no equivalent in mammalian mitochondria and was less readily distinguished from the inner pathway, since it utilized endogenous cytochromes and was sensitive to antimycin A. There were, however, three strong lines of circumstantial evidence that the antimycin A-sensitive pathway for the oxidation of exogenous NADH was distinct from that used to oxidize internal NADH. Firstly, it was important to have intact mitochondria (CUNNINGHAM 1964), thus it was unlikely that the NADH gained access through membrane damage. Secondly, the P/O for the oxidation of external NADH was significantly lower than for internal NADH (BAKER and LIEBERMAN 1962). Finally, the oxidation of external NADH was stimulated by ions, while the oxidation of internal NADH was not (HACKETT 1961). Despite this evidence, it was widely thought that the antimycin A-sensitive component of external NADH oxidation was due to added NADH gaining access to the matrix space as the result of artifacts induced during isolation. LIEBERMAN and BAKER (1965) considered that NADH might enter through small holes in the mitochondrial membrane, while HODGES and HANSON (1967) felt that the oxidation of external NADH was not necessarily a normal physiological activity of the mitochondria, and suggested that it might be an artifact induced during isolation. This view was reiterated by PACKER et al. (1970), who felt that the rigorous grinding necessary to release the mitochondria from the plant cells resulted in damage to the organelle, thus allowing external NADH to gain access to the matrix space and oxidation to occur by the internal pathway. CHANCE et al. (1968) added fuel to this controversy by stating that poorly isolated mitochondria were permeable to exogenous NAD^+, which accounted for earlier observations that the oxidation of NAD^+-linked acids could be stimulated by NAD^+ (DAVIES 1953 and LATIES 1953).

Meanwhile, studies on mitochondria isolated from baker's yeast (OHNISHI et al. 1966) showed that they were similar to plant mitochondria in being able to oxidize external NADH. However, they differed from both plant and mammalian mitochondria in being able to oxidize endogenous NADH via a pathway coupled to only two sites of ATP synthesis. They also observed that the oxidation of both endogenous and exogenous NADH was not inhibited by rotenone, a compound known to block electron flow between NADH and cytochrome b in mammalian mitochondria. SCHATZ et al. (1966) showed that yeast mitochondria could, under certain growth conditions, have a modified NADH dehydrogenase which was not inhibited by rotenone and by-passed the first site of ATP synthesis. Furthermore, it did not contain the characteristic electron paramagnetic spin signal at $g = 1.94$, which is now associated with the iron-sulfur centers. These results led WILSON and HANSON (1969) to investigate the influence of rotenone on the oxidation of both exogenous and endogenous NADH. In the absence of ADP (State 4) they observed that rotenone partially inhibited the oxidation of malate, but had no effect on the exogenous NADH oxidase. In the presence of ADP (State 3) they found that rotenone caused the complete inhibition of malate oxidation and a partial inhibition of external NADH oxidation. WILSON and HANSON (1969) concluded that the difference in inhibitor sensitivity was good proof that external and internal NADH were oxidized by different dehydrogenase systems.

VON JAGOW and KLINGENBERG (1970) used ^{14}C-labeled NADH as a substrate and potassium ferricyanide as a nonpenetrating electron acceptor to show conclusively that yeast mitochondria could oxidize exogenous NADH via an antimycin A-sensitive pathway without the NADH actually entering the matrix space. They postulated that this was achieved by a flavoprotein associated with the external surface of the inner membranes. Palmer and Passam (1971) argued that a similar external NADH dehydrogenase might feed electrons directly into the cytochrome chain in plant mitochondria; if the enzyme was located on the outer surface it would by-pass the first proton pumping system and thus fail to make ATP at the first coupling site, thereby explaining the lower P/O ratios. Coleman and Palmer (1971) attempted to use potassium ferricyanide as a nonpenetrating electron acceptor; ferricyanide is a polar molecule and can only accept electrons from redox components situated on the outer surface of the inner membrane. In mammalian mitochondria, cytochrome c is the usual point at which ferricyanide is reduced, thus ferricyanide reduction is normally sensitive to inhibition by antimycin A. If there was an external NADH dehydrogenase, then ferricyanide could interact directly with the flavoprotein and this reduction would not be inhibited by antimycin A. The results of Coleman and Palmer (1971) showed that no antimycin A-sensitive NADH-ferricyanide reductase could be detected, this suggesting that the dehydrogenase involved was located on the outer surface and accessible to ferricyanide. Coleman and Palmer (1971) also showed that the oxidation of exogenous NADH was strongly inhibited by EGTA and stimulated by Ca^{2+}; since neither of these two reagents can cross the inner membrane, it follows that their site of action is likely to be on the external surface of the inner membrane.

Moreau and Lance (1972) separated the outer and inner membranes of plant mitochondria and showed that the outer membrane contained an antimycin A-resistant NADH-cytochrome c reductase. It appears that this dehydrogenase was responsible for the antimycin A-resistant, cytochrome c-dependent, oxidation of exogenous NADH. The existence of this dehydrogenase complicated the interpretation of the NADH-ferricyanide results obtained by Coleman and Palmer (1971). Douce et al. (1973) showed that the NADH dehydrogenase associated with the external membrane was specific for the removal of the α-hydrogen from NADH and had a low K_m for ferricyanide (5 µM), while a soluble dehydrogenase, probably dissociated from the outer surface of the inner membrane was specific for the β-hydrogen of NADH and had a much greater K_m for ferricyanide (300 µM). They investigated the response of NADH-ferricyanide reductase to antimycin A and to varying concentrations of ferricyanide and concluded that in intact mitochondria all the antimycin A-resistant NADH-ferricyanide reductase was attributable to the dehydrogenase associated with the outer membrane, while the NADH-ferricyanide reductases associated with the inner membrane was completely sensitive to antimycin A. This is in marked contrast to data obtained by Coleman and Palmer (1971) and is not consistent with the external location of the NADH dehydrogenase. Douce et al. (1973) showed that swelling the mitochondria in 20 mOsm medium allowed ferricyanide to gain access to the flavoprotein, possibly by causing it to dissociate from the membrane, and concluded that the use of ferricyanide did not help in locating the position of the dehydrogenase. They suggested that the biphasic stimulation of the NADH-cytochrome c reductase which accompanied the rupture of the outer membrane followed by the rupture of the inner membrane during osmotic swelling was the best evidence for the external nature of the dehydrogenase responsible for the oxidation of exogenous NADH. The results of Douce et al. (1973) are widely accepted as constituting definitive proof for an outward-facing dehydrogenase feeding electrons directly into the cytochrome chain. However, they merely provide another form of circumstantial evidence in favor of the proposal, and conclusive proof remains to be obtained.

During the latter stages of the work on the location of the external NADH dehydrogenases, experimental evidence accumulated concerning the complexity of the internal NADH dehydrogenase. Ikuma and Bonner (1967) studied the response of the oxidation of NAD^+-linked Krebs cycle acids to the respiratory inhibitors rotenone and amytal, which were known to inhibit the respiratory chain-linked NADH dehydrogenases in mammalian mitochondria. They noted that rotenone inhibited 50% of the malate oxidase, while amytal caused a 90% inhibition. The partial inhibition by rotenone suggested that there may be more than one pathway for the oxidation of internal NADH. Brunton and Palmer (1973) showed that piericidin A, an inhibitor that acts at the same site as rotenone, caused a more severe inhibition of pyruvate oxidation than other NAD^+-linked substrates, which suggested that different Krebs cycle substrates have variable levels of access to the different dehydrogenases. They also demonstrated that the effectiveness of piericidin A in inhibiting malate oxidation was increased in the presence of oxaloacetate and the inhibition was rapidly relieved if the oxaloacetate was removed. This data suggested that the relative activity of the different NADH dehydrogenases is influenced by the ratio of malate to oxaloacetate. Using the antimycin A sensitivity of the malate-ferricyanide reductase as a guide, Brunton and Palmer (1973) suggested the piericidin A-resistant dehydrogenase responsible for the oxidation of endogenous NADH was located inside the inner membrane and was distinct from the rotenone-resistant dehydrogenase believed to be located on the outer surface of the inner membrane (Douce et al. 1973). Day and Wiskich (1974) also studied the rotenone-resistant oxidation of malate, which responds strongly to exogenous NAD^+, and argued that in the presence of NAD^+ all the rotenone-resistant oxidation can be accounted for by implicating a transmembrane dehydrogenase, similar to that proposed by Humphreys and Conn (1956); they envisaged a transmembrane dehydrogenase working in reverse to allow the transfer of reducing equivalents from internal NADH to external NAD^+. The suggestion raised the important question of whether the rotenone-resistant oxidation of NAD^+-linked substrates is catalyzed by the same dehydrogenase responsible for the oxidation of external NADH, or whether there is a second dehydrogenase located on the inner surface, as proposed by Brunton and Palmer (1973). Marx and Brinkman (1978) investigated this problem and concluded that the rotenone-resistant dehydrogenase active in the oxidation of internal NADH was different from that located on the exterior surface of the inner membrane. In a later paper Marx and Brinkman (1979) showed that the temperature dependencies of the external and internal rotenone-resistant dehydrogenases were different. Møller and Palmer (1982) showed that, using submitochondrial particles of mixed vectorial polarity, it was possible to distinguish two types of rotenone-resistant NADH dehydrogenases, one sensitive to EGTA and the other not; this was taken to indicate that the external rotenone-resistant dehydrogenase had a divalent ion requirement not shared by the internal enzyme. Thus it seems reasonable to suggest that NADH can be oxidized by four separate pathways; two responsible for the oxidation of exogenous NADH, one associated with the outer membrane, and one associated with the inner membrane. The other two pathways are responsible for the oxidation of endogenous NADH, one sensitive to inhibition by rotenone and the other resistant to rotenone. In the following section the current appreciation of each system will be discussed.

2 NADH Dehydrogenases Oxidizing Exogenous NADH

The current evidence suggests that there are two systems capable of oxidizing external NADH. One associated with the outer membrane, which is not considered to play a central role in respiratory metabolism, and one associated with the inner membrane, that seems able to feed electrons directly to the respiratory chain.

2.1 The Outer Membrane NADH Dehydrogenase

The operation of the pathway is characterized by being insensitive to rotenone and antimycin A, and dependent on added soluble cytochrome c. SOTTOCASA et al. (1967), working with rat liver mitochondria, separated the outer and inner membranes and demonstrated that the bulk of the antimycin A-resistant NADH-cytochrome c reductase was associated with the outer membrane and was specific for removing the 4α-hydrogen from NADH. This enzyme was not coupled to ATP formation. Characterization of the enzyme on the external membrane of mammalian mitochondria has shown it to be similar to the enzyme found in the microsomal fraction which has been more thoroughly characterized (STRITTMATTER 1963, 1966). The microsomal NADH-cytochrome c reductase is composed of two main subunits, an NADH-cytochrome b reductase, which is an FAD-containing flavoprotein with a molecular weight of 40,000, and a cytochrome b component known as b_5, which has absorption peaks at 552 and 557 nm at 77 K. In the mammalian system there are some differences between the microsomal and mitochondrial systems (PALMER and COLEMAN 1974), the mitochondrial enzyme is more sensitive to inhibition by dicumarol and trypsin than is the microsomal enzyme. BÄCKSTRÖM et al. (1973, 1978) have reported the presence of a ferrodoxin type iron-sulfur center closely associated with the mitochondrial enzyme that is not found in the microsomal enzyme complex. In the mammalian enzyme, the NADH reduces the FAD, which is then oxidized by cytochrome b_5. Under normal conditions the natural oxidant of the cytochrome b_5 is not known (cytochrome P-450 is not present in the mitochondrial membrane; MOREAU 1978). It has been observed that cytochrome b_5 can be slowly autooxidized, which could account for the low levels of cyanide and antimycin A-resistant oxidation some times seen in animal mitochondria (ARCHOKOV et al. 1974). If soluble cytochrome c is added to the mito-chondria it can become reduced by the cytochrome b_5 and re-oxidized by the cytochrome c oxidase on the inner membrane to yield a cyanide sensitive pathway of NADH oxidation.

MOREAU and LANCE (1972) were the first to separate the inner and outer membranes of plant mitochondria and showed that the antimycin A-insensitive NADH-cytochrome c reductase was located on the outer membrane together with a b-type cytochrome. DOUCE et al (1973) studied the NADH-cytochrome c reductase associated with the outer membranes isolated from gradient-purified mitochondria, and showed that the enzyme resembled the mammalian system in being specific for the removal of the α-hydrogen from NADH, whilst the dehydrogenase associated with the inner membrane was specific for the removal of the β-hydrogen from NADH. The membranes also contained a flavoprotein and a b-type cytochrome that had a single α absorption peak at 555 nm at room temperature which divided into a double peak at 551 and 557 nm at 77 K. Both MOREAU and LANCE (1972) and DOUCE et al. (1973) used swelling techniques to isolate the outer membranes, whereas DAY and WISKICH (1975) used digitonin to separate the inner and outer membranes. They confirmed the presence of an antimycin A-resistant NADH-cytochrome c reductase on the outer membrane. This enzyme had very low activity towards NADPH and could be distinguished from the NADH dehydrogenase associated with the outer surface of the inner membrane because it was not inhibited by EGTA or by concentrations of dicumarol below 10 µM, both of which inhibit the antimycin A-sensitive oxidation of exogenous NADH in turnip mitochondria (DAY and WISKICH 1975). High concentrations of dicumarol (above 10 µM) did inhibit the outer membrane dehydrogenase. DAY and WISKICH (1975) showed that both the external dehydrogenases (i.e., that on the outer membrane and inner membrane) were inhibited by p-chloromercuribenzoate. Incubation of the enzymes with NADH before exposing them to the inhibitors conferred protection against inhibition, which suggested that $-$SH groups may be involved in binding the NADH to the dehydrogenase.

MOREAU (1978) carried out an extensive investigation of the NADH dehydrogenase associated with the outer membrane of cauliflower mitochondria and compared this with the microsomal system. The data showed the enzyme to be highly specific for NADH and not able to oxidize NADPH. Kinetic data showed that the mitochondrial enzyme had a higher V_{max} than the microsomal enzyme and that it also had a greater affinity for NADH. The K_m(app) for NADH was 2.4 µM for the mitochondrial enzyme and 12 µM for the microsomal system. Inhibitor studies on the mitochondrial enzyme showed mersalyl to be an efficient inhibitor, confirming that $-$SH groups play an important role in enzyme function. MOREAU (1978) showed that thenoyl-trifluoroacetone (an iron chelator) caused a 15% inhibition and concluded iron-sulfur centers play a role in the enzyme sequence. There have been no reports of the detection of an iron-sulfur center by electron paramagnetic resonance techniques in the outer membrane fraction of plant mitochondria. MOREAU (1978) showed the outer membrane contained a flavoprotein with an absorption band at 440–460 nm and fluorescent emission at 515 nm which could be abolished by reduction with dithionite. MOREAU suggested that the flavin may be FAD because the intensity of the fluorescence was greatly enhanced upon acidification. Three b type cytochromes b-559, b-558 and b-554 were identified in the outer membrane preparation and MOREAU suggested these may be the counterparts of the three types of cytochrome b found in plant microsomes (RICH and BENDALL 1975). Examination of his spectra show that there is some cytochrome a present, also 20% of the NADH-cytochrome c reductase was reported to be sensitive to antimycin A, consequently this raises doubts over the purity of the outer membrane fraction used.

The natural electron acceptor from the outer membrane NADH-cytochrome c reductase is not known. The outer membrane of the mitochondrion does not contain any cytochrome P-450 (MOREAU 1978), whereas cytochrome P-450 is known to be present in the microsomal system (RICH and BENDALL 1975) and can bring about the final step in NADH oxidation. MOREAU (1978) proposed that soluble cytochrome c could accept electrons from the cytochrome b-555 on the outer membrane and become oxidized by the cytochrome c oxidase associated with the inner membrane. There are few data available concerning the physiological significance of such a pathway, the operation of such a system would depend on the level of soluble cytochrome c and the physical location of the sites of cytochrome c reduction and oxidation. It is generally accepted that the outer membrane of an intact plant mitochondria is impermeable to cytochrome c (DOUCE et al. 1972). Therefore, for the proposed intermembrane electron transport pathway to work, the soluble cytochrome c must be located in the intramembrane space which would necessitate the site of cytochrome c reduction to be located on the inner surface of the external membrane. The evidence concerning the site of cytochrome c reduction on the outer membrane is mixed. PALMER and COLEMAN (1974) reported that osmotic swelling greatly enhanced (400%) the activity of the antimycin A-resistant NADH-cytochrome c reductase in both rat liver and Jerusalem artichoke mitochondria, suggesting that cytochrome c needs to penetrate the outer membrane before it can be reduced. In contrast DOUCE et al. (1973) reported that the activity of the outer membrane NADH-cytochrome c reductase, in mung bean mitochondria, was not influenced by osmotic stress and concluded that cytochrome c reduction took place on the outer surface of the outer membrane. If the latter view is correct, then it is unlikely that reduced cytochrome c can cross the outer membrane fast enough to provide a meaningful rate of electron transport in intact organelles.

2.2 The Inner Membrane NADH Dehydrogenase

2.2.1 Location of the Dehydrogenase

HUMPHREYS and CONN (1956) were the first to obtain experimental evidence that external NADH could be oxidized by a pathway that did not depend on the presence of cytochrome c and was inhibited by antimycin A. These properties clearly distinguish this pathway from the oxidation of exogenous NADH by the outer membrane enzyme system described in the previous section. HUMPHREYS and CONN (1956) proposed that external NADH could not pass through the inner membrane, and therefore had to be oxidized via an external pathway.

WILSON and HANSON (1969) showed that the oxidation of external NADH was not sensitive to inhibition by rotenone and this feature distinguished the external pathway from that responsible for the oxidation of endogenous NADH. Unfortunately, two later developments complicate the situation; firstly IKUMA and BONNER (1967) reported that malate oxidation was only partially sensitive to rotenone, suggesting that endogenous NADH could have access to a rotenone-resistant dehydrogenase. Subsequently BRUNTON and PALMER (1973) reported the presence of a piericidin A-resistant NADH dehydrogenase capable of oxidizing internal NADH, which is only coupled to two sites of ATP synthesis. MARX and BRINKMAN (1978, 1979) confirmed the presence of a rotenone-resistant internal NADH dehydrogenase. Therefore the argument that the external pathway differs from the internal pathway on the basis of rotenone resistance and two coupling sites is no longer valid. Secondly, more attention has been given to the well-known ability of NAD^+ to stimulate the oxidation of NAD^+-linked substrate (DAVIES 1953). NEUBURGER and DOUCE (1978) were the first to suggest that the added NAD^+ crossed the inner membrane. PALMER et al. (1982) also concluded that many observations could be explained on the basis that the inner membrane of the mitochondria was permeable to NAD^+. Thiamine pyrophosphate can also stimulate the oxidation of pyruvate or α-ketoglutarate (LATIES 1953) and therefore appears able to cross the inner membrane. Since modern techniques produce highly purified, intact mitochondria it seems unlikely that coenzyme permeability occurs through membrane damage and the interesting possibility of regulated coenzyme translocation must be considered. Thus the concept of rigid compartmentation of coenzymes may not apply in the plant system; if NAD^+ can be translocated it is not impossible that external NADH may move across the membrane. Therefore, we cannot entirely dismiss the view that the different responses to ions (HACKETT 1961, MILLER et al. 1970, MØLLER et al. 1981) may be due to the modification of an NADH translocator.

From the foregoing discussion it is apparent that the weight of experimental data favors the view that there is an external NADH dehydrogenase that can feed electrons directly into the ubiquinone of the respiratory chain. However, the direct proof is not yet available, and there is a possibility that reducing equivalents may gain access to the rotenone-resistant internal NADH dehydrogenase. For the purposes of the remainder of this section it will be assumed that there is an external NADH dehydrogenase; if subsequent research shows this to be incorrect, then many of the characteristics attributed to it should apply to the translocator involved in its entry to the matrix. Much of the current understanding concerning the characteristics of this system have recently been reviewed (PALMER and MØLLER 1982).

2.2.2 Nature of the Redox Components and Relationship with the Respiratory Chain

VON JAGOW and KLINGENBERG (1970) showed that the external, respiratory chain linked NADH dehydrogenase in yeast was readily solubilized. DOUCE et al. (1973) obtained an easily solubilized NADH-dehydrogenase from mung bean mitochondria which was identified as a flavoprotein because it showed a bleaching at 450 and 480 nm when reduced by NADH. It was different from the flavin isolated from the outer membrane (Douce et al. 1973, Moreau 1978) in being specific for removing the β-hydrogen of NADH. Storey (1971) studied the spectral changes that accompany the oxidation of NADH in skunk cabbage mitochondria and identified a flavin (Fp.1a) as being directly involved in the external NADH dehydrogenase. The insensitivity of the external NADH dehydrogenase to rotenone and piericidin A (Wilson and Hanson 1969, Brunton and Palmer 1973) coupled with the lack of ATP synthesis between the flavoprotein and ubiquinone suggest that the iron-sulfur proteins play no role in the electron transport sequence to the quinone pool. Cammack and Palmer (1977) examined the iron-sulfur proteins in *Arum* mitochondria and described the presence of a single iron-sulfur protein that is not found in mammalian mitochondria which could be reduced by external NADH, and they tentatively suggested that it may play a role in the external NADH dehydrogenase system.

It is generally assumed that ubiquinone plays an important role in the oxidation of external NADH (Moore and Rich 1980, Palmer and Møller 1982), however the experimental data concerning the involvement of ubiquinone is rather complex. Storey (1971) measured the kinetics of ubiquinone reduction in skunk cabbage mitochondria and reported that external NADH reduced the flavoprotein, cytochromes b-557 and c-549 more rapidly than the ubiquinone. He concluded the kinetic response was not entirely consistent with the ubiquinone playing a central role; he favored it acting as a side chain storage system.

In contrast Huq and Palmer (1978) extracted quinones from *Arum* mitochondria and showed that the oxidation of external NADH ceased; it could be re-started by adding ubiquinone, suggesting that in this tissue quinone played an important role. However, there is difficulty in understanding the role of quinone in the oxidation of external NADH and this becomes especially clear when working with mitochondria which contain both a cyanide-sensitive and cyanide-resistant oxidase. Hackett (1957) using skunk cabbage mitochondria, showed that whilst endogenous NADH could be oxidized by both oxidases, exogenous NADH was only oxidized by the cyanide-sensitive oxidase. Tomlinson and Moreland (1975) described a similar situation in sweet potato mitochondria in which succinate could be oxidized by the cyanide-resistant oxidase, whilst external NADH could not. Huq and Palmer (1978) made the same observation using *Cassava* mitochondria. Recently Rustin et al. (1980) and Gardeström and Edwards (1983) have described situations in which only the endogenous NADH could be oxidized by the cyanide-resistant oxidase. This data makes it difficult to consider the respiratory chain in plant mitochondria as a number of NADH dehydrogenases feeding electrons into a single homogenous pool of ubiquinone, which then distributes them to the two oxidase systems. Palmer (1979) suggested that the data could be explained by assuming that the quinone was organized into more than one functional pool. Moore et al. (1980) observed that dibutylchloromethylitin chloride(DBCT) inhibited the oxidation of succinate and malate but not external NADH, and they concluded that DBCT interacted with a functionally distinct pool of quinone associated with the oxidation of succinate and malate whilst not affecting the quinone involved in the oxidation of external NADH. COOK and CAMMACK (1983) have shown that the quinone analogue 5-(n-undecyl)-6-hydroxy-4,7-dioxobenzothiazole (UHDBT) inhibits the oxidation of succinate and not external NADH via the cyanide-sensitive oxidase; these authors concluded that UHDBT could completely inhibit the re-oxidation of ubiquinone associated with succinate oxidation, but only partly inhibit the re-oxidation of NADH-reduced ubiquinone. COTTINGHAM and MOORE (1983) showed that when NADH was oxidized by either the cyanide-sensitive or cyanide-resistant oxidase, the process exhibited "pool" kinetics, i.e., a diffusable species plays an important role, and ubiquinone was assumed to be the diffusable species involved. When both oxidases were working simultaneously, the "pool" kinetics did not hold, suggesting that different diffusable steps were involved in each pathway. Recently, KAY and PALMER (unpublished results) have shown that 2(n-heptyl)-4-hydroxyquinoline N-oxide (HQNO) caused a more severe inhibition of succinate oxidation than NADH oxidation, whilst antimycin A was equally inhibitory towards both substrates. These observations suggest that the route of electron transport through respiratory complex 3 differs depending on the nature of the donor. These results are most easily explained by assuming that in the presence of HQNO a diffusable pool of quinone mediates electron transport when succinate is the donor, whilst bound quinones mediate the oxidation of external NADH.

2.2.3 Inhibitors of the External Dehydrogenase

The most obvious characteristics of the external NADH dehydrogenase associated with the inner membrane are its resistance to inhibition by rotenone and piericidin A and sensitivity to antimycin A. The most studied inhibitors of the external NADH dehydrogenase are the divalent ion chelators EDTA and EGTA (COLEMAN and PALMER 1971) and citrate (COWLEY and PALMER 1978). These chelators result in the extensive inhibition of the antimycin A-sensitive NADH-cytochrome c reductase, but have no effect on the NADH-ferricyanide reductase (COLEMAN and PALMER 1971). The inhibition can be reversed by the addition of Ca^{2+} (COLEMAN and PALMER 1971, MØLLER et al. 1981, MOORE and ACKERMAN 1982). The precise role that calcium plays in the NADH-cytochrome c reductase is unclear, but it has been suggested that Ca^{2+} may bind the flavin to the correct site on the outer surface of the inner membrane (COLEMAN and PALMER 1971, STOREY 1980). The degree of inhibition caused by the chelators is greatly influenced by experimental conditions. The addition of chelator before the NADH resulted in a high level of inhibition (COWLEY and PALMER 1978, MØLLER et al. 1981, MØLLER and PALMER 1981 b), which became partially reversed as the experiment progressed. If the chelator was added after the NADH, then the chelator took up to 90 s before establishing maximal inhibition. The final rate of oxidation obtained in both sequences was eventually the same for any given concentration of chelator. It has been suggested that when the chelator was added before NADH the Ca^{2+} could be more easily removed than when the chelator was added after the NADH, when the calcium appeared to be more firmly "locked in" (COWLEY and PALMER 1978). The partial recovery of NADH oxidation that occurred when the NADH was added after the chelator was thought to be the result of movement of Ca^{2+} from the matrix into some of the dehydrogenases which are not accessible to the chelator (MØLLER et al. 1981).

The effectiveness of chelators as inhibitors was greatly reduced when the pH was lowered from pH 7.5 (60% inhibition) to pH 6.5 (10% inhibition). EGTA has pK values at 2, 2.7, 8.9 and 9.5, therefore it would be expected to have unchanged chelating abilities over the pH range 6.0 to 7.5. Thus it has been speculated that the different response to EGTA, as the pH is lowered, is the result of changes in the degree of ionization of components on the surface of the membrane (MØLLER and PALMER 1981b). EDTA and EGTA are polar molecules that cannot penetrate the inner membrane and it seems that they must exert their inhibitory influence by removing Ca^{2+} close to the outer surfaces. Chlorotetracycline, a chelator of calcium which fluoresces strongly in an apolar medium, is without effect on the NADH oxidase until Ca^{2+} is added (MØLLER et al. 1983), when a strong inhibition develops which responds to pH changes in the same way as inhibition by EGTA (MØLLER and PALMER 1981b). The mechanism by which Ca^{2+} inhibits in the presence of chlorotetracycline has not been elucidated, but it is possible that the lipid-soluble chlorotetracycline-Ca^{2+} complex might attach itself to the Ca^{2+}-binding site in the NADH-cytochrome c reductase.

DAY and WISKICH (1975) reported that p-chloromercuribenzoate was a potent inhibitor and that incubation of the mitochondria with NADH protected the system against inhibition. ARRON and EDWARDS (1979) surveyed the sensitivity of the NADH and NADPH dehydrogenases to sulfhydryl inhibitors and reported that p-chloromercuribenzoate, n-ethyl maleimide and mersalyl were good inhibitors of NADPH oxidation whilst they were less effective in inhibiting the oxidation of NADH. MØLLER and PALMER (1981b) also investigated the influence of mersalyl on the oxidation of both NADPH and NADH. In Jerusalem artichoke mitochondria mersalyl was found to inhibit the oxidation of NADH at pH 7.2 but was without effect at pH 5.2, whereas the oxidation of NADPH, which is not oxidized at pH 7.2, was strongly inhibited by mesalyl at pH 5.2. In *Arum* mitochondria, at pH 6.7, the oxidation of NADH was not inhibited by mersalyl, whilst the oxidation of NADPH was strongly inhibited. The greater sensitivity of NADPH oxidation to mersalyl has been taken by ARRON and EDWARDS (1979), MØLLER and PALMER (1981b) and NASH and WISKICH (1983) to indicate that exogenous NADH and NADPH were oxidized by different dehydrogenases linked to the respiratory chain.

DAY and WISKICH (1975) found that low concentrations of dicumarol strongly inhibited the oxidation of exogenous NADH and recently RAVANEL et al. (1981) reported that flavones selectively inhibited the oxidation of external NADH.

2.2.4 Specificity of the External NADH Dehydrogenase for the Nicotinamide Adenine Dinucleotide

KOEPPE and MILLER (1972) were the first to investigate the capacity of plant mitochondria to oxidize NADPH and reported that corn mitochondria could oxidize exogenous NADPH. The accompanying oxidative phosphorylation equalled the efficiency that accompanied the oxidation of exogenous NADH. They also found that amytal inhibited the oxidation of NADPH and not NADH, and concluded that the NADPH was oxidized by a separate dehydrogenase.

Mammalian mitochondria are known to contain a transhydrogenase which transfers reducing equivalents between NAD^+ and NADPH (RYDSTROM 1977), such an enzyme would enable NADPH to reduce NAD^+. HACKETT (1963) reported the presence of a transhydrogenase in mitochondria isolated from pea stems, WILSON and BONNER (1970) also reported the presence of a NAD^+-NADPH transhydrogenase in submitochondrial particles made from mung bean mitochondria. The $NADH-NADP^+$ transhydrogenation in plants was studied by HASSON and WEST (1971), who showed that it depended on ATP as an energy source; however, they found it was located in the microsomal fraction and not in the mitochondria. There have been several reports that plant mitochondria do not contain NADP as a co-enzyme (HAWKER and LATIES 1963, HARMEY et al. 1966, IKUMA 1972), which suggests that a mitochondrial transhydrogenase is unnecessary. ARRON and EDWARDS (1979) re-investigated the oxidation of NADH and NADPH by isolated mitochondria, they found they were both oxidized by a variety of mitochondria. When using corn mitochondria they detected no difference between the two donors and their sensitivity to amytal and rotenone as reported by KOEPPE and MILLER (1972). ARRON and EDWARDS reported that the oxidation of NADPH was more sensitive to inhibition by chelators (1979) and mersalyl (1980) and concluded that different flavoprotein dehydrogenases were responsible for the oxidation of each coenzyme. ARRON and EDWARDS (1979) also studied the oxidation of pyridine nucleotide coenzymes in red beetroot mitochondria; in contrast to reports by DAY et al. (1976), they observed that mitochondria from fresh beetroot tissue oxidized both NADH and NADPH. However, after washing discs of beetroot tissue for 24 h before isolating the mitochondria, it was possible to obtain an enhanced capacity to oxidize NADH and no change in the capacity to oxidize NADPH. This also supports their conclusion that separate dehydrogenases were responsible for the oxidation of NADH and NADPH.

MØLLER and PALMER (1981 b) observed that the pH greatly influenced the relative rate of NADH and NADPH oxidation. The pH optimum for the oxidation of NADH was near 7.0, whilst the pH optimum for the oxidation of NADPH was between 6.5 and 6.0, depending on the source of the mitochondria. At pH values below 6.0 the rate of oxidation of NADH and NADPH was identical. MØLLER and PALMER (1981 a) found no evidence that a phosphatase acted directly on the NADPH to convert it to NADH. Neither could evidence be found for a $NADPH-NAD^+$ transhydrogenase, either by direct assay or by the use of palmitoyl-CoA as an inhibitor of the transhydrogenase (RYDSTROM 1977). It was therefore concluded that oxidation of NADPH occurred directly, either by an unspecific NAD(P)H dehydrogenase or by two separate enzyme systems. The influence of pH on the relative rate of NADPH oxidation could be explained on the basis that NADPH contains a 2'-phosphate group that has a pK of 6.2–6.3 (THEORELL 1935). The presence of this group is the principal difference between NADPH and NADH, the phosphate would become protonated as the pH was decreased from 7.0 to 6.0, making the electrostatic properties of NADPH more like NADH, resulting in the loss of specificity at the level of the dehydrogenase. However, Møller and Palmer (1981 b) concluded that the differences in the sensitivity of oxidation of NADPH and NADH to chelators and mersalyl (Arron and Edwards 1979, 1980) were difficult to explain on the basis of one rather unspecific dehydrogenase and consequently the current evidence favors the existence of two separate dehydrogenases (Nash and Wiskich 1983).

2.2.5 Regulation of Electron Flux Through the External NADH Dehydrogenase

In mammalian cells, NADH in the cytosol cannot be directly oxidized by the respiratory chain; the reducing power must be translocated into the matrix via the malate-aspartate shuttle or oxidized indirectly by the glycerolphosphate-dihydroxyacetone cycle. These systems are not able to maintain an equilibrium between the redox potentials of the cytosolic and mitochondrial pools of pyridine nucleotides. Thus in animal cells, under aerobic conditions, the pyridine nucleotides in the mitochondria are more oxidized than those in the cytosol (Chance and Theorell 1959).

In plant cells the external NADH can be oxidized directly by the mitochondria. A similar situation has been described in yeast, where the introduction of oxygen to anaerobic cells results in the simultaneous oxidation of pyridine nucleotides both inside and outside of the mitochondria. It is reasonable to assume that a similar situation occurs in plant cells. The oxidation of cytosolic NADH by the outward-facing dehydrogenase will favor the conversion of hexoses to trioses by the forward direction of glycolysis. Thus electron flux through the external dehydrogenase can have important consequences for cellular metabolism.

There are currently two ways by which electron transport from exogenous NADH can be regulated, firstly by the ionic composition of the assay medium and secondly by competitive electron flux from Krebs cycle acids.

Regulation by Cations. Isolated plant mitochondria suspended in an assay medium containing low ionic concentrations (<1 mM K^+) oxidize exogenous NADH only slowly; this can be stimulated by the addition of a variety of cations (Hackett 1961, Johnston et al. 1979, Møller et al. 1981). Trivalent ions were the most effective, divalent were intermediate and monovalent ions the least effective. This general response to cations has been interpreted as being due to the screening of fixed negative charges associated with the lipid and protein components of the membrane structure. Screening of the charges decreases the electrostatic repulsion of the negatively charged NADH, resulting in an apparent increase in affinity between the enzyme and substrate (i.e., a lowering of the K_m). Screening of the charges also leads to an increase in the V_m (Møller and Palmer 1981 c), which may be attributed to an increased lateral mobility of membrane protein complexes, resulting in an increased collision frequency and higher rates of electron transport (SCHNEIDER et al. 1980).

In addition to regulation by the general phenomenon of charge screening, there is evidence that calcium ions may play a specific role in activating the oxidation of external NADH (HACKETT 1961, MØLLER et al. 1981). The mechanism by which calcium activates the NADH oxidation remains unresolved. There have been two proposals to account for the role of calcium. COLEMAN and PALMER (1971) observed that calcium stimulated the oxidation of NADH by oxygen but not by ferricyanide, and proposed that calcium may bind the flavoprotein to the main respiratory chain, whilst PALMER and MØLLER (1982) speculated that the presence of calcium changed the conformation of the enzyme such that it interacted more efficiently with NADH.

Control by Competitive Electron Flow. There have been several observations that when mitochondria are supplied with more than one substrate, the resulting rate of oxygen uptake is lower than predicted by adding the rates of oxidation of both substrates measured separately (DAY and WISKICH 1977, COWLEY and PALMER 1980, DRY et al. 1983 and BERGMAN and ERICSON 1983). Calculations from the data of DAY and WISKICH (1977) suggest that when malate and NADH were mixed, most of the decrease in expected rate comes from an inhibition of NADH oxidation, whilst when succinate and NADH were used 65% of the reduction was due to loss of NADH oxidation and 35% from succinate. COWLEY and PALMER (1980) showed that under State 4 conditions[1], the lack of additivity between NADH oxidation and succinate oxidation could all be attributed to inhibition of NADH oxidation, and succinate oxidation proceeded normally. The inhibition of NADH oxidation required the active oxidation of succinate and was prevented by adding malonate to inhibit succinate dehydrogenase. This clearly distinguishes this type of inhibition from that caused by citrate, which results from the chelation of calcium (COWLEY and PALMER 1978). The degree of inhibition of NADH oxidation by succinate depended on the metabolic state of the mitochondria. It was greatest when the proton motive force was high, i.e., under State 4 conditions, intermediate under State 3 and least in the presence of weak acid uncouplers. Recently studies on leaf mitochondria (DRY et al. 1983, BERGMAN and ERICSON 1983) have shown that the oxidation of glycine takes precedence over the oxidation of external NADH and Krebs cycle intermediates. Clearly, the regulation of oxidation by competitive electron flux is likely to have important consequences in the integration of cellular metabolism.

The mechanism responsible for this interaction remains unknown. It could be associated with the processes involved in the re-oxidation of the ubiquinol pools by Complex 3. This is supported by the observation that the rates of succinate-ubiquinone I and NADH-ubiquinone I reduction are additive (KAY and PALMER, unpublished observation) and that the redox poise of cytochrome c_1 may determine the extent of interaction between NADH and succinate oxidase (COWLEY and PALMER 1980).

---

1 Definition of the different stages: see Chap. 10, this volume, p. 283.

2.2.6 The Physiological Significance of Regulation
of the NADH Dehydrogenase

The importance of regulation by general charge screening is difficult to gauge since the concentration of cations in the cytosol in plant cells is difficult to measure, but it seems likely that the natural ion concentration is sufficient to bring about maximal levels of charge screening.

Control by modulation of calcium levels is more likely to be physiologically significant (Palmer and Møller 1982). Studies concerning the removal of calcium by citrate (Cowley and Palmer 1978) show that about 1 μM calcium is needed in the cytosol to maintain a maximum rate of NADH oxidation and that 1 mM citrate will cause a 50% reduction in the rate of NADH oxidation. The effective concentrations of both of these modulators are in the same order of magnitude as they may occur in the cytosol, and therefore they could potentially play a role in regulating the rate of oxidation of cytosolic NADH in an in vivo system.

Regulation of NADH oxidation by flow of electrons from succinate also seems to have a potential physiological significance. Inhibition of NADH oxidation would only occur when the phosphate potential is high and under these conditions the mitochondria would be using the nonphosphorylating oxidase to interconvert acids for biosynthetic purposes and would therefore need to keep the pyridine nucleotides in the cytosol reduced to bring about the anabolic reactions (Palmer and Møller 1982). If the phosphate potential were to fall, then the inhibition of external NADH oxidation would be reduced and all electron donors, both in the mitochondrial matrix and in the cytosol would be readily oxidized to restore the ATP level.

3 NADH Dehydrogenases Oxidizing Endogenous NADH

Millerd et al. (1951) found that isolated plant mitochondria were able to oxidize NAD^+-linked Krebs' cycle acids without the need to add coenzymes. Humphreys and Conn (1956) repeated this observation and concluded that isolated mitochondria contained NAD^+ in the matrix space. Humphreys and Conn speculated that the endogenous NAD^+ may be bound to the membrane structure making it difficult for added "soluble" NAD^+ to gain access to the sites of oxidation and reduction. Lance and Bonner (1968) showed that several different types of plant mitochondria contained endogenous NAD^+ and that the concentration varied between 2 and 5 nmol of pyridine nucleotide mg^{-1} mitochondrial protein. Neuburger and Douce (1978) reported that the level of endogenous NAD^+ in potato mitochondria varied and appeared to depend on the physiological state of the original tuber; in this study they estimated the matrix volume to be very approximately 1 μl mg^{-1} protein. Therefore, the estimates of the concentration of NAD^+ in the matrix space varies between 1 and 5 mM.

The endogenous NAD^+ is located in the matrix space, since the enzymes involved in oxidizing the Krebs cycle NAD^+-linked substrates are located in the matrix space (DOUCE et al. 1973, DAY et al. 1979). In addition, the oxidation of malate and other substrates has been reported to be sensitive to inhibition by compounds that prevent the operation of the dicarboxylic acid translocator located in the inner membrane (PHILIPS and WILLIAMS 1973, DAY and WISKICH 1974), indicating that accumulation into the matrix space preceeds oxidation.

Extensive data are available to show that the NADH produced in the matrix space is oxidized via a dehydrogenase not available to oxidize exogenous NADH. BAKER and LIEBERMAN (1962) and CUNNINGHAM (1964) showed that the oxidation of endogenous NADH was coupled to three sites of ATP, whilst the oxidation of external NADH was only coupled to two sites of ATP formation. CUNNINGHAM (1964) suggested that the extra site of ATP formation was associated with electron flux between the NADH and ubiquinone. IKUMA and BONNER (1967) identified another difference between the external NADH dehydrogenase and the internal enzyme when they showed that the oxidation of internal NADH was sensitive to inhibition by amytal and partially sensitive to rotenone, whilst the external pathway was resistant to both inhibitors. The oxidation of endogenous NADH also differs from the oxidation of exogenous NADH in that it is not sensitive to inhibition by divalent ion chelators such as EDTA or EGTA and shows no dependency on calcium (COLEMAN and PALMER 1971).

The partial inhibition of the internal NADH dehydrogenase by rotenone reported by IKUMA and BONNER (1967) has also been reported by other authors (BRUNTON and PALMER 1973, WISKICH and DAY 1982). BRUNTON and PALMER (1973) published the first evidence that there was a rotenone-resistant dehydrogenase responsible for the oxidation of endogenous NADH which was not accessible to external NADH. This inhibitor resistance makes this enzyme similar to the external dehydrogenase, although no evidence for a calcium requirement has been found. MØLLER and PALMER (1982) reported that in submitochondrial particles, of mixed vectorial polarity, two rotenone-resistant pathways for NADH oxidation existed. One was sensitive to chelating agents and stimulated by calcium, this was equated with the external dehydrogenase, and the other was not affected by chelators, which was considered to be the enzyme responsible for catalyzing rotenone-resistant oxidation of endogenous NADH. MARX and BRINKMAN (1978) also reported evidence for a rotenone-resistant NADH dehydrogenase capable of oxidizing endogenous NADH which they considered to be different from the external enzyme. They published evidence that the temperature dependency of the two rotenone-resistant dehydrogenases were different (MARX and BRINKMAN 1979). In a recent paper WISKICH and DAY (1982) also presented evidence which suggests that there is a rotenone-resistant NADH dehydrogenase capable of oxidizing endogenous NADH.

3.1 Rotenone-Sensitive Oxidation of Endogenous NADH

3.1.1 Redox Components Associated with the Dehydrogenase

There are similarities between this enzyme and the rotenone-sensitive NADH dehydrogenase present in mammalian mitochondria (RAGAN 1980). The principal differences between this enzyme and the external NADH dehydrogenase are that it is coupled to the first site of oxidative phosphorylation and is sensitive to inhibition by rotenone and amytal. The plant dehydrogenase is less sensitive to rotenone than the mammalian enzyme, requiring about 15 nmol rotenone mg^{-1} protein (Chauveau and Lance 1971) to cause complete inhibition, whilst the mammalian mitochondria require only 30 pmol mg^{-1} protein (Ernster et al. 1963). In contrast to the mammalian enzyme, the plant enzyme has not been isolated and few definitive data are available concerning the individual redox components involved. The study of flavoproteins and particularly the differentiation between different types of flavoproteins and iron-sulfur centers is difficult. Storey (1971) has examined the flavoproteins associated with plant mitochondria and has recognized five different types, only one of which, a low potential ($E_{m_{(7.2)}} = 155$ mV) fluorescent flavoprotein, follows the redox state of the endogenous pyridine nucleotide pool. This flavoprotein is designated FP_{lf} and is probably involved in the oxidation of α-lipoate. In earlier studies Storey described FP_M, which was a pool of flavin reduced by malate, however, in a recent review Storey (1980) discusses the difficulty in identifying flavoproteins and it is no longer clear which flavin is reduced by internal NADH.

In studies on yeast mitochondria, it has become clear that there is a close relationship between the operation of the first site of ATP synthesis and sensitivity to inhibition by rotenone and piericidin A with the involvement of iron-sulfur centers in the dehydrogenase system (Palmer and Coleman 1974). Since it is known that the rotenone-sensitive dehydrogenase is coupled to the first site of oxidative phosphorylation (Baker and Lieberman 1962) and sensitive to inhibition to piericidin A (Brunton and Palmer 1973), it seems likely that iron-sulfur centers are involved. Cammack and Palmer (1977) studied the iron-sulfur centers in *Arum* mitochondria and published evidence for three components associated with the NADH dehydrogenase; these were center N-1b with $E_m = -240$ mV, center N-2 with $E_m = -110$ mV and a more complex signal believed to be from center N-3 plus N-4 with an $E_m = -275$ mV. Similar results have been found by Rich and Moore (see Storey 1980). It is believed that the iron-sulfur components of the internal NADH dehydrogenase feed electrons directly into the ubiquinone pool, from where they can gain access to either the cyanide sensitive or cyanide resistant oxidase.

3.1.2 Regulation of Electron Flow Through the Rotenone-Sensitive Dehydrogenase

Møller and Palmer (1982) measured the K_m of the rotenone-sensitive dehydrogenase for NADH to be 8 µM, which is lower than for any other NADH dehydrogenase associated with the inner membrane. It is clear that all NAD^+-linked substrates that are oxidized in the matrix space can contribute electrons for oxidation via this internal NADH dehydrogenase. Brunton and Palmer (1973), studying wheat mitochondria, observed that the oxidation of both malate, in the presence of oxaloacetate, and pyruvate supplied electrons which were almost exclusively oxidized via the rotenone sensitive NADH dehydrogenase.

The precise mechanisms by which the operation of the dehydrogenase is regulated are unclear. Because the dehydrogenase is coupled to the first site of oxidative phosphorylation, it is certain that the magnitude of the proton motive force across the inner membrane will have a strong regulatory influence, a fact which is reflected in numerous reports that rotenone will cause stronger inhibition under State 3 conditions, when the proton motive force is less inhibitory, than under State 4 conditions, when the proton motive force prevents the operation of the dehydrogenase.

Evidence exists to support the view that the maximum rate of oxidation of NAD^+-linked substrates cannot be attained in the presence of uncoupling agents unless AMP is added. Under these conditions AMP stimulates the simultaneous oxidation of NADH and reduction of cytochrome b in a rotenone-sensitive manner, suggesting that AMP relieved a rate-limiting step associated with the rotenone-sensitive NADH dehydrogenase (SOTTHIBANDHU and PALMER 1975). The AMP activation occurred in the presence of atractyloside or bongkrekic acid, ruling out the possibility that the AMP gained entry to the matrix space using the ADP/ATP exchange system. JUNG and HANSON (1975) and ABOU-KHALIL and HANSON (1977) have described the presence of an additional ADP or ATP/P_i exchange process in plant mitochondria which results in the net accumulation of adenine nucleotides. This process is not reported to transport AMP and is dependent on a proton motive force to provide the energy necessary and would not be operational in the presence of uncouplers. Therefore it is unlikely that AMP was entering the matrix space and stimulating the activity of NAD^+-linked dehydrogenases or acting as a phosphate acceptor for residual oxidative phosphorylation, resistant to weak acid uncouplers. These observations form the basis for a meaningful control mechanism by which the AMP concentration in the cytosol regulates the flow of electrons through the phosphorylating, rotenone-sensitive, NADH dehydrogenase (PALMER and MØLLER 1982).

3.2 Rotenone-Resistant Oxidation of Endogenous NADH

Little is understood concerning the characteristics of this pathway. IKUMA and BONNER (1967) first studied the action of rotenone and showed that it caused only a partial inhibition of oxidation of NAD^+-linked Krebs cycle acids, whilst amytal caused a much greater inhibition. Rotenone is known to be a very specific inhibitor of the NADH dehydrogenase, whilst amytal is known to inhibit a broader spectrum of reactions (ERNSTER et al. 1963). BRUNTON and PALMER (1973) were the first to obtain experimental data consistent with the presence of a rotenone-resistant dehydrogenase, inaccessible to external NADH, and capable of oxidizing internal NADH. When studying this enzyme, great care must be taken to distinguish it from the external dehydrogenase. MARX and BRINKMAN (1978, 1979) also found evidence for a rotenone-resistant dehydrogenase capable of oxidizing endogenous NADH, which they showed to be different from the external dehydrogenase on the basis of response to temperature. The

concept of an internal rotenone-resistant dehydrogenase has recently been rapidly gaining ground and together with the nonphosphorylating, cyanide-resistant
oxidase, provides a nonphosphorylating pathway for the reoxidation of NADH
(see PALMER 1979, WISKICH and DAY 1982, LAMBERS 1980).

Little is known concerning the nature of the redox components involved
in rotenone-resistant internal dehydrogenase. BRUNTON and PALMER (1973)
showed that the ADP/O ratio decreased when piericidin A was added and
concluded that the dehydrogenase by-passed the first site of ATP synthesis.
The lack of sensitivity to rotenone and piericidin A and lack of coupling at
site I phosphorylation suggests that iron-sulfur proteins play no role. BRUNTON
and PALMER (1973) considered that both the rotenone-sensitive and -resistant
pathways might be catalyzed by the same flavoprotein. However, more recent
work by MØLLER and PALMER (1982) has determined that the K_m for NADH
of the rotenone resistant component is 80 µM, which is ten times greater than
for the rotenone-sensitive dehydrogenase. Such a difference in the relative affinity of the two dehydrogenases for NADH suggests that the same flavoprotein
is not likely to be involved.

3.2.1 Relationship to the Terminal Oxidases

The relationship between the rotenone-resistant internal NADH dehydrogenase
and the two terminal oxidases is complex. PALMER (1979) has suggested that
it would be physiologically desirable to couple electron flux through the nonphosphorylating NADH dehydrogenase with the cyanide-resistant oxidase,
thereby providing a completely nonphosphorylating route for the oxidation
of pyridine nucleotides. Such a system would be able to re-oxidize pyridine
nucleotides under conditions of high phosphorylation potential, enabling mitochondrial metabolism to supply carbon precursors for anabolic metabolism,
without being constrained by a high proton motive force. LIPS and BIALE (1966)
first obtained experimental evidence that electrons passing through the rotenone-
resistant dehydrogenase may be preferentially made available to the cyanide-
resistant oxidase. CAMMACK and PALMER (1973) provided further evidence for
the operation of a totally nonphosphorylating pathway of endogenous NADH
oxidation when they showed that the induction of the cyanide-resistant oxidase,
by washing thin slices of potato tuber tissue, was accompanied by an increase
in the piericidin A-resistant NADH dehydrogenase. More recently, RUSTIN et al.
(1980) have concluded that there is a preferential kinetic relationship between
the rotenone-resistant NADH dehydrogenase and the cyanide-resistant oxidase,
a pathway which is particularly active when malate was supplied as the substrate.
GARDESTRÖM and EDWARDS (1983) have also obtained evidence that mitochondria isolated from the leaves of C_4 plants oxidize malate preferentially via the
cyanide-resistant oxidase, thus adding weight to the desire to establish that
there is a favorable kinetic relationship between the rotenone-resistant dehydrogenase and cyanide-resistant oxidase. However, WISKICH and DAY (1982)
have investigated such a possibility and concluded that there was insufficient
evidence to support the view that there was a direct relationship between the
rotenone-resistant dehydrogenase and cyanide-resistant oxidase.

3.2.2 Relationship to the NAD^+-Linked Krebs Cycle Dehydrogenases

The relationship between the rotenone resistant NADH dehydrogenase and the Krebs cycle NAD^+-linked dehydrogenases, and malate oxidizing enzymes in particular, has proved particularly puzzling and remains incompletely understood. BRUNTON and PALMER (1973) carried out the first analysis of how different acids were oxidized by the piericidin A-resistant dehydrogenase. They observed that, although all substrates could be oxidized via the piericidin A-resistant NADH dehydrogenase, the rate of oxidation of pyruvate, or malate in the presence of oxaloacetate, was very slow in the presence of piericidin A. This led to the conclusion that different donors had different access to the internal NADH dehydrogenase system. DAY and WISKICH (1974, 1978) argued that there was no need to implicate differential access to the two dehydrogenases to explain the kinetic data obtained by BRUNTON and PALMER (1973), and suggested that all the data could be accounted for by the influence of oxaloacetate. The role of oxaloacetate in regulating the contribution made by the two internal NADH dehydrogenases was proposed by BRUNTON and PALMER (1973) and PALMER and ARRON (1976), who showed that when studying malate oxidation, if oxaloacetate was allowed to accumulate, then piericidin A caused almost complete inhibition for 1–2 min, followed by recovery of rotenone insensitivity. If oxaloacetate was removed, then piericidin A caused only a partial inhibition. LANCE et al. (1967) had already reported a similar occurrence when avocado mitochondria attained the State 4 condition. As soon as the ADP was exhausted, the respiratory rate was very low and oxaloacetate high, when the oxaloacetate level decreased and the velocity of State 4 eventually increased as the nonphosphorylating dehydrogenase (rotenone-resistant) became activated. TOBIN et al. (1980) showed that the addition of rotenone to mitochondria oxidizing malate and accumulating oxaloacetate immediately stopped oxygen uptake and diverted the NADH produced by the malic enzyme to the malate dehydrogenase where it reduced the oxaloacetate to malate (see WISKICH and DRY Chap. 10, this Vol. for details of malate oxidation). MØLLER and PALMER (1982) measured the relative K_m for both internal NADH dehydrogenases toward NADH; they found a value of 8 µM for the rotenone-sensitive and 80 µM for the rotenone-resistant dehydrogenase. PALMER et al. (1982) re-examined factors that regulate electron flux through the two internal dehydrogenases. They concluded that since the equilibrium constant for the malate dehydrogenase was 2×10^{-12} and that the enzyme was present in ten times larger amounts than was necessary to saturate the respiratory chain, it was reasonable to assume that the malate dehydrogenase reaction was always in equilibrium, and that unless the oxaloacetate was continually removed, all the NADH oxidized came from the malic enzyme. If oxaloacetate was allowed to accumulate, then the level of NADH would be low and only the high-affinity, rotenone-sensitive dehydrogenase would operate. If NADH oxidation was prevented (e.g., by adding rotenone) then the NADH produced by the malic enzyme would become available to reverse the malate dehydrogenase and remove the oxaloacetate, thus allowing the concentration of NADH to rise and engage the rotenone-resistant dehydrogenase (PALMER and MØLLER 1982). An important consequence of the need for a low concentration of oxaloacetate and high concentration of NADH to engage the rotenone-resistant, nonphosphorylating, NADH dehydrogenase is that the concentration of oxaloacetate falls below the concentration necessary for the synthesis of citrate. Thus acetyl-CoA cannot be oxidized under such conditions, and therefore the rotenone-resistant dehydrogenase can only be active in oxidizing NADH produced by the isocitrate, α-ketoglutarate dehydrogenases and malic enzyme. This emphasizes that the main role of this NADH dehydrogenase is likely to be in enabling the Krebs cycle to provide carbon precursors for biosynthetic processes under conditions when ATP synthesis does not occur.

There are, however, reasons to believe that regulation via the level of oxaloacetate cannot completely replace the concept of compartmentation of the NADH pool. Brunton and Palmer (1973, Fig. 5) showed that when mitochondria became anaerobic whilst metabolizing malate, the pyridine nucleotide pool became reduced in a bi-phasic manner. The first pool became reduced in the presence of oxaloacetate whilst the second pool became reduced only after the oxaloacetate had been removed. Goonewardena and Wilson (1979) also produced evidence that the pyridine nucleotide acts as if it were organized in at least two separate pools. Rustin et al. (1980) came to a similar conclusion, and finally Gardeström and Edwards (1983) also suggest that different NAD^+-linked dehydrogenases produce NADH that can only be oxidized by one of the internal pathways. Therefore, in conclusion, further studies need to be undertaken to prove whether the endogenous NADH produced by different Krebs cycle acids has differential access to different dehydrogenases. It is unlikely that the NAD^+ pool is physically compartmented, but it is possible to speculate that there is a precise spatial relationship between different enzymes producing NADH and the enzymes responsible for the oxidation of that NADH.

References

Abou-Khalil S, Hanson JB (1977) Net adenosine diphosphate accumulation in mitochondria. Arch Biochem Biophys 183:581–587

Archokov AI, Karyakin AV, Skulachev VP (1974) Intermembrane electron transfer in mitochondrial and microsomal systems. FEBS Lett 39:239–242

Arron GP, Edwards GE (1979) Oxidation of reduced adenine dinucleotide phosphate by plant mitochonria. Can J Biochem 57:1392–1394

Arron GP, Edwards GE (1980) Oxidation of reduced nicotinamide adenine dinucleotide phosphate by potato mitochondria. Inhibition by sulfhydryl reagents. Plant Physiol 65:591–594

Bäckström D, Hoffström I, Gustafsson I, Ehrenberg A (1973) An iron-sulphur protein in the mitochondrial outer membrane, reducible by NADH and NADPH. Biochem Biophys Res Commun 53:596–602

Bäckström D, Lorusso M, Anderson K, Ehrenberg A (1978) Characterisation of the iron-sulphur protein of the mitochondrial outer membrane partially purified from beef kidney cortex. Biochim Biophys Acta 502:276–288

Baker JE, Lieberman M (1962) Cytochrome components and electron transfer in sweet potato mitochondria. Plant Physiol 37:90–97

Beevers H (1954) The oxidation of reduced diphosphopyridine nucleotide by an ascorbate system from cucumber. Plant Physiol 29:265–269

Berger J, Avery GS (1943) Dehydrogenases of the *Avena* coleoptile. Am J Bot 30:290–297

Bergman A, Ericson I (1983) Efects of pH, NADH, succinate and malate on the oxidation of glycine in spinach leaf mitochondria. Physiol Plant 59:421–427

Bhagvat K, Hill R (1951) Cytochrome oxidase in higher plants. New Phytol 50:112–120

Bonner WD, Voss DO (1961) Some characteristics of mitochondria extracted from higher plants. Nature (London) 191:682–684

Boswell JG, Whiting GC (1940) Oxidase systems in the tissues of higher plants. New Phytol 39:241–265

Brummond DO, Burris RH (1953) Transfer of C^{14} by lupin mitochondria through reactions of the tricarboxylic acid cycle. Proc Natl Acad Sci USA 39:754–759

Brunton CJ, Palmer JM (1973) Pathways for the oxidation of malate and reduced pyridine nucleotide by wheat mitochondria. Eur J Biochem 39:283–291

Cammack R, Palmer JM (1973) EPR studies of iron-sulphur proteins of plant mitochondria. Ann NY Acad Sci 222:816–823

Cammack R, Palmer JM (1977) Iron-sulphur centres in mitochondria from *Arum maculatum* spadix with very high rates of cyanide resistant respiration. Biochem J 166:347–355

Chance B, Theorell B (1959) Localisation and kinetics of reduced pyridine nucleotides in living cells by microfluorimetry. J Biol Chem 234:3044–3050

Chance B, Bonner WD, Storey BT (1968) Electron transport in respiration. Annu Rev Plant Physiol 19:295–320

Chauveau M, Lance C (1971) Mitochondria of cauliflower influorescences. 2. Effects of electron transport inhibitors. Physiol Veg 9:353–359

Coleman JOD, Palmer JM (1971) Role of Ca^{2+} in the oxidation of exogenous NADH by plant mitochondria. FEBS Lett 17:203–208

Conn E, Vennesland B, Kraemer LM (1949) Distribution of a triphosphopyridine nucleotide specific enzyme catalysing the reversible oxidative decarboxylation of malic acid in higher plants. Arch Biochem 23:179–197

Cook ND, Cammack R (1983) Effects of the quinone analogue 5-n-undecyl-6-hydroxy-4,7-dioxobenzothiazole (UHDBT) on cyanide-sensitive and cyanide-insensitive plant mitochondria. Biochem Soc Trans 11:785

Cottingham IR, Moore AL (1983) Ubiquinone pool behavior in plant mitochondria. Biochim Biophys Acta 724:191–200

Cowley RC, Palmer JM (1978) The interaction of citrate and calcium in regulating the oxidation of exogenous NADH in plant mitochondria. Plant Sci Lett 11:345–350

Cowley RC, Palmer JM (1980) The interaction between exogenous NADH oxidase and succinate oxidase in Jerusalem artichoke (*Helianthus tuberosus*) mitochondria. J Exp Bot 31:199–207

Crane FL (1957) Electron transport and cytochromes of sub-cellular particles from cauliflower buds. Plant Physiol 32:619–625

Cunningham WP (1964) Oxidation of externally added NADH by isolated corn root mitochondria. Plant Physiol 39:699–703

Davies DD (1953) The Krebs cycle enzyme system of pea seedlings. J Exp Bot 4:173–183

Day DA, Wiskich JT (1974) The effect of exogenous nicotinamide adenine dinucleotide on the oxidation of NAD-linked substrates by isolated plant mitochondria. Plant Physiol 54:360–363

Day DA, Wiskich JT (1975) Isolation and properties of the outer membrane of plant mitochondria. Arch Biochem Biophys 171:117–123

Day DA, Wiskich JT (1977) Factors limiting respiration by isolated cauliflower mitochondria. Phytochemistry 16:1499–1502

Day DA, Wiskich JT (1978) Pyridine nucleotide interactions with isolated plant mitochondria. Biochim Biophys Acta 501:396–404

Day DA, Rayner JR, Wiskich JT (1976) Characteristics of external NADH oxidation by beetroot mitochondria. Plant Physiol 58:38–42

Day DA, Arron GP, Laties GG (1979) Enzyme distribution in potato mitochondria. J Exp Bot 30:539–549

Douce R, Christensen EL, Bonner JR WD (1972) Preparation of intact plant mitochondria. Biochim Biophys Acta 275:148–160

Douce R, Manella CA, Bonner WD Jr (1973) The external NADH dehydrogenases of intact plant mitochondria. Biochim Biophys Acta 292:105–116

Dry IB, Day DA, Wiskich JT (1983) Preferential oxidation of glycine by the respiratory chain of pea leaf mitochondria. FEBS Lett 158:154–158

Dubuy HG, Woods MW, Lackey MD (1950) Enzymatic activities of isolated normal and mutant mitochondria and plastids of higher plants. Science 111:572–574

Ernster L, Dallner G, Azzone GF (1963) Differential effects of rotenone and amytal on mitochondrial electron and energy transfer. J Biol Chem 238:1124–1131

Gardeström P, Edwards GE (1983) Isolation of mitochondria from leaf tissue of *Panicum miliaceum*, a NAD-malic enzyme type C_4 plant. Plant Physiol 71:24–29

Goonewardena H, Wilson SB (1979) The oxidation of malate by isolated turnip (*Brassica napus* L.) mitochondria. III The effects of inhibitors. J Exp Bot 30:889–903

Hackett DP (1955a) Recent studies on plant mitochondria. Int Rev Cytol 4:143–196

Hackett DP (1955b) A pathway of terminal oxidation in potato mitochondria. Plant Physiol 30

Hackett DP (1957) Respiratory mechanisms in the aroid spadix. J Exp Bot 8:157–171

Hackett DP (1961) Effect of salts on NADH oxidase activity and structure of sweet potato mitochondria. Plant Physiol 36:445–452

Hackett DP (1963) Respiratory mechanisms and control in the higher plant tissues. In: Wright B (ed) Control mechanisms in respiration and fermentation. Ronald, New York, pp 105–127

Harmey MA, Ikuma H, Bonner WD (1966) Near ultra violet spectrum of white potato mitochondria. Nature (London) 209:174–175

Hasson EP, West CA (1971) Properties of a higher plant pyridine nucleotide transhydrogenase. Fed Proc Am Soc Exp Biol 30:1189 Abs

Hawker JS, Laties GG (1963) Nicotinamide adenine dinucleotide in potato tuber slices in relation to respiratory changes with age. Plant Physiol 38:498–500

Hill R, Bhagvat K (1939) Cytochrome oxidase from flowering plants. Nature (London) 143:726

Hodges TK, Hanson JB (1967) Energy-linked reactions of plant mitochondria. Curr Top Bionenerg 2:65–98

Humphreys TE, Conn EE (1956) Oxidation of reduced diphosphopyridine nucleotide by lupin mitochondria. Arch Biochem Biophys 60:226–243

Huq S, Palmer JM (1978) The involvement and possible role of quinone in cyanide-resistant respiration. In: Ducet G, Lance C (eds) Plant mitochondria. Elsevier/North Holland Biomedical Press, Amsterdam, New York, pp 225–232

Ikuma H (1972) Electron transport in plant respiration. Annu Rev Plant Physiol 23:419–436

Ikuma H, Bonner WD (1967) Properties of higher plant mitochondria. III Effects of respiratory chain inhibitors. Plant Physiol 42:1535–1544

Jagow von G, Klingenberg M (1970) Pathways of hydrogen in mitochondria von *Saccharomyces carlsbergensis*. Eur J Biochem 12:583–592

James WO (1953) The terminal oxidases of plant respiration. Biol Rev 28:245–260

Johnston SP, Møller IM, Palmer JM (1979) The stimulation of exogenous NADH oxidation in Jerusalem artichoke mitochondria by screening of charges on the membranes. FEBS Lett 108:28–32

Jung DW, Hanson JB (1975) Activation of 2,4-dinitrophenol-stimulated ATP-ase activity in cauliflower and corn mitochondria. Arch Biochem Biophys 168:358–368

Kenefick DG, Hanson JB (1966) The contracted state as an energy source for Ca^{2+} binding and Ca^{2+} inorganic phosphate accumulation by corn mitochondria. Plant Physiol 41:1601–1609

Koeppe DE, Miller RJ (1972) Oxidation of reduced nicotinamide adenine dinucleotide phosphate by isolated corn mitochondria. Plant Physiol 49:353–357

Kubowitz F (1938) Spaltung und Resynthese der Polyphenoloxydase und des Hämocyanins. Biochem Z 299:32–57

Lambers H (1980) The physiological significance of cyanide-resistant respiration in higher plants. Plant Cell Physiol 3:293–303

Lance C, Bonner WD (1968) The respiratory chain components of higher plant mitochondria. Plant Physiol 43:756–766

Lance C, Hobson GE, Young RE, Biale JB (1967) Metabolic processes in cytoplasmic particles of the avocado fruits. IX The oxidation of pyruvate and malate during the climacteric cycle. Plant Physiol 42:471–478

Laties GG (1953) The dual role of adenylate in the mitochondrial oxidations of the higher plant. Physiol Plant 6:199–214

Lehninger AL (1951) Phosphorylation coupled to oxidation of dihydrodiphosphopyridine nucleotides. J Biol Chem 190:345–359

Lehninger AL (1955) Oxidative phosphorylation. Harvey Lect Ser 49:176–215

Levy H, Schade AL (1948) Terminal oxidase systems of potato tuber respiration. Arch Biochem 19:273–286

Lieberman M, Baker JE (1965) Respiratory electron transport. Annu Rev Plant Physiol 16:343–382

Lips SH, Biale JB (1966) Stimulation of oxygen uptake by electron transfer inhibitors. Plant Physiol 41:797–802

Lockhart EE (1939) Diaphorase (coenzyme factor). Biochem J 33:613–617

Martin EM, Morton RK (1956) Enzymatic properties of microsomes and mitochondria from silver beet. Biochem J 62:696–704

Marx R, Brinkman K (1978) Characteristics of rotenone-insensitive oxidation of matrix NADH by broad-bean mitochondria. Planta 142:83–90

Marx R, Brinkman K (1979) Effect of temperature on the pathways of NADH oxidation in broad-bean mitochondria. Planta 144:359–365

Miller RJ, Dumford SW, Koeppe DE, Hanson JB (1970) Divalent cation stimulation of substrate oxidation by corn mitochondria. Plant Physiol 45:649–653

Millerd A, Bonner J, Axelrod B, Bandurski R (1951) Oxidative and phosphorylative activity of plant mitochondria. Proc Natl Acad Sci USA 37:855–862

Møller IM, Palmer JM (1981a) Properties of the oxidation of exogenous NADH and NADPH by plant mitochondria. Evidence against a phosphatase or nicotinamide nucleotide transhydrogenase being responsible for NADPH oxidation. Biochim Biophys Acta 681:225–233

Møller IM, Palmer JM (1981b) The inhibition of exogenous NAD(P)H oxidation in plant mitochondria by chelators and mersalyl as a function of pH. Physiol Plant 53:413–420

Møller IM, Palmer JM (1981c) Charge screening by cations affects the conformation of the mitochondrial inner membrane. A study of exogenous NAD(P)H oxidation in plant mitochondria. Biochem J 195:583–588

Møller IM, Palmer JM (1982) Direct evidence for the presence of a rotenone-resistant NADH dehydrogenase on the inner surface of the inner membrane of plant mitochondria. Physiol Plant 54:267–274

Møller IM, Johnston SP, Palmer JM (1981) A specific role for Ca^{2+} in the oxidation of exogenous NADH by Jerusalem artichoke (*Helianthus tuberosus*) mitochondria. Biochem J 194:487–495

Møller IM, Palmer JM, Johnston SP (1983) Inhibition of exogeous NADH oxidation in plant mitochondria by chlorotetracycline in the presence of calcium ions. Biochim Biophys Acta 125:289–291

Moore AL, Ackerman KEO (1982) Ca^{2+} stimulation of the external NADH dehydrogenase in Jerusalem artichoke (*Helianthus tuberosus*) mitochondria. Biochem Biophys Res Commun 109:513–517

Moore AL, Rich PR (1980) The bioenergetics of plant mitochondria. Trends Biochem Sci 5:284–288

Moore AL, Linnett PE, Beechey RB (1980) Dibutylchloromethyltin chloride, a potent inhibitor of electron transport in plant mitochondria. J Bioenerg Biomembr 12:309–322

Moreau F (1978) The electron transport system of the outer membranes of plant mitochondria. In: Ducet G, Lance C (eds) Plant mitochondria. Elsevier/North Holland Biomedical Press, Amsterdam New York, pp 77–84

Moreau F, Lance C (1972) Isolément et propriétés des membranes externes et internes de mitochondries végétales. Biochemie 54:1335–1348

Nash D, Wiskich JT (1983) Properties of substantially chlorophyll-free pea leaf mitochondria prepared by sucrose density gradient separation. Plant Physiol 71:627–634

Neuberger M, Douce R (1978) Transport of NAD$^+$ through the inner membrane of plant mitochondria. In: Ducet G, Lance C (eds) Plant mitochondria. Elsevier/North Holland Biomedical Press, Amsterdam New York, pp 109–116

Newcomer EH (1940) Mitochondria in plants. Bot Rev 6:85–147

Ohnishi T, Kawaguchi K, Hagihara B (1966) Preparation and some properties of yeast mitochondria. J Biol Chem 241:1797–1806

Packer L, Murakami S, Mehard CW (1970) Ion transport in chloroplasts and plant mitochondria. Annu Rev Plant Physiol 21:271–304

Palmer JM (1979) The "uniqueness" of plant mitochondria. Biochem Soc Trans 7:246–252

Palmer JM, Arron GP (1976) The influence of exogenous nicotinamide adenine dinucleotide on the oxidation of malate by Jerusalem artichoke mitochondria. J Exp Bot 27:418–430

Palmer JM, Coleman JOD (1974) Multiple pathways of NADH oxidation in the mitochondrion. Horizons Biochem Biophys 1:220–260

Palmer JM, Møller IM (1982) Regulation of NAD(P)H dehydrogenases in plant mitochondria. Trends Biochem Sci 7:258–261

Palmer JM, Passam HC (1971) The oxidation of NADH by plant mitochondria. Biochem J 122:16–17p

Palmer JM, Schwitzguebel JP, Moller IM (1982) Regulation of malate oxidation in plant mitochondria. Response to rotenone and exogenous NAD. Biochem J 208:703–711

Philips ML, Williams GR (1973) Effects of 2-butylmalonate, 2-phenylsuccinate, benzylmalonate and p-iodobenzylmalonate on the oxidation of substrates by mung bean mitochondria. Plant Physiol 51:225–228

Ragan CI (1980) The molecular organisation of NADH dehydrogenase. Subcellular Biochemistry Vol 7. Plenum, New York, pp 267–307

Ravanel P, Tissut M, Douce R (1981) Effects of flavone on the oxidative properties of intact plant mitochondria. Phytochemistry 20:2101–2103

Rich PR, Bendall DS (1975) Cytochrome components of plant microsomes. Eur J Biochem 55:333–341

Rustin P, Moreau F, Lance C (1980) Malate oxidation in plant mitochondria via malic enzyme and the cyanide-insensitive electron transport pathway. Plant Physiol 66:457–462

Rydstrom J (1977) Energy-linked nicotinamide nucleotide transhydrogenases. Biochem Biophys Acta 463:155–184

Schatz G, Racker E, Tyler DD, Gonze J, Estabrook RW (1966) Studies of the DPNH-cytochrome b segment of the respiratory chain of bakers yeast. Biochem Biophys Res Commun 22:585–590

Schneider H, Lemasters JJ, Hochli M, Hackenbrook CR (1980) Liposome-mitochondrial inner membrane fusion. Lateral diffusion of integral electron transfer components. J Biol Chem 255:3748–3756

Smillie RM (1955) Enzymatic activity of particles isolated from various tissues of the pea plant. Aust J Biol Sci 8:186–195

Sottibandhu R, Palmer JM (1975) The activation of non-phosphorylating electron transport by adenine nucleotides in Jerusalem artichoke (Helianthus tuberosus) mitochondria. Biochem J 152:637–645

Sottocasa GL, Kuylenstierna B, Ernster L, Bergstrand A (1967) An electron transport system associated with the outer membrane of liver mitochondria. J Cell Biol 32:415–438

Storey BT (1971) The respiratory chain of plant mitochondria. X Oxidation-reduction potentials of the flavoproteins of skunk cabbage mitochondria. Plant Physiol 48:493–497

Storey BT (1980) Electron transport and energy coupling in plant mitochondria. In: Davies D (ed) The biochemistry of plants, vol II. A comprehensive treatise. Academic Press, London New York, pp 125–197

Strittmatter P (1963) Microsomal cytochrome b_5 and cytochrome b_5 reductase. In: Boyer PD, Lardy H, Myrbäck K (eds) The enzymes, vol VIII. Academic Press, London New York, pp 113–145

Strittmatter P (1966) NADH-cytochrome b_5 reductase. In: Slater EC (ed) Flavins and flavoproteins. Elsevier, Amsterdam, pp 325–329

Theorell H (1935) Reines Cytochrom C. Vorläufige Mitteilung. Biochem Z 279:463–464

Tobin A, Djerdjour B, Journet E, Neuberger M, Douce R (1980) Effect of NAD^+ on malate oxidation in intact plant mitochondria. Plant Physiol 66:225–229

Tomlinson PF, Moreland DE (1975) Cyanide-resistant respiration of sweet potato mitochondria. Plant Physiol 55:365–369

Warburg O, Christian W (1932) Über ein neues Oxydationsferment und sein Absorptionsspektrum. Biochem Z 254:438–458

Whatley FR (1951) Coenzymes in plants. New Phytol 50:244–257

Wilson RH, Hanson JB (1969) The effect of respiratory inhibitors on NADH, succinate and malate oxidation in corn mitochondria. Plant Physiol 44:1335–1341

Wilson SB, Bonner WD (1970) Preparation and some properties of submitochondrial particles from tightly coupled mung bean mitochondria. Plant Physiol 46:25–30

Wiskich JT, Day DA (1982) Malate oxidation, rotenone resistance and alternative path activity in plant mitochondria. Plant Physiol 70:959–964

Zeltich I, Barber GA (1960) Oxidative phosphorylation and glycolate oxidation by particles from spinach leaves. Plant Physiol 35:205–209

8 The Cyanide-Resistant Pathway of Plant Mitochondria

C. LANCE, M. CHAUVEAU, and P. DIZENGREMEL

1 Introduction

Plant respiration has long been described to be resistant to cyanide, since the first observation was made in 1929 by GENEVOIS on sweet pea (*Lathyrus odoratus*) seedlings. Soon after, VAN HERK and BADENHUIZEN (1934) and VAN HERK (1937a, b, c) showed that the respiration of the spadix of the *Sauromatum guttatum* was highly resistant to cyanide. Respiration in this group of plants (Araceae) is known to be extremely high and linked to heat production, particularly during pollination (see LANCE 1972, MEEUSE 1975). In 1939, OKUNUKI observed that the respiration of Sauromatum pollen was resistant to carbon monoxide, and a similar observation was also made by MARSH and GODDARD (1939) on carrot leaves. These pioneering works established the concept that plant respiration differed from that of animals by its behavior toward respiratory inhibitors. Cyanide could even stimulate respiration, as in potato tubers (HANES and BARKER 1931).

In 1955, JAMES and ELLIOTT, with *Arum maculatum* spadix, brought the demonstration that the site of cyanide resistance was located in the mitochondria. The coexistence of two electron transfer pathways in plant mitochondria was first indicated by OKUNUKI (1939) and then by YOCUM and HACKETT in 1957. Since then, many papers have been published, dealing mainly with the occurrence, characterization, and functioning of this particular pathway which, after some problems of semantics (SIEDOW 1982), became universally known as the cyanide-resistant pathway or the alternative pathway, thus indicating that electrons are offered a choice of two pathways to reach molecular oxygen.

As progress took place, it became evident that the alternative pathway was not associated with the production of ATP. Apparently, therefore, this pathway was useless and wasteful. The crucial question was then asked: what were the distribution of electrons between the two pathways and its incidence on the regulation of energy metabolism in plants?

A breakthrough in the field occurred in 1971 when SCHONBAUM et al. discovered that substituted hydroxamic acids were specific inhibitors of the alternative pathway, just as cyanide or carbon monoxide were specific inhibitors of the main respiratory chain. With two specific tools available it became possible to tackle the question of electron distribution between the two pathways.

Fifty years after its discovery, cyanide resistance in plant respiration has become a widely accepted concept. Its main features have been reviewed periodically. An exhaustive review on the state of the question was produced by HENRY and NYNS in 1975. Over the years comprehensive or more specific reviews have appeared which should be referred to for a complete overview of the subject (HACKETT 1959, BONNER 1965, IKUMA 1972, MEEUSE 1975, PALMER 1976, SOLO-MOS 1977, DAY et al. 1980, LAMBERS 1980, STOREY 1980, LANCE 1981, DIZENGRE-MEL and LANCE 1982, DUCET 1982a, LAMBERS 1982, LATIES 1982, SIEDOW 1982).

There are many facets to the problem of cyanide resistance in plants. Cyanide resistance in plant tissues or mitochondria is related to the transport and oxidation of substrates, to the composition and organization of the electron carriers in the mitochondrial membrane, to the process of energy conservation, and to the regulation of energy metabolism within the cell. All these points will receive specific and extensive coverage in several Chapters of this Volume. Particularly, the significance of cyanide resistance in plant tissues and organs will be dealt with in details by LAMBERS (Chap. 14, this Vol.). For these reasons, it seemed advisable to restrict this topic exclusively to the characterization and operation of the alternative pathway in isolated mitochondria, i.e., outside the regulatory context of the cell. Even with such a restricted field one has to deal with several hundreds of published papers. It was therefore outside the scope of this review to cover the subject exhaustively, and we have to apologize for not mentioning all the papers that have contributed to the progress in this field. Finally, considering isolated mitochondria only, if one limits the coverage of the alternative pathway to what is known, thought to be known, believed, and unknown, one ends up with ample material rather to raise many more questions than to bring final answers.

2 The Measure of Cyanide Resistance

One of the most delicate points concerning cyanide resistance is the measure of the extent of this resistance. When plant mitochondria are given KCN (in the mM range) while oxidizing a respiratory substrate, they are deemed to be cyanide-resistant if the rate of O_2 uptake is not brought to nil, and percent inhibition is computed from the oxidation rates before and after cyanide addition. It is evident that percent inhibition is a parameter that should be used with caution. It will depend on the efficiency of cyanide inhibition of electron transport, on the rate of oxidation prior to cyanide addition and, as will be seen in the next section, on the nature of the substrate being oxidized.

A necessary condition to demonstrate cyanide resistance is that cyanide should be present, and there are in the literature (particularly on organs or tissues) a number of reports of cyanide resistance due to the mere fact of cyanide being absent. Before the technique of measurement of O_2 uptake by the O_2 electrode was introduced, all studies on mitochondrial (and tissue slices) oxidations were carried out using the manometric technique of Warburg. Rather long reaction times (10–30 min) were needed in order to observe significant measurements, compared to an average time of 1 min or so with the O_2 electrode. At usual experimental pH's ($\simeq 7$), cyanide, whose pK is 8.9, is mainly present as its protonated form. Therefore it escapes from the reaction medium as volatile HCN, a situation which is not encountered when using NaN_3 (pK

4.7). Therefore, with the manometric technique, one can only trust the results when the method of Robbie (1948) has been used. With this method, constant concentrations of cyanide are maintained in the medium by placing appropriate mixtures of KCN-KOH or $Ca(CN)_2 - Ca(OH)_2$ in the center well of the Warburg flask. Care should also be taken that cyanide is not eliminated in the presence of ketoacids or sugars through cyanhydrin formation (Laties 1982).

Such a drawback is not encountered with the O_2 electrode, which allows rapid measurements so that the cyanide concentration does not significantly change during the course of the experiment. However, in that case, one faces another danger: the measurement is so fast that it may occur that cyanide has not enough time to produce its full effect. This occurs when too low concentrations of cyanide are used (Chauveau 1976a). In that case, the inhibition is progressive and should be measured several minutes after the establishment of a constant rate. Similarly, incubation of mitochondria for several minutes (3–5 min) in the presence of a low concentration of cyanide yields higher inhibition values than the same concentration added during the course of an experiment (Chauveau 1976a). With plant mitochondria, 1 mM KCN is usually the standard concentration used to observe an immediate and full effect.

The use of percent inhibition values to compare the cyanide resistance of different materials has generally no great meaning unless rigorous experimental conditions are defined. First of all, they will depend on the substrate being oxidized. But, for the same substrate, they will also depend on the rate of oxidation prior to cyanide addition (substrate State, State 3, State 4)[1]. State 3 rate (in the presence of ADP) is always much more strongly inhibited than State 4 rate (in the absence of ADP), but in both instances the rate in the presence of cyanide remains the same. Therefore, the absolute value (nmol O_2 min^{-1} mg^{-1} protein) of the fraction of electron transport that is resistant to cyanide has greater significance, no matter what percentage of the rate prior to cyanide addition it represents. As a consequence, if one wants to compare the cyanide resistance of mitochondria from different origins, some safeguards must be provided. The first of these is the knowledge of the maximal rate of electron transport of the mitochondria under studies. This is an important point: for the same substrate the percent inhibition will decrease if the oxidation rate decreases: this is well illustrated by the different metabolic rates (State 3, State 4, etc.) of mitochondrial oxidations; but these rates can also be modulated by varying the substrate concentration or by adding an inhibitor (malonate in the case of succinate). Any decrease in the rate brings about a decrease in percent inhibition: however, as above mentioned, within certain limits, the rate of the cyanide-resistant electron transport remains constant (Chauveau 1977).

There is no absolute way of knowing the maximal rate of electron transport. However, one may approach this value by releasing all constraints that can impede electron transport: a substrate (succinate) that is actively oxidized by most plant mitochondria should be used; a cocktail of substrates (succinate, malate, NADH, glycine), could also be used, although for reasons of competition for, or saturation of, coenzymes, the rates are not necessarily additive

---

1 Definition of the different states see Chap. 10, this Vol., p. 283

(DAY and WISKICH 1977, LAMBERS et al. 1983); finally, an uncoupler should also be present to release the control exerted by oxidative phosphorylation. Only under such conditions is it possible to measure percent inhibitions adequately and, most of all, the extent of the electron flow that is resistant to cyanide.

An extensive survey of the degrees of cyanide resistance throughout the plant kingdom can be found in HENRY and NYNS (1975). There are very large variations, even when dealing with the same plant material. For instance, for the typical cyanide-resistant mitochondria from *Arum maculatum* spadix, the reported effects of cyanide range between 85% inhibition and 30% stimulation, i.e., the "cyanide resistance" is between 15 and 130% ! (JAMES and ELLIOTT 1955, SIMON 1957, BENDALL 1958, PASSAM and PALMER 1972, WEDDING et al. 1973, 1975, CHAUVEAU 1976a, 1977). Probably the most striking result of these studies is that there are very few plant mitochondria that are 100% cyanide-sensitive. Most of them display some degree of cyanide resistance with some substrate. For instance NADH oxidation can be 100% inhibited by 1 mM cyanide in mitochondria isolated from fresh potato slices (DIZENGREMEL 1975). However, with another substrate such as malate, some degree of resistance will be observed (DIZENGREMEL 1980, RUSTIN et al. 1980). From all the studies on plant mitochondria carried out so far, one gains the impression that cyanide resistance is a general feature of mitochondrial oxidations in plants. In other words, it could be a "constitutive" property of plant mitochondria which could become expressed more or less extensively, depending on internal or external conditions.

3 The Dependence on Respiratory Substrates

Respiratory substrates are oxidized at very different rates by plant mitochondria. Some mitochondria will also oxidize specific substrates: leaf mitochondria, for instance, oxidize glycine (DOUCE et al. 1977, NEUBURGER and DOUCE 1977, JACKSON et al. 1979, DAY and WISKICH 1981). As a rule, succinate and NADH are oxidized at the fastest rates, then comes malate. Pyruvate and α-ketoglutarate are oxidized at much slower rates, probably because of the complexity of the reaction which is dependent on a number of cofactors (NAD^+, thiamine pyrophosphate, etc). Citrate is generally oxidized at a slower rate than malate. All those substrates feed electrons in the respiratory chain at the level of various NADH dehydrogenases for NAD-linked substrates (malate, α-ketoglutarate, etc) or at the level of succinate dehydrogenase. An artificial substrate, ascorbate, generally coupled to TMPD (tetramethyl-p-phenylene diamine), is also used. This substrate yields its electrons to cytochrome c and therefore uses only the terminal part of the respiratory chain (TYLER et al. 1966). Using the old terminology of "phosphorylation sites", quite adequate for the present purpose, the number of sites used by the electron flow is 3 for NAD-linked substrates, 2 for succinate and 1 for ascorbate. The situation is more complex in the case of NADH oxidation (see below).

Table 1. Dependence of cyanide inhibition on the nature of the substrate being oxidized

Plant species	Percent inhibition[a]				Reference
	Malate	Succinate	Ascorbate	NADH	
Arum maculatum	5–56	0–55	80–95	30–55	1, 2, 3, 4, 5, 6, 7
Symplocarpus foetidus	25–45	17–50	86–95	17–50	7, 8, 9, 10, 11
Sauromatum guttatum	51	0–52	92–97	55	7, 12, 13
Zea mays	10–75	80–100	80–100	86–100	14, 31
Ipomea batatas	39	50–63	90–96	87	7, 11, 15, 16
Triticum aestivum	81	78	–	96	17
Cicer arietinum	37	60	79	–	18
Acer pseudoplatanus	51	40	–	90	19
Phaseolus aureus	75–80	64–85	71	88	7, 8, 20, 31
Brassica oleracea	78–90	84–90	80–90	90	21, 22
Helianthus tuberosus	95	95–98		97	22, 23, 24
Solanum tuberosum					
Fresh slices	78–94	86–99	84	86–99	7, 23, 25, 26, 27, 28, 29, 31
Aged slices	47	60–77	82–84	89–94	25, 26, 27, 28, 29, 30

[a] The KCN concentrations used were in the mM range

(1) Bendall 1958, (2) Chauveau 1976a, (3) James and Elliott 1955, (4) Lance 1979, (5) Passam and Palmer 1972, (6) Wedding et al. 1975, (7) Bahr and Bonner 1973a, (8) Bendall and Bonner 1971, (9) Hackett 1957, (10) Hackett and Haas 1958, (11) Wiskich and Bonner 1963, (12) Chauveau 1980, (13) Bonner et al. 1972, (14) Yang et al. 1980, (15) Hackett et al. 1960b, (16) Tomlinson and Moreland 1975, (17) Pomeroy 1975, (18) Burguillo and Nicolas 1977, (19) Wilson 1971, (20) Ikuma and Bonner 1967, (21) Chauveau and Lance 1971, (22) Dizengremel et al. 1973, (23) Dizengremel 1980, (24) Coleman and Palmer 1972, (25) Dizengremel 1975, (26) Dizengremel and Lance 1976, (27) Dizengremel and Lance 1982, (28) Hackett et al. 1960a, (29) Van der Plas and Verleur 1976, (30) Lance and Dizengremel 1978, (31) Neuburger and Douce 1977

A number of data have been gathered in Table 1. They deal with the oxidation of four typical substrates by mitochondria from 12 different plant species. No matter what the plant species, some general remarks can be made. Malate oxidation generally displays the highest degree of resistance. Succinate oxidation, except in corn, Jerusalem artichoke, and fresh potato slices, is also cyanide-resistant, though less so than malate oxidation. By contrast, ascorbate oxidation, which uses cytochrome oxidase only, appears strongly sensitive to cyanide. As to NADH oxidation, it appears to be as sensitive to cyanide as ascorbate oxidation in most mitochondria, except in three instances (*A. maculatum*, *S. foetidus*, *S. guttatum*). This behavior is related to the particular mode of oxidation of NADH in plant mitochondria, which mainly uses a dehydrogenase located on the outer surface of the inner membrane (Von Jagow and Klingenberg 1970, Douce et al. 1973) and to the organization of the respiratory chain in plant mitochondria (Rustin et al. 1980).

It is evident that the degrees of cyanide resistance are highly variable, depending on substrates and plant species (Table 1). As mentioned above, it is difficult to establish which type of mitochondria is the most cyanide-resistant. However, among the plant species listed in Table 1, mitochondria from aroid spadices (*A. maculatum, S. foetidus, S. guttatum*) are deemed to be typically cyanide-resistant. This is also the case with mitochondria from aged potato slices, though to a lesser degree. By contrast, mitochondria from potato tubers or from fresh potato slices are generally considered as the most cyanide-sensitive plant mitochondria. Other types of mitochondria show intermediary levels of cyanide resistance.

A special mention should be made of mitochondria from leaves, which have not been included in Table 1. Only recently has it been possible to eliminate contamination by broken thylakoids, and to obtain rather clean mitochondrial preparations. A general feature of these mitochondria is their rather high cyanide or antimycin resistance (ARRON et al. 1979, RUSTIN et al. 1980, DAY and WISKICH 1981, RUSTIN 1981, GARDESTRÖM and EDWARDS 1983). Generally, malate oxidation is highly resistant, and glycine oxidation is more resistant than succinate oxidation (DOUCE et al. 1977).

From the data of Table 1, one can select several specific features of substrate oxidations in plant mitochondria. Considering the three major substrates (succinate, malate, NADH), these features are: (a) malate oxidation always displays the highest degree of cyanide-resistance; (b) NADH oxidation is strongly cyanide-sensitive. When potato mitochondria become cyanide-resistant upon aging of the slices, succinate and malate oxidations become cyanide-resistant, but NADH oxidation remains very strongly cyanide-sensitive; (c) NADH oxidation is cyanide-resistant only in mitochondria from the aroid spadices. As will be seen later, these features are related to the mechanisms of malate oxidation and to the organization of the respiratory chain.

4 The Inhibition of Electron Transport

Strictly speaking, only the inhibitors of the alternative pathway should be considered here. However, as inhibitors of both the cyanide-resistant and cyanide-sensitive pathways have to be used to study the path of electrons in cyanide-resistant plant mitochondria, an overview of both types of inhibitor will be presented.

According to current knowledge (see TZAGOLOFF 1982), the respiratory chain is thought to consist of four main multienzyme complexes and of two smaller-sized components (ubiquinone and cytochrome c). Two complexes are responsible for NAD-linked substrate (Complex I) and succinate (Complex II) oxidations. Both complexes consist of a flavoprotein associated to several iron-sulfur centers (BEINERT and PALMER 1965); they constitute a "flavoprotein pathway" (CHANCE et al. 1968). Ubiquinone plays a central role as an obligate intermediate carrier for electrons coming from the substrates and on their way to oxygen.

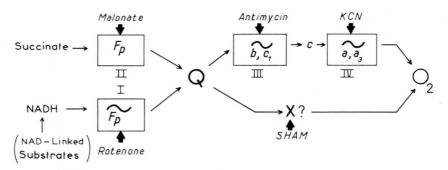

Fig. 1. Scheme of electron transport in plant mitochondria. This scheme shows the organization of the respiratory chain into four complexes, the sites of energy transduction, and the sites of action of specific inhibitors. a, a_3, b, c, c_1 cytochromes; Q ubiquinone; X alternative pathway oxidase (?)

Three of these complexes (I, III, IV) are the sites of energy transduction (CHANCE and WILLIAMS 1965). Moreover, in plant mitochondria, two accesses to oxygen are offered to the electrons. One is made by Complexes III and IV plus cytochrome c. Consisting mainly of cytochrome components, it will be called the cytochrome pathway and is sensitive to cyanide and antimycin. The other one, starting from ubiquinone, is the alternative pathway, resistant to cyanide. This is summarized in Fig. 1.

The electrons are fed to ubiquinone by Complexes I and II. Complex I is related to many NAD-linked dehydrogenases and is a site for energy transduction. Complex II, on the other hand, is linked only to succinate dehydrogenase and is not a site for energy transduction. Quantitatively and qualitatively, Complex I is more important than Complex II. Therefore, electron transport in plant mitochondria can be conveniently depicted as a branched system consisting of three main segments: the flavoprotein, the cytochrome, and the alternative pathways, ubiquinone being the branching point of the three pathways. This picture is the "classical" concept of electron transport in plant mitochondria (STOREY 1980). The important point to be emphasized at the moment is that specific inhibitors can interrupt electron transport at specific sites on these pathways (Fig. 1). Attention will be focused on the inhibitors of the alternative pathway since the inhibitors of the two other pathways will be covered in detail elsewhere (DUCET, Chap. 4, PALMER and WARD, Chap. 7, this Vol.).

4.1 Inhibitors of the Flavoprotein Pathway

The two flavoprotein dehydrogenases of this pathway have very different properties. Succinate dehydrogenase, a long-known enzyme, is competitively inhibited by malonate and noncompetitively inhibited by oxaloacetate (WEBB 1966). It is also sensitive to reagents having strong affinity toward iron, such as thenoyltrifluoroacetone (REDFEARN et al. 1965). Its full activity depends upon prior activation by ATP. One of the possible reasons for discrepancies in the extent of inhibition of succinate oxidation by cyanide is often to be found in an incomplete activation of the enzyme by ATP produced by oxidative phosphorylation. On the whole, succinate dehydrogenase in plant mitochondria behaves like its counterpart in animal mitochondria.

The situation is more complicated with NADH dehydrogenase, the second flavoprotein of this pathway. In contrast with the situation in animal mitochondria, two, and possibly three, different NADH dehydrogenases are to be found in the inner membrane of plant mitochondria (PALMER and WARD, Chap. 7, this Vol.), in addition to the NADH dehydrogenase located on the outer mitochondrial membrane (MOREAU and LANCE 1972).

One of the NADH dehydrogenases is part of Complex I. It is thought to be analogous to the similar enzyme of animal mitochondria. It is basically associated with a site of energy transduction and is inhibited by rotenone, amytal, and piericidin (GARLAND et al. 1969) and, incidentally, by cytokinins (CHAUVEAU et al. 1983). This dehydrogenase is responsible for the oxidation of endogenously generated NADH, and is located on the inner surface of the inner membrane, in close vicinity with the many NAD-linked dehydrogenases of the matrix compartment.

Another NADH dehydrogenase is located on the outer surface of the inner membrane (Von JAGOW and KLINGENBERG 1970, DOUCE et al. 1973). It is responsible for the oxidation of externally supplied NADH. It is not sensitive to rotenone or amytal, does not participate in energy transduction and appears to be strongly linked to the cytochrome pathway. The oxidation of exogenous NADH appears indeed as being highly cyanide-sensitive (Table 1). The activity of this dehydrogenase is also dependent on the presence of Ca^{2+} ions (PALMER 1976). In some mitochondria this dehydrogenase could also carry out the oxidation of NADPH (ARRON and EDWARDS 1979). A separate dehydrogenase for NADPH could also be involved (ARRON and EDWARDS 1979).

Finally, there are presently good arguments to think that another NADH dehydrogenase, distinct from that of Complex I, could also be present on the inner surface of the inner membrane (PALMER 1976, WISKICH and DAY 1979, RUSTIN et al. 1980, MØLLER and PALMER 1982). This dehydrogenase is not linked to energy production and is not sensitive to rotenone or amytal. It constitutes a by-pass around the rotenone-sensitive site and its implication in cyanide resistance (RUSTIN et al. 1980) will be discussed later.

4.2 Inhibitors of the Cytochrome Pathway

By definition this segment of the respiratory chain is sensitive to cyanide. The mechanisms of cyanide inhibition are outside the scope of this article (see DUCET, Chap. 4, this Vol.), but this effect is general and bears on the sensitivity of Complex IV to this inhibitor. Cyanide is thought to bind to cytochrome a_3, thus impeding its reoxidation by molecular oxygen (YONETANI and RAY 1965). Other inhibitors are also active on Complex IV, namely carbon monoxide and sodium azide (NaN_3). NaN_3 is often used instead of cyanide because it is not volatile at usual pH's (7–7.5). It is about ten times less active than KCN (IKUMA and BONNER 1967, CHAUVEAU and LANCE 1971).

The second complex (Complex III) of the cytochrome pathway is specifically inhibited by antimycin A. Antimycin is supposed to bind stoichiometrically to cytochrome b (ESTABROOK 1962, STOREY 1972) or to a specific protein component of Complex III (DAS GUPTA and RIESKE 1973, RINAUDO 1979). Since Complexes III and IV are disposed sequentially in the cytochrome pathway, the inhibitors of either complex should have the same effect. This is actually observed: when used at appropriate concentrations the extents of the inhibitions of succinate or malate oxidation by antimycin are the same as by cyanide (BAHR and BONNER 1973b, HENRY and NYNS 1975, CHAUVEAU 1976a, LANCE and DIZENGREMEL 1978, DIZENGREMEL and LANCE 1982). Another inhibitor of complex III, HOQNO (2, n-heptyl-4-hydroxyquinoline-N-oxide), is sometimes used instead of antimycin. It gives identical results (HACKETT et al. 1960b).

4.3 Inhibitors of the Alternative Pathway

The first indications that the alternative pathway could be specifically inhibited came from the study of a number of iron-chelating compounds such as thiocyanate, α,α'-dipyridyl, 8-hydroxyquinoline and o-phenanthroline. However, the specificity of these compounds is low and very high concentrations have to be used that can also inhibit the cytochrome pathway (LANCE 1969, BENDALL and BONNER 1971, PALMER 1972, BAHR and BONNER 1973a, CHAUVEAU 1976b, VAN DER PLAS and VERLEUR 1976). These inhibitors are no longer used nowadays, but they strongly contributed to establish the idea that iron could be implicated in the cyanide-resistant pathway, possibly as part of its terminal oxidation step.

A breakthrough in the study of cyanide resistance in plant mitochondria occurred in 1971 following the demonstration by SCHONBAUM et al. that hydroxamic acid derivatives were specific inhibitors of the alternative pathway. A variety of derivatives were assayed and SHAM (salicylhydroxamic acid) (Fig. 2) became universally recognized as the inhibitor of the cyanide-resistant

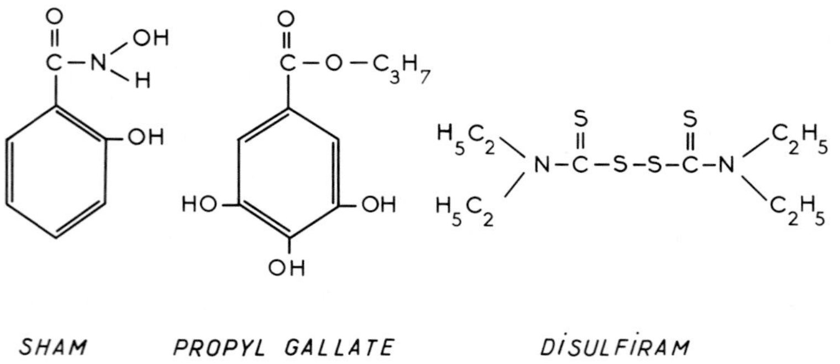

SHAM   PROPYL GALLATE   DISULFIRAM

Fig. 2. Major specific inhibitors of the alternative pathway

pathway although another derivative, m-CLAM (m-chlorobenzhydroxamic acid) is more active than SHAM. SHAM concentrations in the mM range are sufficient to totally inhibit the cyanide-resistant electron transport of plant mitochondria.

More recently, two other compounds have been found to have a strong and specific inhibitory effect on the alternative pathway: These are propyl gallate (Fig. 2) (PARRISH and LEOPOLD 1978) and disulfiram (Fig. 2) (GROVER and LATIES 1978). Both compounds are more than 100 times as effective as SHAM, since 50% inhibition is achieved at 2–5 µM concentrations for propyl gallate (SIEDOW and GIRVIN 1980) and 5–10 µM concentrations for disulfiram (GROVER and LATIES 1981). Some of these compounds are also strong inhibitors of lipoxygenase (MILLER and OBENDORF 1981, PETERMAN and SIEDOW 1983), and this property has contributed to give lipoxygenase a putative role in cyanide resistance of plant mitochondria (PARRISH and LEOPOLD 1978, GOLDSTEIN et al. 1980).

Besides these three main inhibitors, several other compounds have been reported to be active on the alternative pathway. They belong to various chemical families. Terpene derivatives such as β-pinene (DOUCE et al. 1978), kaempferol (RAVANEL et al. 1982), chloroquine (JAMES and SPENCER 1982), or herbicides (GAUVRIT 1978, MORELAND and HUBER 1978, MORELAND 1980) have uncoupling effects on the cytochrome pathway and inhibit the alternative pathway. Among these inhibitors, various cytokinins, when used at concentrations well above their usual physiological concentrations, have also a very strong effect on the alternative pathway (MILLER 1979, 1980, DIZENGREMEL et al. 1982). Benzyladenine, a synthetic cytokinin, appears to be the most active. Apparently there is no direct relationship between the biological activity of these compounds and their abilities to inhibit the alternative pathway; some anticytokinins can even be potent inhibitors of the alternative pathway (DIZENGREMEL et al. 1982).

4.4 Interactions Between Inhibitors

Two types of interaction have to be considered: interactions between inhibitors acting on the same pathway or between inhibitors acting on different pathways.

Rotenone and amytal have similar effects; they are strongly efficient in inhibiting malate oxidation, but have no effect on the oxidation of succinate or exogenous NADH (CARMELI and BIALE 1970, TOMLINSON and MORELAND 1975, DIZENGREMEL 1980). However, amytal can inhibit the NADH oxidation of A. maculatum mitochondria (CHAUVEAU 1976a). It should be pointed out that NADH oxidation is cyanide-resistant in these mitochondria (Table 1). There is, however, some difference between the effects of rotenone and amytal. Generally, cyanide does not increase the extent of rotenone inhibition. In contrast, cyanide used in combination with amytal strongly inhibits the oxidation of malate (CARMELI and BIALE 1970, TOMLINSON and MORELAND 1975, CHAUVEAU 1980, DIZENGREMEL 1980). With cyanide-resistant mitochondria, the cyanide-resistant fraction of succinate oxidation is strongly inhibited by amytal whereas it is not by rotenone. These observations indicate that amytal, contrary to rotenone, has an effect – probably unspecific – on the alternative pathway. Benzyladenine inhibits Complex I activity at the same site as rotenone (CHAUVEAU et al. 1983).

Considering the cytochrome pathway per se, antimycin and cyanide generally give the same values of percent inhibition. Since they act sequentially on a linear pathway, they present no additive effects when used together. On practical grounds, the extents of cyanide and antimycin resistances are the same (BAHR and BONNER 1973b, DIZENGREMEL and LANCE 1982).

The situation is much more complicated with the specific inhibitors of the alternative pathway. SHAM and propyl gallate act on the same site (SIEDOW and GIRVIN 1980). SHAM and disulfiram have different sites of action (GROVER and LATIES 1978, 1981). Finally, benzyladenine and SHAM, on the one hand, and benzyladenine and disulfiram, on the other hand, act on different sites (DIZENGREMEL et al. 1982). From these observations one can conclude that the alternative pathway can be inhibited at three different sites – or in three different ways – at least. These sites are respectively sensitive to: (a) SHAM (and analogs) and propyl gallate; (b) disulfiram and (c) benzyladenine (and analogs). The first group of inhibitors is characterized by the presence of a phenolic hydroxyl group (Fig. 2), which would appear as a minimal structural feature to inhibit the alternative pathway (SIEDOW and BICKETT 1981, SIEDOW 1982). All the inhibitors are lipophilic and the degree of lipophilicity plays an important role for the activity of the inhibitors of the third group (DIZENGREMEL et al. 1982). Disulfiram is the only inhibitor without a ring structure, and could act by reacting with -SH groups in the alternative pathway (GROVER and LATIES 1981).

When the inhibitors of the cytochrome and the alternative pathways are used together, electron transport in cyanide-resistant mitochondria is strongly repressed. However, only in very few instances is electron transport brought to nil. A part of the oxidative process remains resistant to cyanide and SHAM. When dealing with intact organs or tissues, it constitutes the so-called "residual respiration". Residual electron transport is also observed with mitochondria. In some plant materials, it can be very important, as in *Arum maculatum* mitochondria (CHAUVEAU 1980), where it increases markedly as the spadix develops and reaches the respiratory crisis (Fig. 3).

The explanation for the residual electron transport is not simple. One point which should be emphasized is that the inhibitors, at the concentrations ordinarily used in these experiments, are not 100% efficient. If one assumes a K_i of 30 µM for cyanide inhibition of electron transport in mitochondria (IKUMA and BONNER 1967, CHAUVEAU 1976b), this means that a 1 mM concentration will inhibit only 97% of the electron flow. Since the K_i of SHAM for the inhibition of the alternative pathway is about 700 µM (SCHONBAUM et al. 1971, CHAUVEAU 1976b), it results that a 2 mM concentration of SHAM (the concentration generally used in such studies) will inhibit no more than 75% of the alternative pathway activity. Therefore, there are theoretical reasons for the occurrence of a residual electron transport activity resistant to both cyanide and SHAM. However, in some instances, the extent of the cyanide and SHAM resistance is such that it cannot be accounted for by the above considerations

Fig. 3. Effects of inhibitors of electron transport on mitochondria from *A. maculatum* spadix during development. V_T: rate in the absence of inhibitor. V_{alt} rate in the alternative pathway, in the presence of 1 mM KCN; V_{cyt} rate in the cytochrome pathway, in the presence of 2.5 mM SHAM; V_{res} residual electron transport in the presence of both 1 mM KCN and 2.5 mM SHAM. Substrate was 10 mM succinate. (CHAUVEAU 1980)

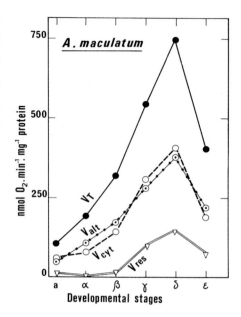

(CHAUVEAU 1976b). Then one generally assumes that a true residual respiration exists in these plant tissues and mitochondria isolated therefrom. In some specific situations it has been shown that azide could inhibit this SHAM- and cyanide-resistant electron transport, thus adding a new pathway to an already rather complicated situation (EDWARDS 1978).

5 The Link with Energy Transduction

The study of cyanide resistance made its fastest progress with plant tissues presenting high metabolic rates. The spadices of plants belonging to the Araceae family display high respiratory rates, which are strongly resistant to cyanide. In these organs this phenomenon is associated with pollination and is characterized by an intense heat production (MEEUSE 1975, CHAUVEAU and LANCE 1982). (Thermogenesis, dissipating energy as heat). The alternative pathway has been considered as a wasteful process, not involved in the production of energy available to the plant. This is a general belief, although it should be pointed out that this pathway is present in a wide variety of mitochondria isolated from nonthermogenic tissues. There are, however, some people who have the opinion that this pathway is linked to some form of energy transduction (WILSON 1980). Nonetheless, the arguments in favor of the nonparticipation of this pathway in energy transduction seem to be overwhelming.

5.1 Oxidative Phosphorylation

It is a common observation that plant mitochondria that are resistant to cyanide display ADP/O values markedly lower than those of cyanide-sensitive mitochondria. This has been shown with a large variety of plant materials: sweet potato (Hackett et al. 1960b), skunk cabbage (Storey and Bahr 1969b), wild-type and poky mutant of *Neurospora* (Lambowitz et al. 1972), *A. maculatum* (Passam and Palmer 1972, Passam 1974, Lance 1974, Lance and Chauveau 1975), aged potato slices (Dizengremel 1975). The ADP/O values can be extremely low: 0.2 to 0.5 for succinate (instead of 2) or 0.8 to 1.2 for malate instead of 3. It should be noticed, however, that these values are highly dependent on the technique used to measure oxidative phosphorylation. The values where direct measurements of ATP synthesis have been made should be given preference. With the oxygen electrode technique, the disappearance of respiratory control does not necessarily mean that oxidative phosphorylation is absent or that ADP/O is nil (Ducet 1982b).

The first hint that the low yield of ATP in cyanide-resistant mitochondria was not due to a low efficiency of the process of oxidative phosphorylation itself but to the occurrence of a nonphosphorylating electron transport pathway was given as early as 1960 by Hackett et al. (1960b). Along the same line, a decrease in the yield of ATP is also observed when the alternative pathway becomes predominant under certain experimental or natural conditions. This is the case with the aging of potato slices (Dizengremel 1975), the growth phases of *Neurospora* (Lambowitz et al. 1972) and the developmental stages of the spadix of *A. maculatum* (Lance 1974, Lance and Chauveau 1975).

A direct demonstration that the alternative pathway is a nonphosphorylating pathway can be given by using the specific inhibitors SHAM (Fig. 2) and KCN. With succinate and malate (Table 2), it is quite clear that the mitochondria present ADP/O ratios well below the average values generally observed in plant mitochondria. The addition of cyanide, by curtailing the cytochrome pathway and diverting the electrons to the alternative pathway, causes a marked decrease in the efficiency of oxidative phosphorylation. One can deduce that no phosphorylation site is associated with the oxidation of succinate, whereas one site

Table 2. The effect of inhibitors of the cytochrome and alternative pathways on ADP/O values of cyanide-resistant mitochondria

Material		ADP/O ratios	
		Succinate	Malate
A. maculatum[a]	Control	0.32	0.82
	+ Antimycin (5 µg mg^{-1} protein)	0.00	0.46
	+ m-Clam (0.3 mM)	0.76	1.23
S. tuberosum[b]	Control	1.02	1.76
(aged slices)	+ KCN (0.4 mM)	0.00	0.63
	+ SHAM (1 mM)	1.24	1.96

[a] Passam (1974)   [b] Dizengremel (1980)

remains associated with the oxidation of malate. This site is the Complex I site, and this observation is quite coherent with a branch point located at the ubiquinone level. On the other hand, the addition of an inhibitor of the alternative pathway considerably increases the efficiency of oxidative phosphorylation by diverting the electrons to the phosphorylating cytochrome pathway. These experiments, which have been repeated with several other plant mitochondria (STOREY and BAHR 1969b, LAMBOWITZ et al. 1972, PASSAM 1974, DIZENGREMEL 1980), elegantly demonstrate that the major function of the alternative pathway is not a production of ATP associated with electron transport.

Opinions have been very clear-cut about this question: the alternative pathway is universally considered as being nonphosphorylating, but does this mean really that it is not linked to any form of energy production? There is one discordant voice in this general consensus: In apparently uncriticizable experiments, using adequate concentrations of cyanide to produce a maximal inhibition of the cytochrome pathway, WILSON (1970, 1978, 1980) has consistently reported that some production of ATP, sensitive to SHAM and oligomycin, could be observed.

5.2 Membrane Potential and Proton Gradient

When speaking of oxidative phosphorylation, the use of the classical term phosphorylation site has the advantage of answering questions by yes or no. Accordingly, the alternative pathway does not include a phosphorylation site and therefore is not concerned with ATP production. The mechanisms by which ATP is produced in mitochondrial membranes are now rather well known. According to the chemiosmotic theory, ATP is synthesized by an ATP synthetase that uses the proton motive force built-up as a result of electron transport along the respiratory chain (see NICHOLS 1982). Adopting this view, the question of the nonphosphorylating character of the alternative pathway must be asked again.

By measuring membrane potential ($\Delta\psi$) and proton extrusion (ΔpH), the two components of the proton motive force, MOORE et al. (1978) observed that the oxidation of succinate by the alternative pathway was not coupled to the building up of a proton motive force. In contrast, the oxidation of malate by the alternative pathway could conserve the capacity to synthesize ATP, presumably through the operation of the coupling site of Complex I. Similar conclusions have been reached using safranine as an optical probe to measure membrane potentials (MOORE and BONNER 1982). An interesting observation is that malate, in the presence of cyanide, can maintain a membrane potential as long as rotenone is absent from the medium (MOORE and BONNER 1982, MOREAU and ROMANI 1982). This indicates that the oxidation of malate through Complex I is associated with energy transduction, whereas it is not when malate oxidation bypasses Complex I. Using similar techniques and avoiding interaction with phosphorylation site I in Complex I, WILSON (1980) nevertheless came to the conclusion that there was some additional evidence for a site of energy conservation in the alternative pathway.

6 The Structure of the Alternative Pathway

If one refers to the scheme of the alternative pathway as depicted in Fig. 1, several questions must be asked: what are the components of this pathway? What is the link with the main respiratory chain? What is the component reacting with oxygen?

Since the beginning of the studies on this pathway it has constantly been assumed that, like the cytochrome pathway, the alternative pathway should react with oxygen through a terminal oxidase, sometimes called the alternative (pathway) oxidase. As this point will be examined in detail later, this section will be mainly concerned with the first and intermediary components of this pathway.

It should be pointed out at once that the alternative pathway has always been considered to be extremely short, consisting of very few elements, as will be seen from the examination of the many hypotheses on the nature of its terminal oxidase. In many instances it consisted only of the terminal oxidase itself. However, some proposals have also been made on the existence of some intermediary components, but the crucial point remains the nature of the branch point which represents the first component of this pathway. By definition, it must be common to the cytochrome and the alternative pathways, and establish the link whith the flavoprotein dehydrogenases.

6.1 Branch Point of the Alternative Pathway

It is rather easy to attribute a rough localization to the point of divergence of the alternative pathway. On the one hand, the degrees of cyanide resistance are the same with cyanide or antimycin. This indicates that the electrons must diverge to the alternative pathway at a point located before the site of antimycin inhibition, namely before the electrons enter Complex III. However, if one admits that antimycin may act on another compound rather than just cytochrome b, this leaves one of the many cytochromes b of plant mitochondria as a potential candidate for the common intermediate. This hypothesis was indeed put forward for a time (BENDALL and HILL 1956).

On the other hand, the fact that, in cyanide-resistant mitochondria, the P/O ratios measured in the presence of cyanide are close to 0 for succinate and close to 1 for malate, strongly suggests that the branch point could be situated after Complexes I and II. Since this region of the respiratory chain, the flavoprotein pathway, is known to be extremely complicated (CHANCE et al. 1968, STOREY 1980), this leaves, as a potential site for the branch point, a segment of the respiratory chain which for a long time was not very well defined in plant mitochondria, i.e., the site of interaction between Complexes I, II, and III. Several electron carriers of this region have been proposed as the branch points, namely a flavoprotein, an iron-sulfur protein, and a cytochrome b. As their function was also that of a terminal oxidase, they will be dealt with later.

To make a long story short, ubiquinone, a rather small hydrophobic molecule (M.W. 863), has been recognized as the branch point of the three pathways

(STOREY 1976), although the situation is by no means simple. Some would tend to consider that ubiquinone could act as a terminal oxidase (RICH and BONNER 1978a, b, c), others would distribute ubiquinone into various pools of different reactivities (RUSTIN et al. 1980, RICH 1981). A first step in this direction was made by STOREY (1973), who clearly demonstrated that ubiquinone was a component of the electron transport pathway of plant mitochondria, located on the main stream of electrons and not acting as a by-pass around cytochrome b. With skunk cabbage mitochondria under anaerobic conditions and in the presence of carbon monoxide, STOREY (1976) demonstrated that ubiquinone (and a flavoprotein as well) was rapidly reoxidized if a pulse of oxygen was given. This reoxidation was strongly inhibited by SHAM, indicating that ubiquinone was a component of the alternative pathway. Since the reoxidation of ubiquinone can also be observed upon giving a pulse of O_2 to anaerobic mitochondria of fresh potato tissue which have no alternative pathway, this clearly establishes that ubiquinone is a carrier shared by both the cytochrome and the alternative pathways.

A further confirmation of a key role for ubiquinone in plant electron transport is given by the observation that extraction of quinones by pentane inhibits cyanine-sensitive and cyanide-resistant electron transport as well. Addition of natural quinones restores the activity of both electron pathways (Von JAGOW and BOHRER 1975, HUQ and PALMER 1978a). Finally, dibromothymoquinone, an analog of ubiquinone, has an inhibitory effect on both pathways (DRABI-KOWSKA 1977, SIEDOW et al. 1978). All these observations point to the fact that ubiquinone plays a central role in electron transport in both pathways, and there is presently a general agreement on the fact that ubiquinone represents the branch point of these pathways. It should be pointed out, however, how scarce the reports on direct determinations of ubiquinone in plant mitochondria are.

6.2 Other Components

Apart from the terminal oxidase, what other components can be present in the alternative pathway? The answer to this question is uncertain. Most people would agree that there is no such component between ubiquinone and the oxidase, not to say oxygen. As a matter of fact, there is only one report indicating the presence of such a component. STOREY (1976) has identified a low-potential flavoprotein (Fp_{ma}) thought to be a possible intermediate. However, there has been no further support for such a flavoprotein (STOREY 1980).

7 The Functional Organization of the Alternative Pathway

In what follows it will be assumed that the alternative pathway is a nonphosphorylating process, that it diverges from the main respiratory chain at the level of ubiquinone, and that it consists of two main components: ubiquinone and the alternative oxidase. This organization was summarized in Fig. 1. This, of course, is an oversimplified view, which will be gradually modified as further questions are examined. The first point to be considered is the topographical localization of this pathway within the mitochondrial membrane. The second is how such a simple scheme as that of Fig. 1 can accommodate a variety of situations as different as those presented in Table 1. In other words, how can different substrates be oxidized in so many different ways by cyanide-sensitive or cyanide-resistant mitochondria?

7.1 Topographical Organization

Very early the site of the cyanide resistance of tissue respiration was determined to be the mitochondrion (JAMES and ELLIOTT 1955, HACKETT 1957). With the improvement of techniques, it became possible to dissociate the outer and the inner mitochondrial membranes, for instance with an osmotic shock (MOREAU and LANCE 1972). The separation of the two membranes of the cyanide-resistant mitochondria from aged potato slices indicates that the cyanide-resistant pathway is firmly associated with the inner membrane (DIZENGREMEL 1980). The removal of the outer membrane has no effect on the operation of this pathway.

A number of assumptions have to be made. They stem from similar studies carried out on animal mitochondria (TZAGOLOFF 1982) and have been found to hold true for plant mitochondria. In short, NADH dehydrogenase, succinate dehydrogenase, cytochrome oxidase and ATP synthetase face the matrix compartment, whereas cytochrome c faces the intermembrane space. Krebs cycle dehydrogenases are thought to be localized in the matrix compartment (TZAGOLOFF 1982). In plant mitochondria, a specific dehydrogenase for exogenous NADH also faces the intermembrane space (DOUCE et al. 1973). It should be realized that ubiquinone, supposed to be a small molecule, contains 10 isoprene units ($= 50$ C) which give ubiquinone enough length (56 Å) to stretch across the whole inner membrane (TRUMPOWER 1981).

After specific inhibitors of the alternative pathway were discovered, and assuming that they combine with the oxidase, it became possible to gain some insight on the location of this oxidase. Considering that hydroxamic acids, known to be good reagents for P-containing molecules, could inhibit the alternative pathway while not affecting oxidative phosphorylation, SCHONBAUM et al. (1971) concluded that the inner mitochondrial membrane was impermeable to hydroxamic acids, since they did not affect ATP synthesis, and that they were acting by complexing a transition metal (Fe ?) present in the oxidase itself. From this it was concluded that the alternative oxidase was located on the outer surface of the inner mitochondrial membrane.

Another view was developed by MOORE et al. (1976), who placed the oxidase on the inner surface of the inner membrane. Their argument was that a close link could be established between an iron-sulfur center of succinate dehydrogenase (on the inner surface) and the alternative pathway. This was consistent with the observation that bathophenanthroline, a chelator of Fe, had no effect on the alternative pathway at concentrations below those active on the cytochrome pathway (RICH et al. 1977).

Fig. 4. Effect of Triton X-100 on the rate of electron transport in the cytochrome (V_{cyt}) and alternative (V_{alt}) pathways measured in the presence of 1 mM SHAM and 1 mM KCN, respectively. Mitochondria from aged potato slices. Substrate was 10 mM succinate. (DIZENGREMEL 1983)

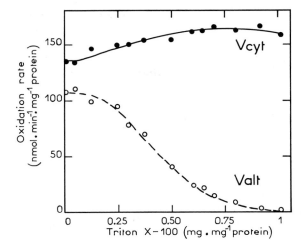

Triton X-100, a nonionic detergent, differentially affects the activities of the cytochrome and alternative pathways (Fig. 4). The activity of the alternative pathway gradually disappears and is totally destroyed for a concentration of detergent that has no effect on the cytochrome pathway. This experiment has been interpreted as indicating that a component of the alternative pathway (alternative oxidase or a ubiquinone pool) was located on the outer surface of the inner membrane and easily accessible to the detergent (DIZENGREMEL 1983). However, it cannot be entirely excluded that the alternative pathway oxidase could be more sensitive to membrane perturbation than the cytochrome pathway.

7.2 Compartmentation

The next crucial point to be explained is the behavior of the different substrates with respect to the alternative pathway (Table 1) and the peculiarities of malate oxidation by plant mitochondria. To sum up, this has led to a controversial situation: a model of organization has been proposed on which not everybody agrees, because this model implies some type of compartmentation for several key intermediates, namely ubiquinone and pyridine nucleotides.

7.2.1 Ubiquinone

The proposal for the compartmentation of ubiquinone is based on several fundamental assumptions, the first one being that ubiquinone is an obligate carrier for electron transport in plant mitochondria: all electrons, therefore, will have to pass through ubiquinone. The second assumption is that substrates showing similar behaviors toward cyanide will be considered as using the same pool of ubiquinone, whereas different pools will have to be used if the behaviors are different.

For plant mitochondria which are highly cyanide-sensitive or resistant, such as the mitochondria from fresh potato tissue or from aroid spadices, a single pool of ubiquinone (or several interconnected pools) could explain the distribution of electrons between the two pathways, although an exception should be made for malate oxidation, as will be seen later.

Then there is the specific case of two moderately cyanide-resistant types of mitochondria: those from aged potato slices or from fresh or aged sweet potato slices (DIZENGREMEL 1975, 1980, TOMLINSON and MORELAND 1975). In these mitochondria, succinate and malate oxidations behave in the same manner: they are inhibited by about 50–60% by cyanide, but NADH oxidation is, or remains, strongly cyanide-sensitive ($\geq 90\%$). In that case, one has to admit that the ubiquinone pool receiving the electrons from the external NADH dehydrogenase has a strong connection with the cytochrome pathway and very little connection with the alternative pathway. This pool is evidently different from the one that is associated with the succinate dehydrogenase and the internal NADH dehydrogenase and which distributes the electrons from succinate and malate between the two pathways.

Finally, one has to take into account the peculiarities of malate oxidation in plant mitochondria. This substrate is oxidized by two enzyme systems located in the matrix compartment: malate dehydrogenase and malic enzyme. The balance between the activities of the two enzymes depends on internal and external factors: lowering the pH to 6.5 favors malic enzyme activity, raising the pH to 7.8 favors malate dehydrogenase activity (MACRAE 1971, TOBIN et al. 1980). At usual experimental pH's (7.0–7.2), both enzymes generally participate in malate oxidation (TOBIN et al. 1980). The addition of NAD^+ to mitochondria oxidizing malate at pH 7.2 stimulates the activity of malic enzyme (MACRAE and MOORHOUSE 1970). An important observation was made: malate oxidation at pH 7.2, coupled to the synthesis of ATP, could be strongly inhibited by rotenone and cyanide and yet an addition of NAD^+ could induce a nonphosphorylating, rotenone- and cyanide-resistant, but SHAM-sensitive electron transport (RUSTIN and MOREAU 1979, RUSTIN et al. 1980). From the analysis of the substrates produced in the course of the reaction, it was concluded that the oxidation of malate by malate dehydrogenase was mediated by a rotenone-sensitive NADH dehydrogenase (Complex I), ubiquinone, and the phosphorylating cytochrome pathway. On the other hand, the oxidation of malate by malic enzyme could be mediated by a rotenone-insensitive dehydrogenase and a ubiquinone pool connected to the oxidase of the alternative pathway. The latter pathway was not phosphorylating. This observation was made on a variety of plant mitochondria and, in this regard, the typically cyanide-sensitive mitochondria from fresh potato slices were indeed cyanide-sensitive as long as malate was oxidized through malate dehydrogenase, but they had to be considered as cyanide-resistant when the oxidation of malate was experimentally shifted to malic enzyme (RUSTIN et al. 1980). These results have been the heart of a controversy (LATIES 1982). There have been reports that the experiments could not be repeated, particularly the stimulation by NAD^+ of malic enzyme activity in the presence of cyanide (PALMER et al. 1982). A recent paper by WISKICH and DAY (1982) reports results that are almost the opposite of those of the paper by RUSTIN et al. (1980). However, two independent sources of support came from two different fields. Basically, they showed that the production of pyruvate from malate is a cyanide-resistant process, whereas the production of oxaloacetate from malate is a cyanide-sensitive process. This was observed in avocado mitochondria during the climacteric rise (MOREAU and ROMANI 1982) and in mitochondria from the mesophyll or bundle-sheath cells of a C_4 plant (GARDESTRÖM and EDWARDS 1983).

From the above considerations one has to conclude that ubiquinone does not behave as a homogenous pool. Three of them can be distinguished: one major pool associated with succinate dehydrogenase and Complex I, another associated with the NADH dehydrogenase of the outer surface of the inner membrane, and a third associated with a rotenone-insensitive NADH dehydrogenase on the inner surface of the inner membrane and accounting for the cyanide-resistant malate oxidation by malic enzyme. It should be recognized that the definition of a quinone pool is not easy, but this notion is becoming familiar (GUTMAN 1977, RICH and BONNER 1978a, DAY et al. 1980). Compartmentation of ubiquinone does not necessarily mean physical compartmentation, though some proteins which can bind ubiquinone have been isolated (TRUMPOWER 1981, YU and YU 1981). Differences in reactivity or affinity could also represent some sort of compartmentation (GUTMAN 1977).

7.2.2 Pyridine Nucleotides

Another major question that should be raised bears on an eventual compartmentation of the pyridine nucleotides at the border line of the matrix substrate dehydrogenases and membrane NADH dehydrogenases. If one accepts the views developed in the previous section (RUSTIN and MOREAU 1979, RUSTIN et al. 1980, MOREAU and ROMANI 1982, GARDESTRÖM and EDWARDS 1983), it appears that the NADH produced by malic enzyme and the NADH produced by malate dehydrogenase are not in equilibrium and freely exchangeable. Therefore, they also show some form of compartmentation. In the former case, NADH could be reoxidized in a cyanide-resistant manner through a specific rotenone-insensitive NADH dehydrogenase. In the latter case, it is reoxidized by the rotenone-sensitive NADH dehydrogenase of Complex I. This scheme, of course, implies the existence of two NADH dehydrogenases on the inner surface of the inner membrane. Such an opinion is by no means new, since the existence of two dehydrogenases has been postulated by several authors (PALMER 1976, WISKICH and DAY 1979, RUSTIN et al. 1980, MØLLER and PALMER 1982).

Once again, it must be recognized that there is no general agreement on the above views. A long time ago, it had been shown that oxaloacetate could inhibit oxygen uptake in mitochondria oxidizing various Krebs cycle intermediates (DOUCE and BONNER 1972). Although oxaloacetate can inhibit succinate dehydrogenase directly, in that instance the inhibition of oxygen uptake was not by a direct effect on the Krebs cycle dehydrogenases but rather by a reversal of the malate dehydrogenase reaction, producing malate in the presence of added oxaloacetate and endogenously generated NADH. Such an effect implies that the NADH produced by the various Krebs cycle dehydrogenases is accessible to malate dehydrogenase, and therefore is not compartmented. Similar views were expressed later (TOBIN et al. 1980, WISKICH and DAY 1982). Even with NADH produced by malic enzyme, it has been shown that it could be accessible to malate dehydrogenase (TOBIN et al. 1980, PALMER et al. 1982). However, it must be noticed that in the presence of rotenone, the kinetics of oxaloacetate disappearance are relatively slow (TOBIN et al. 1980). The same result is seen with glycine as substrate (DAY and WISKICH 1981). No further comment will be made here since this topic will be dealt with in detail elsewhere (see PALMER and WARD, this Volume).

7.3 Organization of the Alternative Pathway

All the views that have been exposed above can be summarized in a rather simple scheme (adapted from RUSTIN et al. 1980). This scheme is far from being unanimously accepted, but it is worth presenting if only for the reason that it is controversial, and can therefore stimulate new research resulting in a better understanding of electron transport in plant mitochondria and that ... one needs God to define an atheist!

Leaving aside the outer mitochondrial membrane and its specific NADH dehydrogenase (MOREAU 1978), the scheme in Fig. 5 summarizes the main fea-

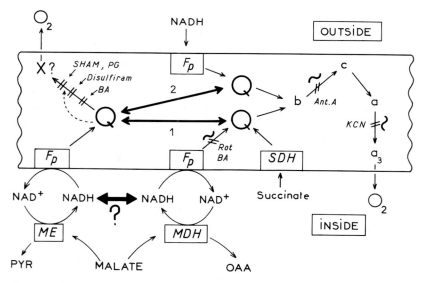

Fig. 5. Scheme of electron transport in the inner membrane of plant mitochondria. F_p flavoproteins (various NADH dehydrogenases); *SDH* succinate dehydrogenase; *MÉ* malic enzyme; *MDH* malate dehydrogenase; *a*, a_3, *b*, *c* cytochromes; *Q* ubiquinone pools; *PYR* pyruvate; *OAA* oxaloacetate. The sites of energy transduction ($\sim$) and of action of the major inhibitors are also indicated. *Ant.A* antimycin A; *Rot* rotenone; *BA* benzyladenine (cytokinin); *PG* propyl gallate; *SHAM* salicylhydroxamic acid. (After RUSTIN et al. 1980)

tures that have just been described. These are the presence of three ubiquinone pools, three NADH dehydrogenases and compartmentation of pyridine nucleotides at the levels of malate dehydrogenase and malic enzyme. In addition, the various pools of ubiquinone can eventually become interconnected. The dehydrogenases linked to the other Krebs cycle intermediates have not been represented; they are supposed to interact with the respiratory chain at the level of the rotenone-sensitive NADH dehydrogenase (Complex I). Finally, the alternative pathway oxidase has been located on the outer surface of the inner membrane, since there seem to be stronger arguments for such a location (cf. above). The merit of such a scheme is its ability to explain the various situations encountered in plant mitochondria (a further advantage will be presented later).

In highly cyanide-sensitive mitochondria (e.g., mitochondria from fresh potato tissue), connections 1 and 2 are absent. NADH, succinate and all Krebs cycle substrate oxidations are all highly cyanide-sensitive, with the exception of the oxidation of malate when it is mediated by malic enzyme. When connection 1 is established, all substrate oxidations become cyanide-resistant except for the oxidation of extramitochondrial NADH which remains highly cyanide-sensitive (e.g., mitochondria from aged potato slices). Finally, when connection 2 becomes established, all substrate oxidations are cyanide-resistant (e.g., aroid spadix mitochondria). As mentioned before, this scheme is probably an oversimplified view of what really occurs within mitochondrial membranes. It is not excluded that interactions could exist between the pools of NADH, which could reconcile antagonistic points of view on NADH compartmentation, but it should be stressed that the conditions of those interactions or of noninteractions are not very well understood at the present time, and the important point remains that NADH produced from malate by malate dehydrogenase or by malic enzyme behaves, with respect to cyanide resistance, in a way that can be most easily interpreted in terms of compartmentation.

Another point of discussion bears on the terminal oxidase of the alternative pathway (cf. next section) and on the effects of various inhibitors on this pathway. To summarize, hydroxamic acids and propyl gallate competitively act on the same site; disulfiram has another site of inhibition and cytokinins have two effects: one at the same site as rotenone, and another on the alternative pathway, but at a site that is different from those of SHAM and disulfiram (cf. Sect. 4.4). Therefore there are at least three different sites of action for these inhibitors of the alternative pathway. An important remark is that with malate oxidation through malic enzyme a part of the electron flow can bypass the cytokinin-sensitive site (CHAUVEAU et al. 1983). This would mean that some bypass could exist around the cytokinin sensitive-site on the alternative pathway (Fig. 5). This indeed makes the situation really very complicated in the region of the alternative pathway oxidase!

8 The Alternative Pathway Oxidase

The question of the terminal oxidation step in the alternative pathway remains one of the most obscure and unsolved problem. Over the years, a number of hypotheses have been proposed. Depending on one's inclination, one can satisfy oneself with a wide variety of opinions ranging from a well-defined oxidase: it has been isolated, its K_m for O_2 is well known, it has specific inhibitors, it is encoded by nuclear genes, etc... to the mere negation of the very existence of such an oxidase. This is the reason why we have chosen to deal with this oxidase separately and at this place, because the sections on the structure and organization of the alternative pathway would have appeared very confusing had all this new material been taken into consideration. Related to the problem of the nature of the terminal oxidase is the problem of the terminal product of oxygen reduction by this pathway. Here again, a variety of answers have been proposed: water (H_2O), hydrogen peroxide (H_2O_2) or superoxide $O_2^{\div}$, but there now seems to be a consensus for H_2O. An interesting approach to the question of the alternate oxidase is to go through all the hypotheses that have been proposed or abandoned: each of them probably bears a part of truth or at least emphasizes a characteristic aspect of the structure or functioning of the alternative pathway.

8.1 Flavoprotein Hypothesis

From the very beginning of the studies on the cyanide-resistant pathway, Van Herk (1937a, b, c) suggested that an auto-oxidizable flavoprotein was the terminal oxidase in the respiration of the spadix of *S. guttatum*. James and Beevers (1950) came to the same conclusion with the spadix of *A. maculatum*, observing that the flavoprotein content was increasing as the spadix was developing and the respiratory rate increasing. The flavoprotein hypothesis was refuted on the ground that the affinity for O_2 it too low and the kinetics of reoxidation too slow (Storey and Bahr 1969a, Bendall and Bonner 1971, Solomos 1977, Siedow 1982). Besides, auto-oxidizable flavoproteins produce H_2O_2 that the cell has to dispose of. Related to this hypothesis, the fact remains that mitochondria from highly cyanide-resistant tissues (aroid spadices) have relatively high contents of flavoprotein (Lance and Bonner 1968, Erecinska and Storey 1970) compared to those of mitochondria from other tissues. Finally, it should also be recalled that a flavoprotein (Fp_{ma}) has been proposed as a component, distinct from the oxidase, of the alternative pathway (Storey 1976). Its oxidation is inhibited by m-chlorobenzhydroxamic acid (m-CLAM), but is not inhibited by cyanide or antimycin.

8.2 Excess Oxidase Hypothesis

Another early hypothesis was that cyanide-resistant tissues had such an excess of cytochrome oxidase that cyanide could not inhibit the electron transfer totally (Ducet and Rosenberg 1962). However, when it became possible to measure

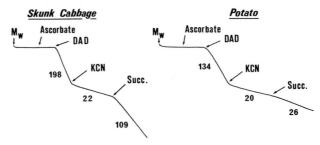

Fig. 6. Demonstration that cytochrome oxidase is functional in cyanide-resistant mitochondria. *Left* mitochondria from skunk cabbage (*Symplocarpus foetidus*) oxidizing ascorbate, which feeds electrons into the respiratory chain at the level of cytochrome c through
an auxiliary carrier DAD (2,3,5,6-tetramethyldiaminobenzene). KCN inhibits the oxidation of ascorbate by acting on cytochrome oxidase. An addition of succinate induces
an oxygen uptake through the alternative pathway. *Right* similar experiment carried
out with mitochondria from fresh potato tissue which do not possess an alternative
pathway. *Numbers* are rates of O_2 uptake in nmol min^{-1} mg^{-1} protein. (After BENDALL
and BONNER 1971)

the cytochrome contents of mitochondria with sophisticated spectrophotometric
techniques, it appeared that the cytochrome contents, and particularly cytochrome oxidase, were very uniformly distributed in plant mitochondria (LANCE
and BONNER 1968, LANCE 1981). There is no excess of cytochrome oxidase
in cyanide-resistant mitochondria. Inaccessibility of cyanide to cytochrome oxidase (SIMON 1957) or incomplete reduction of cytochrome oxidase in the presence of cyanide (CHANCE and HACKETT 1959, YOCUM and HACKETT 1957, HACK-
ETT and HAAS 1958) have also been suggested as possible explanations. The
fact that cytochrome oxidase is indeed a functional enzyme in cyanide-resistant
mitochondria was demonstrated very elegantly by BENDALL and BONNER (1971)
(Fig. 6).

8.3 Cytochrome b_7 Hypothesis

In aroid spadix mitochondria it was observed that a b type cytochrome, that
became known as cytochrome b_7, could stay oxidized in the presence of cyanide
(BENDALL and HILL 1956, YOCUM and HACKETT 1957, BENDALL 1958, HACKETT
and HAAS 1958). For a time this cytochrome was considered as both the branch
point and the oxidase of the alternative pathway (HACKETT 1957). However,
antimycin is able to fully reduce the cytochromes b in cyanide-resistant mitochondria (BENDALL and BONNER 1971) and their kinetics of oxidation-reduction
appears to be too sluggish to account for the high respiratory rates observed
in some tissues (STOREY and BAHR 1969a). Nevertheless, it should be mentioned
that the content of cytochrome b increases in the mitochondria of aging potato
slices, in parallel which the development of cyanide resistance (HACKETT et al.
1960a, COTTE-MARTINON et al. 1969, DIZENGREMEL 1975). However, a close
examination of the different cytochromes b shows that the one that increases

most during aging (cytochrome b-565) is also the one that becomes fully reduced by antimycin under aerobic conditions (DIZENGREMEL 1980). Similarly, the content in cytochrome b increases during the development of the *A. maculatum* spadix, in which the concentrations of the different b cytochromes are above the average values found in other plant mitochondria (CHAUVEAU 1980). Cytochrome b_7 has been identified either as cytochrome b-560 (CHANCE et al. 1968), or with the two cytochromes b-556 and b-560 (CHAUVEAU 1980). As with the flavoproteins, mitochondria from aroid spadices are characterized by high levels of cytochrome b, but the relationship with the alternative pathway is uncertain.

8.4 Nonheme Iron Protein Hypothesis

At the same time as they criticized various prior hypotheses, BENDALL and BONNER (1971) made the suggestion that the alternative pathway oxidase could belong to a family of then newly discovered electron carriers: the nonheme iron proteins (or iron-sulfur proteins), whose typical components are nonheme iron and labile sulfide. The evidence was that the alternative pathway could be selectively inhibited by a variety of iron-binding reagents (thiocyanate, α,α'-dipyridyl, 8-hydroxyquinoline) (LANCE and BONNER 1969, LANCE 1969, BENDALL and BONNER 1971, PALMER 1972). This hypothesis was reinforced by the fact that hydroxamic acid derivatives also have similar properties (SCHONBAUM et al. 1971). This hypothesis became very popular and received strong support (SHARPLESS and BUTOW 1970, WILSON 1971, SHERALD and SISLER 1972, PALMER 1972, DIZENGREMEL et al. 1973, BAHR and BONNER 1973a, CHAUVEAU 1976b, VAN DER PLAS and VERLEUR 1976). In *A. maculatum* mitochondria, the disappearance of the alternative pathway could be related to a release of SH_2, corresponding to the labile sulfide of the oxidase and representing only 20% of the total mitochondrial labile sulfide (CHAUVEAU et al. 1978). Lack of sulfate or iron in the medium during the growth of microorganisms could decrease the operation of the alternative pathway (HADDOCK and GARLAND 1971, HENRY et al. 1978), whereas an addition of Fe^{3+} could stimulate its development (HENRY et al. 1977). This pathway also appeared less stable than the cytochrome pathway: a temperature treatment at 40 °C for 30 min could destroy all alternative pathway activity while leaving the cytochrome pathway activity intact in mitochondria of *A. maculatum* or aged potato slices (CHAUVEAU et al. 1978, DIZENGREMEL and CHAUVEAU 1978). Finally, with *Neurospora crassa*, it could be shown that the biosynthesis of this oxidase was under the control of two nuclear genes (EDWARDS et al. 1974, EDWARDS and ROSENBERG 1976, EDWARDS 1978).

All these observations were strong arguments to support an iron-sulfur protein as the terminal oxidase of the alternative pathway. However, it was observed that the distribution of nonheme iron or labile sulfide were the same in cyanide-sensitive or cyanide-resistant mitochondria (DIZENGREMEL et al. 1973, LANCE 1981). Finally, an end to the nonheme iron protein hypothesis apparently came when it was demonstrated by EPR spectroscopy that no clear signal from the nonheme iron could be observed in relation with the functioning of the alternative pathway (CAMMACK and PALMER 1973, 1977, MOORE et al. 1976, RICH and

BONNER 1978d) – unless this oxidase is an EPR-silent nonheme iron protein (RICH and BONNER 1978b)!

8.5 Ubiquinone (Q-Cycle) Hypothesis

Subsequently, ubiquinone itself became involved in the terminal reaction step of the alternative pathway. In 1976, RICH and MOORE proposed a model of electron transport in plant mitochondria in which the proton motive ubiquinone cycle described by MITCHELL (1975) was implicated. In that scheme (Fig. 7), ubiquinone was both the branch point of the alternative pathway and the alternative pathway itself, since the latter was described as a reversal of a step of this cycle: the reduction by succinate of $QH^{\cdot}$ to QH_2. This step was sensitive to SHAM, yielding H_2O as the terminal product of oxygen reduction. A basically similar scheme was published by RICH and BONNER (1978a), establishing a closer relationship between center S-3 of succinate dehydrogenase and the Q cycle. It differed, however, from the previous scheme in that the terminal product was H_2O_2, not H_2O. The question of the terminal product, whether related to the ubiquinone cycle or to the oxidase itself, has always been puzzling. It has been reported that H_2O_2 could be generated by the oxidase (RICH et al. 1976, HUQ and PALMER 1978c) but the rate was too slow to account for the O_2 uptake by the alternative pathway and it is now clear that the product is H_2O. However, it could not be excluded that H_2O_2 could be transiently produced at higher rate and then destroyed by catalase or peroxidase (RICH et al. 1976). The involvement of superoxide dismutase in the production of H_2O_2 from the superoxide anion $O_2^{\cdot}$, generated by the alternative oxidase or ubiquinone, has also been proposed as an alternative hypothesis (MOORE et al. 1976).

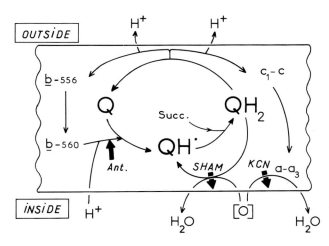

Fig. 7. The proton motive ubiquinone cycle scheme of electron transport in the inner membrane of plant mitochondria. (RICH and MOORE 1976)

8.6 Quinol Oxidase Hypothesis

In the above hypothesis a specific pool of quinone (the branch point) was playing the role of the "oxidase", but no oxidase was really required. The first hint that a specific protein could be associated with the oxidation of this specific quinone pool was made by Rich and Bonner (1978a). Soon after two reports that the oxidase had been isolated appeared in the literature (Huq and Palmer 1978b, Rich 1978). In both papers the oxidase had been solubilized by the use of detergents (deoxycholate, lubrol) and the substrates used were reduced ubiquinone (ubiquinol) or related compounds (duroquinol, menadiol). Protein preparations whose properties were close to those of a putative alternative oxidase were obtained, but these preliminary studies have not been so far followed by detailed reports on the properties of these proteins. Recently, another report appeared on the isolation of a copper-containing quinol oxidase from *A. maculatum* mitochondria, but similar attempts were unsuccessful with other cyanide-resistant materials (Bonner and Rich 1983). Whether this protein is the alternative pathway oxidase is not clear at the present time, but the basic result is that the alternative pathway oxidase activity can be easily removed from plant mitochondria by detergent treatments (Dizengremel 1983) (Fig. 4).

8.7 Lipoxygenase Hypothesis

Following the finding of Parrish and Leopold (1978) that the oxygen uptake of germinating soybean seeds was sensitive to KCN and SHAM and that a high lipoxygenase activity could be observed in this material, the idea that the alternative pathway oxidase could be confounded with lipoxygenase spread rapidly. Indeed these two activities appeared to be sensitive to SHAM and propyl gallate and resistant to cyanide (Siedow and Girvin 1980, Peterman and Siedow 1983). Then the question arose of the presence of lipoxygenase in plant mitochondria. Although this activity seemed to be present (Dupont 1981), it appeared that careful purification of mitochondria could remove the activity from the mitochondrial fraction (Goldstein et al. 1980, Siedow and Girvin 1980). To help solve the question, disulfiram was also found to be an inhibitor of the alternative pathway oxidase, but not of lipoxygenase (Miller and Obendorf 1981). At the present time, it seems generally agreed that lipoxygenase, a cytosolic or plastid enzyme, can be a contaminant of mitochondria and mimic the alternative pathway oxidase (Dupont et al. 1982). It should also be noted that the substrate of lipoxygenase (linoleic acid) is not a main foodstuff for plant mitochondria, which excludes a major participation of lipoxygenase in the oxygen uptake of mitochondria.

8.8 Free Radical Hypothesis

The merit of the lipoxygenase hypothesis was to draw attention to the unsaturated fatty acids present in mitochondrial membranes and on the peroxidation

Fig. 8. The free radical hypothesis of the alternative pathway in plant mitochondria. AH_2 substrate; Q, $QH^·$, QH_2 ubiquinone in various states of reduction; RH unsaturated fatty acid; $R^·$ unsaturated fatty acid radical; $ROO^·$ fatty acid peroxy radical; ROH hydroxy fatty acid; $SHAM$ salicylhydroxamic acid; PG propyl gallate. *Heavy arrows* indicate generation (*1*) and scavenging (*2*) of free radicals. (RUSTIN et al. 1983 b)

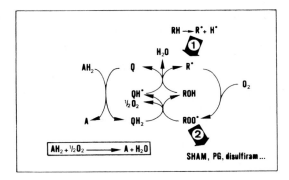

reactions they can undergo. An explanation for the possible implication of lipoxygenase in the alternative pathway was given when it was observed that plant mitochondria, lipoxygenase, and pure cytochrome c could catalyze an oxygen uptake in the presence of linoleic acid (DUPONT et al. 1982). However, the details of the reactions were quite different, depending on the mode of preparation of the linoleic acid emulsions (aerobic or anaerobic). The direct lipoxygenase assay, by measurement of the disappearance of linoleic acid rather than by measurement of O_2 uptake (RUSTIN et al. 1982), made it clear that linoleic acid hydroperoxides were key intermediary components in these reactions, being responsible for the O_2 uptake observed in the presence of mitochondria and cytochrome c.

Considering these results and the fact that unsaturated fatty acids are quantitatively well represented in the phospholipids of mitochondrial membranes (BEN ABDELKADER and MAZLIAK 1971) and are easily released upon damaging the membranes, a new hypothesis was proposed in which peroxy radicals of unsaturated fatty acids play a major role (RUSTIN et al. 1983a, b). This hypothesis is based on the fact that fatty acid radicals or fatty acid peroxy radicals can be spontaneously generated either in aqueous emulsions or in biological membranes. These radicals can be generated by lipoxygenase (GALLIARD and CHAN 1980) but also in the presence of UV radiations, high temperatures, or superoxide. On the other hand, it has been shown that fatty acid peroxy radicals could easily react with reduced quinones (RUSTIN et al. 1983b, 1984).

This being given, a scheme of the alternative pathway in plant mitochondria can be presented (Fig. 8), although the detailed reaction mechanisms of some steps are probably open to further examination. In this scheme, the electrons of all respiratory substrates contribute to the production of reduced ubiquinone (QH_2). On the other hand, fatty acid radicals ($R^·$) are produced (naturally or by contaminating lipoxygenase) and react with O_2 to give peroxy radicals ($ROO^·$). These peroxy radicals can then react with QH_2 in a two-step process, with the transitory formation of $QH^·$ and ROH, to yield Q and H_2O. This scheme is supported by rather good experimental evidence (RUSTIN et al. 1983b) and possesses several advantages over previous hypotheses. First, it could be more general than it appears in Fig. 8. In fact, radicals can be generated in multiple points (flavin, ubiquinone) in the respiratory chain (FORMAN and Bo-

VERIS 1982) resulting in the production of superoxide $O_2^{\bar{\cdot}}$ in the presence of oxygen, and finally of ROO˙ in the presence of membrane unsaturated fatty acids. Therefore the reactions of Fig. 8 could arise at different levels along the respiratory chain, and not only with R˙ or ROO˙. A further advantage is that this alternate pathway scheme, like the ubiquinone hypothesis, does not need an oxidase to carry out the oxygen uptake, thus solving a rather unsolvable question. The oxidase is replaced by the reaction: $R˙ + O_2 \rightarrow ROO˙$. It has also been observed that the effects of inhibitors on the alternative pathway are by no means simple (cf. Sect. 4.3, 4.4, and 7.3). At least three sites of action could be identified. In this scheme, the effects of the inhibitors could be explained simply by assuming that they act as free radical scavengers (RABEK and RANBY 1978). It should also be mentioned that these inhibitors, acting as free radical scavengers, could also interfere at the level of QH˙. Apparently, this is not exactly the case, since these inhibitors are not active on the cytochrome pathway, of which Q/QH_2 is also a part. If ubiquinone is compartimented in the mitochondrial membrane (cf. Sect. 7.2.1), this would not be a problem.

One major question arising from this model is the problem of its regulation in the plant cell. This concerns both the induction of the alternative pathway under various internal or external conditions (LAMBERS, Chap. 14, this Vol.) and the regulation of its functioning within the mitochondrial membrane. The activity of the respiratory chain itself, by producing free radicals, could eventually control the extent of activity of the alternative pathway. Finally, a sequence of reactions like those of Fig. 8 could as well take place in any membrane system in which a source of electrons (AH_2), a quinone and free unsaturated fatty acids would be present together. It is therefore not surprising that such a "classical" cyanide-resistant electron transport has been described ... in the chloroplast, and named chlororespiration (BENNOUN 1982).

8.9 Conclusions

As it stands, the problem of the alternative pathway oxidase is not definitively solved. Over the years, a variety of hypotheses have been proposed which have two characters in common: (a) they have all been "fashionable" hypotheses, following, over a period of 50 years, the progress in the knowledge of the respiratory chain (autooxidizable flavoprotein, cytochrome oxidase, cytochrome b, nonheme iron protein, ubiquinone, free radicals); (b) it is probable that they all hide a part of the truth that, at the present time, has not been fully apprehended by specialists in the field. Summing up, the alternative pathway oxidase had a glorious past and promises to have a brilliant future.

9 The Distribution of Electrons Between the Two Pathways

The distribution of electrons between the two pathways is a fundamental problem, since it has strong consequences on the yield of ATP associated with the mitochondrial electron transport. The consequences are still more important for the regulation of energy metabolism in plants. As this question will be dealt with elsewhere (LAMBERS, Chap. 14, this Vol.), here the problem will be examined at the mitochondrial level only. Moreover, we will come back to a scheme even simpler than that of the alternative pathway presented in Fig. 1, leaving aside the peculiarities relating to the organization of this pathway or to its terminal oxidation step.

9.1 Some Definitions and Remarks

Figure 9 is a simple scheme summarizing the situation of electron transport in plant mitochondria. Joining together succinate and internal NADH oxidations, it consists of three main pathways (flavoprotein, cytochrome, alternative). For each pathway a distinction has to be made between its activity, i.e., the actual electron flow (v), and its capacity, i.e., the maximal electron flow it can accomodate (V). The total electron flow in the flavoprotein pathway (v_T, V_T) is distributed between the cytochrome (v_{cyt}, V_{cyt}) and the alternative pathways (v_{alt}, V_{alt}) [Eqs. (1) and (2)]. If the alternative and the cytochrome pathways do not operate at full capacity, their actual activity represents a fraction (ρ, ρ') of that capacity [Eqs. (3) and (4), with $0 \leq \rho, \rho' \leq 1$]. Therefore to express the actual activity of the respiratory chain, one is confronted with a set of three equations: in Eq. (5) both pathways operate at a fraction of their capacities (normally, this should represent the most general situation) while in Eqs. (6) and (7), one of the pathways is privileged, working at full capacity, and the other one at reduced capacity. When both pathways operate at full capacity one returns to Eq. (2), with the implication that the factors limiting the electron flow (V_T) are the capacities of the cytochrome and alternative pathways and not the capacities or activities of the initial dehydrogenases. For reasons which will become apparent later, Eq. (7) is never used, Eq. (5) has been severely

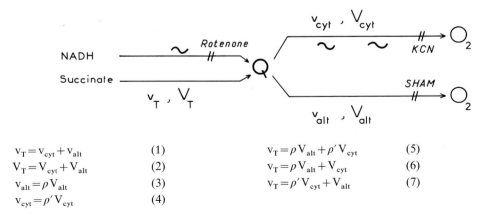

$$v_T = v_{cyt} + v_{alt} \quad (1) \qquad\qquad V_T = \rho\,V_{alt} + \rho'\,V_{cyt} \quad (5)$$
$$V_T = V_{cyt} + V_{alt} \quad (2) \qquad\qquad V_T = \rho\,V_{alt} + V_{cyt} \quad (6)$$
$$v_{alt} = \rho\,V_{alt} \quad (3) \qquad\qquad V_T = \rho'\,V_{cyt} + V_{alt} \quad (7)$$
$$v_{cyt} = \rho'\,V_{cyt} \quad (4)$$

Fig. 9. Quantitative repartition of electrons between the cytochrome and the alternative pathways (see text)

criticized and only a version of Eq. (6) is universally used; it is known as the BAHR and BONNER's (1973a) equation.

None of the above parameters, however, is easy to determine. The access to V_T would imply both the use of an uncoupler to release the control of oxidative phosphorylation and that of a "cocktail" of substrates to saturate the initial dehydrogenases and both pathways. The situation of multisubstrate oxidation is by no means simple (DAY and WISKICH 1977, LAMBERS et al. 1983). An FAD-linked substrate (succinate) can increase the rate of oxidation in the presence of several NAD-linked substrates, but the same is true of glycine oxidation, which implies some separate access (compartmentation?) to the NADH dehydrogenase of Complex I (WALKER et al. 1982). In practice, one is mostly interested in the oxidation of a single substrate. A maximal rate is obtained by using substrate concentrations several times higher than the K_m for the mitochondrial oxidation of this substrate. Therefore, it results that different V_T are measured for different substrates.

The determinations of ρ or ρ' imply the ability to measure V_{cyt} and V_{alt}, the capacities of the pathways. These quantities are not easy to measure either, as will be seen below. The activities v_{cyt} and v_{alt} can be measured from the effects of the specific inhibitors KCN and SHAM used at appropriate efficient concentrations. However, there is a fundamental proviso: in the presence of one specific inhibitor, the electrons should not be rerouted toward the other noninhibited pathway. This situation again is complex and only sophisticated tricks or assumptions can be used to solve the problem.

Finally, it should also be mentioned that Eqs. (1) and (2) are sometimes presented in a more general form, such as: $v_T = v_{cyt} + v_{alt} + v_{res}$, v_{res} representing the residual electron flow that is resistant to both SHAM + KCN used together. Whatever the meaning of this v_{res} (cf. Sect. 4.4), this quantity is usually substracted from v_T, and one then uses a simplified equation: $v'_T = v_{cyt} + v_{alt}$. With tissues v_{res} can be important, but it is negligable with mitochondria, except with *A. maculatum* mitochondria (Fig. 3).

9.2 Distribution of Electrons Between Pathways

Practically, given an actual rate of electron transport in mitochondria (or respiratory rate for a tissue), one is interested in knowing the distribution of electrons between the two pathways. The major trouble is that, except in one instance (which applies only to mitochondria but not to tissues), one has to make use of inhibitors, and therefore inhibit the system, to know how things do occur in the non-inhibited system. Basically, three types of approaches have been used to solve the problem.

9.2.1 Bahr and Bonner's Method

In a most cited paper, BAHR and BONNER (1973a) published a method that was largely used afterwards not only with mitochondrial preparations, but also with tissue and organ respiration (LAMBERS, Chap. 14, this Vol.).

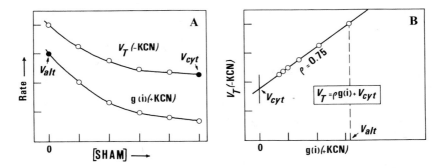

Fig. 10A, B. The BAHR and BONNER's method of measuring the distribution of electrons between the cytochrome and the alternative pathways (see text). (After BAHR and BONNER 1973a)

Roughly speaking, in this method, the rate of a substrate oxidation is measured in the presence of a range of concentrations of SHAM (or of an analog of SHAM) both in the absence or in the presence of a concentration of cyanide sufficient to fully inhibit the cytochrome pathway. One therefore gets a set of two curves (Fig. 10A). Then, (a) if one postulates that the rate in the cytochrome pathway is not affected by the changes intervening in the alternative pathway rate (in the presence of SHAM) and (b) if the maximal contribution of the alternative pathway at each concentration of SHAM is represented by a function $g(i)$, the equation of BAHR and BONNER (1973a) is expressed by the formula (in the authors' original terminology and using *italicized* symbols whose significance is not the same as for the symbols defined above):

$$V_T = \rho g(i) + V_{cyt} \tag{8}$$

where V_T is the observed rate, V_{cyt} is the rate in the cytochrome pathway (not affected by fluctuation of the rate in the alternative pathway) and ρ a factor comprised between 0 and 1, $\rho g(i)$ representing the actual contribution of the alternative pathway. To have an expression of Eq. (8) for each concentration of SHAM one plots the rate in the absence of cyanide (V_T) versus the rate in the presence of cyanide [$g(i)$]. If the above assumptions are correct, one should obtain a straight line (Fig. 10B). This is usually observed, and therefore justifies the assumptions that were made. From this figure, one can obtain V_{cyt} by extrapolation, ρ is measured graphically and $g(i)$ (when SHAM is absent, $g(i)$ is maximum and equals V_{alt}) can be calculated, yielding the final formulation of the BAHR and BONNER's equation, as it is usually known:

$$V_T = \rho V_{alt} + V_{cyt}. \tag{9}$$

It is clear that Eq. (9) does not describe maximal rates as defined above. The parameters apply to particular situations: for instance, V_T and V_{cyt} are different for State 3 and State 4 rates; only V_{alt} is the same for both states and represents "a capacity" of the alternative pathway. V_T and V_{cyt} are in fact v_T and v_{cyt} as defined above, while V_{alt} is not really V_{alt} (see below). It should noted that V_{cyt} (measured by extrapolation, Fig. 10 B) in Eqs. (8) and (9) can be different from the rate one can measure in the presence of a high concentration of SHAM (in the absence of KCN). Similarly, V_{alt} can be different from the rate one can measure in the presence of cyanide and in the absence of SHAM (black symbols on Fig. 10 A). As a matter of fact, the true equation describing any particular situation would be (using the symbols defined previously):

$$v_T = \rho V_{alt} + v_{cyt} \tag{10}$$

with the hypothesis that v_{cyt} is always the highest possible, but variable and not necessarily maximal, depending on the experimental conditions (State 3, State 4, uncoupled State), as it is justified by current observations.

This method has been widely used and, in accordance with the hypotheses made, has contributed to the idea that the alternative pathway operates as an overflow mechanism when the cytochrome pathway is inhibited (State 4) or cannot accommodate the overall electron flow (uncoupling) (LAMBERS 1982, LATIES 1982). A simplified version of the BAHR and BONNER's equation is often used when dealing with tissues and organs. In that case, one measures the rates of O_2 uptake in the absence of inhibitors, in the presence of either SHAM or KCN, or of both inhibitors. After substracting the residual respiration, the SHAM-sensitive rate is considered as measuring the activity of the alternative pathway (v_{alt}) while the cyanide-resistant rate is considered as a measure of the capacity of this pathway (V_{alt}). The ratio between the two parameters (ρ) gives an estimation of the participation of the alternative pathway (LAMBERS et al. 1983).

Though widely used, the method of BAHR and BONNER has sometimes some failures: (a) values of ρ higher than 1 can sometimes be obtained (CHAUVEAU 1980, LAMBERS et al. 1983). This means that SHAM has probably lost its specificity for inhibiting the alternative pathway. (b) In a similar manner to the method described in Fig. 10, one can also titrate the activity of electron transport against cyanide, instead of SHAM. This allows one to determine a ρ' factor measuring the extent of the engagement of the cytochrome pathway (CHAUVEAU 1980, DIZENGREMEL 1980). This, of course, is contrary to the basic assumption of the BAHR and BONNER's equation. Although there is no theoretical reason to favor the cytochrome pathway over the cyanide-resistant pathway, this assumption has been justified by the observations that have been made. It is also satisfactory in that it favors energy production (ATP synthesis) over energy dissipation (heat). (c) With the same mitochondrial preparation, one can measure the oxidation rates of several substrates (succinate, malate, NADH, for instance). For each substrate one can also derive an equation like Eq. (9). In each equation, there will be a different V_{cyt}, in accordance with the original hypothesis, but there will be also a different V_{alt}. It is clear therefore that those V_{alt} do not represent the true capacity of the alternative pathway (V_{alt}). Consequently, ρV_{alt} represents an undefined contribution of the alternative pathway: it does not represent the "engagement" of the alternative pathway (ρV_{alt}), as it is usually understood.

As a matter of fact, the only things interesting to know are the respective contributions, either absolute (nmol O_2) or relative (%), of each pathway in any given rate of substrate oxidation. The "engagement" of each pathway, i.e., the ratio activity/capacity, is really of minor importance. Surprisingly enough, everybody cares about the "engagement" of the alternative pathway but not at all about the "engagement" of the cytochrome pathway, which is deemed to be "maximal" under any set of circumstances. This notion of "engagement" is of secondary importance when dealing with isolated mitochondria, but it has strong physiological implications when dealing with tissue or organ respiration (LATIES 1982, LAMBERS, Chap. 14, this Vol.). A reappraisal of this concept could eventually clarify some situations.

9.2.2 De Troostembergh and Nyns's Method

This method (DE TROOSTEMBERGH and NYNS 1978) was proposed to account for the electron transport pattern in the yeast *Saccharomycopsis lipolytica*. The basic assumption of this method is that ubiquinone feeds the electrons to the alternative and the cytochrome pathways proportionally to their respective electron transport capacities (V_{cyt}, V_{alt}). This model was an adaptation to plant cyanide-resistant mitochondria of a model developed by KRÖGER and KLINGEN-BERG (1973) for electron transport in animal mitochondria. The principle of a competition between the two pathways at the level of ubiquinone was initially rejected by BAHR and BONNER (1973b). This method has also been severely criticized by LATIES (1982). Probably, its major theoretical drawback is that it implies that a fraction of the alternative pathway (i.e., not producing ATP) is constantly used in any type of cyanide-resistant mitochondria: a situation still more uncomfortable than assuming that the activity of the cytochrome pathway is always maximal, since, deliberately, a part of the energy is lost with no possibility of modulating this loss.

9.2.3 ADP/O Ratio Method

LAMBOWITZ et al. (1972) have also published a method to evaluate the participation of the alternative pathway that is based on the yield of ATP in cyanide-resistant mitochondria. According to this method (two examples are given in Table 2), one measures the ADP/O ratio in the absence of inhibitors and then in the presence of inhibitors of the cytochrome (antimycin, cyanide) and alternative (SHAM, m-CLAM) pathways. The first group of inhibitors decreases the yield of ATP while the second group increases this yield (cf. Sect. 5.1, Table 2).

Given the theoretical yields of ATP (2 for succinate, 3 for malate), the yield in the presence of a hydroxamic acid allows the determination of the efficiency (eff) of the oxidative phosphorylation process in these mitochondria. The yield with malate in the presence of an inhibitor of the cytochrome pathway also gives an idea of this efficiency, since site I is functioning. The efficiency of oxidative phosphorylation can also be measured with a substrate, such as ascorbate, whose electrons are not diverted to the alternative pathway. For instance, the efficiency is about 40% in *A. maculatum* mitochondria (Table 2). For each substrate, one can then establish an equation relating the observed yield of ATP and the proportion of the electron flow carried by each pathway:

$$ADP/O_{(measured)} = [2(1-x)+0(x)] \times eff \quad \text{for succinate}$$
$$= [3(1-x)+1(x)] \times eff \quad \text{for malate}$$

(the terminology is not that of Lambowitz et al. 1972), x being the fraction of the electron flow which is mediated by the alternative pathway, and $(1-x)$ the complementary flow through the cytochrome pathway. It should be remembered that the cytochrome pathway associates the synthesis of 3 ATP for malate and 2 for succinate, while the corresponding values are 1 and 0 for the alternative pathway (Figs. 1 and 9). In the above equations, theoretical values are obtained when eff = 1 and x = 0.

This method has not been widely used (Lambowitz et al. 1972, Passam 1974, Lance 1979, Dizengremel 1980) although it is one of the best, since it does not make use of inhibitors to measure the actual rates of electron transport in both pathways. It should be said, however, that its range of application is limited: it applies only to cyanide-resistant mitochondria oxidizing a substrate in State 3; it does not apply to the State 4 condition nor to cells or tissues.

9.3 Mechanism of Electron Distribution

This point has been considered in detail by Bahr and Bonner (1973b) and discussed by Storey (1980). It largely depends on the idea one favors about the alternative pathway oxidase. Assuming that it exists, the best explanation would be an equilibrium between two components, respectively linked to one of each pathway, in the flavoprotein region of the respiratory chain. Due to particular redox potentials, the second component could become reduced only when the first one has reached a relatively high level of reduction (Bahr and Bonner 1973b). For a time, ubiquinone and Fp_{ma} (cf. Sect. 6) seemed to meet the requirement (Storey 1976). The situation appears less certain now (Storey 1980).

10 The Biogenesis of the Alternative Pathway

Conditions can be found that make the cyanide-resistant pathway display or increase its activity (aging, treatment with ethylene, chloramphenicol, changes in the composition of growth medium, natural stresses, etc.) (Henry and Nyns 1975, Lance and Dizengremel 1978, Laties 1982, Lambers 1982). However, when one examines the events occurring at the mitochondrial level, it is very difficult to observe changes in the composition of the mitochondrial membranes that could be related to the functioning of the alternative pathway. For instance, the composition in cytochromes (DIZENGREMEL 1975, LANCE 1981), in nonheme iron proteins (DIZENGREMEL et al. 1973, DIZENGREMEL 1980, LANCE 1981) or in membrane polypeptides (DIZENGREMEL and KADER 1980) shows only insignificant variations. The proposal has therefore been made that the cyanide-resistant pathway is present in all plant mitochondria; its functioning, however, would depend on certain external or internal conditions (DIZENGREMEL and LANCE 1976). In short, it should be a matter of regulation rather than one of biogenesis. Different studies, however, have shown that the biosynthesis of a protein (the oxidase or some intermediary component) should be necessary for the pathway to become functional. These experiments have been carried out mostly with microorganisms: *Neurospora crassa* (EDWARDS et al. 1974, EDWARDS and ROSENBERG 1976) or *Saccharomycopsis lipolytica* (DE TROOSTEMBERGH et al. 1978). Two molecular models of the control of the biosynthesis of this protein by one or several nuclear genes have been produced (EDWARDS et al. 1974, DE TROOSTEMBERGH et al. 1978). Finally, it should also be recalled that the appearance of the cyanide-resistant pathway in potato slices (DIZENGREMEL and LANCE 1976) or during the germination of *Cicer arietinum* seeds (BURGUILLO and NICOLÁS 1977) is inhibited by cycloheximide but not by chloramphenicol. This again points to the biosynthesis of a protein on cytoplasmic ribosomes, under the control of nuclear genes.

11 The Significance of the Alternative Pathway

Every recent review article on cyanide resistance has a terminal section on the significance of the alternative pathway (PALMER 1976, SOLOMOS 1977, LAMBERS 1980, 1982, LANCE 1981, LATIES 1982, DIZENGREMEL and LANCE 1982). This gives the authors an opportunity to express their own personal views on the subject. This article will be no exception. Rather than covering the field exhaustively (LAMBERS, Chap. 14, this Vol.), a biased point of view (it is the authors' privilege) will be presented here concerning the scheme of the alternative pathway as it appears in Fig. 5. The main feature of this scheme is that malate can be oxidized in two ways: one that produces ATP, through malate dehydrogenase; another, through malic enzyme, that can produce or not produce ATP, depending on the fate of the NADH resulting from the reaction (Fig. 5). This property, whose control mechanism is not well understood at the present time, gives malate a key role in the regulation of energy metabolism in the plant cell (LANCE and RUSTIN 1984).

Malate is generally the quantitatively most represented organic acid of the plant cell. Many enzymes, in many compartments of the cell, are concerned with malate metabolism (malate dehydrogenase, malic enzyme, phosphoenolpyruvate carboxylase, phosphoenolpyruvate carboxykinase, etc.). It is not at all irrational to imagine that the product of glycolysis in many plant cells is not pyruvate but malate (by carboxylation and reduction of phosphoenolpyruvate) and that the substrate of the Krebs cycle is malate (through malic enzyme activity) instead of pyruvate. By undergoing the following reactions:

$$\text{phosphoenol pyruvate} + CO_2 \rightleftharpoons \text{oxaloacetate}$$
$$\text{oxaloacetate} + NADH + H^+ \rightarrow \text{malate} + NAD^+$$
$$\text{malate} + NAD^+ \rightleftharpoons \text{pyruvate} + CO_2 + NADH$$

catalyzed by phosphoenolpyruvate carboxylase, malate dehydrogenase, and malic enzyme, malate appears as a storage molecule for both CO_2 and NADH, not to speak of its regulatory function in controlling the cell pH (two of the above equations involve equilibrium between dicarboxylic and monocarboxylic acids) (DAVIES 1977).

   As a storage form of reducing power, malate moves easily across cell membranes, and, by a variety of shuttle mechanisms, can equilibrate the reducing power of the different cell compartments (cell membranes are well known to be rather impermeable to NADH and NADPH). Malate, therefore appears as a mobile reducing power. An excess of reducing power in the cell can be removed easily, without any control by the phosphate potential or energy charge, via malic enzyme and the nonphosphorylating cyanide-resistant pathway.

   The same mechanism can also play a role in CAM and C_4 plants, and this indeed has been shown to occur (RUSTIN 1981, GARDESTRÖM and EDWARDS 1983). In these plants, photosynthesis is a two-step process as far as CO_2 fixation is concerned. First, CO_2 is accumulated as malate, during the night (CAM plants) or in mesophyll cells (C_4 plants). Malate is then used, in the day time (CAM plants) or in the bundle-sheath cells (C_4 plants), as a source of CO_2. CO_2 is produced by the malic enzyme which also generates NADH. It is essential that this NADH be readily reoxidized in order to allow more production of CO_2 for photosynthesis. This reoxidation should also escape the control by the energy charge: a decarboxylation of malate by the mitochondrial malic enzyme solves the problem.

   While we are at it, why not also implicate malate, malic enzyme, and the alternative pathway in the regulation of the intracellular pH of the plant cell, which has such a large vacuole?

   These types of reaction, which integrate the functioning of the alternative pathway into the general metabolism of the plant cell, probably give the alternative pathway a greater significance than its participation in some spectacular events, such as thermogenesis in aroid spadix for instance. Therefore, the main advantage of the scheme of electron transport in Fig. 5 is its implication in the regulation of the plant cell metabolism (LANCE and RUSTIN 1984).

12 Conclusion

This overview of the current knowledge on the cyanide-resistant electron transport in plant mitochondria has shown both the progress that has been made in this field but also the many uncertainties that still remain. In particular, the problem of the alternative pathway oxidase is still intriguing. The fate of the many hypotheses that have been produced may be changing; however, the unquestionable experimental facts on which these hypotheses were built remain. No doubt one day all these dispersed observations will become integrated in a general scheme of the cyanide-resistant electron transport, unless more than one type of cyanide resistance exists in plant mitochondria – a question which has never been raised. New experiments, moderate imagination, serendipity, and modesty are the key ingredients to solve the problem.

References

Arron GP, Edwards GE (1979) Oxidation of reduced nicotinamide adenine dinucleotide phosphate by plant mitochondria. Can J Biochem 57:1392–1399

Arron GP, Spalding MH, Edwards GE (1979) Isolation and oxidative properties of intact mitochondria from the leaves of *Sedum praealtum*. A crassulacean acid metabolism plant. Plant Physiol 64:182–186

Bahr JT, Bonner WD (1973a) Cyanide-insensitive respiration. I The steady states of skunk cabbage spadix and bean hypocotyl mitochondria. J Biol Chem 248:3441–3445

Bahr JT, Bonner WD (1973b) Cyanide-insensitive respiration. II Control of the alternate pathway. J Biol Chem 248:3446–3450

Beinert H, Palmer G (1965) Contribution of EPR spectroscopy to our knowledge of oxidative enzymes. Adv Enzymol 27:105–198

Ben Abdelkader A, Mazliak P (1971) Renouvellement des lipides dans diverses fractions cellulaires de parenchyme de Pomme de terre ou de Chou-fleur. Physiol Vég 9:227–240

Bendall DS (1958) Cytochromes and some respiratory enzymes in mitochondria from the spadix of *Arum maculatum*. Biochem J 70:381–390

Bendall DS, Bonner WD (1971) Cyanide-insensitive respiration in plant mitochondria. Plant Physiol 47:236–245

Bendall DS, Hill R (1956) Cytochrome components in the spadix of *Arum maculatum*. New Phytol 55:206–212

Bennoun P (1982) Evidence for a respiratory chain in the chloroplast. Proc Natl Acad Sci USA 79:4352–4356

Bonner WD (1965) Mitochondria and electron transport. In: Bonner J, Varner JE (eds) Plant biochemistry. Academic Press, London New York, pp 89–123

Bonner WD, Rich PR (1983) p-Quinol: oxygen oxidoreductase, a new copper oxidase in plant mitochondria. Plant Physiol 72 Suppl:19

Bonner WD, Christensen EL, Bahr JT (1972) Cyanide- and antimycin-insensitive respiration. In: Azzone GF, Carafoli E, Lehninger AL, Quagliariello E, Siliprandi N (eds) Biochemistry and biophysics of mitochondrial membranes. Academic Press, London New York, pp 113–119

Burguillo PF, Nicolás G (1977) Appearance of an alternate pathway cyanide-resistant during germination of seeds of *Cicer arietinum*. Plant Physiol 60:424–427

Cammack R, Palmer JM (1973) EPR studies of iron-sulphur proteins of plant mitochondria. Ann NY Acad Sci 222:816–823

Cammack R, Palmer JM (1977) Iron-sulphur centres in mitochondria from *Arum maculatum* spadix with very high rates of cyanide-resistant respiration. Biochem J 166:347–355

Carmeli C, Biale JB (1970) The nature of the oxidation states of sweet potato mitochondria. Plant Cell Physiol 11:65–81

Chance B, Hackett DP (1959) The electron transfer system of skunk cabbage mitochondria. Plant Physiol 34:33–49

Chance B, Williams GR (1965) The respiratory chain and oxidative phosphorylation. Adv Enzymol 17:65–134

Chance B, Bonner WD, Storey BT (1968) Electron transport in respiration. Annu Rev Plant Physiol 19:295–320

Chauveau M (1976a) La chaîne respiratoire des mitochondries d'*Arum*. I Inhibition de la voie d'oxydation cytochromique. Physiol Vég 14:309–323

Chauveau M (1976b) La chaîne respiratoire des mitochondries d'*Arum*. II Inhibition de la voie d'oxydation insensible au cyanure. Physiol Vég 14:325–337

Chauveau M (1977) Variabilité de la résistance apparente à l'action inhibitrice du cyanure dans les mitochondries d'*Arum*. CR Acad Sci Sér D 284:1281–1284

Chauveau M (1980) Etude comparative des mécanismes de transfert d'électrons dans les mitochondries végétales sensibles ou résistantes au cyanure. Ph D Thes, Univ Pierre et Marie Curie, Paris

Chauveau M, Lance C (1971) Les mitochondries de l'inflorescence de Chou-fleur. II Action des inhibiteurs de transport des électrons. Physiol Vég 9:353–372

Chauveau M, Lance C (1982) Respiration et thermogenèse chez les Aracées. Bull Soc Bot Fr 129, Actual Bot (2):123–134

Chauveau M, Dizengremel P, Lance C (1978) Thermolability of the alternative electron transport pathway in higher plant mitochondria. Physiol Plant 42:214–220

Chauveau M, Dizengremel P, Roussaux J (1983) Interaction of benzylaminopurine with electron transport in plant mitochondria during malate oxidation. Plant Physiol 73:945–948

Coleman JOD, Palmer JM (1972) The oxidation of malate by isolated plant mitochondria. Eur J Biochem 26:499–509

Cotte-Martinon M, Diano M, Meunier D, Ducet G (1969) Étude de diverses fractions cellulaires dans le tubercule de Pomme de terre au cours de la survie. I Oxydations respiratoires et composition en cytochromes. Bull Soc Fr Physiol Vég 15:279–295

Davies DD (1977) Control of pH and glycolysis. In: Smith H (ed) Regulation of enzyme synthesis and activity in higher plants. Academic Press, London New York, pp 41–62

Day DA, Wiskich JT (1977) Factors limiting respiration by isolated cauliflower mitochondria. Phytochemistry 16:1499–1502

Day DA, Wiskich JT (1981) Glycine metabolism and oxalacetate transport by pea leaf mitochondria. Plant Physiol 68:425–429

Day DA, Arron GP, Laties GG (1980) Nature and control of respiratory pathway in plants: the interaction of cyanide-resistant respiration with the cyanide-sensitive pathway, Vol II. In: Davies DD (ed) The biochemistry of plants. Academic Press, London New York, pp 197–241

Dizengremel P (1975) La voie d'oxydation insensible au cyanure dans les mitochondries de tranches de Pomme de terre (*Solanum tuberosum* L) maintenues en survie. Physiol Vég 13:39–54

Dizengremel P (1980) Modifications des transferts d'électrons mitochondriaux au cours de la survie de tranches de tubercule de Pomme de terre (*Solanum tuberosum* L). Ph D Thes, Univ Pierre et Marie Curie, Paris

Dizengremel P (1983) Effect of Triton X-100 on the cyanide-resistant pathway in plant mitochondria. Physiol Vég 21:743–752

Dizengremel P, Chauveau M (1978) Mechanism of temperature action on electron transport in cyanide-sensitive and cyanide-resistant plant mitochondria. In: Ducet G, Lance C (eds) Plant mitochondria. Elsevier/North Holland Biomedical Press, Amsterdam, pp 267–274

Dizengremel P, Kader JC (1980) Effect of aging on the composition of mitochondrial membranes from potato slices. Phytochemistry 19:211–214

Dizengremel P, Lance C (1976) Control of changes in mitochondrial activities during aging of potato slices. Plant Physiol 58:147–151

Dizengremel P, Lance C (1982) La respiration insensible au cyanure chez les végétaux. Bull Soc Bot Fr 129, Actual Bot (2):19–36

Dizengremel P, Chauveau M, Lance C (1973) Résistance au cyanure et teneur en ferrosulfoprotéines des mitochondries végétales. CR Acad Sci 277, Sér D:239–242

Dizengremel P, Chauveau M, Roussaux J (1982) Inhibition by adenine derivatives of the cyanide-insensitive electron transport pathway of plant mitochondria. Plant Physiol 70:585–589

Douce R, Bonner WD (1972) Oxalacetate control of Krebs cycle oxidation in purified plant mitochondria. Biochem Biophys Res Commun 47:619–624

Douce R, Manella CA, Bonner WD (1973) The external NADH dehydrogenases of intact plant mitochondria. Biochim Biophys Acta 292:105–116

Douce R, Moore AL, Neuburger M (1977) Isolation and oxidative properties of intact mitochondria isolated from spinach leaves. Plant Physiol 60:625–628

Douce R, Neuburger M, Bligny R, Pauly G (1978) Effects of β-pinene on the oxidative properties of purified intact plant mitochondria. In: Ducet G, Lance C (eds) Plant mitochondria. Elsevier/North Holland Biomedical Press, Amsterdam, pp 207–214

Drabikowska AK (1977) Dibromothymoquinone inhibition of the cyanide-sensitive and -insensitive respiration of MI-1 mutant of Neurospora. Life Sci 21:667–674

Ducet G (1982a) Les mécanismes respiratoires chez les végétaux. Bull Soc Bot Fr 129, Actual Bot (2):7–18

Ducet G (1982b) Respiratory coupling in potato tuber mitochondria. Effect of mitochondrial aging. Physiol Vég 20:187–199

Ducet G, Rosenberg AJ (1962) Leaf respiration. Annu Rev Plant Physiol 13:171–200

Dupont J (1981) Lipoxygenase-mediated cleavage of fatty acids in plant mitochondria. Physiol Plant 52:225–232

Dupont J, Rustin P, Lance C (1982) Interaction between mitochondrial cytochromes and linoleic acid hydroperoxide. Possible confusion with lipoxygenase and alternative pathway. Plant Physiol 69:1308–1314

Edwards DL (1978) Cyanide-insensitive respiratory systems in Neurospora. In: Degn H, Lloyd D, Hill GC (eds) Functions of alternative terminal oxidases, Vol 49. Pergamon Press, Oxford New York, pp 21–29

Edwards DL, Rosenberg E (1976) Regulation of cyanide-insensitive respiration in Neurospora. Eur J Biochem 62:217–221

Edwards DL, Rosenberg E, Maroney PA (1974) Induction of cyanide-insensitive respiration in Neurospora crassa. J Biol Chem 249:3551–3556

Erecinska M, Storey BT (1970) The respiratory chain of plant mitochondria. VII Kinetics of flavoprotein oxidation in skunk cabbage mitochondria. Plant Physiol 46:618–624

Estabrook RW (1962) Observations on the antimycin A inhibition of biological oxidations. I Stoichiometry and pH effects. Biochim Biophys Acta 60:236–248

Forman HJ, Boveris A (1982) Superoxide radical and hydrogen peroxide in mitochondria. In: Pryor WA (ed) Free radicals in biology, Vol V. Academic Press, London New York, pp 65–90

Galliard T, Chan HWS (1980) Lipoxygenases. In: Stumpf PK (ed) The biochemistry of plants, vol IV. Academic Press, London New York, pp 131–161

Gardeström P, Edwards GE (1983) Isolation of mitochondria from leaf tissue of Panicum miliaceum, a NAD-malic enzyme type C_4 plant. Plant Physiol 71:24–29

Garland PB, Clegg RA, Light PA, Ragan CI (1969) Mechanisms and inhibitors acting on electron transport and energy conservation between NADH and the cytochrome chain. In: Bücher T, Sies H (eds) Inhibitors tools in cell research. Springer, Berlin Heidelberg New York, pp 217–246

Gauvrit C (1978) Effects of some substituted urea herbicides on oxidative phosphorylation in potato tuber mitochondria (Solanum tuberosum L.). In: Ducet G, Lance C (eds) Plant mitochondria. Elsevier/North Holland Biomedical Press, Amsterdam, pp 199–206

Genevois ML (1929) Sur la fermentation et sur la respiration chez les végétaux chlorophylliens. Rev Gén Bot 41:252–271

Goldstein AH, Anderson JO, Mc Daniel RG (1980) Cyanide-insensitive and cyanide-sensitive O_2 uptake in wheat. I Gradient-purified mitochondria. Plant Physiol 66:488–493

Grover SD, Laties GG (1978) Characterization of the binding properties of disulfiram, an inhibitor of cyanide-resistant respiration. In: Ducet G, Lance C (eds) Plant mitochondria. Elsevier/North Holland Biomedical Press, Amsterdam, pp 259–266

Grover SD, Laties GG (1981) Disulfiram inhibition of the alternative respiratory pathway in plant mitochondria. Plant Physiol 68:393–400

Gupta Das U, Rieske JS (1973) Identification of a protein component of the antimycin-binding site of the respiratory chain by photoaffinity labeling. Biochem Biophys Res Commun 54:1247–1254

Gutman M (1977) Functional compartmentation of the mitochondrial quinone during simultaneous oxidation of two substrates. In: Packer L, Pappageorgiou GC, Trebst A (eds) Bioenergetics of membranes. Elsevier/North Holland Biomedical Press, Amsterdam, pp 165–175

Hackett DP (1957) Respiratory mechanisms in the aroid spadix. J Exp Bot 8:157–171

Hackett DP (1959) Respiratory mechanisms in higher plants. Annu Rev Plant Physiol 10:113–146

Hackett DP, Haas DW (1958) Oxidative phosphorylation and functional cytochromes in skunk cabbage mitochondria. Plant Physiol 33:27–32

Hackett DP, Haas DW, Griffiths SK, Niederpruem DJ (1960a) Studies on the development of cyanide-resistant respiration in potato tuber slices. Plant Physiol 35:8–19

Hackett DP, Rice B, Schmid C (1960b) The partial dissociation of phosphorylation from oxidation in plant mitochondria by respiratory chain inhibitors. J Biol Chem 235:2140–2144

Haddock BA, Garland PB (1971) Effect of sulphate-limited growth on mitochondrial electron transfer and energy conservation between reduced nicotinamide-adenine dinucleotide and the cytochromes in *Torulopsis utilis*. Biochem J 124:155–170

Hanes CS, Barker J (1931) The effects of cyanide on the respiration and sugar content of the potato at 15° C. Proc R Soc London Ser B 108:95–118

Henry MF, Nyns EJ (1975) Cyanide-insensitive respiration. An alternative mitochondrial pathway. Sub-Cell Biochem 4:1–65

Henry MF, Bonner WD, Nyns EJ (1977) Involvement of iron in the biogenesis of the cyanide-insensitive respiration in the yeast *Saccharomycopsis lipolytica*. Biochim Biophys Acta 460:94–100

Henry MF, Troostembergh De JC, Nyns EJ (1978) Biogenesis and properties of the mitochondrial cyanide-insensitive alternative respiratory pathway in the yeast *Saccharomycopsis lipolytica*. In: Degn H, Lloyd D, Hill GC (eds) Functions of alternative terminal oxidases. Pergamon Press, Oxford New York, pp 55–62

Herk Van AWH (1937a) Die chemischen Vorgänge im *Sauromatum*-Kolben. I. Die Fragestellung. Rec Trav Bot Néerl 34:69–156

Herk Van AWH (1937b) Die chemischen Vorgänge im *Sauromatum*-Kolben. II. Mitteilung. Proc Kon Akad Wetensch 40:607–614

Herk Van AWH (1937c) Die chemischen Vorgänge im *Sauromatum*-Kolben. III. Mitteilung. Proc Kon Akad Wetensch 40:709–719

Herk Van AWH, Badenhuizen NP (1934) Über die Atmung und Katalasewirkung im *Sauromatum*-Kolben. Proc Kon Akad Wetensch 37:99–105

Huq S, Palmer JM (1978a) The involvement and possible role of quinone in cyanide-resistant respiration. In: Ducet G, Lance C (eds) Plant mitochondria. Elsevier/North Holland Biomedical Press, Amsterdam, pp 225–232

Huq S, Palmer JM (1978b) Isolation of a cyanide-resistant duroquinol oxidase from *Arum maculatum* mitochondria. FEBS Lett 95:217–220

Huq S, Palmer JM (1978c) Superoxide and hydrogen peroxide production in cyanide-resistant *Arum* mitochondria. Plant Sci Lett 11:351–358

Ikuma H (1972) Electron transport in plant respiration. Annu Rev Plant Physiol 23:419–436

Ikuma H, Bonner WD (1967) Properties of higher plant mitochondria. III. Effects of respiratory inhibitors. Plant Physiol 42:1535–1544

Jackson C, Dench JE, Hall DO, Moore AL (1979) Separation of mitochondria from contaminating subcellular structures utilizing silica sol gradient centrifugation. Plant Physiol 64:150–153

Jagow Von G, Bohrer C (1975) Inhibition of electron transfer from ferrocytochrome b to ubiquinone, cytochrome c_1 and duroquinone by antimycin. Biochim Biophys Acta 387:409–424

Jagow Von G, Klingenberg M (1970) Pathways of hydrogen in mitochondria of *Saccharomyces carlsbergensis*. Eur J Biochem 12:583–592

James TW, Spencer MS (1982) Inhibition of cyanide-resistant respiration in pea cotyledon mitochondria by chloroquine. Plant Physiol 69:1113–1115

James WO, Beevers H (1950) The respiration of *Arum* spadix. A rapid respiration, resistant to cyanide. New Phytol 49:353–374

James WO, Elliott DC (1955) Cyanide-resistant mitochondria from the spadix of an *Arum*. Nature (London) 175:89

Kröger A, Klingenberg M (1973) The kinetics of the redox reactions of ubiquinone related to the electron transport activity in the respiratory chain. Eur J Biochem 34:358–368

Lambers H (1980) The physiological significance of cyanide-resistant respiration. Plant Cell Environ 3:293–302

Lambers H (1982) Cyanide-resistant respiration: A non-phosphorylating electron transport pathway acting as an energy overflow. Physiol Plant 55:478–485

Lambers H, Day DA, Azcón-Bieto J (1983) Cyanide-resistant respiration in roots and leaves. Measurements with intact tissues and isolated mitochondria. Physiol Plant 58:148–154

Lambowitz AM, Smith EW, Slayman CW (1972) Oxidative phosphorylation in *Neurospora* mitochondria. Studies on wild-type, *poky* and chloramphenicol induced wild type. J Biol Chem 247:4859–4865

Lance C (1969) Données récentes sur la composition et l'activité de la chaîne respiratoire chez les végétaux supérieurs. Bull Soc Fr Physiol Vég 15:259–278

Lance C (1972) La respiration de l'*Arum maculatum* au cours du développement de l'inflorescence. Ann Sci Nat Bot, 12. Sér 13:477–495

Lance C (1974) Respiratory control and oxidative phosphorylation in *Arum maculatum* mitochondria. Plant Sci Lett 2:165–171

Lance C (1979) La respiration insensible au cyanure chez les végétaux. Agron Lusit 39:131–152

Lance C (1981) Cyanide-insensitive respiration in fruits and vegetables. In: Friend J, Rhodes MJC (eds) Recent advances in the biochemistry of fruits and vegetables. Academic Press, London New York, pp 63–87

Lance C, Bonner WD (1968) The respiratory chain components of higher plant mitochondria. Plant Physiol 43:756–766

Lance C, Bonner WD (1969) Cyanide and antimycin resistance in plant mitochondria. XI Int Bot Congr, Seatle, p 121

Lance C, Chauveau M (1975) Évolution des activités oxydatives et phosphorylantes des mitochondries de l'*Arum maculatum* L. au cours du développement de l'inflorescence. Physiol Vég 13:83–94

Lance C, Dizengremel P (1978) Slicing-induced alterations in electron transport systems during aging of storage tissues. In: Kahl G (ed) Biochemistry of wounded plant tissues. De Gruyter, Berlin, pp 467–501

Lance C, Rustin P (1984) The central role of malate in plant metabolism. Physiol Vég 22:625–641

Laties GG (1982) The cyanide-resistant, alternative path in higher plant respiration. Annu Rev Plant Physiol 33:519–555

Macrae AR (1971) Effect of pH on the oxidation of malate by isolated cauliflower bud mitochondria. Phytochemistry 10:1453–1458

Macrae AR, Moorhouse R (1970) The oxidation of malate by mitochondria isolated from cauliflower buds. Eur J Biochem 16:96–102

Marsh PB, Goddard DR (1939) Respiration and fermentation in the carrot *Daucus carota*. I Respiration. Am J Bot 26:724–728

Meeuse BJD (1975) Thermogenic respiration in Aroids. Annu Rev Plant Physiol 26:117–126

Miller CO (1979) Cytokinin inhibition of respiration by cells and mitochondria of soybean, *Glycine max* (L.) Merrill. Planta 146:503–511

Miller CO (1980) Cytokinin inhibition of respiration in mitochondria from six plant species. Proc Natl Acad Sci USA 77:4731–4735

Miller MG, Obendorf RL (1981) Use of tetraethylthiuram disulfide to discriminate between alternative respiration and lipoxygenase. Plant Physiol 67:962–964

Mitchell P (1975) Protonmotive redox mechanism of the cytochrome b-c_1 complex in the respiratory chain: protonmotive ubiquinone cycle. FEBS Lett 56:1–6

Møller IM, Palmer JM (1982) Direct evidence for the presence of a rotenone-resistant NADH dehydrognase on the inner surface of the inner membrane of plant mitochondria. Physiol Plant 54:267–274

Moore AL, Bonner WD (1982) Measurements of membrane potentials in plant mitochondria with the safranine method. Plant Physiol 70:1271–1276

Moore AL, Rich PR, Bonner WD, Ingledew WJ (1976) A complex EPR signal in mung bean mitochondria and its possible relation to the alternate pathway. Biochem Biophys Res Commun 72:1099–1107

Moore AL, Bonner WD, Rich PR (1978) The determination of the proton-motive force during cyanide-insensitive respiration in plant mitochondria. Arch Biochem Biophys 186:298–306

Moreau F (1978) The electron transport system of outer membranes of plant mitochondria. In: Ducet G, Lance C (eds) Plant mitochondria. Elsevier/North Holland Biomedical Press, Amsterdam, pp 77–84

Moreau F, Lance C (1972) Isolement et propriétés des membranes externes et internes des mitochondries végétales. Biochimie 54:1335–1348

Moreau F, Romani R (1982) Malate oxidation and cyanide-insensitive respiration in avocado mitochondria during the climacteric cycle. Plant Physiol 70:1385–1390

Moreland DE (1980) Mechanisms of action of herbicides. Annu Rev Plant Physiol 31:597–638

Moreland DE, Huber SC (1978) Fluidity and permeability changes induced in the inner mitochondrial membrane by herbicides. In: Ducet G, Lance C (eds) Plant mitochondria. Elsevier/North Holland Biomedical Press, Amsterdam, pp 191–198

Neuburger M, Douce R (1977) Oxydation du malate, du NADH et de la glycine par les mitochondries de plantes en C_3 et C_4. CR Acad Sci, Sér D 285:881–884

Nichols DG (1982) Bioenergetics. An introduction to the chemiosmotic theory. Academic Press, London New York

Okunuki K (1939) Über den Gaswechsel der Pollen. II. Acta Phytochim 11:27–64

Palmer JM (1972) Inhibition of electron transport in *Arum maculatum* mitochondria by potassium thiocyanate. Phytochemistry 11:2957–2961

Palmer JM (1976) The organization and regulation of electron transport in plant mitochondria. Annu Rev Plant Physiol 27:133–157

Palmer JM, Schwitzguebel JP, Møller IM (1982) Regulation of malate oxidation in plant mitochondria. Response to rotenone and exogenous NAD^+. Biochem J 208:703–711

Parrish DJ, Leopold AC (1978) Confounding of alternate respiration by lipoxygenase activity. Plant Physiol 62:470–472

Passam HC (1974) The oxidative phosphorylation and ATPase activity of *Arum* spadix mitochondria in relation to heat generation. J Exp Bot 25:653–657

Passam HC, Palmer JM (1972) Electron transport and oxidative phosphorylation in *Arum* spadix mitochondria. J Exp Bot 23:366–374

Peterman TK, Siedow JN (1983) Structural features required for inhibition of soybean lipoxygenase-2 by propyl gallate. Evidence that lipoxygenase activity is distinct from the alternative pathway. Plant Physiol 71:55–58

Plas Van Der LHW, Verleur JD (1976) CN-resistant respiration in potato (*Solanum tuberosum* L.). Changes after storage of the tuber. Plant Sci Lett 7:149–154

Pomeroy MK (1975) The effect of nucleotides and inhibitors on respiration in isolated wheat mitochondria. Plant Physiol 55:51–58

Rabek JF, Ranby B (1978) The role of singlet oxygen in the photooxidation of polymers. Photochem Photobiol 28:557–570

Ravanel P, Tissut M, Douce R (1982) Effect of kaempferol on the oxidative properties of intact plant mitochondria. Plant Physiol 69:375–378

Redfearn ER, Whittaker PA, Burgos J (1965) The interaction of electron carriers in the mitochondrial $NADH_2$ and succinate oxidase systems. In: King TE, Morrison M (eds) Oxidases and related redox systems, Vol II. John Wiley, New York, pp 943–959

Rich PR (1978) Quinol oxidation in *Arum maculatum* mitochondria and its application to the assay, solubilisation and partial purification of the alternative oxidase. FEBS Lett 96:252–256

Rich PR (1981) A generalised model for the equilibration of quinone pools with their biological donors and acceptors in membrane-bound electron transfer chain. FEBS Lett 130:173–178

Rich PR, Bonner WD (1978a) EPR studies of higher plant mitochondria. II. Center S-3 of succinate dehydrogenase and its relation to alternative respiratory oxidations. Biochim Biophys Acta 501:381–395

Rich PR, Bonner WD (1978b) An EPR analysis of cyanide-resistant mitochondria isolated from the mutant *poky* strain of *Neurospora crassa*. Biochim Biophys Acta 504:345–363

Rich PR, Bonner WD (1978c) The nature and location of cyanide- and antimycin-resistant respiration in higher plants. In: Degn H, Lloyd D, Hill GC (eds) Functions of alternative terminal oxidases. Pergamon Press, Oxford New York, pp 149–158

Rich PR, Bonner WD (1978d) Paramagnetic centers associated with higher plant and *poky Neurospora crassa* mitochondria. In: Ducet G, Lance C (eds) Plant mitochondria. Elsevier/North Holland Biomedical Press, Amsterdam, pp 61–68

Rich PR, Moore AL (1976) The involvement of the proton motive ubiquinone cycle in the respiratory chain of higher plants and its relation to the branchpoint of the alternative pathway. FEBS Lett 65:339–344

Rich PR, Boveris A, Bonner WD, Moore AL (1976) Hydrogen peroxide generation by the alternate oxidase of higher plants. Biochem Biophys Res Commun 71:695–703

Rich PR, Moore AL, Bonner WD (1977) The effects of bathophenanthroline, bathophenanthrolinesulphonate and 2-thenoyltrifluoroacetone on mung bean mitochondria and submitochondrial particles. Biochem J 162:205–208

Rinaudo JB (1979) Oxydation anaérobic du cytochrome b-566. Nouveau modèle pour le complexe III de la chaîne respiratoire. Physiol Vég 17:525–534

Robbie WA (1948) Use of cyanide in tissue respiration studies. Meth Med Res 1:307–316

Rustin P (1981) Spécificité de l'oxydation mitochondriale du malate pendant l'acquisition du fonctionnement CAM induit par la photopériode. Physiol Vég 19:464

Rustin P, Moreau F (1979) Malic enzyme activity and cyanide-insensitive electron transport in plant mitochondria. Biochem Biophys Res Commun 88:1125–1131

Rustin P, Moreau F, Lance C (1980) Malate oxidation in plant mitochondria via malic enzyme and the cyanide-insensitive electron transport pathway. Plant Physiol 66:457–462

Rustin P, Dupont J, Lance C (1982) Specific and rapid measurement of lipoxygenase activity in biological materials. Physiol Vég 20:721–727

Rustin P, Dupont J, Lance C (1983a) A role for fatty acid peroxy radicals in the cyanide-insensitive pathway of plant mitochondria. Trends Biochem Sci 8:155–157

Rustin P, Dupont J, Lance C (1983b) Oxidative interactions between fatty acid peroxy radicals and quinones: possible involvement in cyanide-resistant electron transport in plant mitochondria. Arch Biochem Biophys 225:630–639

Rustin P, Dupont J, Lance C (1984) On the involvement of lipid peroxy radicals in the cyanide-resistant pathway. Physiol Vég 22:643–663

Schonbaum GR, Bonner WD, Storey BT, Bahr JT (1971) Specific inhibition of the cyanide-insensitive respiratory pathway in plant mitochondria by hydroxamic acids. Plant Physiol 47:124–128

Sharpless TK, Butow RA (1970) An inducible terminal oxidase in *Euglena gracilis* mitochondria. J Biol Chem 245:58–70

Sherald JL, Sisler HD (1972) Selective inhibition of antimycin A-insensitive respiration in *Ustilago maydis* and *Ceratocystis ulmi*. Plant Cell Physiol 13:1039–1052

Siedow JN (1982) The nature of cyanide-resistant pathway in plant mitochondria. Rec Adv Phytochem 16:47–84

Siedow JN, Bickett DM (1981) Structural features required for inhibition of cyanide-insensitive electron transfer by propyl gallate. Arch Biochem Biophys 207:32–39

Siedow JN, Girvin ME (1980) Alternative respiratory pathway. Its role in seed respiration and its inhibition by propyl gallate. Plant Physiol 65:669–674

Siedow JN, Huber JC, Moreland DE (1978) Inhibition of mung bean mitochondrial electron transfer by dibromothymoquinone. In: Ducet G, Lance C (eds) Plant mitochondria. Elsevier/North Holland Biomedical Press, Amsterdam, pp 233–240

Simon EW (1957) Succinoxidase and cytochrome oxidase in mitochondria from the spadix of *Arum*. J Exp Bot 8:20–35

Solomos T (1977) Cyanide-resistant respiration in higher plants. Annu Rev Plant Physiol 28:279–297

Storey BT (1972) The respiratory chain of plant mitochondria. XIII. Redox state changes of cytochrome b-562 in mung bean seedling mitochondria treated with antimycin A. Biochim Biophys Acta 267:48–64

Storey BT (1973) The respiratory chain of plant mitochondria. XV. Equilibration of cytochromes c-549, b-553, b-557 and ubisemiquinone in mung bean mitochondria: placement of cytochrome b-557 and estimation of the midpoint potential of ubiquinone. Biochim Biophys Acta 292:592–603

Storey BT (1976) Respiratory chain of plant mitochondria. XVIII. Point of interaction of the alternate oxidase with the respiratory chain. Plant Physiol 58:521–525

Storey BT (1980) Electron transport and energy coupling in plant mitochondria. In: Davies DD (ed) The biochemistry of plants, Vol II. Academic Press, London New York, pp 125–195

Storey BT, Bahr JT (1969a) The respiratory chain of plant mitochondria. I. Electron transport between succinate and oxygen in skunk cabbage mitochondria. Plant Physiol 44:115–125

Storey BT, Bahr JT (1969b) The respiratory chain of plant mitochondria. II. Oxidative phosphorylation in skunk cabbage mitochondria. Plant Physiol 44:126–134

Tobin A, Djerdjour B, Journet E, Neuburger M, Douce R (1980) Effect of NAD$^+$ on malate oxidation in intact plant mitochondria. Plant Physiol 66:225–229

Tomlinson PF, Moreland DE (1975) Cyanide-resistant respiration of sweet potato mitochondria. Plant Physiol 55:365–369

Troostembergh De JC, Nyns EJ (1978) Kinetics of the respiration of cyanide-insensitive mitochondria from the yeast *Saccharomycopsis lipolytica*. Eur J Biochem 85:423–432

Troostembergh De JC, Ledrut-Damanet MJ, Nyns EJ (1978) Acetate, a trigger for the induction of the alternative, cyanide-insensitive respiration in the yeast *Saccharomycopsis lipolytica*. In: Ducet G, Lance C (eds) Plant mitochondria. Elsevier/North Holland Biomedical Press, Amsterdam, pp 331–338

Trumpower BL (1981) New concepts on the role of ubiquinone in the mitochondrial respiratory chain. J Bioenerg Biomembr 13:1–24

Tyler DD, Estabrook RW, Sanadi DR (1966) Studies of oxidative phosphorylation and energy-linked cytochrome b reduction with tetramethyl-p-phenylenediamine as substrate. Arch Biochem Biophys 114:239–251

Tzagoloff A (1982) Mitochondria. Plenum Press, New York London

Walker GH, Oliver DJ, Sarojini G (1982) Simultaneous oxidation of glycine and malate by pea leaf mitochondria. Plant Physiol 70:1465–1469

Webb JL (1966) Enzyme and metabolic inhibitors, Vol II. Academic Press, London, New York, pp 1–244

Wedding RT, McCready CC, Harley JL (1973) Inhibition and stimulation of the respiration of *Arum* mitochondria by cyanide and its relation to the coupling of oxidation and phosphorylation. New Phytol 72:1–13

Wedding RT, McCready CC, Harley JL (1975) Changing patterns in the response to cyanide of *Arum* mitochondria oxidizing malate and other substrates. New Phytol 74:1–17

Wilson SB (1970) Energy conservation associated with cyanide-insensitive respiration in plant mitochondria. Biochim Biophys Acta 223:383–387

Wilson SB (1971) Studies on the cyanide-insensitive oxidase of plant mitochondria. FEBS Lett 15:49–52

Wilson SB (1978) Cyanide-insensitive oxidation of ascorbate + NNN'N'-tetramethyl-p-phenylene-diamine mixture by mung bean (*Phaseolus aureus*) mitochondria. Biochem J 176:129–136

Wilson SB (1980) Energy conservation by the plant mitochondrial cyanide-insensitive oxidase. Some additional evidence. Biochem J 190:349–360

Wiskich JT, Bonner WD (1963) Preparation and properties of sweet potato mitochondria. Plant Physiol 38:594–604

Wiskich JT, Day DA (1979) Rotenone-insensitive malate oxidation by isolated plant mitochondria. J Exp Bot 30:99–107

Wiskich JT, Day DA (1982) Malate oxidation, rotenone-resistance, and alternative path activity in plant mitochondria. Plant Physiol 70:959–964

Yang F, Xing Q, Wang S (1980) Cyanide-insensitive respiration in corn mitochondria. I. Comparison between sensitivities of α-ketoglutarate and succinate oxidations to cyanide. Sci Sin 23:774–784

Yocum CS, Hackett DP (1957) Participation of cytochromes in the respiration of the aroid spadix. Plant Physiol 32:186–191

Yonetani T, Ray GS (1965) Studies on cytochrome oxidase. VI. Kinetics of the aerobic oxidation of ferrocytochrome b by cytochrome oxidase. J Biol Chem 240:3392–3398

Yu CA, Yu L (1981) Ubiquinone-binding proteins. Biochim Biophys Acta 639:99–128

9 Membrane Transport Systems of Plant Mitochondria

J.B. HANSON

1 Introduction

Mitochondria are loci of intense oxidative activity in the plant cell, accounting for the bulk of the oxygen consumption. Operation of the TCA cycle (Chap. 10) in the matrix oxidized substrate organic acids to CO_2 and H_2O with the production of NADH and reduced flavoprotein, which in turn are oxidized by O_2 through the respiratory chain of the inner-membrane, producing ATP (Chap. 6). In addition, partial operation of the TCA cycle or transamination or oxidative deamination produces products utilized in intermediary metabolism of the cell. Both functions, ATP formation and participation in intermediary metabolism, require regulated transport of ions and metabolites between the cytosol and mitochondrial matrix. Mitochondria also carry out transcription-translation in the matrix, which requires the uptake of nucleotides and amino acids from the cytosol. Plant mitochondria may also participate in the regulation of free Ca^{2+} levels in the cytosol by the uptake and release of Ca^{2+}. In general terms, selective and regulated transport between the cytosol and mitochondrial matrix functions in the complex web of plant metabolism.

It is these transport functions which form the subject of this chapter. They are localized in the bounding osmotic barrier of the mitochondrion, the inner membrane. (The outer membrane is not an osmotic barrier, being permeable to solutes up to 4,000–5,000 molecular weight; see Chap. 5). As with all membrane transport investigations, the principal concerns are with permeability, selectivity, energy-linkage, and regulation. Ultimately, these bear on understanding the integration of mitochondrial metabolism with that of the rest of the cell.

Investigations of inner membrane transport are incomplete, especially when compared to the extensive investigations of animal mitochondria. Indeed, much of the work on plant mitochondria has been derived from observations initially made with animal mitochondria. Sufficient investigation has been made to show that the basic principles of mitochondrial transport are held in common among all eukaryotes. However, plant mitochondria do differ in certain details which appear to be linked to autotrophic metabolism, although these are not necessarily manifest, or equally developed, in all species and tissues. Not all plant mitochondria are alike in their transport functions, and distinction based on species, tissues, developmental stage, environments, and techniques are frequently evident. Critical comparisons are not often made, however, and little can be done other than to call attention to the existence of these differences. They may be inherent, or reflect exposure or masking of transport avenues during isolation.

Several recent reviews and chapters have discussed transport in plant mitochondria (HANSON and KOEPPE 1975, WILSON and GRAESSER 1976, WISKICH
1977, HANSON and DAY 1980).

2 Structural and Osmotic Properties

Certain structural and osmotic properties of plant mitochondria are relevant
to investigation of transport.

Figure 1 (ÖPIK 1974) provides a graphic representation of a typical plant
mitochondrion as seen in electron micrographs of plant cells. The matrix, which
contains the solutes of concern here, is delimited by the inner membrane, which
is invaginated to produce saccate or tubular cristae. The mitochondrion is
bounded by the permeable and more rigid outer membrane. Inner membrane
surface area is governed by the amount of invagination. The lumen of the
cristae is part of the intermembrane space.

Figure 2 (POMEROY 1977) shows mitochondria in cells of wheat seedlings
fixed with glutaraldehyde-osmic acid, which preserves matrix structure, and
$KMnO_4$, which accentuates membrane structure. Figure 3 (POMEROY 1977)
shows wheat mitochondria isolated and suspended in buffered 0.4 M sucrose,
which is impermeant, or 0.2 M KCl, which penetrates into the matrix. These
media are hypertonic, as were the media used in isolation and washing, and
there is evident expansion of the intermembrane space at the expense of the
matrix and alteration of the cristae (cf. Fig. 2). Practically all transport work
has been done with isolated mitochondria like those of Fig. 3, but generally
transferred to media of 250–300 mOsM where they appear more like they do
in vivo (BAKER et al. 1968). The assumption is made that the changes accompanying isolation do not alter transport properties, but this is not certain; respiratory activity is reduced by isolation (BLIGNY and DOUCE 1976). Isolation in

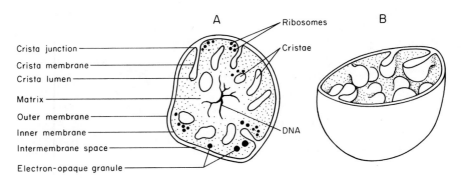

Fig. 1A, B. Illustrations of a typical plant mitochondrion. (ÖPIK 1974). The number
of cristae is variable depending on cell development and respiratory demand, but in
general serves to approximately double the inner membrane surface area compared to
the outer. (HANSON and DAY 1980)

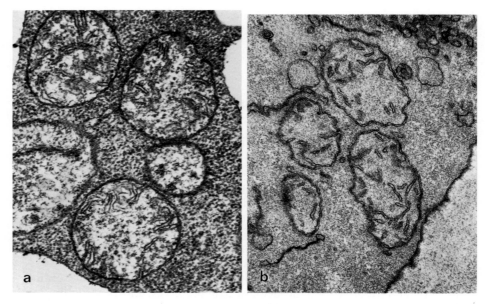

Fig. 2a, b. Electron micrographs of wheat (*Triticum aestivum* L.) seedling mitochondria in situ. **a** Fixed with glutaraldehyde-osmic acid. **b** Fixed in $KMnO_4$. (POMEROY 1977)

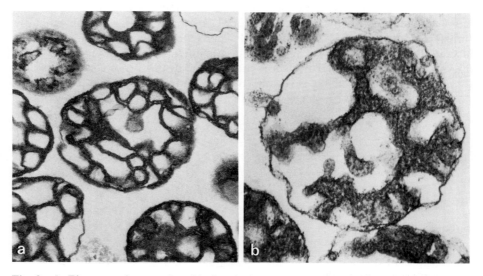

Fig. 3a, b. Electron micrographs of isolated wheat mitochondria fixed in glutaraldehyde-osmic acid. **a** Suspended in 0.4 M sucrose. **b** Suspended in 0.2 M KCl for 3.5 min. Note the shrinkage of the matrix (cf. Figs. 2a and 3a) and the lessened matrix density as swelling occurs in KCl. (POMEROY 1977)

hypertonic media shrinks the matrix, setting up steep concentration gradients for efflux of ions and metabolites. Probably most isolated mitochondria are depleted of these.

Inner membranes are characterized by a high protein/lipid ratio, a phospholipid fraction uniquely enriched in diphosphatidyl glycerol (BLIGNY and DOUCE 1980) and a high proportion of unsaturated fatty acids which are believed to be responsible for their plasticity (MOREAU et al. 1974). Most of the membrane protein is accounted for by the electron transfer chain and coupling ATPase, but intrinsic proteins acting as ion carriers presumably contribute.

The inner membrane forms a nearly perfect osmometer (LONGO and ARRIGONI 1964, YOSHIDA and SATO 1968, LORIMER and MILLER 1969, DeSANTIS et al. 1975). Water penetrates very rapidly and most solutes very slowly, so that a change in external solute concentration produces an osmotic adjustment in matrix volume in a few seconds. A contributing factor to the rapidity is the large surface/volume ratio. MASSARI et al. (1972) report for rat liver mitochondria a water permeability coefficient of 5.3×10^{-3} cm s^{-1}, a value similar to that of other biological membranes. Permeability to O_2 and CO_2 is nonlimiting to metabolism.

Inner membranes are relatively impermeable to protons. Isolated potato mitochondria have a H$^+$ conductance of 0.31 nmol s^{-1} pH unit^{-1} mg^{-1} protein, a value about three fold higher than that of liver mitochondria (DUCET 1979).

The volume of the matrix at different levels of osmotic swelling can be estimated from packed volume less the osmotic "dead space" (i.e., space not osmotically responsive). For corn mitochondria the matrix volume varies from 0.8 µl mg^{-1} protein at 8.4 atm external osmotic pressure to 4.1 µl at 1.8 atm (Fig. 2 in LORIMER and MILLER 1969). Estimates based on volumes accessible to H$_3$O but inaccessible to ^{14}C sucrose or sorbitol range from about 0.9 µl mg^{-1} protein (TOBIN et al. 1980) to 2 µl mg^{-1} protein (KIRK and HANSON 1973, MOORE and BONNER 1981) to about 12 µl mg^{-1} protein (HAMPP 1979). Much of the variation may lie with techniques of recovery by centrifugation through silicone oil (HAMPP, personal communication).

When mitochondria are transferred to solutions containing a solute that penetrates the inner membrane, the initial rapid adjustment is followed by a slower swelling. In this case the increased osmotic potential created by solute entry into the matrix is continuously adjusted by the accompanying influx of an osmotic equivalent of water. Conversely, contraction (shrinkage) of the matrix occurs if solute is lost to the solution.

The swelling or shrinkage that results from the osmotic volume adjustment can be monitored by measurements of absorbancy or percent transmission usually at 520–540 nm. A decrease in absorbancy (or an increase in percent transmission) indicates swelling due to an osmotic influx of water. This has been verified by determinations of water content (STONER and HANSON 1966) and packed volume (YOSHIDA and SATO 1968, LORIMER and MILLER 1969). There is a linear relationship between increases in packed volume and increases in percent transmission (LORIMER and MILLER 1969); between packed volume and the reciprocal of the osmotic potential of the medium (LORIMER and MILLER 1969); and between the reciprocals of the absorbancy and the osmolarity (YOSHIDA and SATO 1968, DeSANTIS et al. 1975). These osmotic relationships are like those of animal mitochondria (TEDESCHI and HARRIS 1955) and are expected from osmotic law. Correlations between ion uptake, increase in matrix volume, and decreases in absorbancy have been observed (KIRK and HANSON 1973).

It was once considered that light absorption detects changes in light scattering due to changes in particle size, but electron microscopy reveals decreases in matrix density with swelling which are of equal or greater importance (POMEROY 1977). For wheat mitochondria transferred from 0.4 M sucrose to 0.2 M KCl (which penetrates) there was an increase in average mitochondrial diameter as determined by Coulter counter from 0.86 μm to 1.27 μm in 3.5 min of swelling (POMEROY 1977; see Fig. 3). However, further swelling, as detected by a decline in 520 nm absorbance, did not produce additional increase in size, and POMEROY concluded that continued modification of the cristae and matrix was responsible. The outer membrane is more rigid than the inner, and unless the outer membrane bursts, it sets a limit to outward expansion of the inner. MANNELLA and BONNER (1975) found the breaking strength of the outer membrane to be directly correlated with its galacturonate and divalent cation content, and osmotic potential differences of 5 to 7 bar between mitochondria and medium were required for rupture. Thus, once the size limit is reached, continued swelling of the matrix space takes place at the expense of the intermembrane space. Dilution of the matrix brings its refractive index closer to that of the medium, decreasing light scattering (NICHOLLS 1982). OVERMAN et al. (1970) have analyzed mathematically the reciprocal transfer of intermembrane and matrix space.

Swelling can affect permeability properties of both inner and outer membranes. EARNSHAW and TRUELOVE (1968) report swelling to cause release of malic dehydrogenase from the matrix and to increase oxidation of reduced cytochrome c (damaged outer membranes). (See Chap. 1).

3 Techniques of Measuring Transport

Determining transport by absorbancy change is useful primarily where a single permeating salt or solute is being followed. It is possible to quantify the transport such that changes in salt content can be calculated from changes in absorbancy (BRIERLEY et al. 1971), but careful standardization of techniques is required. Absorbancy is widely used as a very sensitive qualitative measure of net transport of osmotically active solutes. Determinations are simply and rapidly made, and can be carried out simultaneously with measurements of related parameters (e.g., O_2 consumption, pH).

Quantitative techniques incorporate some type of analysis for solute entering (or leaving) the mitochondria, either by measuring loss (or gain) from solution or by reisolating and analyzing the mitochondria. Changes in solution can only be followed from dilute concentrations of the traced solute. Chemical analysis of mitochondrial content is not always readily performed on small samples, and many of the data have been obtained with radioisotope labeling. Tracing radioisotopes actually measures influx or efflux, not net transport, and may incorporate a measure of exchange for an unlabeled solute. There are experiments where determination of exchange is desired.

Re-isolation of mitochondria after a period of uptake or exchange introduces problems of separating them from contaminating reaction medium. The medium can be washed away after collecting the mitochondria on filters, or by resuspension and recentrifugation, but if the membrane is permeable to the solute there will be at least some leaching from the matrix. (Leakage is greater in mitochondria with low respiratory and phosphorylative activity; MALHOTRA and SPENCER 1970). In what has come to be the accepted technique, the reacted mitochondria are layered over silicone oil of the proper density and centrifuged through this into a receiving solution, usually acidic for the extraction of the solute. For some transport reactions there are known inhibitors of the ion carriers which will block other than nonspecific efflux (e.g., the phosphate transporter can be blocked with mercurials). In closely timed transport studies, addition of the inhibitor followed by rapid centrifugation through silicone oil allows for exacting kinetic analyses (e.g., HAMPP 1979).

It is possible to make a qualitative determination on penetration of respiratory substrates by determining increases in O_2 consumption. This technique becomes quantitative when ^{14}C-labeled substrates are used and analyses made for CO_2 and/or other TCA cycle acids produced. Substrate exchange mechanisms have been investigated by preloading mitochondria with labeled substrates and determining short-term exchange for nonlabeled substrates added to the medium (DESANTIS et al. 1976).

These techniques have been discussed in detail by WISKICH (1977). In what follows they will be brought in when relevant to the experimental findings.

4 Transport of Inorganic Ions and Acetate

4.1 Background

The earliest studies of transport in plant mitochondria were patterned after the passive swelling and energized contraction in KCl determined by absorbancy as reported for liver mitochondria (LEHNINGER 1962, AZZI and AZZONE 1967a, b). It was found that cauliflower (LYONS and PRATT 1964), pea (LONGO and ARRIGONI 1964), corn (STONER et al. 1964, STONER and HANSON 1966), bean (EARNSHAW and TRUELOVE 1968), and castor bean (YOSHIDA and SATO 1968, YOSHIDA 1968a) mitochondria would swell passively in buffered solutions of 100–200 mM KCl. Addition of ATP + Mg^{2+} or a respiratory substrate produced a rapid contraction of the swollen mitochondria, or if supplied initially prevented swelling. LYONS and PRATT (1964) compared rat liver and cauliflower mitochondria; the plant mitochondria swelled five times more rapidly in KCl, and did not show rapid swelling on addition of 1 mM $CaCl_2$, a characteristic of liver mitochondria.

ATP-energized contraction was linked to ATPase activity and was inhibited by oligomycin (STONER et al. 1964, STONER and HANSON 1966); a linear relationship was found between contraction and P/O ratio (EARNSHAW and TRUELOVE 1968). Respiration-energized contraction was inhibited by CN^- and antimycin (STONER and HANSON 1966, YOSHIDA 1968 a). Uncouplers such as DNP (2,4-dinitrophenol) and Cl-CCP (carbonyl cyanide m-chlorophenyl hydrazone) slightly promoted swelling and inhibited contraction with either energy source, but concentrations in excess of those required for uncoupling oxidative phosphorylation were needed (STONER et al. 1964, STONER and HANSON 1966, YOSHIDA 1968 a). Gramicidin, an ionophore which facilitates penetration of the membrane by monovalent cations, greatly accelerated swelling and eliminated energized contraction (YOSHIDA 1968 a). Addition of phosphate, sulfate or acetate to the KCl swelling medium gave additional passive swelling, but lowered the rate and level of active contraction (STONER and HANSON 1966, YOSHIDA 1968 a). YOSHIDA (1968 a, b) correctly ascribed this response to active uptake of these anions, thus reducing the net salt efflux.

Energy-linked contraction is only found with electrolytes, not with penetrating nonpolar solutes such as ribose (LORIMER and MILLER 1969).

Substitution of other salts for KCl gave the following series of increasing effectiveness in passive swelling: for K^+ salts, $F^- < Cl^- < Br^- < NO_3^- < I^- < CNS^-$; for Cl^- salts, $Ca^{2+} < Mg^{2+} \ll NH_4^+ < Li^+ < Na^+ < K^+$ (YOSHIDA and SATO 1968). The degree of energized contraction decreased as passive swelling increased in the series from Cl^- to CNS^- (YOSHIDA 1968 a); i.e., net efflux pumping of salt is best realized under conditions of limited passive influx.

Passive swelling in KCl was found to be very pH-sensitive (STONER and HANSON 1966, YOSHIDA 1968 a). Minimal swelling occurs at about pH 6.5, with an exponential rise in rate as the pH increases. JUNG and BRIERLY (1979) used pH 8.1 and 37 °C to initiate spontaneous swelling of potato mitochondria in KCl and KNO_3. Energized contraction is optimal at pH 7–7.5, and is eliminated at pH 8.5 where passive swelling is extremely rapid. The presence of Mg^{2+} retards the swelling rate, especially at high pH (JUNG and BRIERLY 1979, FLUEGEL and HANSON 1981), but is without effect on respiration-energized contraction (STONER and HANSON 1966). Aging of isolated mitochondria increases passive swelling and decreases energized contraction (POMEROY 1976).

Subsequently, use was made of the ammonium salt swelling technique of CHAPPELL and HAAROFF (1967) to follow proton-linked anion uptake. Part of this centered on transport of substrate acids (PHILLIPS and WILLIAMS 1973 b, WISKICH 1974), but acetate and phosphate transport were also studied (DESANTIS et al. 1975). Passive phosphate uptake from NH_4^+ salts proved to be like that driven by respiration in sensitivity to SH binding reagents (mersalyl, N-ethylmaleimide), indicating carrier-mediated phosphate transport.

4.2 Mechanisms of Passive Transport

Figure 4 presents transport models for passive swelling, energy-linked swelling and energy-linked contraction. All ion flux is passive down thermodynamic gradients, except for energy-linked proton efflux. Passive swelling is driven by

Passive: Potential generated by diffusion

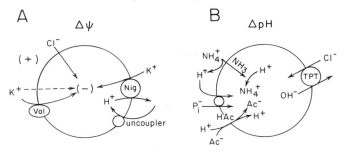

Energy linked: Potential generated by H^+ pumping.

$$\Delta \bar{\mu}_{H^+} = \Delta \psi - Z \Delta pH$$

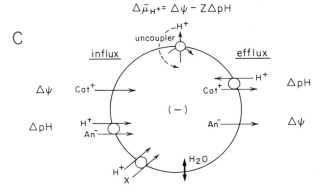

Fig. 4A–C. Schematic diagrams of passive and active transport in plant mitochondria. **A** Passive swelling in KCl driven by the electrogenic influx of Cl^- down its concentration gradient creating a membrane potential ($\Delta \psi$). K^+ influx is accelerated by valinomycin or by nigericin + uncoupler. **B** Passive swelling in NH_4^+ salts of proton-conducting anions driving the electroneutral influx of NH_3, creating a proton gradient (ΔpH). Anions enter with H^+ either on an intrinsic transporter (phosphate) or as the undissociated acid (acetic) or on an anion/OH^- exchanging ionophore such as tripropyltin (chloride). **C** Transport linked to metabolic energy derived from H^+ efflux pumping by the respiratory chain or coupling ATPase. Influx of salts is by cation uniport plus anion/H^+ symport at the expense of $\Delta \psi$ and ΔpH, respectively. Conversely, efflux is by cation/H^+ antiport plus anion uniport. Electroneutral molecules (X) may be cotransported with H^+ bound to the same carrier at the expense of $\Delta \bar{\mu}_{H^+}$

concentration gradients, while energy-linked swelling and contraction arise from the proton gradient created by active H^+ pumping.

Passive swelling in KCl is driven by the imposed concentration gradient of salt, generally 100–150 mM KCl externally, and is an electrogenic system based on creation of a diffusion potential (Fig. 4A). Since isolated plant mitochondria contain 120–140 mM K^+, there is little or no K^+ concentration gradient, and with negligible endogenous respiration (addition of antimycin or cyanide has no effect on passive swelling; JUNG and BRIERLEY 1979, FLUEGEL and Hanson 1981) the electrochemical gradient of K^+ must be very small. However, matrix Cl^- is on the order of 10 mM (Fluegel and Hanson 1981) and passive swelling thus is driven by $\Delta\bar{\mu}_{Cl}$ across a membrane with a relatively high permeability for Cl^- (P_{Cl}), producing a diffusion potential for K^+ influx. Based on the rapid swelling produced by the ionophores gramicidin and valino-mycin, a general conclusion is that spontaneous swelling is rate-limited by per-meability to K^+; i.e., $P_{Cl} > P_K$ (Hanson and Koeppe 1975, Wilson and Graesser 1976). In some preparations of mitochondria there is little KCl swelling until valinomycin is added to increase P_K (turnip, Moore and Wilson 1977; mung bean, Huber and Moreland 1979, potato, Jung and Brierley 1979). One cannot conclude that mitochondria are anion-impermeable without subjecting them to this test. Huber and Moreland (1979) observed passive swelling by mung bean mitochondria in 100 mM phosphate, sulfate or acetate when K^+ flux was activated by valinomycin (see also Day and Wiskich 1980); sulfhydryl reagents promoted rather than inhibited transport. In corn mitochondria an electrogenic influx of acetate can be seen at pH 7.5, but is scarcely detectable at pH 6.0 (Fluegel and Hanson 1981).

The increased swelling rates found with NO_3^-, I^-, and CNS^- compared to Cl^- can be attributed to their greater lipophilicity and higher permeability constants, which in turn permit generation of higher diffusion potentials for driving electrophoretic K^+ influx. Swelling in KCNS is extremely rapid with no need for ionophores (Yoshida and Sato 1968, Moore and Wilson 1977), demonstrating that K^+ influx rises dramatically with increased $\Delta\psi$. Apparently, those anions and cations which are effective in electrogenic swelling must pene-trate the lipid bilayer of the inner membrane. An early observation was that saturation of the KCl solution with CCl_4 or C_2H_4 doubled the KCl swelling rate (Lyons and Pratt 1964), suggesting that permeability is governed by the organization of the lipid bilayer. Swelling is increased in mitochondria with a high content of unsaturated fatty acids (Lyons et al. 1964), and by free acids (Earnshaw et al. 1970). There appears to be no evidence indicative of specific membrane carriers.

The sharp rise in KCl swelling with increased pH is attributed to increased chloride permeability in liver mitochondria (Azzi and Azzone 1967a) since Cl^- exchange is promoted at high pH but K^+ exchange is not. This is supported for corn mitochondria by the finding that at higher pH there is less stimulation of swelling by the Cl^-/OH^- exchanging ionophore tripropyltin (TPT^+); i.e., Cl^- permeability is less limiting (Fluegel and Hanson 1981). However, it has not been established that there is no effect of pH on K^+ permeability.

Figure 4A also illustrates another potential avenue for K^+ entry. Both mam-malian (Brierley 1976) and plant (Hanson and Koeppe 1975) mitochondria have an intrinsic mechanism for H^+/K^+ exchange disclosed by studies of ener-gized contraction (see below). There is no theoretical reason why this "antiport" should not function in KCl swelling via H_{out}^+/K_{in}^+ exchange if the pH gradient produced can be dissipated by H^+-mobilizing uncouplers (protonophores). Un-coupler-stimulated swelling of corn mitochondria is explained by this mechanism (Fluegel and Hanson 1981). Nigericin is a lipid-soluble antibiotic which acts as a synthetic H^+/K^+ antiport, augmenting endogenous activity. Nigericin alone has no effect on KCl swelling, but greatly increases it in the presence of un-coupler (MOORE and WILSON 1977, FLUEGEL and HANSON 1981), the net result of which is an electrophoretic influx of K^+ (Fig. 4A).

Nigericin alone produces rapid swelling with salts of weak acids, such as acetate, which can penetrate as the undissociated acid (MOORE and WILSON 1977, JUNG and BRIERLEY 1979). A combination of valinomycin plus uncoupler gives the same response as nigericin. This type of swelling is based on a diffusion-created pH gradient rather than an electrical gradient. The most widely used technique for passive swelling of the ΔpH type is based on the ready penetration of biological membranes by NH_3, but not by NH_4^+ (CHAPPELL and HAAROFF 1967). As illustrated in Fig. 4B, NH_4^+ can penetrate down a concentration gra-dient by losing a proton externally and regaining it internally. This is an electri-cally neutral transfer of NH_4^+ because an equivalent proton gradient with oppo-site polarity is established. In turn, the pH gradient drives the uptake of those anions whose transfer requires a cotransport (*symport*) of H^+, or an exchange transport (*antiport*) with OH^-, producing an electrically neutral ammonium salt influx. Thus there is little electrogenic swelling in NH_4Cl due to low perme-ability of the membrane to NH_4^+ and H^+, but if an alkyl tin, such as tripropyltin (TPT), is introduced to effect Cl^-/OH^- exchange, a very rapid swelling ensues based on creation and utilization of a pH gradient (Fig. 4B; HUBER and MORE-LAND 1979, JUNG and BRIERLEY 1979). TPT is a lipid-soluble cation that can form membrane-mobile complexes with OH^- or halides.

Ammonium swelling studies have been particularly useful in establishing the existence of anion transport by H^+ symport or OH^- antiport free from any ambiguities arising from respiratory metabolism (CHAPPELL 1968). In plant mitochondria it is demonstrated that phosphate (and arsenate) uptake is driven by ΔpH via an intrinsic carrier sensitive to SH-binding reagents (PHILLIPS ad WILLIAMS 1973b, DESANTIS et al. 1975, WICKES and WISKICH 1976). The same SH-sensitivity is seen in respiration-driven phosphate (and arsenate and sulfate) uptake, giving assurance that the normal operation of the phosphate carrier requires a pH gradient (see below).

4.3 Energy-Linked Transport

Discussion of energy-linked transport starts with the recognition that ATP hy-drolysis via the coupling ATPase, or electron transfer in the respiratory chain, produces an electrochemical gradient of protons, or protonmotive force (see MOORE and RICH, Chap. 6, this Vol.). Expressed in millivolts, $pmf = \Delta\bar{\mu}_{H^+}/F = \Delta\psi - Z\Delta pH$, where $Z = 2.3$ RT/F (MITCHELL 1966). Energy-linked extrusion of H^+ leaves the matrix space negatively charged and alkaline with respect to the external phase, and provides both the electrical ($\Delta\psi$) and chemical (ΔpH) potentials that are formed separately by diffusion (cf. Fig. 4A and 4B with 4C). With salts both electrophoretic *uniport* and proton *symport* or *antiport* are required to discharge $\Delta\bar{\mu}_{H^+}$; electrophoretic *cotransport* of H^+ with a neutral solute will also discharge $\Delta\bar{\mu}_{H^+}$ (Fig. 4C, X/H^+ transport).

It is important to recognize that the energized condition is the normal condi-tion, and that energy-linked and equilibrated transport processes act to maintain ion and substrate contents suitable for the metabolic requirements. Figure 4C schematically presents the basis for energy-linked influx and efflux transport, and the means of maintaining ion balance. Note the following changes compared to the nonenergized state.

a) Although passive diffusion of ions across the membrane still occurs, it is no longer regulated by concentration gradients alone. Respiration imposes electrical potentials on the order of 140 mV, interior negative, and pH gradients equivalent to 30–40 mV (MOORE et al. 1978, MOORE and BONNER 1981), which have a pronounced effect on the electrochemical gradient down which passive flux occurs.

b) Since the salt gradient is no longer the sole source of the potential, a different relationship arises between them. Instead of salt concentrations generating the potential, potentials are giving rise to salt concentrations. In short, the mobile ion content of the mitochondrion is a function of $\Delta\bar{\mu}_{H^+}$; if $\Delta\bar{\mu}_{H^+}$ falls, as with uncoupling or inhibition of respiration, so does salt content (assuming isotonic osmotic support).

c) Salt content is maintained in steady-state equilibrium by influx and efflux pumping. In the absence of H^+ leaks and oxidative phosphorylation this pumping regulates respiration. Note that a minimum of an equivalent of H^+ pumping is required for an equivalent of salt influx or salt efflux. Also, recycling of an equivalent of ion can consume an equivalent of H^+ pumping (e.g., K_{in}^+ at the expense of $\Delta\psi$, K_{out}^+ at the expense of ΔpH). Cyclic pumping has been demonstrated under State 4 conditions with heart mitochondria (JUNG et al. 1977, CHÁVEZ et al. 1977) and is proposed as a principal cause of high State 4 rates in plant mitochondria (FLUEGEL and HANSON 1981)[1].

d) The relative contributions of $\Delta\psi$ and ΔpH to $\Delta\bar{\mu}_{H^+}$ can be varied by variable resistance to cation or anion flux. For example, if anion/H^+ symport meets less resistance ($P_{an} > P_{cat}$) there will be hyperpolarization of the membrane potential at the expense of the pH gradient. Rapid phosphate/H^+ symport has been proposed as a major component in maintaining high $\Delta\psi$ (LEHNINGER 1974).

4.3.1 Salt Efflux

As mentioned above (Sect. 4.1), energy-linked salt transport was first observed in the respiration- or ATP-energized contraction of mitochondria passively swollen in KCl. The generally accepted mechanism for this is that shown in Fig. 4C (efflux). In KCl the only avenue for recycling extruded H^+ is through H_{in}^+/K_{out}^+ exchange, with charge balance maintained by efflux of Cl^- driven by $\Delta\psi$. During the rapid phase of net salt efflux, the respiration rate is increased (JUNG and BRIERLEY 1979), demonstrating the coupling of H^+ pumping to the efflux.

This mechanism deserves closer study than it has received. It is not altogether clear why the introduction of a large $\Delta\psi$ should drive Cl^- efflux rather than K^+ influx (i.e., not drive a futile cycle of K_{in}^+ at the expense of $\Delta\psi$, followed by K_{out}^+ at the expense of ΔpH). Presumably the relative resistance to electrophoretic conduction of K^+ remains higher than that of Cl^-: as already noted (Sect. 4.2), during passive swelling $P_{Cl} > P_K$. JUNG and BRIERLEY (1979) report

---

1 Definition of the different stages, see this Vol., Chap. 10, p. 283

that energy-linked contraction of potato mitochondria in KCl is promoted by nigericin (i.e., supplements endogenous H^+/K^+ antiport), while valinomycin produces swelling and high respiration (i.e., increases P_K, KCl uptake, and futile cycling of K^+). Energy conservation as $\Delta\bar{\mu}_{H^+}$ clearly requires not only high resistance to H^+ backflow, but also high resistance to K^+ uniport and/or H^+/K^+ antiport.

The H^+/K^+ antiport does seem to be under some form of control. Presumably it is an intrinsic proteinaceous carrier, but no specific inhibitor for it has been identified. In potato mitochondria it is activated by pH 8, (JUNG and BRIERLEY 1979), which also induces spontaneous KCl or KNO_3 swelling, increases respiration, and decreases the efficiency of respiration in driving contraction (which can be related to a K^+ or Cl^- futile cycle, a form of uncoupling). In mung bean mitochondria the antiport is activated by respiration (HUBER and MORELAND 1979). In corn mitochondria a low concentration of uncoupler *increases* contraction as if membrane lipids limited H^+ access to the antiport (STONER and HANSON 1966, HENSLEY and HANSON 1975); similarly, valinomycin acts to increase the accessibility of the antiport to K^+ (HENSLEY and HANSON 1975). Efflux pumping is favored by high concentrations of K^+ (HENSLEY and HANSON 1975). Usually the antiport functions only in cation efflux (JUNG and BRIERLEY 1979), but in corn mitochondria it can function in KCl or NaCl swelling in the presence of uncoupler (FLUEGEL and HANSON 1981). In potato mitochondria the rates of $H^+/cation^+$ exchange in NO_3^- efflux pumping are $Li^+ > Na^+ > K^+$ (JUNG and BRIERLEY 1979). This is the opposite of the rates of uniport influx under concentration gradients (Sect. 4.2).

In rat liver mitochondria, lowering of free matrix Mg^{2+} appears to be required for activation of the K^+/H^+ antiport (GARLID 1980). However, external Mg^{2+} also has a role, since it retards passive swelling (Sect. 4.2) and accelerates the rate and extent of energy-linked contraction (JUNG and BRIERLEY 1979); i.e., it appears to "tighten" ion channels.

By pulsing respiration with limited additions of O_2 or of NADH (as substrate) cyclic contraction and re-swelling can be obtained (YOSHIDA 1968a). It is observed that the rate and respiratory efficiency of contraction increases with repeated cycles (JUNG and BRIERLEY 1979). There is no explanation for this increase, but re-swelling is not much affected, so presumably there is progressive activation of the $cation^+/H^+$ antiport.

Anions, such as phosphate, which enter the mitochondria via H^+ symport (or OH^- antiport) are competitors of the H^+/K^+ antiport for the transport potential available in ΔpH (Fig. 4C). Phosphate, sulfate (which enters on the phosphate transporter; ABOU-KHALIL and HANSON 1979a) and acetate in mM concentrations act to reduce the rate and level of energy-linked contraction in KCl media (Sect. 4.1) by producing swelling which offsets the contraction. However, it is possible to demonstrate energy-linked contraction of K phosphate-loaded mitochondria by blocking further influx with mersalyl, a nonpenetrating sulfhydryl-binding mercurial (HENSLEY and HANSON 1975). As mentioned above, this contraction is accelerated by low concentrations of uncoupler and high concentrations of K^+ in the presence of valinomycin, all of which appear to facilitate H^+/K^+ antiport activity.

4.3.2 Salt Influx

Major attention in energy-linked ion transport has been directed to the phosphate transporter. It exhibits sulfhydryl sensitivity and saturation kinetics, indicating that it is an intrinsic membrane protein. It is very active, and the rapid influx of phosphate at the expense of the pH gradient contributes to $\Delta\psi$. It is essential to oxidative phosphorylation, which occurs with matrix phosphate. It is required in substrate acid transport (Sect. 5.1).

Sulfhydryl sensitivity of respiration-linked phosphate transport has been universally observed. Nonpenetrating mercurials such as mersalyl, p-hydroxymercuribenzoate, and methyl mercuric iodide are more effective than N-ethylmaleimide (NEM), and their inhibition can be reversed by sulfhydryl-protecting agents such as cysteine or mecaptoethanol (HANSON et al. 1972, DESANTIS et al. 1975, HAMPP 1979). The effectiveness of the mercurials is sometimes suggested to lie with inhibition of both Pi/H^+ symport (or Pi/OH^- antiport) and Pi/dicarboxylate exchange (Sect. 5.1) while NEM inhibits only the former; however, inhibition of dicarboxylate exchange should not affect net solute uptake and swelling. DESANTIS et al. (1975) report almost complete inhibition of respiration-driven phosphate swelling in bean mitochondria at 20 nmol mersalyl mg^{-1} protein; which compares favorably with results from animal mitochondria (this was not true for NEM inhibition which required much higher concentrations than reported for rat liver mitochondria). HAMPP (1979), using oat leaf mitochondria and determining initial rates of ^{32}Pi influx into the matrix, found mersalyl to be tenfold less effective than with animal mitochondria. The reasons for the discrepancy are not obvious; they may lie with species or techniques or the assumption that the mitochondria are responding to the total amount of mercurial added rather than solution concentration.

Saturation kinetics for K phosphate uptake were first observed with corn mitochondria by the swelling technique (HANSON et al. 1972). A biphasic curve was obtained with half-saturation of the initial phase at about 0.25 mM, the second phase at about 5 mM. It is possible that the second phase is introduced by enhanced K^+ influx, since the addition of valinomycin showed that the rate and extent of swelling are limited by K^+ permeability. This work also demonstrated that the amount of ATP that can be formed in phosphate-loaded, mersalyl-blocked mitochondria is limited to the amount of accumulated phosphate (nonpenetrating mersalyl does not inhibit the coupling ATPase).

HAMPP (1979) has made an exacting kinetic study by determining phosphate uptake into the matrix over the initial 4 s, using oat leaf mitochondria isolated at various periods after illuminating etiolated plants. After 3 h there was about four fold increase over initial rates of ATP formation, and with phosphate concentrations up to 1 mM a similar increase in V_{max} of phosphate uptake reaching 17 μmol mg^{-1} protein h^{-1}. By 24 h the activity had declined below that of mitochondria from the initial etiolated tissue. However, the K_m for phosphate influx remained constant at 0.23 mM. This work demonstrates that metabolic controls exist which can vary the activity (or activation?) of the phosphate transporter without altering its affinity for phosphate. HAMPP suggests that this may reflect the pH gradient across the inner membrane; that is, control by demand. An alternative of control through the exposed SH-groups was less attractive because the amount of SH-inhibition was largely independent of the stage of development; however, it was not established that the level of functional SH-groups is constant with development.

The phosphate transporter functions in efflux as well as influx. When respiration is inhibited or uncoupled, accumulated phosphate diffuses from the matrix; mersalyl will block this efflux (HENSLEY and HANSON 1975). If mersalyl is added to uncoupled mitochondria hydrolyzing ATP a rapid swelling is produced by the released phosphate, which accumulates in the matrix (DeSANTIS et al. 1975). There is a small, contraction-swelling cycle during the State 4–State 3–State 4 transition upon limited ADP addition (HANSON 1972), which correlates with a small decline and recovery in $\Delta\bar{\mu}_{H^+}$ (MOORE et al. 1978). Clearly matrix phosphate content is a function of $\Delta\bar{\mu}_{H^+}$.

The critical dependence of phosphate transport on ΔpH is illustrated by the increased uptake of phosphate salts when oligomycin is added (MILLARD et al. 1965, BERTAGNOLLI and HANSON 1973). There appears to be some backflow of H^+ through the coupling ATPase in the absence of added ADP which is increased by turnover with matrix phosphate, or especially arsenate, and which constitutes a proton "leak" (HANSON and DAY 1980). Oligomycin, which blocks H^+ transport through the ATPase, stops the leak and increases ion transport.

The rate and extent of mersalyl-sensitive phosphate swelling is greatly increased by the presence of basic proteins, such as protamine, accelerating respiration and lowering ADP/O ratios (HANSON 1972). It appears that screening of negative surface charge by polycations permits rapid uptake and cycling of phosphate, and suggests that coulombic repulsion normally limits the accessibility of phosphate to the transporter.

Arsenate (HANSON et al. 1972, BERTAGNOLLI and HANSON 1973, WICKES and WISKICH 1976) and sulfate (ABOU-KHALIL and HANSON 1979a) are also transported by the phosphate transporter. DeSANTIS et al. (1976) found exchange transport of sulfate with the dicarboxylate transporter in bean mitochondria, but not with the phosphate transporter (Sect. 5.1). These results were from short-term passive loading and exchange in KCl media. In corn mitochondria respiring NADH (i.e., no substrate uptake) both K_2SO_4 swelling and ^{35}S-sulfate uptake showed transport characteristics paralleling those of phosphate: strong inhibition by mersalyl and NEM, weak inhibition by n-butylmalonate (an inhibitor of the dicarboxylate transporter), poor transport in KCl media, and promotion by Mg^{2+} and oligomycin (ABOU-KHALIL and HANSON 1979a). There was clear indication for phosphate/sulfate exchange, and sulfate + phosphate swelling was additive. Respiration-linked swelling is much more rapid with phosphate than sulfate, and addition of valinomycin causes additional swelling with phosphate, but contraction with sulfate (KIMPEL and HANSON 1978). This is explained on the relative resistance of the H^+/sulfate symport and the H^+/K^+ antiport (Fig. 4C); mobilization of K^+ with valinomycin makes H^+/K^+ antiport the pathway of least resistance to reentry of extruded H^+.

Energy-linked acetate swelling is rapid and extensive, and completely uncouples corn mitochondria (WILSON et al. 1969). Acetate enters as acetic acid (i.e., an acetate$^-$/H^+ symport) and the cation penetrates electrophoretically. There is little discrimination between cations; Na^+, Mg^{2+}, $Tris^+$, and tetraethylammonium$^+$ will all serve (WILSON et al. 1969), indicating that the cation channel is not very selective.

Respiration-driven phosphate uptake can be greatly increased by Ca^{2+}, Sr^{2+} or Ba^{2+} with massive deposition of insoluble phosphate precipitates in the matrix (reviewed by WILSON and GRAESSER 1976). The Ca^{2+}/Pi uptake ratio is 1.6 in corn mitochondria (ELZAM and HODGES 1968), which suggests that the deposits are calcium hydroxyapatite as determined for animal mitochondria (LEHNINGER 1970). In bean mitochondria the Sr^{2+}/Pi ratio is 1.4 (JOHNSON and WILSON 1972). Beet mitochondria show extensive phosphate uptake with Mg^{2+} in the presence of oligomycin (MILLARD et al. 1965), but this has been difficult to confirm with other mitochondria, probably because phosphate is not removed by deposition. HODGES and HANSON (1965) found Mg^{2+} uptake with phosphate in the presence of Ca^{2+} under conditions where Mg^{2+} seems to compete for Ca^{2+} transport sites. ABOU-KHALIL and HANSON (1978, 1979c) report phosphate and Mg^{2+} uptake in corn mitochondria accumulating ADP, with a Pi:ADP:Mg^{2+} ratio of approximately 1:1:1. However, phosphate uptake is only 4 to 5% of that secured with Ca^{2+}.

The critical question is how Ca^{2+} (or related Sr^{2+} and Ba^{2+}) is actively transported. This matter appears to be resolved for animal mitochondria (LEHNINGER 1974, SCARPA 1979, CARAFOLI 1981, NICHOLLS and ÅKERMAN 1982). Rapid Ca^{2+} influx is electrophoretic through a Ca^{2+} channel which is probably proteinaceous and is specifically inhibited by ruthenium red. The resulting collapse of $\Delta\psi$ is countered by a proportional rise in $Z\Delta pH$ as respiratory H^+ pumping continues, and in the presence of an H^+ conducting anion (e.g., phosphate, acetate) to restore $\Delta\psi$ continued and massive uptake of Ca^{2+} can be secured. Without such anions Ca^{2+} uptake is limited by the level of endogenous phosphate. Free Ca^{2+} in the matrix is in the micromolar range, maintained dynamically by Ca^{2+} efflux through Ca^{2+}/Na^+ or Ca^{2+}/H^+ antiports.

There is a great deal of detail here which is unexplored in plant mitochondria. The investigations that have been made with plants show one striking difference: it has not been possible to demonstrate that Ca^{2+} uptake in plant mitochondria is by simple uniport. If the uniport exists as it does in animal mitochondria, there should be active Ca^{2+} swelling with acetate, and there is not. Calcium uptake is only secured in the presence of phosphate (HODGES and HANSON 1965, JOHNSON and WILSON 1972, DIETER and MARMÉ 1980, FUKUMOTO and NAKAI 1982), with arsenate in the presence of oligomycin a weak substitute (KENEFICK and HANSON 1967). This is not due to failure to transport acetate, which with K^+ salts gives more rapid and extensive uptake than phosphate (WILSON et al. 1969, JOHNSON and WILSON 1972). The high affinity binding sites associated with Ca^{2+} transport in animal mitochondria are with one possible exception (sweet potato mitochondria) lacking in plant mitochondria (CHEN and LEHNINGER 1973). Animal mitochondria show a pronounced transient release of respiration and H^+ efflux on addition of Ca^{2+}, but this is either nonexistent (MOORE and BONNER 1977, JOHNSON and WILSON 1972) or very small and dependent on the level of endogenous phosphate (DAY et al. 1978) in plant mitochondria. An early comparative study of passive KCl swelling showed Ca^{2+}

to be promotive in rat liver mitochondria (i.e., a permeant cation) but slightly inhibitory in cauliflower mitochondria (LYONS and PRATT 1964). Some preparations of mung bean mitochondria show no Ca^{2+} uptake or depression of $\Delta\psi$ upon Ca^{2+} addition (MOORE and BONNER 1977, MOORE et al. 1978), while others show pronounced Sr^{2+}, Ca^{2+} and Ba^{2+} uptake in the presence of phosphate (WILSON and MINTON 1974). MOORE and BONNER (1977) did not actually measure Ca uptake in the absence of phosphorylating conditions, which are known to inhibit Ca^{2+} phosphate uptake (HANSON and MILLER 1967).

The phosphate requirement has led to suggestions that Ca^{2+} uptake might be through transport of a Ca phosphate complex (WILSON and MINTON 1974, DAY et al. 1978), which was once suggested for animal mitochondria (MOYLE and MITCHELL 1977) but now appears to be discredited (NICHOLLS and ÅKERMAN 1982). However, if Ca^{2+} uptake is by simple uniport in plant mitochondria it must be through a channel with very high resistance.

It is possible that this is the case. In a study with corn mitochondria suspended in buffered 100 mM KCl or K acetate with 1 mM $^{45}CaCl_2$ but without phosphate and oxidizing NADH, EARNSHAW et al. (1973) found a rapid and sustained uptake of Ca^{2+} in KCl medium which was dependent on mersalyl-sensitive transport of readily leached endogenous phosphate, and which was associated with typical KCl efflux pumping and contraction. When the NADH was exhausted there was rapid passive efflux of Ca^{2+} and endogenous phosphate; addition of 1 mM ^{32}P-labeled phosphate strongly retarded the loss of Ca^{2+} and phosphate, but permitted ^{32}Pi influx by mersalyl-insensitive exchange (EARNSHAW and HANSON 1973). The initial uptake of Ca^{2+} was accompanied by the ejection of H^+, with an average H^+/Ca^{2+} ratio of 0.9; this was attributed to formation of $Ca_3(PO_4)_2$ in the matrix.

In contrast, the addition of NADH to the acetate medium produced rapid and extensive swelling. Calcium uptake accompanied the swelling for less than 30 s, after which the Ca^{2+} was spontaneously lost. There was no detectable uptake of leached endogenous phosphate as in the KCl medium. Further investigation showed that oxidative phosphorylation was active in the KCl medium, but uncoupled in the acetate medium (also WILSON et al. 1969). These results may signify that Ca^{2+} was entering by uniport with acetate, but that uptake could not be maintained due to uncoupling by way of an acetic acid/acetate futile cycle. Possibly phosphate transport differs in producing a sufficiently large and sustained $\Delta\psi$ to electrophoretically drag Ca^{2+} through an uniport of high resistance. In this case the difference between corn and liver mitochondria would lie with uniport resistance. However, adding the divalent cation ionophore A23187 to mung bean mitochondria did not introduce Ca^{2+} uptake (MOORE and BONNER 1977).

As matters stand, Ca^{2+} uptake by plant mitochondria is specifically dependent upon SH-sensitive phosphate transport with no firm evidence for a Ca^{2+} uniport of the animal type. The transport has been interpreted as Ca^{2+}-activated phosphate transport on the grounds that the Ca^{2+}/Mg^{2+} ratio determines whether the available energy is used for ATP formation or transport (HANSON and MILLER 1967); plant mitochondria do not show strong Ca^{2+} uncoupling of ATP formation and ATP formation is a good inhibitor of Ca^{2+} uptake and retention (HODGES and HANSON 1965, JOHNSON and WILSON 1973). Empirically, Ca^{2+}, Sr^{2+}, and Ba^{2+} appear to activate phosphate transport and accompany it. It is noted that these divalent cations, unlike Mg^{2+}, are characterized by high rates of water substitution in complex formation (MILLER et al. 1970), but the significance of this for transport is obscure. These cations may have their greatest effect by forming insoluble phosphate salts, thus maintaining low chemical activities in the matrix.

Recently, FUKUMOTO and NAGAI (1982) reported Ca^{2+} uptake and Ca^{2+}-ATPase in apple mitochondria to be inhibited by chlorpromazine, a calmodulin inhibitor, and they suggest Ca^{2+} transport may be due to a Ca^{2+}-ATPase similar to that of microsomal membranes. Although DIETER and MARMÉ (1980) did not find that added calmodulin stimulated mitochondrial Ca^{2+} uptake, it is possible that tenaciously bound calmodulin is present (or that chlorpromazine has other deleterious effects).

5 Transport of Organic Metabolites and Cofactors

5.1 General Characteristics

Mitochondrial metabolites and cofactors are predominately anions or zwitterions – organic acids, amino acids, nucleotides, etc. – largely neutralized by K^+ and Mg^{2+}. The potential available for transport is that described above for inorganic ions and acetate, $\Delta\psi$ and ΔpH (Fig. 4). Since $\Delta\psi$ is 120–140 mV (MOORE et al. 1978), anions are charge-repelled and must enter either by neutral exchange for another anion or be coupled to ΔpH through neutral H^+ symport (or OH^- antiport) as in Fig. 4C. Furthermore, the transport processes must be selective for those metabolites required from the cytosol, or required by the cytosol (or cell organelles) from the mitochondria.

CHAPPELL and colleagues, using liver mitochondria, provided the initial description of the exchange transport system for citric acid cycle anions (CHAPPELL and CROFTS 1966, CHAPPELL and HAARHOFF 1967, CHAPPELL 1968). Much subsequent research has confirmed and extended the original observations (MEIJER and VAN DAM 1974, LANOUE and SCHOOLWERTH 1979, SCARPA 1979). To the extent investigated the animal system has been confirmed in plant mitochondria (WISKICH 1977). Figure 5 (reactions 2, 3, 4, and 5) illustrates the primary exchanges.

It is generally believed that metabolite transport is carried out by intrinsic membrane proteins specific for the transported ion or a group of related ions (WISKICH 1977, LANOUE and SCHOOLWERTH 1979). In animal mitochondria nearly all transport is inhibited by some form of SH-reagent or by substrate analogs which compete for binding sites (LANOUE and SCHOOLWERTH 1979). As far as investigated, this also appears to be true for plant mitochondria (WISKICH 1977, HANSON and DAY 1980), and a reasonable first assumption is that selective metabolite transport lies with the characteristics of intrinsic transport proteins.

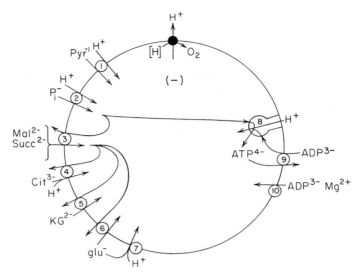

Fig. 5. Schematic diagram of substrate uptake and/or oxidative phosphorylation in respiring plant mitochondria. Numbered transport pathways are discussed in the text. *Cit* citrate; *glu* glutamate; *KG* α-ketoglutarate; *Mal* malate; P_i inorganic phosphate; *Pyr* pyruvate; *Succ* = succinate. ○ Represents a metabolite carrier; ● represents proton translocation by the respiratory chain. In addition to the transport systems shown, there is evidence for a separate carrier which catalyzes the transport of oxaloacetate

However, it does not follow that all transport observed in vitro is necessarily mediated by carrier protein. Recall that even phosphate has been observed to penetrate electrogenically from 100 mM K phosphate in the presence of valinomycin (HUBER and MORELAND 1979). In what follows it will be seen that some mitochondrial preparations exhibit passive swelling in high concentrations of metabolites. The significance of this is uncertain, but it apparently demonstrates the presence of "channels" of low specificity for diffusive flux. In general, the less ambiguous work is done with osmotically supported mitochondria with low physiological levels of transported metabolites, a condition more nearly approaching that in vivo. Passive influx is reduced and energy-linked transport mechanisms are more securely identified.

Lastly, the transport activities of mitochondria can reflect specialized metabolic demands. Mitochondria from green leaves transport and oxidize glycine (WOO and OSMOND 1976) and formate (OLIVER 1981), introducing transport mechanisms additional to those found in nongreen tissues.

5.2 Monocarboxylate Transport

As we have seen, K acetate or propionate are rapidly transported by respiring mitochondria, producing massive swelling and uncoupling. A comparison made with corn mitochondria suspended in 100 mM K propionate, lactate or pyruvate showed that α-hydroxy or α-keto substitution eliminated the propionate uncoupling and gave rise to energy-linked contraction with NADH (WILSON et al. 1969), presumably due to the reduction in pK_a. Bean mitochondria in a 310 mM mannitol medium containing 5 mM Mg^{2+}, 20 µM rotenone, and buffered at pH 7 show some passive swelling with 20 mM K acetate, β-hydroxybutyrate, propionate, or pyruvate, presumably by anion uniport (LEE and WILSON 1972). Upon addition of NADH as respiratory substrate there was additional energy-linked swelling, least with pyruvate, greatest with acetate; in theory this should be by H^+-anion symport (Sect. 4.3). Similar pyruvate swelling driven by NADH oxidation (a nonpenetrating substrate) has been reported for corn mitochondria in a sucrose medium (DAY and HANSON 1977b). Thus even at 10 to 20 mM concentrations of anion there is a limited amount of passive electrogenic monocarboxylate influx in some mitochondrial preparations, and there is a mechanism whereby respiration can drive more extensive uptake. In the case of pyruvate ($pK_a = 2.5$), it is difficult to conceive of this mechanism being simple penetration of the undissociated acid.

DAY and HANSON (1977b) investigated pyruvate transport by corn mitochondria using 10 mM or less external pyruvate. As with liver mitochondria, the uptake exhibited saturation kinetics ($K_m = 0.53$ mM) and was sensitive to the SH-inhibitors α-cyano-4-hydroxycinnamic acid (CHCA) and mersalyl. Lactate was transported by the same mechanism. Pyruvate transport rates were one-third those of malate and, unlike malate, were not stimulated by phosphate. PROUDLOVE and MOORE (1982) using green pea leaf and etiolated mung bean shoot mitochondria and measuring ^{14}C-pyruvate uptake also found saturation kinetics and SH-sensitivity. The conclusion is that pyruvate enters by H^+-symport (or OH^- antiport) on a carrier specific for pyruvate and lactate. Transport rates are relatively low ($V_{max} = 22$ to 28 nmol mg^{-1} protein^{-1} h^{-1}; PROUDLOVE and MOORE 1982) and are inadequate to support rapid respiration. It is suggested that malate transport plus malic enzyme furnishes much of the matrix pyruvate under high respiratory demand (DAY and HANSON 1977b).

Pea cotyledons (which do not contain glyoxysomes) yield mitochondria which will oxidize the series of saturated fatty acid salts from butyrate to palmitate in the presence of Mg^{2+}, CoASH, ATP, and 1 mM malate as a sparker (McNEIL and THOMAS 1976, THOMAS and McNEIL 1976). Carnitine will stimulate butyrate and palmitate oxidation by 22 to 28%; malate oxidation is not stimulated. It is proposed that transport of the fatty acids occurs as the acylcarnitine complex is known to function this way in some animal mitochondria. SCARPA (1979) suggests that $\Delta\psi$ may drive the transport if the acylcarnitine becomes protonated upon complexing to the membrane.

5.3 Dicarboxylate Transport

As illustrated in Figure 5 (reaction 3) dicarboxylate substrates (malate, succinate, oxaloacetate) and malonate enter the mitochondria in exchange for previously accumulated phosphate. Initial support for this came from experiments with substrate analogs (2-butylmalonate, 2-phenylsuccinate, benzylmalonate, p-iodobenzylmalonate), which had been demonstrated to inhibit dicarboxylate transport competitively in animal mitochondria. These compounds inhibit dicarboxylate respiration (PHILLIPS and WILLIAMS 1973a, WISKICH 1975) and NH_4^+-driven swelling (PHILLIPS and WILLIAMS 1973b, WISKICH 1974). By use of mersalyl or N-ethylmaleimide (NEM) to inhibit phosphate transport these investigations also showed that dicarboxylate transport is largely (but not absolutely) linked to phosphate transport.

DeSANTIS et al. (1976) used the more sensitive procedure of measuring passive exchange of preloaded labeled metabolites for exogenous metabolites to provide an extensive survey of exchange transport in bean mitochondria. The principal findings can be summarized as follows:

a) Exchange of preloaded phosphate for exogenous phosphate is inhibited by mersalyl or by NEM + butylmalonate, but not by NEM or butylmalonate alone (investigations with animal mitochondria have shown NEM to inhibit the phosphate transporter, but not the dicarboxylate; MEIJER and VAN DAM 1974). Hence, phosphate/phosphate exchange occurs on both the phosphate and dicarboxylate transporters.

b) Preloaded phosphate will exchange with malate, oxaloacetate, succinate, and malonate in processes inhibited by mersalyl and butylmalonate, but not by NEM. No phosphate exchange was found with added citrate, α-ketoglutarate, pyruvate, glutamate, maleate, fumarate, or ADP. Thus uptake by exchange with phosphate on the "dicarboxylate transporter" is limited to a special group of dicarboxylates.

c) Exchange of preloaded sulfate for external sulfate, sulfite, phosphate, and malonate is inhibited by mersalyl and butylmalonate but not by NEM, indicating that sulfate enters only on the dicarboxylate carrier. (This is not true for net sulfate uptake by corn mitochondria which is by the phosphate transporter; see Sect. 4.3.2). Malate and succinate also exchange for sulfate, but α-ketoglutarate, citrate, pyruvate, glutamate, and ADP do not. Hence sulfate acts as a phosphate analog on the dicarboxylate transporter.

d) Preloaded malonate (used as a representative dicarboxylate) is indicated by the inhibitors as exchanging for external malonate, phosphate, sulfate, or sulfite on the dicarboxylate transporter. No exchange is seen with glutamate or pyruvate. However, malonate will exchange with citrate or α-ketoglutarate in processes insensitive to mersalyl but sensitive to 1,2,3-benzenetricarboxylate and butylmalonate, respectively. This result indicates the presence of specialized carriers capable of exchanging citrate and α-keto-glutarate, but not phosphate. Hence there are three distinct carriers capable of dicarboxylate/dicarboxylate exchange involving malate, succinate, malonate, and oxaloacetate, which differ in selectivity for exchange with phosphate, citrate and α-ketoglutarate (respectively, reactions 3, 4 and 5, Fig. 5). From a physiological viewpoint there are multiple avenues for the entry of malate, the principal dicarboxylate substrate.

e) Studies with preloaded α-ketoglutarate and citrate affirm that these are distinct carriers. Endogenous α-ketoglutarate exchanges readily for exogenous α-ketoglutarate, malonate, succinate, malate, and oxaloacetate, but not with phosphate, sulfate, citrate, pyruvate, glutamate, or phosphoenolpyruvate. The exchange is insensitive to mersalyl or benzenetricarboxylate, but it is inhibited by the dicarboxylate analogs butylmalonate and phenylsuccinate. Preloaded citrate exchanges readily for added citrate, cis-aconitate, phosphoenolpyruvate, malonate, and malate, but not for α-ketoglutarate, pyruvate, glutamate, phosphate, or sulfate. The only effective inhibitor is benzenetricarboxylate used as the high concentration of 20 mM compared to 2 mM exchanging substrate.

DeSantis et al. (1976) observe that these results resemble those obtained with animal mitochondria with one major distinction; oxaloacetate can be transported by both the dicarboxylate and α-ketoglutarate transporters. Oxaloacetate is noted for its ready penetration into plant mitochondria (Wiskich 1977). Lack of fumarate exchange should also be noted; fumarate appears to be the one citric acid cycle anion not carrier-transported (Wiskich 1977).

These exchange studies are of great value, but they are limited to one species (*Vigna sinensis*) and measure passive efflux for 1 min at 8 °C with respiration blocked by antimycin A. Determinations of respiration, swelling, and uptake which involve net transport under more physiological conditions reveal some additional characteristics. One of these has been mentioned, the net uptake of sulfate via the phosphate transporter (Abou-Khalil and Hanson 1979a). It is not known why these differences occur, but they should be considered.

Using standard NH_4 swelling techniques, Wiskich (1974) showed that penetration of malate, succinate, and malonate into cauliflower mitochondria required the presence of phosphate, but castor bean and wheat mitochondria with good respiratory control and ADP/O ratios swelled spontaneously in ammonium malate, succinate, and citrate. In a mannitol medium containing rotenone and 20 mM K salts of malate, malonate, maleate, glutarate, α-ketoglutarate, and citrate, Lee and Wilson (1972) found that bean (*Phaseolus vulgaris*) mitochondria swell somewhat upon addition of NADH. In similar experiments with 10 mM malate, corn mitochondria showed NADH-driven swelling in the absence of phosphate, and although swelling was increased in the presence of phosphate, neither mersalyl nor butylmalonate completely prevented it (Day and Hanson 1977a). It was confirmed that labeled malate uptake occurred in the absence of phosphate, although uptake was doubled in its presence. Under comparable conditions at pH 6.3, corn mitochondria show enhanced swelling in 10 mM malate upon addition of nigericin (K^+/H^+ exchange), suggesting malate influx via H^+-symport (Birnberg and Hanson 1983).

Malate/H^+ symport at pH's below 7 should not be dismissed. Mitochondria from cultured tobacco cells have optimal respiration at pH 6.5, and Horn and Mertz (1982) suggest that pH in this range may normally occur in the intermembrane space. In this pH range malic enzyme activity keeps inhibitory oxaloacetate levels to a minimum (Neuburger and Douce 1980).

Thus, on present evidence, there are two mechanisms for dicarboxylate transport; (1) transport via the exchange systems illustrated in Fig. 5, and (2) what appears to be anion/H^+ symport (or anion/OH^- antiport), as illustrated in Fig. 4C. Evidence for the latter is largely ignored, but it does not go away. Perhaps under respiring conditions with high ΔpH and nearly complete substrate oxidation, one or more of the dicarboxylate carriers accommodates dicarboxylate/H^+ symport, or equivalent OH^- or HCO_3^- antiport. Alternatively, under conditions requiring metabolite exchange in intermediary metabolism, the observed exchange system would be favored. In vivo, something of each can be expected.

Questions have arisen about the mode of oxaloacetate transport. Following up a brief report on the inhibitory properties of phthalonate on transport (MOORE et al. 1979), DAY and WISKICH (1981 a, b) found that phthalonate inhibits oxaloacetate transport but not dicarboxylate or α-ketoglutarate transport. They report that cauliflower mitochondria swell spontaneously in NH_4 oxalacetate, implying exchange for OH^- ions, but the swelling is insensitive to phthalonate. JOURNET et al. (1982) found exit of succinate and α-ketoglutarate during oxaloacetate uptake (used to trigger glutamate oxidation), and suggest that different carriers may be involved in influx and efflux. DAY and WISKICH (1981 a) also found phthalonate to inhibit oxaloacetate efflux. They note that the action of phthalonate is quite different from that in liver mitochondria, where it inhibits α-ketoglutarate uptake (MEIJER et al. 1976).

These anomalies may indicate that inhibitions by substrate analogs are manifestations of the level of specific substrate binding to a common carrier, not carrier turnover. If so, it is conceivable that binding of one substrate may be blocked, while another has access. In the shuttle hypothesis (successive reactions 2, 3, 4, or 5, Fig. 5), the carriers bind the exchanging metabolites alternately at the inner and outer faces of the membrane, and it is possible that different conformational changes associated with different substrates provide different susceptibility to competitive inhibitors. The alternative is a very large (and increasing) number of substrate- and inhibititor-specific carrier proteins.

5.4 Tricarboxylate Transport

As indicated above, one of the dicarboxylate exchangers has the distinction of accommodating citrate, cis-aconitate, and phosphoenolpyruvate, but not phosphate, sulfate, α-ketoglutarate, glutamate, or pyruvate (DESANTIS et al. 1976). Citrate exchange is inhibited by 1,2,3-benzenetricarboxylate but not by butylmalonate. Swelling in NH_4^+ citrate requires the presence of phosphate and malate or succinate (PHILLIPS nd WILLIAMS 1973b, WISKICH 1974), a result reasonably interpreted by the cascade of exchanges illustrated in Fig. 5 (reactions 2, 3, and 4). As with the other carriers, the tricarboxylate exchanger works in either direction (i.e., citrate influx or efflux, DESANTIS et al. 1976). Thus the tricarboxylate transporter of plants is at least fundamentally like that of animals (MEIJER and VAN DAM 1974, LANOUE and SCHOOLWOERTH 1979, SCARPA 1979), although very little of the extensive research with animal mitochondria has been repeated and confirmed (WISKICH 1977).

Again there are anomalies. WISKICH (1974) reported that beet root mitochondria swell in NH_4^+ citrate without malate addition, and castor bean and wheat coleoptile mitochondria swelled in the absence of both phosphate and malate. HUBER and MORELAND (1979) found that mung bean mitochondria would not swell in NH_4^+ citrate with or without catalytic additions of phosphate and malate. JUNG and LATIES (1979) reported phosphate and malate to stimulate respiration-driven ^{14}C-citrate uptake by potato mitochondria, but substantial accumulation was found without these additions. Corn mitochondria in a sucrose medium at pH 7.4 and containing mM concentrations of citrate and phosphate (or sulfate) but no dicarboxylate show swelling and citrate oxidation upon lowering the pH to 6.4 (Kimpel and Hanson 1977). Mersalyl strongly inhibited this low pH activation of citrate and phosphate transport.

An extensive investigation of citrate transport in corn mitochondria under passive and active conditions (Birnberg et al. 1982, Birnberg and Hanson 1983) discloses the following:

a) Although added phosphate and malate somewhat promote ^{14}C-citrate uptake and swelling, they are not essential. Only at very high concentrations did butylmalonate or benzenetricarboxylate give any inhibition.

b) Low pH promotes citrate uptake and oxidation. Addition of nigericin (H^+/K^+ antiporter) greatly increases passive influx of K-citrate, indicating that citrate enters by H^+/citrate symport. Although a variety of other organic acids show some nigericin-stimulated uptake, only isocitrate appeared to enter via the H^+/citrate symport.

c) Citrate/malate exchange does exist, but it is not effective in citrate uptake. In contrast, the exchange is very effective in removing citrate from the matrix provided the mitochondria are respiring; i.e., $citrate_{out}^{3-}/malate_{in}^{2-}$ driven by $\Delta\psi$.

In short, such citrate uptake as occurs in corn mitochondria (it is a poor respiratory substrate) is largely driven by H^+ symport. The major function of the tricarboxylate transporter is to furnish citrate (or isocitrate) to the cytosol. An important principle emerges here: swelling and exchange studies can detect the presence of metabolite carriers, but they do not identify physiological function.

Other plant mitochondria are much more effective in oxidizing (and hence transporting) citrate. Journet and Douce (1983) find that potato mitochondria supplemented with thiamine pyrophosphate and NAD^+ rapidly oxidize citrate in the absence of ADP, secreting α-ketoglutarate into the medium. After transamination with aspartate, glutamate and malate are secreted. This might be taken as evidence for citrate exchange with α-ketoglutarate or glutamate, but it does not follow that these are carrier-linked exchanges; the concerted action of two discrete carriers will appear as an exchange.

5.5 Amino Acid Transport

Amino acids are zwitterions, but can carry a net charge depending on pH and carboxyl or amine substitution on the R-groups. In the case of simple amino acids lacking such substitution, cyclization through H-bonding of the amino and carboxyl groups may make them truly neutral (Halling et al. 1973). Thus with some amino acids transport will be electroneutral, in others electrogenic, Transport studies in plant mitochondria have been largely focused on the amino acids which can be utilized by the mitochondria as respiratory substrates – glutamate, aspartate, glycine, and proline.

Cavalieri and Huang (1980) found passive swelling of respiration-blocked mung bean mitochondria in 200 mM solutions of neutral amino acids buffered at pH 7.2 in the order proline > serine > methionine > glycine > threonine > alanine. Swelling was generally greater with the L- than the D-stereoisomers, with D/L ratios of 0.87 with proline and 0.60 with serine. The mercurials p-mercuribenzoate and mersalyl (but not N-ethylmaleimide) inhibited passive swelling, although the effect on methionine was small or nonexistent and 10 min preincubation was required for maximum effect. The inhibition was reversed by dithiothreitol. The authors conclude that, as in rat liver mitochondria (reviewed by LaNoue and Schoolwerth 1979), proteinaceous uniports exist for diffusive entry and exit.

Some studies on glycine swelling, oxidation, and ^{14}C-glycine accumulation confirm that uptake is passive (Day and Wiskich 1980, 1981b, Proudlove and Moore 1982); others describe respiration-linked uptake inhibited by uncoupler, mersalyl, and glycine analogs (Walker et al. 1982). Proudlove and Moore (1982), using mung bean and pea leaf mitochondria, question the existence of carriers in that uptake showed no evidence of saturation up to 25 mM glycine, serine, aspartate, or glutamate, and no inhibition by mersalyl, uncoupler, or valinomycin. Day and Wiskich (1980) show that mersalyl promotes, not inhibits, the passive swelling of pea leaf mitochondria in glycine. Glycine oxidation is restricted to mitochondria from photosynthetic tissue (Gardeström et al. 1980), but mitochondria from etiolated mung bean hypocotyls also absorb it (Proudlove and Moore 1982). There is a steady efflux of serine into the medium during glycine oxidation by spinach leaf mitochondria (Journet et al. 1981) but there is as yet no evidence for a specific glycine/serine antiport or common carrier.

A very recent study by Yu et al. (1983) may help to resolve some of the discrepant results outlined above. These authors made a direct study of the transport of glycine (and other neutral amino acids) into plant mitochondria and found evidence for both carrier-mediated transport and diffusion. The former mechanism was only apparent at concentrations of glycine below 0.5 mM; at higher concentrations, the diffusion pathway was dominant. However, earlier reports of inhibition by SH-reagents of glycine uptake could not be confirmed by Yu et al. (1983).

Oxidation of proline and its oxidation product (Δ^1-pyrroline-5-carboxylate producing glutamate) appear to occur at the inner surface of the inner membrane (Elthon and Stewart 1981), and the products of ^{14}C-labeling show up in organic acids and glutamate (Boggess et al. 1978), indicating a requirement for transport. Beyond the above-mentioned swelling studies, however, there is little evidence for specific proline transporters (see also Yu et al. 1983).

Glutamate transport is not resolved. In animal mitochondria glutamate/H^+ symport (or OH^- antiport) and electrogenic exchange of H^+-neutralized glutamate for charged aspartate have been described (LaNoue and Schoolwerth 1979, Nicholls 1982). Neither of these systems is clearly confirmed in plant mitochondria, and aspartate transport has been scarcely studied although it certainly enters (Journet et al. 1981) and exits (Journet et al. 1982), as shown by participation in matrix transamination. The same can be said for glutamate, and empirically a glutamate/aspartate exchange is observed, although the means of exchange is not known.

The anionic character of the dicarboxylate amino acids suggests that exchange on one or more of the carriers handling dicarboxylate anions might occur, but DeSantis et al. (1976) found no glutamate exchange with phosphate, α-ketoglutarate, or citrate in bean mitochondria (glutamate/aspartate exchange was not determined). On the contrary, Day and Wiskich (1977) found that rapid swelling of cauliflower and beet mitochondria in NH_4 glutamate required additions of phosphate and malate, and that butylmalonate was somewhat inhibitory. These are the same requirements found for citrate and α-keto-glutarate swelling; i.e., malate taken up in exchange for phosphate exchanges for glutamate (Fig. 5, reaction 6). It was later discovered that phthalonate would inhibit the apparent malate/glutamate exchange and glutamate oxidation (Day and Wiskich 1981a). They suggest a special glutamate carrier on the grounds that glutamate did not inhibit citrate/isocitrate exchange.

As noted above, Proudlove and Moore (1982) found no evidence for energy-linked carrier-mediated ^{14}C-glutamate or aspartate uptake. They state that phthalonate has no effect on the influx of glutamate or aspartate, and suggest that fluxes may be determined by movement of the corresponding keto acids (i.e., α-ketoglutarate and oxaloacetate) and by local concentrations of each.

Resolution of glutamate transport will require a patient comparison of the techniques producing these opposed results. For the purpose of this chapter, a H^+/glutamate symport is illustrated in Fig. 5 (reaction 7); if glutamate is passively diffusing via nonspecific channels it must be neutralized. It should be noted that cotransport mechanisms (i.e., H^+/X, Fig. 4C) for amino acid transport are not eliminated; that is, protonation of the carrier, rather than the substrate, with uptake driven by the total protonmotive force ($\Delta\bar{\mu}_{H^+}$). A carrier of this type could conceivably participate in passive flux as well.

5.6 Nucleotide Transport

Most studies of nucleotide transport have been on the ADP/ATP exchange accompanying oxidative phosphorylation, again building on intensive investigations with animal mitochondria (Klingenberg 1970, Vignais 1976). In addition, there have been a few studies of net uptake of nucleotides by plant mitochondria. Since matrix nucleotides partake in the exchange, net transport will be discussed first.

Plant mitochondria contain nucleotides and other cofactors which are required for substrate oxidation, phosphorylation, and polynucleotide synthesis. Osmotic shrinkage and metabolism-inhibiting conditions during isolation can produce mitochondria which have lost substrates and cofactors (Hanson and Day 1980), and the deficiency affects metabolism. An example of this is found in the report of Stoner and Hanson (1966) that the simple addition of 5 mM NAD^+ would activate cyanide-sensitive endogenous respiration and produce contraction of KCl-swollen corn mitochondria. Apparently the swollen mitochondria were deficient in NAD^+, but passive influx from very high external concentrations was sufficient to activate the dehydrogenases. Efflux of NAD^+ from bean cotyledon mitochondria at different development stages correlates with their loss of metabolic activity (MALHOTRA and SPENCER 1970). With ample substrate NAD$^+$ deficiency is not always observed, but TOBIN et al. (1980) found that potato mitochondria require added NAD$^+$ for optimal malate oxidation, and accumulated it in an energy-linked process ($K_m = 0.3$ mM, $V_{max} = 2$ nmol min^{-1} mg^{-1} protein). Mung bean mitochondria had a higher endogenous NAD$^+$ content (5.7 nmol mg^{-1} protein vs. 1.1 for potato), lower V_{max} for uptake (0.5 to 0.7 nmol min^{-1} mg^{-1} protein), and did not require added NAD$^+$ for optimal malate oxidation. The authors speculate that the higher matrix content of NAD$^+$ might affect uptake rates.

Adenine nucleotide (AdN) content of plant mitochondria can also be deficient. Cauliflower mitochondria do not exhibit an uncoupler-activated ATPase activity unless "primed" by a brief period of respiration in the presence of Mg and phosphate during which the AdN content rises two- to three fold above the initial level of 3.2 nmol mg^{-1} protein (JUNG and HANSON 1975). Priming is not needed for corn mitochondria with an initial AdN content of 7 nmol mg^{-1} protein. Priming is believed to effect release of an inhibitor protein of the coupling ATPase complex (TAKEUCHI 1975), but the observation on net ADP or ATP accumulation increasing the matrix pool of functional AdN is still of importance.

ABOU-KHALIL and HANSON (1977, 1979b, c) investigated in greater detail the energy-linked accumulation of AdN in corn mitochondria (reaction 10, Fig. 5). The process is more effective with ADP than ATP, requires both Mg^{2+} and phosphate, and the accumulated AdN are lost when $\Delta\bar{\mu}_{H^+}$ is collapsed with uncoupler. Extensive accumulation requires inhibition of ATP formation by oligomycin (blocks the proton channel of the ATPase, reaction 8, Fig. 5) or by carboxyatractyloside (blocks ADP/ATP exchange, reaction 9, Fig. 5). However, even during normal oxidative phosphorylation the AdN content rises from 4–5 nmol mg^{-1} protein to 16–18 nmol, occurring predominately as ATP. Arsenate will substitute for phosphate (sulfate partially). ADP does not exchange for accumulated ^{32}Pi. Mersalyl is very inhibitory, possibly due to inhibition of phosphate uptake. Mn^{2+}, and to a lesser degree Ca^{2+} and Zn^{2+}, can substitute for Mg^{2+}. Accumulated ADP will rapidly exchange for exogenous ADP through an AdN-specific pathway which is insensitive to carboxyatractyloside. Citrate is a competitive inhibitor of ADP uptake ($K_i = 10$ mM), but it does not exchange for matrix AdN. The empirical observation is that net ADP uptake involves concerted transport of phosphate and Mg^{2+}; it is possible that an extended function of the phosphate transporter is responsible.

The adenine nucleotide translocator (reaction 9, Fig. 5) carries out an electrogenic ADP$^{3-}_{in}$/ATP$^{4-}_{out}$ exchange driven by $\Delta\psi$, which along with the H$^+$/phosphate symport (reaction 2) at the expense of ΔpH utilizes part of the $\Delta\bar{\mu}_{H^+}$ required for oxidative phosphorylation (the balance is utilized in the electrogenic influx of 2 H$^+$ through the coupling ATPase). Most of what is known about the AdN translocator in plant mitochondria is derived from studies utilizing the specific inhibitors atractyloside, carboxyatractyloside, and bongkrekic acid. After some initial uncertainty arising from the relative ineffectiveness of atractyloside, it has become clear that, with minor differences, the transport process is very similar to that in animal mitochondria.

VIGNAIS et al. (1976) note three differences between plant and animal mitochondria with respect to AdN transport: a small AdN pool (see above discussion), low atractyloside sensitivity, and competitive inhibition of ADP transport by carboxyatractyloside. Competitive inhibition of ADP transport as determined by inhibition of State 3 respiration or ^{14}C-ADP exchange has been observed for both atractyloside (JUNG and HANSON 1973, VIGNAIS et al. 1976, EARNSHAW 1977, SILVA LIMA and DENSLOW 1979) and carboxyatractyloside (VIGNAIS et al. 1976, SILVA LIMA and DENSLOW 1979), with carboxyatractyloside 30- to 50-fold more effective. Binding affinities show comparable discrimination; $K_D = 10$ to 20 nM for carboxyatractyloside, about 0.45 µM for atractyloside in potato mitochondria (VIGNAIS et al. 1976). Binding of atractyloside prevents the small, abrupt contraction detected by absorbancy upon addition of ADP (JUNG and HANSON 1973, EARNSHAW and HUGHES 1976); this is thought to result from effectively fixing the carrier in its outer, ADP-binding orientation and preventing the conformational change to an inner, ATP-binding configuration (EARNSHAW and HUGHES 1976).

Bongkrekic acid, a noncompetitive inhibitor of the AdN transporter, is also effective at low concentrations ($K_D = 0.7$ nmol mg^{-1}protein) with the amount bound increased in the presence of ADP (VIGNAIS et al. 1976). Oddly enough, bongkrekic acid at higher concentrations can relieve the atractyloside inhibition of ADP-induced contraction, presumably by reorientation of the carrier to an inside configuration (EARNSHAW and HUGHES 1976). However, only the orientation is altered; transport is blocked.

Measurements of ^{14}C-ADP uptake by corn mitochondria show initial transport to be completed in less than 15 s (EARNSHAW 1977); this contrasts with the net ADP uptake mechanism which requires about 5 min (ABOU-KHALIL and HANSON 1977). With mitochondria from greening oat seedling leaves and measurements of ^{3}H-ADP uptake into the matrix space, about 5 min are required to reach maximum uptake (HAMPP and WELLBURN 1980). In both cases, exchange of preloaded ADP was only activated by added AMP, ADP, or ATP; other nucleotides had little or negligible effect (AMP may exchange after conversion to ADP by adenylate kinase). In oat mitochondria the K_m for ADP uptake was 27 µM, for ATP uptake 100 µM (HAMPP and WELLBURN 1980).

EARNSHAW (1977) determined carrier-bound (atractyloside-removable) ADP to be 0.55 nmol mg^{-1} protein, a value similar to that for animal mitochondria. Exchange of ADP into the matrix saturates at about 5 µM external ADP.

Hence, as far as examined the AdN transporter functions as it does in animal mitochondria by delivering ADP to the matrix and removing the ATP formed. BERTAGNOLLI and HANSON (1973) found that arsenate uncoupling involved cyclic arsenate transport, in on the phosphate transporter, out (as ADP-As) on the AdN transporter. Since ADP-As should hydrolyze if set free in the aqueous matrix, it was suggested that the coupling ATPase might directly exchange nucleotides with the transporter. The alternative is that ADP-As hydrolysis in the matrix is slow compared to transport.

Lastly, DRY and WISKICH (1982) using pea leaf and cauliflower mitochondria have found that State 3 respiration rates are determined more by the concentration of exogenous ADP than the ADP/ATP ratio. This suggests that there are still important things to learn about ADP binding to the carrier, carrier turnover, and matrix concentrations of ATP.

6 Conclusions

It is evident that isolated plant mitochondria have transport systems for metabolites which function in a fashion similar to those of other eukaryotes. It is equally true that they show differences which reflect autotrophic metabolism. One can expect differences in kind as well as degree as more is learned about the mitochondria of green leaves. Plant mitochondria warrant study directed at explaining their participation in plant metabolism, with less dependence on comparisons with rat liver or beef heart. Work on these lines has begun, and transport studies form an integral part of it.

Most disturbing are the diverse and sometimes conflicting results found with mitochondria from different species and tissues. Some of this diversity may lie with techniques of isolation and assay, and investigation of these cannot be considered closed until mitochondrial preparations can be shown to have properties and activities comparable to those in vivo. The difficulty here lies with determining the in vivo standards; however, at the minimum, estimates can be made of respiratory rates in vivo and in vitro, and calculations made on substrate transport in different tissues at different stages of development. In short, the biochemical and biophysical data on transport (and metabolism) can be referred to physiological requirements.

References

Abou-Khalil S, Hanson JB (1977) Net adenosine diphosphate accumulation in mitochondria. Arch Biochem Biophys 183:581–587

Abou-Khalil S, Hanson JB (1978) Net accumulation of adenine nucleotide by corn mitochondria. In: Ducet G, Lance C (eds) Plant mitochondria. Elsevier/North Holland, Biomedical Press, Amsterdam, pp 141–149

Abou-Khalil S, Hanson JB (1979a) Energy-linked sulfate uptake by corn mitochondria via the phosphate transporter. Plant Physiol 63:635–638

Abou-Khalil S, Hanson JB (1979b) Energy-linked adenosine diphosphate accumulation by corn mitochondria. I. General characteristics and effect of inhibitors. Plant Physiol 64:276–280

Abou-Khalil S, Hanson JB (1979c) Energy-linked adenosine diphosphate accumulation by corn mitochondria. II. Phosphate and divalent cation requirement. Plant Physiol 64:281–284

Azzi A, Azzone GF (1967a) Swelling and shrinkage phenomena in liver mitochondria. VI. Metabolism-independent swelling coupled to ion movement. Biochim Biophys Acta 131:468–478

Azzi A, Azzone GF (1967b) Ion transport in liver mitochondria. II. Metabolism-linked ion extrusion. Biochim Biophys Acta 135:444–453

Baker JE, Elfvin LG, Biale JB, Honda SI (1968) Studies on the ultrastructure and purification of isolated plant mitochondria. Plant Physiol 43:2001–2022

Bertagnolli BL, Hanson JB (1973) Functioning of the adenine nucleotide transporter in the arsenate uncoupling of corn mitochondria. Plant Physiol 52:431–435

Birnberg PR, Hanson JB (1983) Mehanisms of citrate transport and exchange in corn mitochondria. Plant Physiol 71:803–809

Birnberg PR, Jayroe DL, Hanson JB (1982) Citrate transport in corn mitochondria. Plant Physiol 70:511–516

Bligny R, Douce R (1976) Les mitochondries de cellules végétales isolées (*Acer pseudoplatanus* L). I. Étude comparée des propriétés oxydatives des mitochondries extraites des mitochondries placées dans leur contexte cellulaire. Physiol Vég 14:499–515

Bligny R, Douce R (1980) A precise localization of cardiolipin in plant cells. Biochim Biophys Acta 617:254–263

Boggess SF, Koeppe DE, Stewart CR (1978) Oxidation of proline by plant mitochondria. Plant Physiol 62:22–25

Brierley GP (1976) The uptake and extrusion of monovalent cations by isolated heart mitochondria. Mol Cell Biochem 10:41–62

Brierley GP, Jurkowitz M, Scott KM, Merola AJ (1971) Ion transport by heart mitochondria. XXII. Spontaneous energy-linked accumulation of acetate and phosphate salts of monovalent cations. Arch Biochem Biophys 147:545–556

Carafoli E (1981) The uptake and release of calcium by mitochondria. In: Lee CP, Schatz G, Dallner G (eds) Mitochondria and microsomes. Addison-Wesley, Reading (Mass), pp 357–374

Cavalieri AJ, Huang AHC (1980) Carrier protein-mediated transport of neutral amino acids into mung bean mitochondria. Plant Physiol 66:588–591

Chappell JB (1968) Systems used for transport of substrates into mitochondria. Br Med Bull 24:150–157

Chappell JB, Crofts AR (1966) Ion transport and reversible volume changes of isolated mitochondria. BBA Libr 7:293–316

Chappell JB, Haarhoff KN (1967) The penetration of the mitochondrial membrane by anions and cations. In: Slater EC, Kaninga Z, Wojtczak L (eds) Biochemistry of mitochondria. Academic Press, London, New York, pp 75–91

Chávez E, Jung DW, Brierley GP (1977) Energy-dependent exchange of K^+ in heart mitochondria. K^+ efflux. Arch Biochem Biophys 183:460–470

Chen C, Lehninger AL (1973) Ca^{2+} transport activity in mitochondria from some plant tissues. Arch Biochem Biophys 157:183–196

Day DA, Hanson JB (1977a) Effect of phosphate and uncouplers on substrate transport and oxidation by isolated corn mitochondria. Plant Physiol 59:139–144

Day DA, Hanson JB (1977b) Pyruvate and malate transport and oxidation in corn mitochondria. Plant Physiol 59:630–635

Day DA, Wiskich JT (1977) Glutamate transport by plant mitochondria. Plant Sci Lett 9:33–36

Day DA, Wiskich JT (1980) Glycine transport by pea leaf mitochondria. FEBS Lett 112:191–194

Day DA, Wiskich JT (1981a) Effect of phthalonic acid on respiration and metabolite transport in higher plant mitochondria. Arch Biochem Biophys 211:100–107

Day DA, Wiskich JT (1971b) Glycine metabolism and oxalacetate transport by pea leaf mitochondria. Plant Physiol 68:425–429

Day DA, Bertagnolli BL, Hanson JB (1978) The effect of calcium on the respiratory responses of corn mitochondria. Biochim Biophys Acta 502:289–297

DeSantis A, Borraccino G, Arrigoni O, Palmieri F (1975) The mechanism of phosphate permeation in purified bean mitochondria. Plant Cell Physiol 16:911–923

DeSantis A, Arrigoni O, Palmieri F (1976) Carrier-mediated transport of metabolites in purified bean mitochondria. Plant Cell Physiol 17:1221–1233

Dieter P, Marmé D (1980) Ca^{2+} transport in mitochondrial and microsomal fractions from higher plants. Planta 150:1–8

Dry IB, Wiskich JT (1982) Role of the external adenosine triphosphate/adenosine diphosphate ratio in the control of plant mitochondrial respiration. Arch Biochem Biophys 217:72–79

Ducet G (1979) Influence of bovine serum albumin on the proton conductance of potato mitochondrial membranes. Planta 147:122–126

Earnshaw MJ (1977) Adenine nucleotide translocation in plant mitochondria. Phytochemistry 16:181–184

Earnshaw MJ, Hanson JB (1973) Inhibition of post-oxidative calcium release in corn mitochondria by inorganic phosphate. Plant Physiol 52:403–406

Earnshaw MJ, Hughes EA (1976) The adenine nucleotide carrier of plant mitochondria: Contraction induced by adenosine diphosphate. Plant Sci Lett 6:343–348

Earnshaw MJ, Truelove B (1968) Swelling and contraction of *Phaseolus* hypocotyl mitochondria. Plant Physiol 43:121–129

Earnshaw MJ, Truelove B, Butler RD (1970) Swelling of *Phaseolus* mitochondria in relation to free fatty acid levels. Plant Physiol 45:318–321

Earnshaw MJ, Madden DM, Hanson JB (1973) Calcium accumulation by corn mitochondria. J Exp Bot 24:828–840

Elthon TE, Stewart CR (1981) Submitochondrial location and electron transport characteristics of enzymes involved in proline oxidation. Plant Physiol 67:780–784

Elzam OE, Hodges TK (1968) Characterization of energy-dependent Ca^{2+} transport in maize mitochondria. Plant Physiol 43:1108–1114

Fluegel MJ, Hanson JB (1981) Mechanisms of passive potassium influx in corn mitochondria. Plant Physiol 68:267–271

Fukumoto M, Nagai K (1982) Effects of calmodulin antagonists on the mitochondrial and microsomal Ca^{2+} uptake by apple fruit. Plant Cell Physiol 23:1435–1441

Gardeström P, Bergman A, Ericson I (1980) Oxidation of glycine via the respiratory chain in mitochondria prepared from different parts of spinach. Plant Physiol 65:389–391

Garlid KD (1980) On the mechanism of regulation of the mitochondrial K^+/H^+ exchanger. J Biol Chem 255:11273–11279

Halling PJ, Brand MD, Chappell JB (1973) Permeability of mitochondria to neutral amino acids. FEBS Lett 34:169–171

Hampp R (1979) Kinetics of mitochondrial phosphate transport and rates of respiration and phosphorylation during greening of etiolated *Avena* leaves. Planta 144:325–332

Hampp R, Wellburn AR (1980) Translocation and phosphorylation of adenine nucleotides by mitochondria and plastids during greening. Z Pflanzenphysiol 98:289–304

Hanson JB (1972) Ion transport induced by polycations and its relationship to loose coupling of corn mitochondria. Plant Physiol 49:707–715

Hanson JB, Day DA (1980) Plant mitochondria. In: Tolbert NE (ed) The plant cell. Biochem Plants 1:315–358

Hanson JB, Koeppe DE (1975) Mitochondria. In: Baker DA, Hall JL (eds) Ion transport in plant cells and tissues. Elsevier/North Holland Biomedical Press, Amsterdam, pp 79–99

Hanson JB, Miller RJ (1967) Evidence for active phosphate transport in maize mitochondria. Proc Natl Acad Sci USA 58:727–734

Hanson JB, Bertagnolli BL, Shepherd WD (1972) Phosphate-induced stimulation of acceptorless respiration in corn mitochondria. Plant Physiol 50:347–354

Hensley JR, Hanson JB (1975) The action of valinomycin in uncoupling corn mitochondria. Plant Physiol 56:13–18

Hodges TK, Hanson JB (1965) Calcium accumulation by maize mitochondria. Plant Physiol 40:101–109

Horn ME, Mertz D (1982) Cyanide-resistant respiration in suspension cultured cells of *Nicotiana glutinosa* L. Plant Physiol 69:1439–1443

Huber SC, Moreland DE (1979) Permeability properties of the inner membrane of mung bean mitochondria and changes during energization. Plant Physiol 64:115–119

278 J.B. HANSON:

Johnson HM, Wilson RH (1972) Sr^{2+} uptake by bean (*Phaseolus vulgaris*) mitochondria.
Biochim Biophys Acta 267:398–408
Johnson HM, Wilson RH (1973) The accumulation and the release of divalent cations
across mitochondrial membranes. Am J Bot 60:858–862
Journet E-P, Douce R (1983) Mechanisms of citrate oxidation by Percoll-purified mito-
chondria from potato tuber. Plant Physiol 72:802–808
Journet E-P, Neuburger M, Douce R (1981) Role of glutamate-oxaloacetate transaminase
and malate dehydrogenase in the regeneration of NAD$^+$ for glycine oxidation by
spinach leaf mitochondria. Plant Physiol 57:467–469
Journet E-P, Bonner WD Jr, Douce R (1982) Glutamate metabolism triggered by oxalo-
acetate in intact plant mitochondria. Arch Biochem Biophys 214:366–375
Jung DW, Brierley GP (1979) Swelling and contraction of potato mitochondria. Plant
Physiol 64:948–953
Jung DW, Hanson JB (1973) Atractyloside inhibition of adenine nucleotide transport
in mitochondria from plants. Biochim Biophys Acta 325:189–192
Jung DW, Hanson JB (1975) Activation of 2,4-dinitrophenol-stimulated ATPase activity
in cauliflower and corn mitochondria. Arch Biochem Biophys 168:358–368
Jung DW, Laties GG (1979) Citrate and succinate uptake by potato mitochondria. Plant
Physiol 63:591–597
Jung DW, Chávez E, Brierley GP (1977) Energy-dependent exchange of K$^+$ in heart
mitochondria. K$^+$ influx. Arch Biochem Biophys 183:452–459
Kenefick DG, Hanson JB (1967) Active accumulation of phosphate by maize mitochon-
dria. In: Fried M (ed) Isotopes in plant nutrition and physiology. Int Atom Energy
Ag, Vienna, pp 271–286
Kimpel JA, Hanson JB (1977) Activation of endogenous respiration and anion transport
in corn mitochondria by acidification of the medium. Plant Physiol 60:933–934
Kimpel JA, Hanson JB (1978) Efflux and influx of potassium salts elicited by valinomycin
in plant mitochondria. Plant Sci Lett 11:329–335
Kirk BI, Hanson JB (1973) The stoichiometry of respiration-driven potassium transport
in corn mitochondria. Plant Physiol 51:357–362
Klingenberg M (1970) Metabolite transport in mitochondria: An example for intracellular
membrane function. Essays Biochem 6:119–159
Klingenberg M (1981) The ADP translocation system of mitochondria. In: Lee CP,
Schatz G, Dallner G (eds) Mitochondria and microsomes. Addison-Wesley, Reading
(Mass), pp 293–316
LaNoue KF, Schoolwerth AC (1979) Metabolite transport in mitochondria. Annu Rev
Biochem 48:871–922
Lee DC, Wilson RH (1972) Swelling in bean shoot mitochondria induced by a series
of potassium salts or organic anions. Physiol Plant 27:195–201
Lehninger AL (1962) Water uptake and extrusion by mitochondria in relation to oxidative
phosphorylation. Physiol Rev 42:467–517
Lehninger AL (1970) Mitochondria and calcium transport. Biochem J 119:129–138
Lehninger AL (1974) Role of phosphate and other proteon-donating anions in respira-
tion-coupled transport of Ca^{2+} by mitochondria. Proc Natl Acad Sci USA
71:1520–1524
Longo C, Arrigoni O (1964) Functional properties of isolated plant mitochondria. High
amplitude swelling. Exp Cell Res 35:572–579
Lorimer GH, Miller RJ (1969) The osmotic behavior of corn mitochondria. Plant Physiol
44:839–844
Lyons JM, Pratt HK (1964) An effect of ethylene on swelling of isolated mitochondria.
Arch Biochem Biophys 104:318–324
Lyons JM, Wheaton TA, Pratt HK (1964) Relationship between the physical nature
of mitochondrial membranes and chilling sensitivity in plants. Plant Physiol
39:262–268
Malhotra SS, Spencer M (1970) Changes in the respiratory, enzymatic, and swelling
and contraction properties of mitochondria from cotyledons of *Phaseolus vulgaris* L.
during germination. Plant Physiol 46:40–44

Mannella CA, Bonner WD Jr (1975) Biochemical characteristics of the outer membranes of plant mitochondria. Biochim Biophys Acta 413:213–225

Massari S, Frigeri L, Azzone GF (1972) Permeability to water, dimension of surface, and structural changes during swelling of rat liver mitochondria. J Membrane Biol 9:57–70

McNeil PH, Thomas DR (1976) The effect of carnitine on palmitate oxidation by pea cotyledon mitochondria. J Exp Bot 27:1163–1180

Meijer AJ, Van Dam K (1974) The metabolic significance of anion transport in mitochondria. Biochim Biophys Acta 346:213–244

Meijer AJ, Woerkom Van GM, Eggelte TA (1976) Phthalonic acid, an inhibitor of α-oxoglutarate transport in mitochondria. Biochim Biophys Acta 430:53–61

Millard DL, Wiskich JT, Robertson RN (1965) Ion uptake and phosphorylation in mitochondria: Effect of monovalent ions. Plant Physiol 40:1129–1135

Miller RJ, Dumford SW, Koeppe DE, Hanson JB (1970) Divalent cation stimulation of substrate oxidation by corn mitochondria. Plant Physiol 45:649–653

Mitchell P (1966) Chemiosmotic coupling in oxidative and photosynthetic phosphorylation. Glenn Res Ltd, Bodmin, Cornwall, England, pp 29–35

Moore AL, Bonner WD Jr (1977) The effect of calcium on the respiratory responses of mung bean mitochondria. Biochim Biophys Acta 460:455–466

Moore AL, Bonner WD Jr (1981) A comparison of the phosphorylation potential and electrochemical proton gradient in mung bean mitochondria and phosphorylating submitochondrial particles. Biochim Biophys Acta 634:117–128

Moore AL, Wilson SB (1977) Translocation of some anions, cations and acids in turnip (Brassica napus L.) mitochondria. J Exp Bot 28:607–618

Moore AL, Rich PR, Bonner WD Jr (1978) Factors influencing the components of the total proton motive force in mung bean mitochondria. J Exp Bot 29:1–12

Moore AL, Jackson C, Deuch J, Morris P, Hall DO (1979) The relationship of glycine decarboxylase to the phosphorylation potential. Plant Physiol 63:S-613

Moreau F, DuPont J, Lance C (1974) Phospholipid and fatty acid composition of outer and inner membranes of plant mitochondria. Biochim Biophys Acta 345:294–304

Moyle J, Mitchell P (1977) The lanthanide-sensitive calcium phosphate porter of rat liver mitochondria. FEBS Lett 77:136–140

Neuburger M, Douce R (1980) Effect of bicarbonate and oxaloacetate on malate oxidation by spinach leaf mitochondria. Biochim Biophys Acta 589:176–189

Nicholls DG (1982) Bioenergetics. An introduction to the chemiosmotic theory. Academic Press, London New York, pp 183

Nicholls D, Åkerman K (1982) Mitochondrial calcium transport. Biochim Biophys Acta 683:57–88

Oliver DJ (1981) Formate oxidation and oxygen reduction by leaf mitochondria. Plant Physiol 68:703–705

Öpik H (1974) Mitochondria. In: Robards AW (ed) Dynamic aspects of plant ultrastructure. McGraw-Hill, London, pp 53–83

Overman AR, Lorimer GH, Miller RJ (1970) Diffusion and osmotic transfer in corn mitochondria. Plant Physiol 45:126–132

Phillips ML, Williams GR (1973a) Effects of 2-butlymalonate, 2-phenylsuccinate, benzylmalonate, and ρ-iodobenzylmalonate on the oxidation of substrates by mung bean mitochondria. Plant Physiol 51:225–228

Phillips ML, Williams GR (1973b) Anion transporters in plant mitochondria. Plant Physiol 51:667–670

Pomeroy MK (1976) Swelling and contraction of mitochondria from cold-hardened and nonhardened wheat and rye seedlings. Plant Physiol 57:469–473

Pomeroy MK (1977) Ultrastructural changes during swelling and contraction of mitochondria from cold-hardened and nonhardened winter wheat. Plant Physiol 59:250–255

Proudlove MO, Moore AL (1982) Movement of amino acids into isolated plant mitochondria. FEBS Lett 147:26–30

Scarpa A (1979) Transport in mitochondrial membranes. In: Giebisch G, Tosteson DC, Ussing HH (eds) Membrane transport in biology, vol II. Springer, Berlin Heidelberg New York, pp 261–355

Silva Lima A, Denslow ND (1979) The effect of atractyloside and carboxyatractyloside on adenine nucleotide translocation in mitochondria of *Vigna sinensis* cv. Serido. Arch Biochem Biophys 193:368–372

Stoner CD, Hanson JB (1966) Swelling and contraction of corn mitochondria. Plant Physiol 4:255–266

Stoner CD, Hodges TK, Hanson JB (1964) Chloramphenicol as an inhibitor of energy-linked processes in maize mitochondria. Nature (London) 203:258–261

Takeuchi Y (1975) Respiration-dependent uncoupler-stimulated ATPase activity in castor bean endosperm mitochondria and submitochondrial particles. Biochim Biophys Acta 376:505–518

Tedeschi H, Harris DL (1955) The osmotic behaviour and permeability to nonelectrolytes of mitochondria. Arch Biochem Biophys 58:52–67

Thomas DR, McNeil PH (1976) The effect of carnitine on the oxidation of saturated fatty acids by pea cotyledon mitochondria. Planta 132:61–63

Tobin A, Djerdjour B, Journet E, Neuburger M, Douce R (1980) Effect of NAD on malate oxidation in intact plant mitochondria. Plant Physiol 66:225–229

Vignais PV (1976) Molecular and physiological aspects of adenine nucleotide transport in mitochondria. Biochim Biophys Acta 456:1–38

Vignais PV, Douce R, Lauquin GJM, Vignais PM (1976) Binding of radioactively labeled carboxyatractyloside, atractyloside and bongkrekic acid to the ADP translocator of potato mitochondria. Biochim Biophys Acta 440:688–696

Walker GH, Sarogini G, Oliver DJ (1982) Identification of a glycine transporter from pea leaf mitochondria. Biochem Biophys Res Commun 107:856–861

Wickes WA, Wiskich JT (1976) Arsenate uncoupling of oxidative phosphorylation in isolated plant mitochondria. Aust J Plant Physiol 3:153–162

Wilson RH, Graesser RJ (1976) Ion transport in plant mitochondria. In: Stocking CR, Heber U (eds) Transport in plants III. Encyclopedia of plant physiology, vol III. Springer, Berlin Heidelberg New York, pp 377–397

Wilson RH, Minton GA (1974) The comparative uptake of Ba^{2+} and other alkaline earth metals by plant mitochondria. Biochim Biophys Acta 333:22–27

Wilson RH, Hanson JB, Mollenhauer HH (1969) Active swelling and acetate uptake in corn mitochondria. Biochemistry 8:1203–1213

Wiskich JT (1974) Substrate transport into plant mitochondria. Swelling studies. Aust J Plant Physiol 1:177–181

Wiskich JT (1975) Phosphate-dependent substrate transport into mitochondria. Oxidative studies. Plant Physiol 56:121–125

Wiskich JT (1977) Mitochondrial metabolite transport. Annu Rev Plant Physiol 28:45–69

Woo KC, Osmond CB (1976) Glycine decarboxylation in mitochondria isolated from spinach leaves. Aust J Plant Physiol 3:771–783

Yoshida K (1968a) Swelling and contraction of isolated plant mitochondria II. Reversible metabolism-dependent contraction. J Fac Sci Univ Tokyo 10:63–82

Yoshida K (1968b) Swelling and contraction of isolated plant mitochondria, III. Reversible metabolism-dependent swelling. J Fac Sci Univ Tokyo 10:83–95

Yoshida K, Sato S (1968) Swelling and contraction of isolated plant mitochondria, I. Passive swelling in sugar and electrolyte solutions. J Fac Sci Univ Tokyo 10:49–62

Yu C, Claybrook DL, Huang AHC (1983) Transport of glycine, serine and proline into spinach leaf mitochondria. Arch Biochem Biophys 227:180–187

10 The Tricarboxylic Acid Cycle in Plant Mitochondria: Its Operation and Regulation

J.T. Wiskich and I.B. Dry

1 Introduction

A major role of plant mitochondria is to generate usable energy required for cellular processes. This is achieved by oxidizing organic acids in the Krebs or tricarboxylic acid (TCA) cycle (Fig. 1) and transferring the reducing equivalents, via the electron transfer chain, to molecular oxygen. During the latter process ADP is phosphorylated to ATP. However, plant mitochondrial function is not as simple as that. First, other substrates can be oxidized [e.g., external (cytoplasmic) NADH, fatty acids and amino acids, glutamate, and glycine.] Second, plant mitochondria possess nonphosphorylating pathways such as the cyanide-insensitive, alternative oxidase and the rotenone-insensitive by-pass for oxidation of intramitochondrial NADH (see Chap. 8 and HANSON and DAY 1980). The combination of these two pathways does not lead to the formation of any ATP. Third, other enzymes can play an important part in mitochondrial function, e.g., NAD-malic enzyme and aspartate aminotransferase.

These features tend to complicate our interpretations because we do not fully understand their activation or de-activation in tissues. However, they do highlight functions, other than maintenance of energy charge (ATKINSON 1977), of plant mitochondria. These functions would require turnover of part or all of the TCA cycle. Although only a single control mechanism need be operating at any one time, different states or conditions of the tissue may impose different controls. The isolation of mitochondria from tissue will usually release them from the controlling influence. Thus the study of isolated mitochondria will, at best, only reveal the potential of their activity or control. Much of the current research into plant mitochondria is aimed at explaining tissue respiration because the TCA cycle plays a central role in plant metabolism.

Quantitative assessments of its in vivo activity have proved difficult (AP REES 1980). Nevertheless, we have analyzed the current information with the view of extrapolating it back to the tissue.

Abbreviations. AMP: adenosine 5′-monophosphate; ADP: adenosine 5′-diphosphate; ATP: adenosine 5′-triphosphate; CHCA: α-cyano-4-hydroxycinnamic acid; CAM: crassulacean acid metabolism; CoA: coenzyme A; NAD: nicotinamide adenine dinucleotide; NADH: nicotinamide adenine dinucleotide (reduced form); PEP: phosphoenolpyruvate; P$_i$: inorganic phosphate; TCA: tricarboxylic acid; TES: N-Tris(hydroxymethyl)methyl-2-aminoethanesulphonic acid.

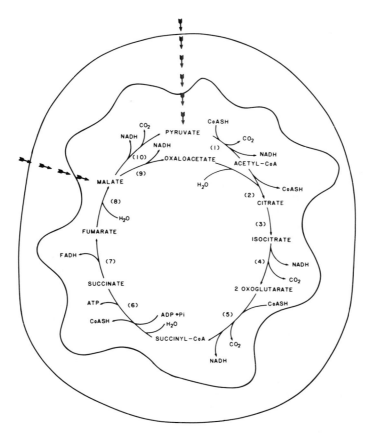

Fig. 1. The tricarboxylic acid cycle. The numbered enzymes are: *1* pyruvate dehydrogenase; *2* citrate synthetase; *3* aconitase; *4* NAD-isocitrate dehydrogenase; *5* oxoglutarate dehydrogenase; *6* succinyl-CoA synthetase; *7* succinate dehydrogenase; *8* fumarase; *9* NAD-malate dehydrogenase; *10* NAD-malic enzyme

2 Control

This topic has been treated recently (WISKICH 1980) and the more general and historic aspects will not be repeated here. For a detailed theoretical discussion of control of metabolic pathways the reader is referred to NEWSHOLME and CRABTREE (1979).

2.1 Adenylate Energy

Mitochondria isolated from most plant tissues demonstrate respiratory control (Fig. 2), in which the rate of oxygen uptake is governed by the presence or absence of phosphate or phosphate acceptor. It took about 10 years for this

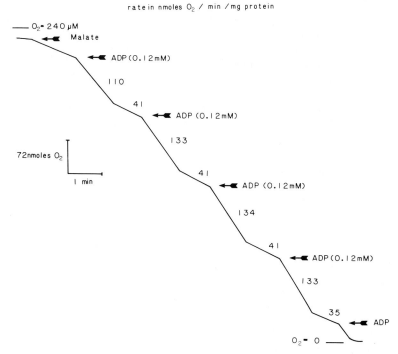

Fig. 2. Oxygen electrode trace of respiring cauliflower mitochondria showing respiratory control. Assayed in 3-ml medium containing 0.3 M sucrose, 10 mM TES pH 7.2, 10 mM phosphate buffer pH 7.2, 5 mM MgCl$_2$, 7 mM malate, 0.3 mM thiamine pyrophosphate and 1.04 mg mitochondrial protein. *Numbers* along the trace represent oxygen uptake as nmol O$_2$ mg^{-1} protein min^{-1}

phenomenon to be translated from animal (LARDY and WELLMAN 1952) to plant (BONNER and VOSS 1961, WISKICH and BONNER 1963) mitochondria. CHANCE and WILLIAMS (1955a, b) developed a terminology to describe various "states" of the mitochondria under these differing conditions.

State 3 is defined as the condition where substrate, oxygen, ADP, and phosphate are present in high concentration and the rate-limiting step lies within the respiratory enzymes. State 4 follows from State 3 after all, or most of the ADP has been phosphorylated. In State 4, the concentration of ADP is limiting, but this may not be directly related to its affinity for the phosphorylation system. Under these conditions the ATP/ADP ratio is high and the phosphorylation potential (ΔG_p) is high.

$$\Delta G_p = -\Delta G'_0 + 2.303 \; RT \frac{\log[ATP]}{[ADP][P_i]}$$

where $\Delta G'_0$ is the standard free energy of ATP hydrolysis, R is the gas constant and T, the absolute temperature.

The phosphorylation potential is the Gibbs free energy of ATP synthesis and clearly it will vary with the ATP/ADP ratio (assuming [P_i] undergoes little change). MOORE and BONNER (1981) have determined "null"- or "equilibrium"-point ΔG_P values of about 64 kJ mol^{-1} for mung bean mitochondria. This would represent an efficiency of 85% in the transfer of energy from endogenous NADH oxidation to ATP synthesis – assuming a P/O ratio of 3. In other words, phosphorylation and oxidation tend to be in thermodynamic equilibrium. This is in contrast to calorimetric measurements of tightly coupled rat liver mitochondria (ESTABROOK and GUTFREUND 1960), which show that heat production during State 3 electron flow (104.5 kJ mol^{-1} substrate) is about two-thirds of that during uncoupled oxidation (146.3 KJ mol^{-1} substrate). These calorimetric measurements need to be verified as they suggest inefficient energy transduction. Consequently, in thermogenic tissues, which increase their temperature significantly above ambient, respiration would need to be rapid, rather than nonphosphorylating or uncoupled.

Although respiratory control can be demonstrated with isolated plant mitochondria, DRY and WISKICH (1982) showed that external ATP/ADP ratios exceeding 20 were required to influence State 3 oxidation rates. These ratios are much greater than those published for ATP/ADP in the cellular compartments of photosynthetic tissues (KEYS and WHITTINGHAM 1969, SELLAMI 1976, HAMPP et al. 1982, STITT et al. 1982) or calculated from energy charge values in plant material generally (PRADET and RAYMOND 1983). The sensitivity of the controlling influence of ATP/ADP ratios was dependent on the total adenylate concentration, suggesting that at lower concentrations the absolute concentration of ADP was more important than the ATP/ADP ratio (DRY and WISKICH 1982).

Mitochondria contain an active adenine nucleotide transporter which in the energized state is quite specific for the outward movement of ATP and inward movement of ADP (WISKICH 1977, LANOUE and SCHOOLWERTH 1979). It is quite possible that oxidation could be controlled by the operation of this adenine nucleotide transporter. In spite of much effort, there is no clear evidence to suggest that this is so (LANOUE and SCHOOLWERTH 1979), and VIGNAIS and LAUQUIN (1979) claim that, in liver cells, it is not the carrier which limits respiration. From the high external ATP/ADP ratios required to reduce the rate of State 3 respiration, DRY and WISKICH (1982) came to a similar conclusion, and suggested that in respiratory control it is the absolute amount of available ADP which regulates the rate of respiration, i.e., the turnover of ADP is the "pace-maker" and determines the rates of transport, phosphorylation, and respiration. This idea can be extended to in vivo respiration, but because of the lack of information it must be applied with caution. Even for animal mitochondria and tissues, where detailed information is readily available, conclusions are no more definite because of the controversies and uncertainties (ERECINSKA and WILSON 1982, JACOBUS et al. 1982) that still persist. Thus, it appears that tissue respiration is controlled by the rate of ADP turnover – this, under normal circumstances, would control the rate of the TCA cycle and place the mitochondria in a condition intermediate between State 3 and State 4. However, it must be remembered that supplying ATP is not the only function of plant mitochondria, and supply of ADP should not always restrict TCA cycle turnover. The operation of nonphosphorylating pathways is one method of avoiding control by low levels of ADP – akin to futile cycles in animal tissues (NEWSHOLME and CRABTREE 1976) – allowing TCA cycle turnover to continue. Another is to reoxidize intramitochondrial NADH by some system other than the electron transfer chain (e.g., reversal of malate dehydrogenase). However, for the TCA cycle to operate completely, such a system would require some ADP turnover to allow substrate-level phosphorylation at succinyl-CoA synthetase, and would need to re-oxidize reduced succinate dehydrogenase, either by oxygen uptake or by reversed electron flow.

2.2 Substrate Supply

The most important substrate of plant respiration is carbohydrate, which generates pyruvate via the glycolytic sequence. Thus, under most circumstances, pyruvate is the major substrate for the TCA cycle, and control of its production could well regulate the rate of TCA cycle turnover. The glycolytic enzymes, phosphofructokinase and pyruvate kinase, have regulatory properties suitable for a control function (TURNER and TURNER 1980). Control is effected by adenine nucleotides, other metabolites, ions, and perhaps by fructose 2,6-bisphosphate (HERS and VAN SCHAFTINGEN 1982).

It has been shown that the respiration of unstarved cells of *Acer pseudoplatanus* L. was stimulated by dinitrophenol and by pyruvate, but not by externally supplied sugars (GIVAN and TORREY 1968). The authors concluded that the respiration rate of these cells was limited by the quantity of substrate supplied to the mitochondria from glycolysis. It has often been observed (BEEVERS 1961) that uncouplers stimulate glycolysis (via the adenine nucleotide balance) more so than the TCA cycle, leading to aerobic fermentation. This suggests that a mitochondrial enzyme is rate-limiting, but the real explanation may be that uncouplers do not permit maximum expression of pyruvate transport. The transporter acts as a pyruvate/H^+ symport (or pyruvate/OH^- antiport) system and would be subject to inhibition by uncouplers (WISKICH 1977).

In fact, it has been suggested (DAY and HANSON 1977) that in corn shoot mitochondria, pyruvate transport becomes rate-limiting under conditions of high energy demand (i.e., in the absence of uncouplers). However, no comparisons were made with the rates of tissue respiration and the required rates of in vitro pyruvate transport. It was further suggested that malate transport (Fig. 3) could help alleviate this limitation. Plant mitochondria contain a NAD-specific malic enzyme which can generate pyruvate internally. Such a system has been advocated in soybean roots and nodules (ADAMS and RINNIE 1981, COKER and SCHUBERT 1981). These schemes have malate (or a combination of malate and pyruvate) as the final products of glycolysis, and can be detected by rapid cycling of $^{14}CO_2$, carbon dioxide being fixed by PEP carboxylase and released by NAD-malic enzyme. Thus, high levels of PEP carboxylase activity are taken to imply a by-pass of pyruvate kinase. Similarly, the thermogenic tissues of Araceae develop very high activities of PEP carboxylase leading to the synthesis of C_4 acids which are rapidly decarboxylated (AP REES et al. 1981).

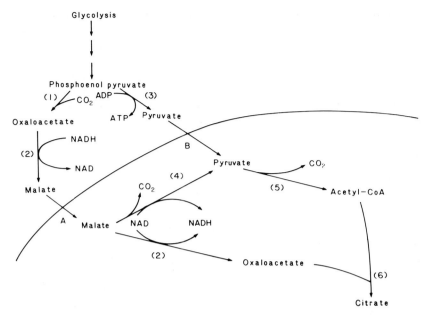

Fig. 3. Pathways of entry of glycolytic carbon into the tricarboxylic acid cycle. *A* malate transporter; *B* pyruvate transporter; the numbered enzymes are: *1* phosphoenolpyruvate carboxylase; *2* NAD-malate dehydrogenase; *3* pyruvate kinase; *4* NAD-malic enzyme; *5* pyruvate dehydrogenase; *6* citrate synthetase

The development of the respiratory "crisis" in spadices of *Arum maculatum* was largely prevented by treatment with 2-N-butylmalonate, suggesting that the mitochondrial substrate was, indeed, cytosolic malate (AP REES et al. 1983). Two possible reasons for terminating the glycolytic sequence with malate rather than with pyruvate were suggested: one was that the mitochondria had a limited capacity for pyruvate transport, and the other that this was a means of ensuring re-oxidation of cytosolic NADH (AP REES et al. 1983). However, the evidence (WISKICH and AP REES, unpublished) for the first possibility is not convincing. Table 1 shows that mitochondria isolated from *Arum maculatum* spadices can show rates of pyruvate transport (as judged by sensitivity to the pyruvate-transport inhibitor, α-cyano-4-hydroxycinnamic acid) sufficient to account for rates of oxygen uptake at thermogenesis (9,000–30,000 µl O_2 h^{-1} g^{-1} f.w., AP REES et al. 1976). The assumptions made in Table 1 are: (a) that α-cyano-hydroxycinnamic acid specifically inhibits pyruvate transport, (b) that the inhibited oxygen uptake represents that due to pyruvate uptake via its carrier and (c) that pyruvate dehydrogenase is responsible for half of the inhibited oxygen consumption and that malate dehydrogenase accounts for the other half. Although most of the oxygen uptake occurs via the alternative oxidase, the in vitro transport of pyruvate would require the maintenance of a proton motive force (MITCHELL 1979). This requirement also applies to net malate influx and does not conflict with current views (LATIES 1982) that the cyanide-insensitive

Table 1. An estimation of pyruvate transport in isolated *Arum maculatum* mitochondria

Fumarase assays:[a]	
Mitochondrial rate	$= 3\ \mu mol\ g^{-1}\ min^{-1}$
Tissue rate	$= 43\ \mu mol\ g^{-1}\ min^{-1}$
Yield of mitochondria	$= 7\%$
Oxygen electrode assays[b]	
Pyruvate + malate oxidation	$= 680\ nmol\ O_2\ g^{-1}\ min^{-1}$
Rate in the presence of 20 μM CHCA	$= 150\ nmol\ O_2\ g^{-1}\ min^{-1}$
CHCA-sensitive rate	$= 530\ nmol\ O_2\ g^{-1}\ min^{-1}$
Tissue rate of pyruvate oxidation	$= 7.6\ \mu mol\ O_2\ g^{-1}\ min^{-1}$
(i.e., corrected for mitochondrial yield)	
Assuming 50% due to pyruvate	$= 3.8\ \mu mol\ O_2\ g^{-1}\ min^{-1}$
(and 50% due to malate → OAA)	
	$= 228\ \mu mol\ O_2\ g^{-1}\ h^{-1}$
Pyruvate transport	$= 456\ \mu mol\ g^{-1}\ h^{-1}$
Equivalent tissue respiration[c]	$= \dfrac{456 \times 6 \times 22.4}{2} = 30{,}640\ \mu l\ O_2\ g^{-1}\ h^{-1}$

[a] Fumarase was measured in both the initial tissue homogenate and the final mitochondrial suspension. Both tissue and mitochondria were disrupted with 0.1% Triton X-100 before dilution and assay. Assayed in 0.1 M phosphate buffer (pH 7.7) containing 30 mM malate, at 240 nm.

[b] Assayed in 0.3 M sucrose, containing 10 mM TES (pH 7.2), 10 mM phosphate (pH 7.2), 5 mM $MgSO_4$, 0.2 mM thiamine pyrophosphate, 0.4 mM ADP, 1 mM NAD, 11 mM pyruvate and 1.0 mM malate. α-cyano-4-hydroxycinnamic acid (CHCA) was present at 20 μM). The mitochondria were purified on a sucrose-density gradient before use.

[c] Pyruvate oxidation accounts for 1/6 of the oxygen uptake by the tissue and 1 mol pyruvate $= 0.5\ mol\ O_2$.

pathway becomes operational after the cytochrome path has been saturated. It is only when mitochondrial data can be accurately extrapolated back to the tissue situation that valid comments can be made about rate-limiting steps in vivo. Isolated mitochondria usually express a capacity in excess of that required to explain whole tissue metabolism (see Table 3). Nevertheless, there appears little doubt that the end-product of plant glycolysis can be regarded as either pyruvate or malate. The advantage of converting PEP to malate is not very clear. Apart from the doubtful suggestion that pyruvate transport may be rate-limiting, the following possibilities are worthy of further consideration: (a) Malate influx into plant mitochondria is anaplerotic. It could lead to higher concentrations of oxaloacetate thereby allowing the TCA cycle to turn over more rapidly. Of course, NAD-malic enzyme would also increase the concentration of intramitochondrial pyruvate. (b) It by-passes the pyruvate kinase reaction and any restrictions applied to it by adenine nucleotides or other effector molecules (TURNER and TURNER 1980). (c) The reduction of oxaloacetate to malate serves to oxidize cytoplasmic NADH and re-cycle NAD for the earlier reaction of glycolysis. It has been shown that the oxidation of external (and by analogy cytoplasmic) NADH by plant mitochondria can be severely restricted if TCA cycle substrates are being oxidized simultaneously (DAY and WISKICH 1977, COWLEY and PALMER 1980). Only one or two substrates of the TCA cycle need be oxidized simultaneously to detect this inhibition, which was more severe in State 4 than in State 3. In most plant tissues the mitochondria are likely to be in some intermediate state, and the re-cycling of cytoplasmic NAD may be more of a problem than appreciated.

2.3 Enzyme Activity

The operation of a biochemical pathway can be regulated by the activity of one of its component enzymes. This limitation may be imposed by the amount of enzyme or by other factors restricting the maximal activity. For example, AP REES et al. (1976) argue that the gradual increase in total phosphofructokinase activity during development of *Arum* spadices acts as a coarse control of glycolysis. Other factors (TURNER and TURNER 1980) exert a fine control, regulating the expression of phosphofructokinase activity.

Table 2 shows the rates of oxidation of various substrates by isolated mung bean mitochondria. Analysis of these results shows that, of the TCA cycle intermediates, citrate and isocitrate are oxidized very slowly indeed. Relatively slow rates of isocitrate oxidation have been reported in cauliflower mitochondria (DAY and WISKICH 1977). In the latter case the oxidation rate was increased by disrupting the mitochondria in the presence of NAD and phenazine methosulphate. The authors concluded that substrate penetration or availability (because of the equilibrium of the aconitase reaction) rather than isocitrate dehydrogenase was rate-limiting. External isocitrate is unlikely to serve as a major mitochondrial substrate in vivo, and its transport is therefore of little importance. However, limited maximum activity of isocitrate dehydrogenase may be of great importance.

An investigation of mitochondrial oxidation rates was made with isolated pea leaf mitochondria. Table 3 shows the mitochondrial rate of oxidation of

Table 2. Rates of oxidation of substrates by mung bean mitochondria. (BOWMAN et al. 1976)

Substrate	State 3 rate[a] (nmol O_2 min^{-1} mg^{-1} protein)	Relative rate[b] (%)
Succinate	206	100
Malate	126	61
Pyruvate[c]	61	30
Citrate	34	17
Isocitrate	22	11
2-Oxoglutarate	76	37
Glutamate	33	16
NADH	212	103

[a] These rates were the second State 3 rates obtained in the presence of NAD and thiamine pyrophosphate.
[b] The rate with succinate is taken to be 100%.
[c] A sparker concentration of malate was present and it has been assumed that only half of the observed rate was due to pyruvate oxidation itself.

Table 3. Some oxidation rates of isolated pea leaf mitochondria

Substrate	Mito-chondrial oxidation[a]	Tissue rate[b] single-step oxidation	Total tissue rate[c] whole TCA cycle	Total tissue rate[d]
	(%)	(μmol O_2 g^{-1} h^{-1})	(μmol O_2 g^{-1} h^{-1})	(μmol CO_2 mg^{-1} chl h^{-1})
A. Mitochondrial oxidations				
Glycine	100 (26)	28.4	–	26
Malate (+ glutamate)	104 (26)	29.5	177	84
Succinate	121 (11)	34.4	206	95
2-Oxoglutarate	50 (5)	14.2	85	39
Isocitrate / citrate	26 (3)	7.4	44	20
Pyruvate (+ malate)	38 (1)	10.8	65	30
B. Isocitrate dehydrogenase	–	9.4	56	26
C. Leaf respiration	–	–	23[e]	11

[a] The rate of glycine oxidation was 125 nmol O_2 min^{-1} mg^{-1} protein. The numbers in parentheses represent the total number of preparations in which the rate of substrate oxidation was compared directly with the rate of glycine oxidation.

[b] These values are the mitochondrial values corrected for the recovery of mitochondria (based on fumarase assays of both tissue and mitochondria) and assume a single-step oxidation.

[c] These values are the total values expected if all oxidative steps of the respiratory process were operating at the same rate as the single-step rate (i.e., single-step oxidations represent 1/6 of the total respiratory oxygen uptake).

[d] These rates have been converted assuming an R.Q. of 1 for the TCA cycle substrates. The pea leaves contained 2.18 ± 0.06 (n = 7) mg chl. g^{-1} fr.w. of tissue.

[e] The measured dark respiration of pea leaf slices was 22.5 ± 2.8 (n = 11) μmol O_2 g^{-1} fr.w. h^{-1}.

glycine and of some TCA cycle intermediates. All of these oxidations are assumed to be single-step oxidations (i.e., the reaction mixture contained malonate with 2-oxoglutarate to prevent succinate oxidation, and half of the pyruvate (+ malate) oxidation was assumed to be due to pyruvate dehydrogenase). These rates have been corrected for the yield of pea leaf mitochondria (based on assays of fumarase recovery – see Table 1) and the individual rates for the TCA cycle-intermediates have been multiplied by 6 to give the total maximum rate of oxygen uptake assuming the rest of the TCA cycle enzymes were operating at the rate of the substrate being oxidized.

Two things become immediately obvious from Table 3. First, the mitochondrial rate of glycine oxidation is barely adequate (DAY and WISKICH 1981) to account for measured rates of 30–40 μmol CO_2 mg^{-1} chlorophyll h^{-1} (LORIMER and ANDREWS 1981) for tissue photorespiration. Second, the mitochondrial rates of oxidation of the TCA cycle intermediates tested exceed that required to explain the rate of dark respiration of leaf slices. The slowest rate of mitochondrial oxidation was observed with isocitrate, but even this extrapolated to a tissue rate of 44.4 μmol O_2 g^{-1} h^{-1} compared to a measured dark respiration rate of 23 μmol O_2 g^{-1} h^{-1}. The measured rate of isocitrate oxidation was almost equal to the maximum rate of isocitrate dehydrogenase activity (Table 3).

Dark respiration of leaves is usually much slower than photorespiration (Canvin et al. 1980), and it appears that total enzymic capacity does not limit the rate of dark respiration under normal circumstances. However, if an enzyme is to limit the maximum rate of turnover of the TCA cycle, isocitrate dehydrogenase may be the most likely candidate.

It should be remembered that (a) the TCA cycle can turn over without reducing oxygen (Graham 1980) if the reducing power from NADH is diverted to other purposes and (b) in healthy tissue some other factors (e.g., allosteric effectors) may control the rate of enzyme turnover. Nevertheless, the general conclusion that enzymic capacity does not limit the TCA cycle appears to hold for most tissues. This is supported by the stimulation of oxygen uptake observed on the addition of uncouplers.

2.3.1 Enzyme Turnover

Although the maximal potential activity of an individual enzyme may exceed that required for metabolism, it is quite possible for it to limit the rate or flux along that particular biochemical pathway. However, it would seem unnecessary for the TCA cycle, operating as a cycle, to regulate any enzyme other than the one reacting with the substrate which feeds carbon into the cycle. If under varying conditions, different parts of the TCA cycle operate at different rates, then other regulating sites become necessary. Thus turnover of the TCA cycle, and other mitochondrial enzymes which interact with so many aspects of metabolism, becomes very complicated.

2.3.2 Pyruvate Dehydrogenase

The pyruvate dehydrogenase complex carries out the following reaction:

$$\text{pyruvate} + NAD^+ + CoA \rightarrow \text{acetyl-CoA} + CO_2 + NADH + H^+.$$

The complex is a multienzyme system of three reactions operating in sequence: pyruvate dehydrogenase, lipoate acetyltransferase, and lipoamide dehydrogenase. The plant enzyme has been isolated and its regulatory properties are quite complex (Rubin and Randall 1977, Thompson et al. 1977a, b, Randall et al. 1981). These have been summarized (Wiskich 1980), and control at this point would effectively regulate turnover of the complete TCA cycle. However, condensation with oxaloacetate to form citrate may not be the only fate for the acetyl-CoA. There is a suggestion (Liedvogel and Stumpf 1982) that in spinach leaves, mitochondrial acetyl-CoA provides acetate for fatty acid synthesis in the chloroplast. The acetyl-CoA is hydrolyzed via mitochondrial acetyl-CoA hydrolase producing acetate which is utilized by the chloroplast to re-generate acetyl-CoA (Murphy and Walker 1982) via acetyl-CoA synthetase. Conversion to the free acetate anion is necessary because acetyl-CoA is thought not to penetrate mitochondrial (Thomas and Wood 1982) or chloroplastic membranes. However, mitochondria may not be a general source of acetate for chloroplasts. Chloroplasts from spinach appear not to contain pyruvate dehydrogenase whereas those from pea leaves do, and can form acetyl-CoA from pyruvate, generated internally, or supplied from an external source (GIVAN 1983). Thus, in those circumstances where mitochondrial pyruvate dehydrogenase fuels both mitochondria and chloroplasts, it would be more appropriate for some other system to regulate TCA cycle turnover. The regulation of acetyl-CoA hydrolase would be crucial for maintaining TCA cycle activity.

2.3.3 Citrate Synthase

In view of the comments above on pyruvate dehydrogenase, it would appear more appropriate for citrate synthase to control movement of carbon around the TCA cycle. The activity of citrate synthase from plant mitochondria is inhibited by ATP (GREENBLATT and SARKISSIAN 1973), whereas that from glyoxysomes is not affected by ATP (AXELROD and BEEVERS 1972). Conditions of high ATP/ADP ratios would restrict flow of carbon at this point. However, the real value of adenylate control here is rather speculative. If electron flow is restricted, one would expect pyridine nucleotide to become more reduced. This would tend to reverse malate dehydrogenase activity and lower the intramitochondrial oxaloacetate. In fact, both substrates for citrate synthase are likely to be present in low amounts, oxaloacetate for reasons mentioned above and acetyl-CoA because of the generally assumed low levels of coenzyme A and complex control pattern of pyruvate dehydrogenase. Thus any further restriction of enzyme activity could lead to a complex, but very fine, control. The purpose of control would be to regulate the supply of carbon to the TCA cycle and its associated reactions. The major fate of citrate in plant tissues appears to be metabolism via the TCA cycle or via the glyoxylate cycle. Where the latter is particularly active (e.g., during germination of fat-containing seeds) citrate synthase is located in glyoxysomes as well as in mitochondria. Under these circumstances carbon from citrate is converted to sugar by gluconeogenesis (BEEVERS 1980).

2.3.4 Isocitrate Dehydrogenase

Isocitrate oxidation is often slow in isolated plant mitochondria (BOWMAN et al. 1976, DAY and WISKICH 1977) and, as seen in Tables 2 and 3, isocitrate dehydrogenase could be rate-limiting for the TCA cycle. However, even this slow rate is sufficient to account for tissue rates of oxygen uptake (Table 3). Further, the stimulation of oxygen uptake by uncouplers shows that none of the enzymes of the TCA cycle was limiting the rate of respiration. However, there appears to be a limited amount of this enzyme present in mitochondria, and it may be well suited for regulation, under some conditions.

DAY and WISKICH (1977) showed that the slow rate of isocitrate oxidation was not due to limiting amounts of isocitrate dehydrogenase and attributed it to slow rates of isocitrate transport and the unfavorable equilibrium with aconitase. (Most of the isocitrate would be converted to citrate). Similar observa-

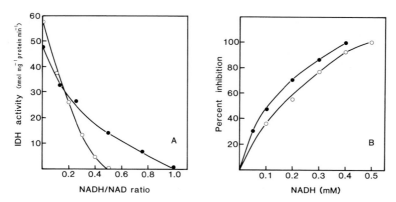

Fig. 4A, B. Inhibition of pea leaf mitochondrial isocitrate dehydrogenase by NADH. The assays contained 50 mM TES (pH 7.4), 0.04% (v/v) Triton X-100, 1 mM MgSO$_4$, 0.1 mM KCN, a constant NAD concentration of either 0.4 mM (●) or 1.0 mM (○) and varying levels of NADH. The reaction was initiated by the addition of 4 mM D,L-isocitrate and the initial rate of NAD reduction measured at 340 nm. The results are plotted in terms of **A** enzyme activity as a function of the NADH/NAD ratio and **B** percent inhibition of enzyme activity produced by a particular NADH concentration in comparison to the maximal activity in the absence of added NADH

tions implicating isocitrate transport as the rate-determining step have now been made with mitochondria isolated from *Neurospora* (SCHWITZGUEBEL et al. 1981). Animal and yeast isocitrate dehydrogenases are allosterically controlled by adenine nucleotides, but the enzyme from higher plants is unaffected (see WISKICH 1980); however, it is inhibited by NADH. DUGGLEBY and DENNIS (1970) have shown that the pea epicotyl enzyme is inhibited 50% when only 10% of the total available NAD (0.4 or 1.0 mM) is reduced (i.e., when the mole fraction of NADH is 0.1). The enzyme is irreversible, but it has a greater affinity for NADH than for NAD, hence the strong inhibition with increasing NADH/NAD ratio. However, the inhibition was reduced in the presence of saturating concentrations of isocitrate or in the presence of the activator, citrate. Intramitochondrial NADH/NAD ratios are not known, but it has been suggested that plant mitochondria contain about 1–5 mM NAD (TOBIN et al. 1980, PALMER et al. 1982). It is unlikely that the total mitochondrial pool of NAD will be available to the enzyme (see Sect. 5). Figure 4A shows that an increasing NADH/NAD ratio (at a constant concentration of NAD) has a more pronounced inhibition at 1 mM NAD that at 0.4 mM NAD. This suggests that NADH concentration may be more important than the NADH/NAD ratio, under saturating conditions. Figure 4B shows the same data replotted as percentage inhibition against NADH concentration. The similarity of these curves also points to NADH concentration being important in restricting isocitrate dehydrogenase activity. NADH is a competitive inhibitor with an inhibition constant (K_i) of 1.9×10^{-4} M (COX and DAVIES 1967). It should be remembered that these enzyme assays are usually performed at high substrate concentrations, whereas the mitochondrial concentration of isocitrate is likely to be low because of the equilibrium of aconitase. However, analyses of TCA cycle intermediates in *Arum* spadices as they develop from α-stage through pre-thermogenesis to thermogenesis (AP REES et al. 1981) reveal little change; during the same period of development there is a dramatic increase in the turnover of the TCA cycle. It is interesting that the concentrations of isocitrate and oxaloacetate were found to be lower than those of the other intermediates. Malate was the only intermediate whose concentration changed dramatically, and this could be accounted for by the increase in PEP carboxylase activity.

Thus the combination of citrate synthase and NAD-isocitrate dehydrogenase warrants further consideration as a control point, especially if the concentration of isocitrate is low, if the total amount of isocitrate dehydrogenase is limited (Tables 2 and 3), and if it exists in two forms (LATIES 1983, TEZUKA and LATIES 1983).

2.3.5 Oxoglutarate Dehydrogenase

Although 2-oxoglutarate dehydrogenase can be activated by adenine nucleotides (WEDDING and BLACK 1971), it is unlikely that this would be an effective control mechanism as far as the TCA cycle is concerned. Activation by AMP may be due to a number of factors which appear to affect the first enzyme of this complex. The enzymic mechanism is very similar to that of pyruvate dehydrogenase (see Sect. 2.3.2). AMP acts as a positive effector on K_m up to pH 8 and as a negative effector above pH 8. Maximum velocity, on the other hand, is affected in a positive manner at all pH values (CRAIG and WEDDING 1980a, b). Despite this, one would expect that some other effect of adenine nucleotides on respiration would predominate. The product of the reaction, succinyl-coenzyme A, is unlikely to leave the mitochondria and 2-oxoglutarate oxidation would be dependent on cycling of coenzyme A. Thus, turnover of 2-oxoglutarate dehydrogenase is vitally dependent on product removal.

2.3.6 Succinyl Coenzyme A Synthetase

A substrate-level phosphorylation requiring ADP (PALMER and WEDDING 1966) ensures that succinyl-CoA synthetase turnover is related to energy demand. The enzymic reaction is reversible, so presumably under conditions of low energy demand coenzyme A is maintained in an acyl form. However, because it is obligatorily coupled to a phosphorylation and because its substrate (succinyl-CoA) is derived solely from the preceding reaction it would seem unnecessary to exert further controls on this enzyme.

2.3.7 Succinate Dehydrogenase

Succinate dehydrogenase can be activated by a number of metabolites, especially anions (OESTREICHER et al. 1973, SINGER et al. 1973). Recently, the enzyme has been partially purified to a multiprotein complex (Complex II) containing two polypeptides, a b-type cytochrome and the dehydrogenase which showed Fe-S-protein type electron paramagnetic resonance signals (BURKE et al. 1982).

However, in spite of its complex kinetics there seems little reason to postulate any regulation of the TCA cycle by succinate dehydrogenase. It is one of the most active enzymes in isolated mitochondria (Tables 2 and 3).

2.3.8 Fumarase

Fumarase carries out the freely reversible hydration of fumarate to malate, and although solutions of it are stabilized by phosphate and other anions (Hill and Bradshaw 1969), there has been no suggestion that it plays a regulatory role. Analyses of enzyme rates in mitochondrial extracts show that fumarase activity should not limit TCA cycle turnover (Bowman et al. 1976) – it usually exceeds that of isocitrate dehydrogenase (Nash et al. 1982).

2.3.9 Malate Dehydrogenase and Malic Enzyme

The equilibrium for malate dehydrogenase lies strongly in favor of NADH oxidation and oxaloacetate reduction. Thus, product removal is essential for maintaining turnover of malate dehydrogenase. NADH oxidation via the electron transport chain will be controlled by ADP levels if the alternative oxidase is not involved. Suggestions for export of reducing power from intramitochondrial NADH to the cytoplasm involve a reversal of malate dehydrogenase. It has been postulated in such systems that malate exchanges for oxaloacetate (Woo et al. 1980) or for 2-oxoglutarate (Journet et al. 1981). Clearly, under such circumstances TCA cycle activity can be maintained (Graham 1980) without malate dehydrogenase generating any reducing power.

Under more usual circumstances it does provide both reducing power and oxaloacetate. Whatever the source of carbon for the TCA cycle (Figs. 1 and 3), oxaloacetate is required for the complete cycle to be maintained. Thus, it seems unnecessary to postulate any other special regulatory property of the enzyme. However, it seems that malate dehydrogenase is spread throughout the mitochondrial matrix and has access to all of the intramitochondrial NAD, because the addition of oxaloacetate inhibits oxidation of all substrates (Douce and Bonner 1972 and Sect. 5).

The role of NAD-malic enzyme is more confusing. The reaction is reversible at relatively high substrate concentrations but the rate of reductive carboxylation of pyruvate is only about one-twentyfifth of the rate of malate decarboxylation (Macrae 1971a). In isolated mitochondria, malate oxidation at pH 7.0 or below is predominantly via NAD-malic enzyme (i.e., pyruvate production) whereas at pH values above 7.0, malate dehydrogenase activity and oxaloacetate formation is favored (Macrae 1971b). The purified enzyme shows a pH optimum of 6.8–6.9 (Macrae 1971a), except when isolated from CAM or C_4 plants, when the pH optimum lies in the range 7.2–7.6 (see Grover et al. 1981). These pH optima must be considered with other factors (e.g., fumarate, malate, coenzyme A) which modulate the kinetic activity of the enzyme. Our current understanding of mitochondrial function (Mitchell 1979), assuming the cytosol has a pH value close to neutrality, would suggest an alkaline matrix within mitochondria in vivo. This means that the enzyme would not experience its optimum pH unless the mitochondria were uncoupled, or the tissue suffered some damage – either obvious disruption of cells or subtle increases in H^+ permeability of membranes. It may be that activation of NAD-malic enzyme is a normal or common consequence of tissue damage and release of vacuolar acids. The enzyme exists in three oligomeric forms (GROVER and WEDDING 1982). The dimer has a high K_m for malate and a low V_{max}. The tetramer has a higher affinity for malate ($K_m = 4.8$ mM) and a high maximum velocity. The octamer appears to have mixed kinetic properties with a low K_m, but also a low maximum velocity. Concentrations of malate which normally saturate an assay system (50–100 mM) convert the enzyme to the tetrameric form. Citrate also converts the enzyme into the tetramer form. It is important to note that low levels of malate (4 mM) failed to cause tetramer formation. This interconversion of oligomeric forms at different malate concentrations may help explain some of the variation in kinetic parameters recorded in the literature.

The purified enzyme is also activated by fumarate, coenzyme A (MACRAE 1971a, VALENTI and PUPILLO 1981, GRISSOM et al. 1983) and sulfate (CANELLAS et al. 1983). Fumarate and coenzyme A appear to bind at different sites and activate the enzyme by reducing the K_m for malate. Both compounds would accumulate under conditions of limiting pyruvate supply and activation of NAD-malic enzyme would help alleviate the problem. However, under conditions of limiting carbon entry into the TCA cycle (be it via pyruvate, malate, other organic acids, or a combination), diversion of malate to pyruvate would eventually deprive the cycle of its carbon pool. The enzyme does not decarboxylate oxaloacetate, and either Mg^{2+} or Mn^{2+} can be used to demonstrate catalysis in vitro, but WEDDING et al. (1981) suggest that Mg^{2+} is the divalent cation used in the plant cell. The high concentration of malate required to keep the enzyme in the active, tetrameric form is of some concern. It is unlikely that such concentrations exist in mitochondria (or cytoplasm) of most plant cells. Thus, NAD-malic enzyme activity may also be controlled by the availability of malate. As more malate becomes available to the mitochondria (e.g., from the vacuole or via carboxylation reactions such as PEP carboxylase), malic enzyme activity would be stimulated, thus maintaining the balance between oxaloacetate and pyruvate production. However, it should be noted that (a) the TCA cycle can continue to operate in the absence of NAD-malic enzyme activity, and (b) even in the relatively inactive form (high K_m and low V_{max}), the enzyme can contribute to metabolism. The need to synchronize NAD-malic enzyme activity with that of malate dehydrogenase cannot be overstressed. Thermogenic tissues of the Araceae are an interesting example. It is suggested that they carboxylate phosphoenolpyruvate to malic acid (AP REES et al. 1981), and that this activity increases during development of the spadix. Although the content of malic acid on a g, fr.w. basis decreases during development, the concentration of malic acid may rise significantly, if the bulk of the increase in fresh weight is due to starch deposition. During the same period of development the activity of the NAD-malic enzyme also increases (AP REES et al. 1983). Hence, both the increase in acidity and the increase in malic acid concentration would tend to activate malic enzyme in vivo. However, such examples are rare, and it is still not clear under what conditions healthy tissues such as cauliflower bud or potato tuber need their mitochondrial NAD-malic enzyme to be fully activated.

3 Fatty Acid Oxidation

Fatty acid degradation in plant tissues can occur via the β-oxidation pathway (GALLIARD 1980). The process has been studied intensively in germinating fat-containing seeds where the pathway is localized in glyoxysomes and associated with the glyoxylate cycle (BEEVERS 1980). β-oxidation may be a more general property of plant microbodies than previously appreciated because it has now been demonstrated in peroxisomes isolated from Jerusalem artichoke tubers (MACEY and STUMPF 1983) and spinach leaves (GERHARDT 1981).

It has been reported (McNEIL and THOMAS 1976, THOMAS and McNEIL 1976) that pea cotyledon mitochondria will oxidize palmitate and other fatty acids. These oxidations were stimulated by the addition of carnitine, and it has been suggested (THOMAS and WOOD 1982) that carnitine acts as an acyl carrier across the mitochondrial membrane. Stimulation by carnitine strongly suggests a mitochondrial β-oxidation system because the peroxisomal and glyoxysomal systems are independent of carnitine (TOLBERT 1981). Figure 5

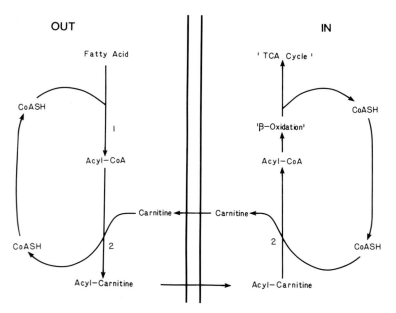

Fig. 5. The operation of the fatty-acid carnitine transporter in mitochondria. (After THOMAS and WOOD 1982). The scheme involves a carnitine/acylcarnitine transporter and acyl-CoA synthetase (*1*) and carnitine acyltransferase (*2*)

shows how fatty acids can be transported across the mitochondrial inner membrane using a carnitine-acyl carnitine carrier. If the acyl group is replaced specifically with acetate, and the acyl transfer is freely reversible (THOMAS and WOOD 1982), this system would not only import acetate into the mitochondria, but also export it (see Sect. 2.3.2). It could be that the carnitine stimulation of fatty acid oxidation in pea cotyledon mitochondria was due to acetate import, i.e., microbody contamination of the mitochondrial pellet could have produced acetate via β-oxidation, and the mitochondria were oxidizing acetate and not fatty acids in general. MACEY (1983) has shown that mitochondrial pellets prepared from pea cotyledons are contaminated with microbodies. However, the "pure" mitochondria, prepared on linear sucrose density gradients, either contained some β-oxidation enzymes or were contaminated with microbodies. Thus, it is still uncertain whether plant mitochondria are capable of β-oxidation or not – and one should not ignore slow, but detectable rates of fatty acid oxidation.

The proposal of a carnitine-acyl carnitine exchange system assumes that coenzyme A or acyl-coenzyme A does not penetrate mitochondrial membranes. Recently, NEUBURGER et al. (1984) have shown that isolated plant mitochondria will accumulate coenzyme A, in an energy-dependent manner. The relevance of this carrier-mediated uptake system to any metabolite exchange system requires further evaluation.

Fatty acids can also be metabolized via an α-oxidation pathway which releases CO_2 and reduces NAD (GALLIARD 1980). This process probably occurs in endoplasmic reticulum membranes. The NADH could be re-oxidized by the external-facing NADH dehydrogenase located on the inner mitochondrial membrane. Results of inhibitor studies on the respiration of uncoupled slices of fresh potato have led WU and LATIES (1983) to propose that oxidation of this NADH occurs via the intra-mitochondrial NADH dehydrogenase. They propose the operation of $malate_{(in)}/phosphate_{(out)}$ and $oxaloacetate_{(out)}^-/OH_{(in)}$ exchanges to transport reducing equivalents into the mitochondria. This seems unnecessarily complicated, but the overall scheme may prove to have some validity. However, an $oxaloacetate_{(out)}/OH_{(in)}^-$ exchange system could not operate in vivo.

3.1 Glyoxysome – Mitochondria Interactions

Studies with germinating castor-oil bean seeds showed that β-oxidation and the glyoxylate cycle were localized in glyoxysomes (BEEVERS 1980). A schematic outline of the cycle is shown in Fig. 6, and the balance sheet of palmitate oxidation in Table 4. It should be noted that carbon from fatty acids effectively ends up as succinate, which must enter the mitochondria for further metabolism. Mitochondrial oxidation of succinate yields oxaloacetate, which is the substrate for gluconeogenesis (BEEVERS 1975). Two features about this process are relevant. First, glyoxysomes produce more NADH than succinate and more than is required for the gluconeogenic process. This excess of NADH is unlikely to be oxidized directly by the mitochondria because castor bean mitochondria

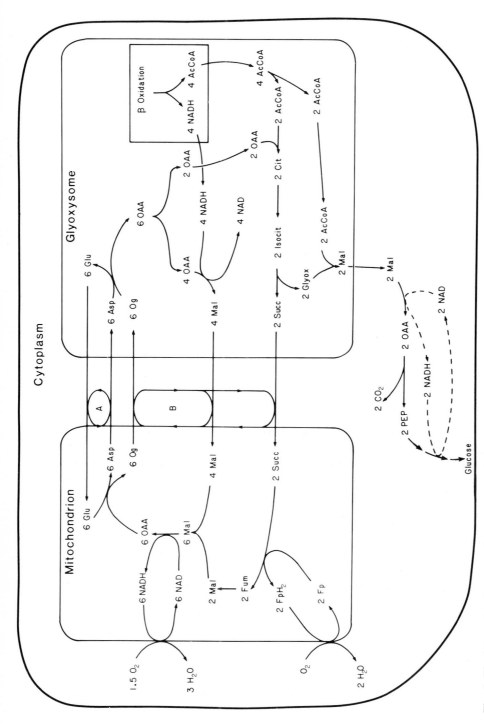

Fig. 6. The interaction between glyoxysomes and mitochondria during germination of castor bean seeds. *A* glutamate/aspartate transporter; *B* 2-oxoglutarate transporter. *AcCoA* acetyl-CoA; *Asp* aspartate; *Cit* citrate; *Fum* fumarate; *Fp* flavoprotein (succinate dehydrogenase); *Glu* glutamate; *Glyox* glyoxylate; *Isocit* isocitrate; *Mal* malate; *OAA* oxaloacetate; *Og* 2-oxoglutarate; *PEP* phosphoenolpyruvate; *Succ* succinate

Table 4. Balance sheet for palmitate oxidation

	Product	ATP equivalent
Glyoxysomes		
1. Activation		
Palmitate + ATP + CoA $\rightarrow$ palmitoyl-CoA + AMP + $P-P_i$	-2 ATP	-2
2. Oxidation		
Palmitoyl-CoA + 7 NAD + 7 CoA $\rightarrow$ 8 acetyl-CoA + 7 NADH	$+7$ NADH(ext)	$+14$
3. Glyoxylate cycle		
8 Acetyl-CoA + 4 NAD $\rightarrow$ 4 succinate + 4NADH	$+4$ NADH(ext)	$+8$
Mitochondria		
4. TCA cycle		
4 Succinate + 4 NAD + 4 fp $\rightarrow$ 4 oxaloacetate + 4 NADH + 4 fpH	$+4$NADH(int) $+4$ fpH(int)	$+12$ $+8$
Cytoplasm		
5. Gluconeogenesis		
4 Oxaloacetate + 8 ATP + 4 NADH $\rightarrow$ 2 glucose + 2CO$_2$ + 8 ADP + 4 NAD	-4 NADH(ext) -8 ATP	-8 -8
Sum	$+7$ NADH (ext) $+4$ NADH (int) $+4$ fpH (int)	$+24$

do not oxidize external NADH rapidly (MILLHOUSE et al. 1983). Second, oxalo-acetate, at low physiological concentrations, does not penetrate the membranes of castor bean mitochondria readily (MILLHOUSE et al. 1983). For these reasons a malate/aspartate shuttle has been postulated (WISKICH 1980, METTLER and BEEVERS 1980). These shuttle systems are shown in Fig. 6. The succinate – oxo-glutarate exchange constitutes the carbon balance in which glyoxysomal succin-ate becomes equivalent to the oxaloacetate used in the cytoplasm for gluconeo-genesis and the malate – oxoglutarate exchange can be regarded as a means of transporting reducing power into the mitochondria (WISKICH 1977). These two exchanges can operate via the 2-oxoglutarate carrier. Glutamate and aspar-tate also need to exchange to maintain balance because of the involvement of aspartate aminotransferase. It has been shown that such an exchange system can maintain glyoxysomal NAD in the oxidized form (METTLER and BEEVERS 1980). The operation of these transporters has been demonstrated in purified castor-bean mitochondria (MILLHOUSE et al. 1983). It was shown that:

1. Internal aspartate will exchange for external aspartate, glutamate, 2-oxo-glutarate, malate or succinate, and

2. internal 2-oxoglutarate will exchange for external 2-oxoglutarate, malate or succinate, but not for external aspartate or glutamate.

Thus, the postulated exchanges (Fig. 6) are present in castor bean mitochondria. The aspartate/glutamate shuttle seemed to be freely reversible, because internal glutamate exchanged for external aspartate. Glutamate efflux was also facilitated by external oxoglutarate, less so by malate, and not at all by succinate. Interestingly, while internal succinate would exchange with all of the organic acids tested, internal malate would not exchange with succinate or 2-oxoglutarate. These latter effects would prevent a short-circuit of carbon flow as depicted in Fig. 6, i.e., it would ensure that malate is oxidized to oxaloacetate.

Endosperm tissue of germinating castor beans is rather different from most tissues. TCA cycle substrates are succinate and malate and the product (oxaloacetate) is exported from the mitochondria. Studies with labeled intermediates suggest that the cycle between pyruvate and succinate does not function (BEEVERS et al. 1966). NAD-malic enzyme activity is quite low in these mitochondria (WISKICH and DAY 1979). Thus, CO_2 is not evolved by the mitochondria during gluconeogenesis, but O_2 is consumed very rapidly. The purified mitochondria yield quite respectable ADP/O and respiratory control ratios (MILLHOUSE et al. 1983), indicating a potential to produce large amounts of ATP rapidly (Table 4). The fate of this ATP needs to be determined because the tissue respiration does not appear under adenylate control and there is no evidence of any mitochondrial cyanide-insensitive respiration.

4 Malate Oxidation:
Malate Dehydrogenase or NAD-Malic Enzyme?

It has been suggested (LANCE 1981) that NADH from malate dehydrogenase is oxidized via the phosphorylating, cytochrome pathway, while that from NAD-malic enzyme is oxidized via the nonphosphorylating, alternative pathway. It is further suggested that oxidation of NADH from malic enzyme to ubiquinone occurs via the nonphosphorylating rotenone-insensitive pathway. Presumably, the NADH from malate dehydrogenase does not have access to the rotenone-insensitive by-pass. It is not intended to discuss the alternative oxidase pathway, but to concentrate on the internal oxidation pathways between NADH and ubiquinone – if these can be shown to be accessible to all intramitochondrial NADH, then the alternative oxidase must also be generally accessible.

Both of the malate-oxidizing enzymes are known to be located in the matrix of the mitochondria (DAY et al. 1979), and separate pathways of NADH oxidation would imply compartmentation of some type. However, the addition of oxaloacetate inhibits the oxidation of all NAD-linked substrates, whether rotenone is present or not (WISKICH and DAY 1982), indicating that malate dehydrogenase has access to all of the intramitochondrial NADH. Although this eliminates any idea of compartmentation of malate dehydrogenase it still allows for separate pools of NAD (see Sect. 5).

Plant mitochondria can oxidize intramitochondrial NADH either via a phosphorylating, rotenone-sensitive path, or via a nonphosphorylating, rotenone-insensitive pathway. The latter by-passes Site I phosphorylation and, in the presence of rotenone, ADP/O ratios are reduced by one-third (BRUNTON and PALMER 1973, WISKICH and RAYNER 1978). Rates of malate oxidation in the presence of rotenone can be quite rapid. Often the addition of rotenone causes a severe inhibition, followed by a gradual increase in the rate of oxidation (PALMER and ARRON 1976, WISKICH and DAY 1979). This was attributed to the accumulation of oxaloacetate and the lag in the recovery of oxygen uptake represented its removal. Indeed, WISKICH and RAYNER (1978) showed that beet-root mitochondria, which rely mainly on malate dehydrogenase, became less sensitive to rotenone, and the apparent inhibition decreased if better oxaloace-tate-removing systems were employed. This has been verified by TOBIN et al. (1980), who analyzed the products and showed that the transient inhibition coincided with a decrease in oxaloacetate concentration. Recently, PALMER et al. (1982) showed that when oxygen uptake recovered, after rotenone addition, there was no net accumulation of oxaloacetate. This is to be expected because the experimental conditions have imposed equilibrium status on malate dehy-drogenase (WISKICH and RAYNER 1978). Thus, the rotenone-inensitive oxygen uptake is stoichiometrically related to pyruvate formation, i.e., NAD-malic en-zyme activity. It is important to remember that malate dehydrogenase is under equilibrium conditions, sustained by NAD reduction via malic enzyme, and therefore cannot contribute to net malate oxidation or to oxygen uptake. The data of TOBIN et al. (1980) and PALMER et al. (1982) show clearly that the rate of pyruvate accumulation did not change (not even during the transient inhibi-tion of oxygen uptake), so that malic enzyme activity was constant, i.e., there was no suggestion of a switch-over of enzymes.

Another problem is associated with the effects of added NAD on malate oxidation. In most cases NAD stimulates the rate of malate oxidation and overcomes any inhibition caused by rotenone. The latter is usually associated with pyruvate accumulation and has led to suggestions that NAD specifically activates NAD-malic enzyme. It has now been shown that plant mitochondria can accumulate NAD from the external medium (TOBIN et al. 1980) and that the effects of NAD are more pronounced on NAD-deficient mitochondria (values of NAD content were 1.1 nmol NAD mg^{-1} protein for potato tuber and 5.7 nmol NAD mg^{-1} protein for mung bean hypocotyls). Stimulation of NAD-malic enzyme could occur because it has a low affinity for NAD [$K_m = 0.9$ mM, for potato (GROVER et al. 1981), but $K_m = 0.04$ mM for pea cotyledon (JOHNSON-FLANAGAN and SPENCER 1981)], or because it has difficulty communi-cating with the membrane-bound NADH-oxidizing systems. However, it has now been shown (MØLLER and PALMER 1982), that the rotenone-insensitive NADH oxidizing system has a lower affinity for NADH (apparent $K_m = 80$ μM) than does the rotenone-sensitive system (apparent $K_m = 8$ μM). Therefore, stimu-lation of malate oxidation by adding NAD to isolated mitochondria is not restricted to NAD-malic enzyme (TOBIN et al. 1980).

Thus, there are two systems for oxidizing intramitochondrial malate and two systems for oxidizing intramitochondrial NADH. Malate oxidation can be manipulated, and misinterpreted, because of the unfavorable equilibrium of malate dehydrogenase (WISKICH and RAYNER 1978, TOBIN et al. 1980).

NADH oxidation via the rotenone-insensitive path is dependent on the intramitochondrial NADH concentration or NADH/NAD ratio, but the path is accessible to all enzymes (Wiskich and Rayner 1978, Palmer et al. 1982), and not just to NAD-malic enzyme.

However, under certain circumstances, or in some tissues, malate oxidation may follow a specific path. This is a reflection of the conditions and does not exclude other oxidations from ever using that same path. It has been claimed that malate oxidation in cauliflower bud (Rustin and Moreau 1979), potato tuber (Rustin et al. 1980), and avocado (Moreau and Romani 1982) mitochondria occurs via NAD-malic enzyme specifically linked to the rotenone-insensitive and the cyanide-insensitive pathway. Thus, malate oxidation would not be subjected to any adenylate control. Rustin et al. (1980) suggest that this malate oxidation has its own exclusive electron transport chain and pool of ubiquinone (see Moore and Cottingham 1983 for a discussion of ubiquinone pools). The need for this degree of specialization is not immediately apparent. Lance (1981) regards this as an elimination of reducing power and describes the operation of this system in CAM plants and in leaves infected with powdery mildew. Certainly, in some CAM plants (Osmond and Holtum 1981), malate is oxidatively decarboxylated via a NAD-malic enzyme system and the pyruvate polymerized to sugar. If the reducing power is not required in such a system and if the enzymic capacity is available (Wiskich and Day 1982), it would be advantageous to oxidize malate via a pathway free of adenylate or other controls. Mitochondria isolated from some CAM plants show rotenone-resistant malate oxidation (Arron et al. 1979). Similarly, it was suggested (Wiskich and Day 1982) that in some C_4 plants it may be necessary to have NAD-malic enzyme operating independently of the TCA cycle and via a nonphosphorylating pathway. Gardeström and Edwards (1983) have shown that malate oxidation by mitochondria isolated from the bundle sheath of *Panicum miliaceum* is without respiratory control and largely cyanide-insensitive, whereas that from the mesophyll is subject to respiratory control and strongly inhibited by cyanide.

In both cases discussed above malate acts as a carrier of CO_2, which is the pertinent product. The production of pyruvate can be regarded as a regeneration of the CO_2 acceptor, and of little consequence to mitochondrial metabolism. A parallel situation may occur during fatty acid synthesis. Plastids do not appear to contain all the machinery to make acetyl-CoA from carbohydrates or trioses (Givan 1983). Pyruvate can be supplied from mitochondria or from the cytoplasm, but acetate, as required by some plants (Liedvogel and Stumpf 1982, Murphy and Walker 1982) needs to be of mitochondrial origin. Whatever the original source of acetate, mitochondria would be generating NADH from pyruvate dehydrogenase, (and possibly from NAD-malic enzyme), and would need to re-oxidize it.

In all of the above there is a specific role or need for pyruvate. However, most plant tissues do not operate in this manner. The mitochondrial pyruvate is destined for the TCA cycle, and oxaloacetate is required. If malate is the source of carbon, oxaloacetate and pyruvate must be produced in a 1:1 ratio. If both malate and pyruvate are carbon sources, then the rate of malate dehydrogenase should exceed that of NAD-malic enzyme. Thus, it would seem most inappropriate for a cell to have one enzyme (NAD-malic enzyme) operating without control, and a parallel, more important, enzyme (malate dehydrogenase) subjected to respiratory control. Furthermore, as explained above (Sect. 2.3.9) the diversion of malate via NAD-malic enzyme to pyruvate could deplete mitochondria of their carbon pool.

It has been shown (WISKICH and DAY 1982, DRY et al. 1983) that oxaloacetate and malate dehydrogense have access to all of the mitochondrial NAD, and postulating separate compartments may be confusing. The reasons for some of the controversy have been (a) the use of high concentrations of malate (see Sect. 2.3.9), (b) forgetting the reversibility of malate dehydrogenase, and (c) measuring single-step oxidation products under conditions which preclude the operation of some pathways or the accumulation of some products.

5 Glycine Oxidation

During photorespiration (TOLBERT 1980, LORIMER and ANDREWS 1981), glycine moves from peroxisomes to mitochondria for conversion to serine. The reaction is:

$$2 \text{ glycine} + NAD^+ + H_2O \rightarrow \text{serine} + NADH + H^+ + CO_2 + NH_3.$$

The NH_3 is re-assimilated in the chloroplast (WOO and OSMOND 1982, DRY and WISKICH 1983), while the serine returns to the peroxisomes for further metabolism toward phosphoglycerate. A reduction step is involved in this metabolism, and it has been suggested that reducing power may come from the mitochondrial glycine oxidation. The NADH would reduce oxaloacetate, which would be transported directly (WOO and OSMOND 1976), or, after transamination (JOURNET et al. 1981), out of the mitochondria. However, the mitochondria can also oxidize the NADH, via their electron transport chain, using oxygen as the electron acceptor (MOORE et al. 1977, DAY and WISKICH 1981). The transport shuttles were invoked in the belief that mitochondrial electron transport does not function in the light – a view that is difficult to sustain (JORDAN and GIVAN 1979, GRAHAM 1980, DRY and WISKICH 1982, REED and CANVIN 1982). Nevertheless, it is clear that for photorespiration to operate continuously, the mitochondrial NAD must be re-cycled.

DAY and WISKICH (1981) noted that the release of NH_3, or of $^{14}CO_2$, from glycine was little affected by the presence of TCA cycle substrates. They concluded that TCA-cycle substrates did not compete with glycine for intramitochondrial NAD. However, competition does exist because the presence of glycine reduces the rate of oxidation of TCA-cycle substrates. This can be determined using ^{14}C-malate (WALKER et al. 1982) for NAD-malic enzyme activity, or by producing balance sheets for any combination of TCA cycle enzymes. DRY et al. (1983) showed that no combination of substrates influenced the rate of glycine oxidation, even under State 4 conditions when one may expect the competition to be most severe. On the other hand, the presence of glycine reduced the rates of oxidation of the other substrates. Thus, glycine is a high-priority substrate for the pea-leaf mitochondrial electron transfer chain. This is quite different to the results observed with glycine oxidation in rat liver mitochondria (Hampson et al. 1983). In the face of probable TCA cycle turnover in the light (Graham 1980), it would be expected that the system used to oxidize NADH in leaf mitochondria should exhibit some form of preference for glycine to ensure the continuous operation of the photorespiratory cycle. The electron transfer chain certainly shows that preference. In contrast, oxaloacetate oxidizes matrix NADH without preference for any NAD-linked substrate (Dry et al. 1983). It is possible, therefore, that NADH from glycine oxidation may be oxidized in vivo by the electron transfer chain.

The effects described by Dry et al. (1983) pose interesting questions about competition for intramitochondrial NAD. For example, why is glycine oxidation so resistant to any competitive effects? It cannot be because its oxidation occurs via glycine-specific pools of NAD or glycine-specific electron transport chains. The inhibition of oxidation of other acids by glycine shows that there are common pools and mutual interactions, and the results with oxaloacetate show that malate dehydrogenase, at least, has access to all of the mitochondrial NADH. The results of Fig. 7 further complicate the issue. Freshly isolated pea leaf mitochondria are slightly NAD-deficient as shown by the stimulation of glycine oxidation by added NAD (compare traces A and B). This dependence of substrate oxidation on added NAD can be dramatically increased by incubating the mitochondria in a large volume of suitable medium at 4 °C. After 1–2 hours of incubation a marked decrease in the rate of rotenone-insensitive glycine oxidation is observed which can be recovered by the addition of NAD (compare traces C and D), indicating that NAD had been lost from the matrix of the mitochondria during the incubation period. The fall in the rotenone-resistant rate of glycine oxidation was much more pronounced than the decrease in the state 3 rate because of the poor affinity of the rotenone-insensitive NADH dehydrogenase for NADH, relative to the rotenone-sensitive dehydrogenase (Møller and Palmer 1982). High intramitochondrial NADH concentrations are required for electron flow via the rotenone-insensitive bypass. The results of Fig. 7 also show that the rate of oxygen consumption in the presence of glycine and rotenone could also be markedly stimulated by the addition of a second NAD-linked substrate – malate. On the basis of all NAD-linked mito-chondrial enzymes having access to a common NAD pool, one might have predicted the addition of malate to have had no effect under these conditions i.e. an NAD-limitation of the rotenone-resistant rate for one enzyme should represent an NAD-limitation for all other enzymes with access to that same NAD pool. The results cannot be explained in terms of glycine decarboxylase having a low affinity for NAD because state 3 rates, in the absence of rotenone in these NAD-depleted mitochondria, are still much faster than in the presence of rotenone. Furthermore, it should be noted that glycine oxidation, in NAD-sufficient mitochondria, is actually less sensitive to rotenone inhibition than malate oxidation (see Fig. 8) indicating that glycine decarboxylase can actually generate a higher NADH/NAD ratio than the malate-oxidizing enzymes in the presence of rotenone. This result would appear to suggest therefore that the malate-oxidizing enzymes have access to some intramitochondrial NAD

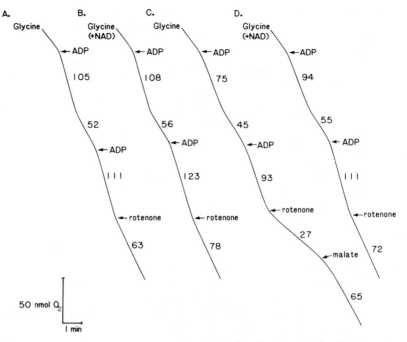

Fig. 7 A–D. Effect of NAD and second-substrate addition on the rotenone-insensitive state 3 rate of glycine oxidation by NAD-depleted pea leaf mitochondria. Mitochondria were depleted of NAD by suspension in a large volume of standard reaction medium containing 2 mM glycine to a concentration of approximately 0.2 mg protein ml^{-1} for 80 min at 4 °C. The reaction medium contained 0.3 M sorbitol, 10 mM Tes, 10 mM KH_2PO_4, 2 mM $MgCl_2$, 1 mM EGTA and 0.1% bovine serum albumen, all adjusted to pH 7.2. Concentrations used: 11 mM glycine, 11 mM malate, 1 mM NAD, 20 μM rotenone, 0.14 mM ADP (first addition); 0.7 mM (second addition). Glutamate (11 mM) and TPP (0.1 mM) were included with the malate addition. Traces *A* and *B* – freshly isolated mitochondria; traces *C* and *D* – NAD-depleted mitochondria. Rates shown are expressed as nmol O_2 consumed mg^{-1} protein min^{-1}

(within these NAD-depleted mitochondria) which is unavailable to glycine decarboxylase and can effectively generate a higher internal NADH concentration under these conditions. Thus, intramitochondrial NAD should not necessarily be regarded as one homogeneous pool common or equally accessible to all NAD-linked enzymes.

Further support for this hypothesis comes from the data in Fig. 8 which shows the state 3 rates (with and without rotenone) of glycine and malate oxidation during NAD-depletion of pea leaf mitochondria. The results indicate that glycine oxidation, in the presence of rotenone, is much more sensitive to NAD loss than is malate oxidation under similar conditions. At time zero it can be seen that the glycine rotenone rate exceeded the malate rate and was clearly much less sensitive to rotenone inhibition than malate oxidation. However, after only a short period of the depletion treatment the glycine rotenone rate fell well below that achieved with malate. The rotenone rates in the

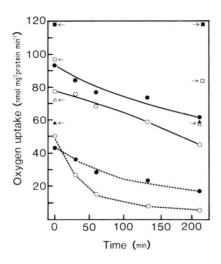

Fig. 8. Changes in malate and glycine oxidation rates during NAD-depletion of pea leaf
mitochondria. Experimental conditions as described in Fig. 7. Samples were withdrawn at
regular intervals, centrifuged at 10,000 g for
20 min and the mitochondria resuspended in
a medium containing 0.3 M sorbitol, 20 mM
Tes (pH 7.5) and 0.1% bovine serum albumen.
Open symbols – glycine; closed symbols – malate; State 3 (——); rotenone-insensitive state
3 (----); State 3 + NAD (■); rotenone-insensitive state 3 + NAD (▲)

presence of NAD were maintained at a reasonably constant level during the
depletion period. Thus, the loss of intramitochondrial NAD during the depletion
treatment appears to have reduced the amount or concentration of NAD available to glycine decarboxylase more significantly than that available to the malate-oxidizing enzymes. Again these results would appear to contradict the model
of a homogenous distribution of enzymes within a common NAD pool and
would lend support for the hypothesis that the glycine decarboxylase and malate
oxidizing enzymes have differential access to NAD within pea leaf mitochondria.
This may arise from differences in the distribution of the respective enzymes
within the mitochondria. The distribution of glycine decarboxylase may be limited by association with the inner mitochondrial membrane (Sarojini and
Oliver 1983) and, consequently, may not have access to the total matrix NAD
pool. The malate-oxidizing enzymes, on the other hand, are most probably
ubiquitous within the mitochondria and thus would have access to a larger total
NAD pool – including that available to glycine decarboxylase (as demonstrated
by the re-oxidation of glycine NADH by MDH in the presence of OAA).

Other results from our laboratory show that not only does malate oxidation
communicate with more of the intramitochondrial NAD than does glycine,
but it is oxidized by more electron transport chains than is glycine. The latter
difference is almost certainly related to the former, i.e. glycine decarboxylase
has a limited distribution within mitochondria and hence functions with only
a portion of the intramitochondrial NAD and the corresponding portion of
electron transport chains. It will be interesting to determine whether other enzymes behave like glycine decarboxylase. This sort of information will help
in constructing an architectural model of the inner mitochondrial membrane.

Glycine oxidation is a single-step process as far as the mitochondria are
concerned. It would seem adequate to have it controlled by the rate of substrate
supply. Glycine oxidation rates are the same whether oxaloacetate or oxygen
is used as electron acceptor (Day and Wiskich 1981), suggesting that the enzyme
is rate-limiting. [Glycine penetration appears to be a diffusive, rather than a
carrier-mediated process (Day and Wiskich 1980) although at concentrations

lower than 0.5 mM an active uptake process has been described (YU et al. 1983)]. Extrapolation of these data to the tissue (Table 3; DAY and WISKICH 1981) shows that the enzymic activity is barely adequate to explain photorespiratory rates (LORIMER and ANDREWS 1981) – so a coarse control by enzymic capacity may apply.

6 Physiological Control of the TCA Cycle

A variety of controlling factors of the TCA cycle have been examined. Transport of substrates into mitochondria is a possibility, but considered an unlikely factor. If transport is not adequate, then the fault will lie with the rate of supply of the substrate, be it pyruvate, malate, succinate, glycine, or any combination of these. Rates of substrate transport required to maintain oxygen uptake in tissue are quite low, and diversion of phosphoenolpyruvate from pyruvate to malate is probably due to problems with oxidation of cytoplasmic NADH rather than problems with pyruvate transport.

Total enzymic capacity is unlikely to act as a control of tissue respiration. Reliable quantitative estimates are not known for all of the TCA cycle enzymes, but of those available, isocitrate dehydrogenase appears to have the lowest maximum rate, but this is ample to explain tissue rates. A number of the TCA cycle enzymes show allosteric properties and would be suitable candidates for control, but the evidence for the in vivo regulation of the TCA cycle by such systems is unfortunately absent.

The adenine nucleotide balance can control mitochondrial oxidations and there seems little doubt that such control operates in tissues whose respiration is totally cyanide-sensitive. Cellular ATP turnover would regulate ADP supply and in so doing act as a "pace-maker" for the TCA cycle both through regulation of oxidative phosphorylation and through the influence of the adenine nucleotide balance on the supply of substrate from glycolysis. However, the important factor in mitochondrial respiratory control is likely to be the absolute concentration of ADP (DRY and WISKICH 1982), more than energy charge or ATP/ADP ratio, which may be more important in regulating glycolysis. This "coarse" control would effectively balance substrate supply with the demands made on the TCA cycle by the oxidative phosphorylation system. Such demands may be communicated to the individual enzymes of the TCA cycle in various "fine" control mechanisms such as NADH/NAD ratios (e.g., isocitrate dehydrogenase, malate dehydrogenase) and substrate-level phosphorylation (e.g., succinyl-CoA synthetase). This "fine" control system would not only maintain balance among the TCA cycle intermediates, but also permit small sections of the cycle to operate independently of the whole. For example, during fat oxidation in germinating castor beans, during photorespiration, and during amplified use of NAD-malic enzyme (as occurs in CAM and C_4 plants), only part of the whole cycle is involved and obviously different controls must come into play.

With respect to tissues with cyanide-insensitive respiration, a model depicting it as an "energy overflow" system has been proposed (LAMBERS 1982). Briefly, it is a mechanism for oxidizing carbohydrate (or other substrates) without adenylate control – a feature of greater importance than heat production (see Sect. 2.1). That correlations can be made between substrate supply and utilization is not surprising – the very nature of the cyanide-insensitive pathway is to process substrates quickly (PALMER 1976, SOLOMOS 1977, WISKICH 1980) and the extent of its contribution is likely to be controlled by substrate. However, it requires that the phosphorylating cytochrome path is first saturated (with respect to the adenylate control applied) before the excess electron flow spills over to the alternative oxidase. There appears to be no physiological device to "switch on" the cyanide-insensitive pathway. Mitochondria isolated from *Arum* spadices are capable of rapid rates of cyanide-insensitive respiration, well, before thermogenesis, and appear fully equipped for thermogenesis. What appears to be lacking in the pre-thermogenic spadix is the supply of substrate. Thus, the real question with cyanide-insensitive respiration may be what controls the supply of substrate.

7 List of Enzymes

NAD-malic enzyme	(EC 1.1.1.39)	Aconitase	(EC 4.2.1.3)
Aspartate aminotransferase	(EC 2.6.1.1)	Pyruvate dehydrogenase	(EC 1.2.4.1)
NAD-malate dehydrogenase	(EC 1.1.1.37)	Lipoate acetyltransferase	(EC 2.3.1.12)
Succinyl-CoA synthetase	(EC 6.2.1.5)	Lipoamide dehydrogenase	(EC 1.6.43)
Succinate dehydrogenase	(EC 1.3.99.1)	Acetyl-CoA hydrolase	(EC 3.1.2.1)
Phosphofructokinase	(EC 2.7.1.11)	Acetyl-CoA synthetase	(EC 6.2.1.1)
Pyruvate kinase	(EC 2.7.1.40)	Citrate synthase	(EC 4.1.3.7)
PEP carboxylase	(EC 4.1.1.31)	2-oxoglutarate dehydrogenase	(EC 1.2.4.2)
Isocitrate dehydrogenase	(EC 6.1.1.41)	Fumarase	(EC 4.2.1.2)

Acknowledgements. The authors are grateful to Dr. JAMES BRYCE for helpful comments, and to CAROL WILKINS, KATHY SOOLE and ANTHONY FOX for supplying their technical expertise. Work performed in the authors' laboratory was supported by the Australian Research Grants Scheme.

References

Adams CA, Rinnie RW (1981) Interactions of phosphoenolpyruvate carboxylase and pyruvic kinase in developing soybean seeds. Plant Cell Physiol 22:1011–1021

ap Rees T (1980) Assessment of the contributions of metabolic pathways to plant respiration. In: Davies D (ed) The biochemistry of plants, Vol II. Metabolism and respiration. Academic Press, London New York, pp 1–29

ap Rees T, Fuller WA, Wright BW (1976) Pathways of carbohydrate oxidation during thermogenesis by the spadix of *Arum maculatum.* Biochim Biophys Acta 437:22–35

ap Rees T, Fuller WA, Green JH (1981) Extremely high activities of phosphoenolpyruvate carboxylase in thermogenic tissues of Araceae. Planta 152:79–86

ap Rees T, Bryce JH, Wilson PM, Green JH (1983) Role and location of NAD-malic enzyme in thermogenic tissues of Araceae. Arch Biochem Biophys 227:511–521

Arron GP, Spalding MM, Edwards GE (1979) Isolation and oxidation properties of intact mitochondria from the leaves of *Sedum praealtum,* a crassulacean acid metabolism plant. Plant Physiol 64:182–186

Atkinson DE (1977) Cellular energy metabolism and its regulation. Academic Press, London New York

Axelrod B, Beevers H (1972) Differential response of mitochondrial and glyoxysomal citrate synthase to ATP. Biochim Biophys Acta 256:175–178

Beevers H (1961) Respiratory metabolism in plants. Row, Peterson, White Plains, NY

Beevers H (1975) Organelles from castor bean seedlings: Biochemical roles in gluconeogenesis and phospholipid biosynthesis. In: Galliard T, Mercer EI (eds) Proceedings of the phytochemical society. No 12. Recent advances in the chemistry and biochemistry of plant lipids. Academic Press, London New York, pp 287–299

Beevers H (1980) The role of the glyoxylate cycle. In: Stumpf PK (ed) The biochemistry of plants, vol IV. Lipids: structure and function. Academic Press, London New York, pp 117–130

Beevers H, Stiller ML, Butt VS (1966) Metabolism of the organic acids. In: Steward FC (ed) Plant physiology, vol IVB. Metabolism: intermediary metabolism and pathology. Academic Press, London New York, pp 119–262

Bonner WD, Voss DO (1961) Some characteristics of mitochondria extracted from higher plants. Nature (London) 191:682–684

Bowman EJ, Ikuma H, Stein HJ (1976) Citric acid cycle activity in mitochondria isolated from mung bean hypocotyls. Plant Physiol 58:426–432

Brunton CJ, Palmer JM (1973) Pathways for the oxidation of malate and reduced pyridine nucleotide by wheat mitochondria. Eur J Biochem 39:283–291

Burke JJ, Siedow JN, Moreland DE (1982) Succinate dehydrogenase. A partial purification from mung bean hypocotyls and soybean cotyledons. Plant Physiol 70:1577–1581

Canellas PF, Grissom CB, Wedding RT (1983) Allosteric regulation of the NAD malic enzyme from cauliflower: Activation by sulfate. Arch Biochem Biophys 220:116–132

Canvin DT, Berry JA, Badger MR, Fock H, Osmond CB (1980) Oxygen exchange in leaves in the light. Plant Physiol 66:302–307

Chance B, Williams GR (1955a) Respiration enzymes in oxidative phosphorylation I. Kinetics of oxygen utilization. J Biol Chem 217:383–393

Chance B, Williams GR (1955b) A method for the localisation of sites for oxidative phosphorylation. Nature (London) 176:250–254

Coker GT, Schubert KR (1981) Carbon dioxide fixation in soybean roots and nodules I. Characterization and comparison with N_2 fixation and composition of xylem exudate during early nodule development. Plant Physiol 67:691–696

Cowley RC, Palmer JM (1980) The interaction between exogenous NADH oxidase and succinate oxidase in Jerusalem artichoke (*Helianthus tuberosus*) mitochondria. J Exp Bot 31:199–207

Cox GF, Davies DD (1967) Nicotinamide-adenine dinucleotide-specific isocitrate dehydrogenase from pea mitochondria. Biochem J 105:729–734

Craig DW, Wedding RT (1980a) Regulation of the 2-oxoglutarate dehydrogenase lipoate succinyltransferase complex from cauliflower by nucleotide. Pre-steady-state kinetics and physical studies. J Biol Chem 255:5769–5775

Craig DW, Wedding RT (1980b) Regulation of the 2-oxoglutarate dehydrogenase lipoate succinyltransferase complex from cauliflower. Steady-state kinetics studies. J Biol Chem 255:5763–5768

Day DA, Hanson JB (1977) Pyruvate and malate transport and oxidation in corn mitochondria. Plant Physiol 59:630–635

Day DA, Wiskich JT (1975) Isolation and properties of the outer membrane of plant mitochondria. Arch Biochem Biophys 171:117–123

Day DA, Wiskich JT (1977) Factors limiting respiration by isolated cauliflower mitochondria. Phytochemistry 16:1499–1502

Day DA, Wiskich JT (1980) Glycine transport by pea leaf mitochondria FEBS Lett 112:191–194

Day DA, Wiskich JT (1981) Glycine metabolism and oxaloacetate transport by pea leaf mitochondria. Plant Physiol 68:425–429

Day DA, Arron GP, Laties GG (1979) Enzyme distribution in potato mitochondria. J Exp Bot 30:539–549

Douce R, Bonner WD (1972) Oxalacetate control of Krebs cycle in purified plant mito-chondria. Biochem Biophys Res Commun 47:619–624

Dry IB, Wiskich JT (1982) Role of the external adenosine triphosphate/adenosine diphos-phate ratio in the control of plant mitochondrial respiration. Arch Biochem Biophys 217:72–79

Dry IB, Wiskich JT (1983) Characterisation of dicarboxylate stimulation of ammonia, glutamine and 2-oxoglutarate-dependent O_2 evolution in isolated pea chloroplasts. Plant Physiol 72:291–296

Dry IB, Day DA, Wiskich JT (1983) Preferential oxidation of glycine by the respiratory chain of pea leaf mitochondria. FEBS Lett 158:154–158

Duggleby RG, Dennis DT (1970) Regulation of nicotinamide adenine dinucleotide-specif-ic isocitrate dehydrogenase from a higher plant. The effect of reduced nicotinamide adenine dinucleotide and mixtures of citrate and isocitrate. J Biol Chem 245:3751–3754

Erecinska M, Wilson DF (1982) Regulation of cellular energy metabolism. J Membr Biol 70:1–14

Estabrook RW, Gutfreund M (1960) Calorimetric measurements during biological oxida-tions. Fed Proc 19:39

Galliard T (1980) Degradation of acyl lipids: Hydrolytic and oxidative enzymes. In: Stumpf PK (ed) The biochemistry of plants, vol IV. Lipids: structure and function. Academic Press, London New York, pp 85–116

Gardeström P, Edwards GE (1983) Isolation of mitochondria from leaf tissue of *Panicum miliaceum* a NAD-malic enzyme type C_4 plant. Plant Physiol 71:24–29

Gerhardt B (1981) Enzyme activities of the β-oxidation pathway in spinach leaf perox-isomes. FEBS Lett 126:71–73

Givan CV (1983) The source of acetyl coenzyme A in chloroplasts of higher plants. Physiol Plant 57:311–316

Givan CV, Torrey JG (1968) Respiratory response to *Acer pseudoplatanus* cells to pyru-vate and 2,4-dinitrophenol. Plant Physiol 43:635–640

Graham D (1980) Effects of light on "dark" respiration. In: Davies DD (ed) The bio-chemistry of plants, vol II. Metabolism and respiration. Academic Press, London New York, pp 525–579

Greenblatt GA, Sarkissian IV (1973) Citrate synthase of plants: Sensitivity to sulfhydryl reagents and molecular weight of the enzyme. Physiol Plant 29:361–364

Grissom CB, Canellas PF, Wedding RT (1983) Allosteric regulation of the NAD malic enzyme from cauliflower: Activation by fumarate and coenzyme A. Arch Biochem Biophys 220:133–144

Grover SD, Wedding RT (1982) Kinetic ramifications of the association – dissociation behaviour of NAD malic enzyme. A possible regulatory mechanism. Plant Physiol 70:1169–1172

Grover SD, Canellas PF, Wedding RT (1981) Purification of NAD malic enzyme from potato and investigation of some physical and kinetic properties. Arch Biochem Bio-phys 209:396–407

Hampp R, Goller M, Ziegler H (1982) Adenylate levels, energy charge and phosphoryla-tion potential during dark-light and light-dark transition in chloroplasts, mitochondria and cytosol of mesophyll protoplasts from *Avena sativa* L. Plant Physiol 69:448–455

Hampson RK, Barron LL, Olson MS (1983) Regulation of the glycine cleavage system in isolated rat liver mitochondria. J Biol Chem 258:2993–2999

Hanson JB, Day DA (1980) Plant mitochondria. In: Tolbert NE (ed) The biochemistry of plants, vol I. The plant cell. Academic Press, London New York, pp 315–358

Hers H-G, Schaftingen van E (1982) Fructose 2,6-bisphosphate 2 years after its discovery. Biochem J 206:1–12

Hill RL, Bradshaw RA (1969) Fumarase. In: Lowenstein JM (ed) Methods of enzymol-ogy, Vol XIII. Citric acid cycle. Academic Press, London New York, pp 91–99

Jacobus WE, Moreadith RW, Vandegaer KM (1982) Mitochondrial respiratory control. Evidence against the regulation of respiration by extramitochondrial phosphorylation potentials or by ATP/ADP ratios. J Biol Chem 257:2397–2402

Johnson-Flanagan AM, Spencer M (1981) The effect of rotenone on respiration in pea cotyledon mitochondria. Plant Physiol 68:1211–1217

Jordan BR, Givan CV (1979) Effects of light and inhibitors on glutamate metabolism in leaf discs of *Vicia faba* L. Sources of ATP for glutamine synthesis and photoregulation of tricarboxylic acid cycle metabolism. Plant Physiol 64:1043–1047

Journet E-P, Neuburger M, Douce R (1981) Role of glutamate-oxaloacetate transaminase and malate dehydrogenase in the regeneration of NAD^+ for glycine oxidation in spinach leaf mitochondria. Plant Physiol 67:467–469

Keys AJ, Whittingham CP (1969) Nucleotide metabolism in chloroplast and non-chloroplast components of tobacco leaves. In: Metzner H (ed) Progress in photosynthetic research, vol I. Laupp, Tübingen, pp 352–358

Lambers H (1982) Cyanide-resistant respiration. A non-phosphorylating electron transport pathway acting as an energy overflow. Physiol Plant 55:478–485

Lance C (1981) Cyanide-insensitive respiration in fruits and vegetables. In: Friend J, Rhodes MJC (eds) Recent advances in the biochemistry of fruits and vegetables. Ann Proc Phytochem Soc Eur No 19. Academic Press, London New York, pp 63–87

Lanoue KF, Schoolwerth AC (1979) Metabolite transport in mitochondria. Annu Rev Biochem 48:871–922

Lardy HA, Wellman H (1952) Oxidative phosphorylations: Role of inorganic phosphate and acceptor systems in control of metabolic rates. J Biol Chem 195:215–224

Laties GG (1982) The cyanide-resistant, alternative respiration path in higher plant respiration. Annu Rev Plant Physiol 33:519–535

Laties GG (1983) Membrane-associated NAD-dependent isocitrate dehydrogenase in potato mitochondria. Plant Physiol 72:953–958

Liedvogel B, Stumpf PK (1982) Origin of acetate in spinach leaf cell. Plant Physiol 69:897–903

Lorimer GH, Andrews TJ (1981) The C_2 chemo- and photorespiratory carbon oxidation cycle. In: Hatch MD, Boardman NK (eds) The biochemistry of plants, vol VIII. Photosynthesis. Academic Press, London New York, pp 329–374

Macey MJK (1983) β-Oxidation and associated enzyme activities in microbodies from germinating peas. Plant Sci Lett 30:53–60

Macey MJK, Stumpf PK (1983) β-Oxidation enzymes in microbodies from tubers of *Helianthus tuberosus*. Plant Sci Lett 28:207–212

Macrae AR (1971a) Isolation and properties of a "malic" enzyme from cauliflower bud mitochondria. Biochem J 122:495–501

Macrae AR (1971b) Effect of pH on the oxidation of malate by isolated cauliflower bud mitochondria. Phytochemistry 10:1453–1458

McNeil PH, Thomas DR (1976) The effect of carnitine on palmitate oxidation by pea cotyledon mitochondria. J Exp Bot 27:1163–1180

Mettler IJ, Beevers H (1980) Oxidation of NADH in glyoxysomes by a malate-aspartate shuttle. Plant Physiol 66:555–560

Millhouse J, Wiskich JT, Beevers H (1983) Metabolite oxidation and transport in mitochondria of endosperm from germinating castor bean. Aust J Plant Physiol 10:167–177

Mitchell P (1979) Compartmentation and communication in living systems. Ligand conduction: a general catalytic principle in chemical, osmotic and chemiosmotic reaction systems. Eur J Biochem 95:1–20

Møller IM, Palmer JM (1982) Direct evidence for the presence of a rotenone-resistant NADH dehydrogenase on the inner surface of the inner membrane of plant mitochondria. Physiol Plant 54:267–274

Moore AL, Bonner WD (1981) A comparison of the phosphorylation potential and electrochemical proton gradient in mung bean mitochondria and phosphorylating sub-mitochondrial particles. Biochim Biophys Acta 634:117–128

Moore AL, Cottingham IR (1983) Characteristics of the higher plant respiratory chain. In: Robb DA, Pierpont S (eds) Metals and micronutrients. Academic Press, London New York, pp 169–204

Moore AL, Jackson C, Halliwell B, Dench JE, Hall DO (1977) Intramitochondrial localization of glycine decarboxylase in spinach leaves. Biochem Biophys Res Commun 78:483–491

Moreau F, Romani R (1982) Malate oxidation and cyanide-insensitive respiration in avocado mitochondria during the climacteric cycle. Plant Physiol 70:1385–1390

Murphy DJ, Walker DA (1982) Acetyl coenzyme A biosynthesis in the chloroplast. What is the physiological precursor? Planta 156:84–88

Nash D, Paleg LG, Wiskich JT (1982) Effect of proline, betaine and some other solutes on the heat stability of mitochondrial enzymes. Aust J Plant Physiol 9:47–57

Neuburger M, Day DA, Douce R (1984) Transport of coenzyme A in plant mitochondria. Arch Biochem Biophys 229:253–258

Newsholme EA, Crabtree B (1976) Substrate cycles in metabolic regulation and in heat generation. Biochem Soc Symp 41:61–109

Newsholme EA, Crabtree B (1979) Theoretical principles in the approaches to control of metabolic pathways and their application to glycolysis in muscle. J Mol Cell Cardiol 11:839–856

Oestreicher G, Hogue P, Singer TP (1973) Regulation of succinate dehydrogenase in higher plants II. Activation by substrates, reduced coenzyme Q, nucleotides and anions. Plant Physiol 52:622–626

Osmond CB, Holtum JAM (1981) Crassulacean acid metabolism. In: Hatch MD, Boardman NK (eds) The biochemistry of plants, vol VIII. Photosynthesis. Academic Press, London New York, pp 283–328

Palmer JM (1976) The organization and regulation of electron transport in plant mitochondria. Annu Rev Plant Physiol 27:133–157

Palmer JM, Arron GP (1976) The influence of exogenous nicotinamide adenine dinucleotide on the oxidation of malate by Jerusalem artichoke mitochondria. J Exp Bot 27:418–430

Palmer JM, Wedding RT (1966) Purification and properties of succinyl-CoA synthetase from Jerusalem artichoke mitochondria. Biochim Biophys Acta 113:167–174

Palmer JM, Schwitzguebel J-P, Møller IM (1982) Regulation of malate oxidation in plant mitochondria. Biohem J 208:703–711

Pradet A, Raymond P (1983) Adenine nucleotide ratios and adenylate energy charge in energy metabolism. Annu Rev Plant Physiol 34:199–224

Randall DD, Williams M, Rapp BJ (1981) Phosphorylation – dephosphorylation of pyruvate dehydrogenase complex from pea leaf mitochondria. Arch Biochem Biophys 207:437–444

Reed AJ, Canvin DT (1982) Light and dark controls of nitrate reduction in wheat (*Triticum aestivum* L.) protoplasts. Plant Physiol 69:508–513

Rubin PM, Randall DD (1977) Regulation of plant pyruvate dehydrogenase complex by phosphorylation. Plant Physiol 60:34–39

Rustin P, Moreau F (1979) Malic enzyme activity and cyanide-insensitive electron transport in plant mitochondria. Biochem Biophys Res Commun 88:1125–1131

Rustin P, Moreau F, Lance C (1980) Malate oxidation in plant mitochondria via malic enzyme and the cyanide-insensitive electron transport pathway. Plant Physiol 66:457–462

Sarojini G, Oliver DJ (1983) Extraction and partial characterization of the glycine decarboxylase multienzyme complex from pea leaf mitochondria. Plant Physiol 72:194–199

Schwitzguebel JP, Møller IM, Palmer JH (1981) The oxidation of tricarboxylate anions by mitochondria isolated from *Neurospora crassa*. J Gen Microbiol 126:297–303

Sellami A (1976) Évolution des adénosine phosphates et de la charge énergétique dans les compartiments chloroplastiques et non chloroplastiques des feuilles de blé. Biochim Biophys Acta 423:524–539

Singer TP, Oestreicher G, Hogue P, Contreiras J, Brandao I (1973) Regulation of succinate dehydrogenase in higher plants I Some general characteristics of the membrane-bound enzyme. Plant Physiol 52:616–621

Solomos T (1977) Cyanide-resistant respiration in higher plants. Annu Rev Plant Physiol 28:279–297

Stitt M, Lilley RMcC, Heldt HW (1982) Adenine nucleotide levels in the cytosol, chloroplasts and mitochondria of wheat leaf protoplasts. Plant Physiol 70:971–977

Tezuka T, Laties GG (1983) Isolation and characterization of inner membrane-associated and matrix NAD-specific isocitrate dehydrogenase in potato mitochondria. Plant Physiol 72:959–963

Thomas DR, McNeil PH (1976) The effect of carnitine on the oxidation of saturated fatty acids by pea cotyledon mitochondria. Planta 132:61–63

Thomas DR, Wood C (1982) Oxidation of acetate, acetyl CoA and acetyl-carnitine by pea mitochondria. Planta 154:145–149

Thompson P, Reid EE, Lyttle CR, Dennis DT (1977a) Pyruvate dehydrogenase complex from higher plant mitochondria and proplastids: Kinetics. Plant Physiol 59:849–853

Thompson P, Reid EE, Lyttle CR, Dennis DT (1977b) Pyruvate dehydrogenase complex from higher plant mitochondria and proplastids: Regulation. Plant Physiol 59:854–858

Tobin A, Djerdjour B, Journet E, Neuburger M, Douce R (1980) Effect of NAD$^+$ on malate oxidation in intact plant mitochondria. Plant Physiol 66:225–229

Tolbert NE (1980) Photorespiration. In: Davies DD (ed) The biochemistry of plants, vol II. Metabolism and respiration. Academic Press, London New York, pp 487–523

Tolbert NE (1981) Metabolic pathways in peroxisomes and glyoxysomes. Annu Rev Biochem 50:133–157

Turner JF, Turner DH (1980) The regulation of glycolysis and the pentose phosphate pathway. In: Davies DD (ed) The biochemistry of plants, vol II. Metabolism and respiration. Academic Press, London New York, pp 279–316

Valenti V, Pupillo P (1981) Activation kinetics of NAD-dependent malic enzyme of cauliflower bud mitochondria. Plant Physiol 68:1191–1196

Vignais PV, Lauquin GJM (1979) Mitochondrial adenine nucleotide transport and its role in the economy of the cell. Trends Biochem Sci 4:90–92

Walker GH, Oliver DJ, Sarojini G (1982) Simultaneous oxidation of glycine and malate by pea leaf mitochondria. Plant Physiol 70:1465–1469

Wedding RT, Black MK (1971) Nucleotide activation of cauliflower α-keto-dehydrogenase. J Biol Chem 246:1638–1643

Wedding RT, Canellas PF, Black MK (1981) Slow transients in the activity of NAD malic enzyme from Crassula. Plant Physiol 68:1416–1423

Wiskich JT (1977) Mitochondrial metabolite transport. Annu Rev Plant Physiol 28:45–69

Wiskich JT (1980) Control of the Krebs cycle. In: Davies DD (ed) The biochemistry of plants, vol II. Metabolism and respiration. Academic Press, London New York, pp 244–278

Wiskich JT, Bonner WD (1963) Preparation and properties of sweet potato mitochondria. Plant Physiol 38:594–604

Wiskich JT, Day DA (1979) Rotenone-insensitive malate oxidation by isolated plant mitochondria. J Exp Bot 30:99–107

Wiskich JT, Day DA (1982) Malate oxidation, rotenone-resistance, and alternate path activity in plant mitochondria. Plant Physiol 70:959–964

Wiskich JT, Rayner JR (1978) Interactions between malate and nicotinamide adenine dinucleotide oxidation in plant mitochondria. In: Ducet G, Lance C (eds) Plant mitochondria. Elsevier/North Holland Biomedical Press, Amsterdam New York, pp 101–108

Woo KC, Osmond CB (1976) Glycine decarboxylation in mitochondria isolated from spinach leaves. Aust J Plant Physiol 3:771–785

Woo KC, Osmond CB (1982) Stimulation of ammonia and 2-oxoglutarate-dependent O$_2$ evolution in isolated chloroplasts by dicarboxylates and the role of the chloroplast in photorespiratory nitrogen recycling. Plant Physiol 69:591–596

Woo KC, Jokinen M, Canvin DT (1980) Reduction of nitrate via a dicarboxylate shuttle in a reconstituted system of supernatant and mitochondria from spinach leaves. Plant Physiol 65:433–436

Wu S, Laties GG (1983) A malate-oxaloacetate shuttle linking cytosolic fatty acid α-oxidation to the mitochondrial electron transport chain in uncoupled fresh potato slices. Physiol Plant 58:81–88

Yu C, Claybrook DL, Huang AHC (1983) Transport of glycine, serine, and proline into spinach leaf mitochondria. Arch Biochem Biophys 227:180–187

11 Leaf Mitochondria (C₃ + C₄ + CAM)

P. GARDESTRÖM and G.E. EDWARDS

1 Introduction

The properties of isolated mitochondria from nongreen tissue have been studied for several years and have been the subject of a number of recent reviews (PALMER 1976, 1979, HANSON and DAY 1980). Mitochondria from leaf tissue have been considerably less studied. The discovery of the photorespiratory glycolate pathway, to which reactions in chloroplasts, peroxisomes, and mitochondria all contribute, have increased the interest in leaf mitochondria. Recent developments in preparation procedures have made more detailed studies on leaf mitochondria possible.

Leaf mitochondria are similar in many respects to mitochondria from nongreen plant tissue. Like other plant mitochondria, they oxidize externally added NADH and NADPH, possibly via two different dehydrogenases on the outer surface of the inner membrane (ARRON and EDWARDS 1979, NASH and WISKICH 1983). As in other plant mitochondria, two malate-oxidizing enzymes (i.e., malate dehydrogenase and NAD-malic enzyme) are found in the matrix of leaf mitochondria (NEUBURGER and DOUCE 1980), and cyanide-resistant oxidation via an alternative oxidase has been shown (RUSTIN et al. 1980, GARDESTRÖM and EDWARDS 1983). Thus, leaf mitochondria possess the same versatility in substrate oxidation as other plant mitochondria, enabling the mitochondria to fulfill complex metabolic functions in the cell (as discussed by PALMER 1979). Some differences between mitochondria from photosynthesizing and nonphotosynthesizing tissues are also known, the best characterized differences being connected with the photorespiratory reactions in C₃ plant mitochondria and decarboxylations in the photosynthetic carbon cycle in CAM and C₄ plants. In this review we shall deal with preparation procedures for leaf mitochondria and the known specializations in leaf mitochondria.

2 Effects of Light on Dark Respiration

In the light there are some fundamental differences between a photosynthetic and a nonphotosynthetic cell. The photosynthetically active cell produces PGA (direct or indirect) from CO_2 for further use in metabolism or export to nonpho-

---

Abbreviations. cyt: cytochrome; Chl: chlorophyll; ME type: malic enzyme type; PAGE: polyacrylamide gel electrophoresis; PEP-CK: PEP carboxykinase; PCA cycle: photosynthetic carbon assimilation cycle; PCO cycle: photosynthetic carbon oxidation cycle; PEP: phosphoenolpyruvate; PGA: 3-phosphoglycerate; RuBP: ribulose bisphosphate.

tosynthetic cells in the form of triose phosphate. There are also two different systems for phosphorylation of ADP: oxidative phosphorylation in the mitochondria and photosynthetic phosphorylation in the chloroplasts. In the dark, the photosynthetic cell depends on carbon reserves and oxidative phosphorylation. The differences in metabolism in the photosynthetic cell in the light and in the dark are likely to affect mitochondrial functions.

The effect of light on dark respiration has recently been reviewed (GRAHAM and CHAPMAN 1979, GRAHAM 1980), but the subject is very complex and not well understood. One general view has been that ATP produced by photosynthetic phosphorylation in the chloroplasts in the light would make oxidative phosphorylation superfluous. Thus, mitochondrial respiration via cytochrome oxidase could be inhibited by a high ATP/ADP ratio in the cytoplasm in the light (triose-phosphate exported from the chloroplast and metabolized to PGA in the cytoplasm would generate ATP). On the other hand, intermediates formed in the mitochondria are needed in the light as well as in the dark. The matter is further complicated by the possible involvement of the alternative oxidase. Electron transport to this oxidase is considered to be nonphosphorylating and thus not regulated by ATP/ADP ratios (DAY et al. 1980). It is clear that leaf mitochondria play an important role in the metabolism of the photosynthesizing cell, not only in photorespiration but also by providing intermediates for further biosynthesis in the cell (CHAPMAN and GRAHAM 1974).

The ATP/ADP ratios in different cell compartments in the light and in the dark have recently been measured using an ingenious technique where protoplasts are broken and fractionated into a chloroplast fraction, a mitochondrial fraction, and a cytosolic fraction (in each case within 0.1 s) using membrane filtration (LILLEY et al. 1982, STITT et al. 1982). Surprisingly, the ATP/ADP ratio in the cytoplasm was found to be higher in the dark (9.2) than in the light (6.4), which is in direct conflict with earlier assumptions (see STITT et al. 1982). In the light the mitochondrial ATP/ADP ratio increased, while the ratio decreased in the cytosol, which is consistent with a decreased mitochondrial energization. The cause of this and the consequences it has for mitochondrial functions are not known. Whether the results reflect the in vivo condition in the light is open to question. It is possible that during isolation of protoplasts the total adenylate pool decreases and influences the change in ATP/ADP ratios in different cellular compartments upon illumination. In addition, the experiments do not reflect the ATP/ADP ratios in vivo under photorespiring conditions as high bicarbonate levels were used in the assay.

DRY and WISKICH (1982) reported that high ATP/ADP ratios outside the mitochondria (up to 20) were required to inhibit respiration. High concentrations of ADP acted as a buffer against inhibition of respiration by a high ATP/ADP ratio. Thus, it is questionable whether the ATP/ADP ratio reaches sufficient levels in the light to control respiration.

There is evidence that dark respiration is higher following a period of illumination than at the end of the night period (AZCON-BIETO and OSMOND 1983, AZCON-BIETO et al. 1983). In the plants in the light, a high level of soluble carbohydrates is available, the respiration in the dark following illumination is stimulated by uncoupler, so respiration may be limited by ADP (i.e., functioning between States 3 and 4)[1]. The higher level of respiration in pre-illuminated plants is correlated with a high leaf carbohydrate content and a higher CO_2 compensation point (Azcon-Bieto and Osmond 1983). It was suggested that the rate of respiration of carbohydrate in the light is comparable to dark respiration, and that the TCA cycle may be modified in the light to function anaplerotically. Of particular interest are recent data which suggest that the higher level of respiration of illuminated leaves may be associated with function of the alternative pathway (Azcon-Bieto et al. 1983, also see Sect. 5.7).

It appears it is no longer a question of whether the TCA cycle runs in the light, but to what extent it functions by shuttling reductant out of the mitochondria (e.g., for nitrate reduction), by engagement of the alternative pathway, and by electron flow coupled to the cytochrome pathway. It is clear that more work is needed on isolated mitochondria, protoplasts, and whole plants to elucidate the details of how photosynthesis influences the function of mitochondria.

3 Preparation of Leaf Mitochondria

3.1 Introduction

The preparation of leaf mitochondria involves, in general, the same problems as preparations of other plant mitochondria. The high mechanical resistance of the cell wall makes it difficult to break the cell without damaging the mitochondria. Harmful vacuolar content might also be released upon homogenization. These problems can be overcome by proper selection of plant material, i.e., leaves of spinach and pea seem to be good in this respect. An additional problem in the purification of mitochondria from leaves is the presence of chloroplasts. In oat leaves, it was estimated that 84% of the cellular protein was associated with chloroplasts and only 2% of the protein associated with the mitochondria (Fuchs et al. 1981). During homogenization, some of the chloroplasts break into fragments of various sizes. These heterogeneous fragments are then difficult to separate from the mitochondria. Table 1 shows the yield of mitochondria from leaf tissue of C_3, C_4, and CAM plants, and the degree of purity relative to chlorophyll from procedures developed by several investigators, which are discussed in more detail below.

Different studies have different requirements for mitochondrial preparation with respect to degree of purity and functional integrity. In some cases, the purity of the preparation is not a major concern (for example studies of respiratory properties). However, other studies are made difficult or impossible by contaminating chloroplast fragments (for example, studies of enzyme compartmentation, composition of the mitochondria and metabolite transport).

---

[1] Definition of the different states, see this Vol., Chap. 10, p. 283.

Table 1. Examples from the literature of yield of purified leaf mitochondria and purity relative to chlorophyll

Sub-group	Species	Fresh weight of material used (g)	Method	Yield of mito-chondrial protein (mg)	Purity[a] mg protein mg^{-1} chl	Reference
C_3	Spinach	50	M	2	100	GARDESTRÖM et al. (1978)
	Spinach	n.s.	M	n.s.	123	JACKSON et al. (1979a)
	Spinach	150	M	4.9	Essentially chl-free	BERGMAN et al. (1980)
	Oats	10	P	1	Essentially chl-free	FUCHS et al. (1981)
	Spinach	n.s.	P	n.s.	Essentially chl-free	NISHIMURA et al. (1982)
	Pea	n.s.	M	n.s.	Substantially chl-free	NASH and WISKICH (1983)
C_4	*Panicum miliaceum*	50	M	1.2 (Ms) 0.8 (Bs)	77 223	GARDESTRÖM and EDWARDS (1983)
CAM	*Sedum praealtum*	55	M	4.9	56	ARRON et al. (1979a)

[a] Thylakoids equivalent to 1 mg chl are estimated to contain about 7 mg of protein. The calculated in vivo mg mitochondrial protein mg^{-1} chlorophyll was 0.4 for spinach (from data of BERGMAN et al. 1980), 0.3 for mesophyll cells and 1.2 for bundle-sheath cells of *Panicum miliaceum* (from data of GARDESTRÖM and EDWARDS 1983), based on recovery of mitochondrial marker enzymes (isocitrate dehydrogenase, cytochrome c oxidase).

n.s. Not stated   Ms Mesophyll
M   Mechanical   Bs Bundle-sheath
P   From protoplasts

3.2 Development of Preparation Procedures

3.2.1 C_3 Plants

OHMURA (1955) showed that a particulate fraction obtained from spinach leaves could carry out oxidative phosphorylation when oxidizing Krebs cycle intermediates, and concluded that the activity was present in chloroplast fragments. Later work revealed a difference in distribution between chloroplasts and oxidative activities (PIERPOINT 1959, 1962), showing that the activities took place in leaf mitochondria. For a long period of time, few studies on isolated leaf mitochondria were made. Preparations with respiratory control were obtained after improvements of the preparation methods were developed (BIRD et al. 1972, WOO and OSMOND 1976).

An important improvement in the preparation procedure for leaf mitochondria was introduced by Douce et al. (1977). The combination of an appropriate grinding medium (0.3 M mannitol, 4 mM cysteine, 1 mM EDTA, 30 mM morpholino propane sulfonic acid (MOPS) buffer, pH 7.5, 0.2% defatted BSA, and 0.6% insoluble polyvinylpyrrolidine (PVP) and a very short grinding time (2 s) resulted in a mitochondrial preparation which, after differential centrifugation, had improved respiratory functions compared with previous preparations. Although it was suitable for respiratory measurements, the preparation was heavily contaminated with chloroplast material. It was estimated that about 40% of the protein in the mitochondrial fraction was associated with thylakoid membranes. Generally the mitochondrial fraction obtained from spinach leaves by differential centrifugation has a ratio of protein to chlorophyll of approximately 15. For many studies it is necessary to have purer preparations. Methods have been developed to obtain these by employing separation methods other than differential centrifugation when isolated mechanically or by isolating mitochondria from leaf protoplasts as summarized in Table 1.

Differences in surface properties between membrane particles can be employed for separation purposes by partitioning in dextran-polyethylene glycol two-phase systems (Albertsson et al. 1982). This method has been used to purify spinach leaf mitochondria (Gardeström et al. 1978). When this method was used, combined with differential centrifugation, the resulting mitochondrial preparation retained respiratory functions, and the enrichment of mitochondria relative to chloroplast material was 200–400 times.

Density gradients based on silica sol were used with limited success for purification of leaf mitochondria (Gronebaum-Turck and Willenbrink 1971). A silica gel coated with polyvinylpyrrolidone in order to reduce the deleterious effects of the silica sol on biological material was introduced under the name Percoll in the mid-1970's (Pharmacia Fine Chemicals, Uppsala, Sweden). Percoll was successfully used to purify the crude mitochondrial fraction obtained by differential centrifugation of a spinach leaf homogenate (Jackson et al. 1979a). The mitochondria were intact, as indicated by respiratory control and ADP/O ratios, and the enrichment relative to chloroplast material was 330. By combining differential centrifugation, phase partition and Percoll density gradient centrifugation, a preparation of functionally intact spinach leaf mitochondria was obtained (Bergman et al. 1980). The preparation was essentially chlorophyll-free and was functionally intact, as shown by respiratory properties. Peroxisomal material was the main contaminant, the enrichment of mitochondria relative to peroxisomes (NAD-isocitrate dehydrogenase/glycolate oxidase) was about 12 compared to the original leaf homogenate. Sucrose density gradient centrifugation in combination with differential centrifugation was used to purify pea leaf mitochndria (Nash and Wiskich 1983). The preparation showed respiratory control and good ADP/O ratios and was stated to be "substantially free of contamination by chlorophyll and peroxisomes"; the enrichment and yield were, however, not specified.

Another approach to the purification problem is taken when isolation of protoplasts is the first step in the purification (Nishimura et al. 1976). Protoplasts are broken and the organelles released by mild mechanical treatment.

Few chloroplasts are broken, and thus a very small amount of chloroplast fragments are produced. Intact chloroplasts can then effectively be removed by differential centrifugation. Very pure mitochondria were obtained from oat leaf protoplasts after differential centrifugation and sucrose density gradient centrifugation (FUCHS et al. 1981). NISHIMURA et al. (1982) isolated mitochondria from spinach leaf protoplasts which showed very high rates of respiration. The disadvantage with the protoplast method is that it is difficult to obtain a high yield of mitochondria. The mechanical methods are much easier to scale up in order to increase the yield.

3.2.2 CAM Plants

Mitochondria have been isolated from leaf tissue of the CAM plants *Mesembryanthemum crystallinum* (VON WILLERT and SCHWÖBEL 1978), *Sedum praealtum* (ARRON et al. 1979a, SPALDING et al. 1980); and *Kalanchoë daigremontiana* (DAY 1980) (all ME types), having good respiratory control and reasonable ADP/O ratios. The isolation media and grinding procedure were similar to those used in isolating spinach leaf mitochondria (Sect. 3.2.1). The mitochondria from *Sedum praealtum*, which were prepared by differential centrifugation and washed once, were largely free of thylakoid contamination (estimate of less than 2% of the protein in the preparation of thylakoid origin). Therefore, unlike most species, with *S. praealtum*, the thylakoids do not tend to co-sediment with the mitochondria. Further purification on a linear sucrose gradient removed other nonmitochondrial proteinaceous material, resulting in a five fold increase in specific activity of cytochrome c oxidase. In comparison to the original leaf homogenate, the purified mitochondria were enriched 250-fold relative to peroxisomes (cytochrome c oxidase/glycolate oxidase activity).

3.2.3 C$_4$ Plants[1]

With the discovery of NAD-ME type C$_4$ plants, KAGAWA and HATCH (1974) isolated mitochondria from bundle-sheath cells and showed that they contained NAD-malic enzyme. Among C$_4$ plants, mitochondria having good coupling of phosphorylation to electron transport were first obtained from maize (NADP-ME type), using a mechanical grinding procedure similar to that used with spinach (NEUBURGER and DOUCE 1977). With the short grinding time used, these mitochondria were probably predominantly of mesophyll origin.

The major difficulties in isolating mitochondria from photosynthetic tissue of C$_4$ plants are first, to get a reasonable separation of mesophyll from bundle-sheath tissue and second, to break the thick-walled bundle-sheath cells without breaking the mitochondria. If adequate precautions are taken, some tissues, like those of *Panicum miliaceum* and *Atriplex spongiosa*, are amenable to differential mechanical grinding, and mesophyll and bundle-sheath fractions can be isolated which have a purity of 80% or greater. The degree of cross-contamination in the original extracts can be assessed by assaying marker enzymes for

---

[1] For definitions of the sub-types see Vol. 3, this Series (C.R. STOCKING and U. HEBER eds.), Chap. 4 by M.D. HATCH and C.B. OSMOND, pp. 150–160 (1976).

the respective cell types like PEP carboxylase (mesophyll enzyme) and RuBP carboxylase (bundle-sheath cell enzyme). Selection of plant material is quite important, since adequate separation by mechanical means is not possible with some species.

Gardeström and Edwards (1983) developed a mechanical procedure for isolating pure, intact mesophyll and bundle-sheath mitochondria from *Panicum miliaceum*, a NAD-ME type species, and studied their respiratory properties. The mesophyll fraction of *P. miliaceum* was isolated using a Polytron (setting 7 for 4 s). Further treatment with the Polytron yielded bundle-sheath strands which were collected by filtration and extracted using a mortar and pestle. A crude mitochondrial preparation was obtained by differential centrifugation and further purified on a Percoll step gradient. In both the mesophyll and bundle-sheath preparations, the enrichment of mitochondria relative to chloroplast material was about 75-fold. It is possible to separate mesophyll and bundle-sheath tissue enzymatically from maize and *P. miliaceum* using digestive enzymes and to obtain functional mitochondria from these preparations (R. Kanai, personal communication), but this is technically more difficult to perform on a routine basis than mechanical procedures.

4 Properties of Isolated Leaf Mitochondria

4.1 Purity and Intactness

Most studies on mitochondria from leaves have been made on preparations from spinach and pea. To be able to compare results, (e.g., on rates of substrate oxidation) with different preparations within and between species, it is important to characterize the preparations with respect to purity and intactness. Also, studies on metabolite transport, where shrinkage/swelling and silicone oil centrifugation techniques are used, are difficult to interpret unless the preparations are pure.

Purity of mitochondria has generally been considered relative to contamination by chloroplasts (particularly thylakoids from broken chloroplasts) and peroxisomes. For example, measurements on the parent tissue in comparison to the mitochondrial preparation of the chlorophyll (chlorolast marker)/cyt c oxidase activity, glycolate oxidase (peroxisome marker)/cyt c oxidase activity and NADPH: cytochrome c oxidoreductase (microsomal marker)/cyt c oxidase activity can be used to determine the degree of purity relative to the original leaf. Usually adherence of cytoplasmic enzymes to the mitochondria will be negligible, although in some tissue this could be a problem (e.g., in some CAM plants, possibly associated with a high phenolic content, see Edwards et al. 1982). Since activity is expressed mg^{-1} mitochondrial protein, comparisons of various mitochondrial activities between different tissues are only possible if the preparations are reasonably pure. As the chloroplasts make up such a large percentage of the protein of the leaf, a high degree of purification is required in order to avoid significant contamination by thylakoid protein. Electron microscopy can be employed to estimate the purity of organelle preparations. The method gives a good general idea about the purity with respect to membranous contamination, but it is very difficult to quantify. Electron microscopy is not a good means for determining degree of mitochondrial intactness.

The most commonly used intactness criteria are respiratory control and latency of enzyme activities. The respiratory control is of limited value as an indicator for intactness of plant mitochondria due to the branching of the respiratory chain to the alternative oxidase as discussed by PALMER (1976). The most widely used intactness assay is to measure the ratio of succinate: cytochrome c oxidoreductase activity in isotonic and hypotonic media (DOUCE et al. 1972). The assay is designed to measure the intactness of the outer membrane, as cytochrome c cannot penetrate the intact outer membrane. The same principle applies for KCN-sensitive ascorbate-cytochrome c-dependent O_2 uptake (NEUBURGER et al. 1982). To assay inner membrane intactness, the permeability of the membrane to malate dehydrogenase (DOUCE et al. 1973, WOO and OSMOND 1977) and the latency of isocitrate dehydrogenase (GARDESTRÖM et al. 1978) have been used. The main disadvantage with the malate dehydrogenase assay is that it measures the permeability of the membrane for protein molecules. NAD-isocitrate dehydrogenase is unstable in solution which makes the method less reliable. Due to these limitations, neither of the two methods is particularly suitable as a general intactness criterion.

4.2 Composition

The recent development of preparation procedures has made studies on composition of leaf mitochondria possible. The cytochrome content as determined by difference spectroscopy is qualitatively the same in spinach leaf mitochondria as in other plant mitochondria. The amount of cytochromes compared to protein was, however, considerably lower in leaf mitochondria than in other plant mitochondria (GARDESTRÖM et al. 1981 a). In these studies, mitochondria from spinach leaves and leaf petioles (stalks), representing mitochondria from photosynthetic and nonphotosynthetic tissue, were compared.

The main phospholipids of mitochondria isolated from oat leaves were phosphatidyl choline, phosphatidyl ethanolamine and diphosphatidyl glycerol (cardiolipin) with C18:2 as the main fatty acid (FUCHS et al. 1981), which is similar to mitochondria from nongreen tissue. Very similar phospholipid composition was also obtained for spinach leaf mitochondria (EDMAN and ERICSON 1984). On a protein basis, spinach leaf mitochondria contained somewhat less phospholipids than leaf petiole mitochondria (GARDESTRÖM et al. 1983) and leaf mitochondria contained relatively more of the fatty acid C18:3 (EDMAN and ERICSON 1984).

Spinach leaf mitochondria differ significantly in the polypeptide composition from mitochondria from the nonphotosynthetic tissues of leaf petioles and roots as determined by SDS-PAGE. In spinach, the matrix fraction of leaf mitochondria contains at least four major polypeptides (15,900; 41,700; 50,700 and 101,000 M.W.), which are absent or present in only small amounts in the matrix fraction from petiole mitochondria (SAHLSTRÖM and ERICSON 1984). Leaf mitochondria also contained relatively more protein in the matrix fraction compared to the membrane fraction than did petiole mitochondria (GARDESTRÖM et al. 1983).

The differences in protein, lipid, and cytochrome composition taken together indicate that leaf mitochondria have fewer respiratory chains in relation to other membrane components than leaf petiole mitochondria (GARDESTRÖM et al. 1983). This difference is possibly due to metabolic differences between mitochondria from photosynthetic and nonphotosynthetic tissue. To speculate, perhaps, in leaf mitochondria there is a tendency to shuttle the NADH out and utilize it as a reductant, and to oxidize it in the alternative pathway rather than via the cytochrome chain. These possibilities have been considered during glycine oxidation in photorespiration (Sect. 5.7).

5 Special Functions of Leaf Mitochondria – Role in Photorespiration

5.1 Photorespiration [1]

In photorespiration, recently fixed carbon is released as CO_2 (LORIMER and ANDREWS 1981, EDWARDS and WALKER 1983); the function of this apparently wasteful process is not known. The photorespiratory cycle involves reactions in chloroplasts, peroxisomes, and mitochondria. It is generally agreed that the first step in photorespiration is oxygenation of RuBP to yield phosphoglycolate in the chloroplasts. Phosphoglycolate is then converted to glycine in the peroxisomes. In the mitochondria, two glycine then form one serine, which is converted to glycerate in the peroxisomes, and then glycerate is converted back to PGA, in the chloroplast (Fig. 1). As illustrated in Fig. 1 there is considerable transport of metabolites between organelles, and the mechanisms are largely unresolved. It was shown by KISAKI et al. (1971a) that the conversion of glycine to serine in the photorespiratory glycolate cycle took place in the mitochondria. The interest in leaf mitochondria increased as their involvement in photorespiration became known and most work on the organelle from leaves has centered on this aspect. The conversion of glycine to serine is specific for mitochondria from photosynthesizing tissue, as the activity was not detected in mitochondria from roots, leaf petioles (stalks), or leaf veins from spinach (GARDESTRÖM et al. 1980). The activity is also very low in mitochondria from etiolated tissue (MOORE et al. 1977, NEUBURGER and DOUCE 1977). The activity is light-induced in etiolated leaf tissue and follows the increase in glycolate oxidase activity, but is independent of chlorophyll synthesis (ARRON and EDWARDS 1980a).

In the mitochondrial reaction, one molecule each of serine, CO_2, and NH_3 is formed from two molecules of glycine (KISAKI et al. 1971b, BIRD et al. 1972) and NAD is also reduced to NADH (BIRD et al. 1972, ARRON et al. 1979b). The reaction is complex and involves two enzymes: glycine decarboxylase and

1 See also in other volumes of this Encyclopedia, in particular: Vol. 6 Photosynthesis II, Chap. 27 and 28 (I. ZELITCH, D.T. CANVIN) pp. 353–396; Vol. 3 Transport in Plants III, Chap. 5 (C. SCHNARRENBERGER, H. FOCK) pp. 185–234.

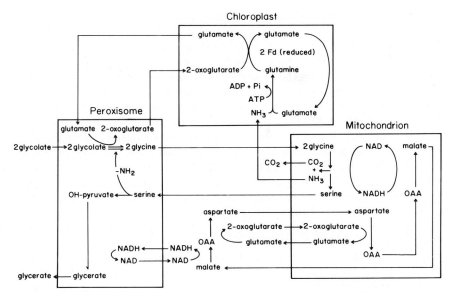

Fig. 1. Possible shuttle of metabolites between mitochondria, peroxisomes, and chloroplasts during metabolism in the glycolate pathway

serine hydroxymethyltransferase. The stoichiometry of CO_2 evolution to O_2 uptake during glycine oxidation is 2:1, as expected if one NADH is formed per CO_2 evolved (ARRON et al. 1979b). Some earlier reports implied a ratio of 1, which may have been due to the inaccuracy of assaying decarboxylation and O_2 uptake separately (see ARRON et al. 1979b).

In C_4 plants, metabolism in the glycolate pathway is considered to occur in bundle-sheath cells, but at a much lower level than in C_3 plants (EDWARDS and WALKER 1983). Although there is some evidence for glycine oxidation by mitochondria from bundle-sheath cells, in some studies the mitochondria had no capacity to oxidize glycine, i.e., in maize (WOO and OSMOND 1977) and in *Panicum miliaceum* (millet: GARDESTRÖM and EDWARDS, unpublished). In the latter case the mitochondria readily oxidized malate. Mitochondria from enzymatically isolated bundle-sheath strands of maize and *P. miliaceum* have some capacity to oxidize glycine (R. KANAI, personal communication). Therefore, the glycine oxidation complex may be particularly labile in some species and its function lost during mitochondrial isolation.

In comparison to mitochondria of C_3 plants, the mitochondria of ME type CAM plants oxidize glycine only slowly (SPALDING et al. 1980, DAY 1980). In CAM plants, little photorespiration occurs during malate metabolism behind closed stomata, due to the relatively high CO_2/O_2 ratio in the leaf, as compared to C_3 plants (see EDWARDS and WALKER 1983). Therefore, the requirement for glycine oxidation under these conditions would be much lower than in C_3 plants. However, many CAM plants fix CO_2 directly from the atmosphere and photorespire during the latter part of the light period. Thus, some capacity for glycine oxidation by the mitochondria is expected.

5.2 Transport of Photorespiratory Metabolites

For the continuous oxidative decarboxylation of glycine in the mitochondria, glycine must be transported into and serine out of the mitochondria. Considering the proposed stoichiometry of carboxylation to oxygenation of RuBP, the required level of glycine transport in mitochondria is about 20 to 30% of the net rate of photosynthesis (e.g., see EDWARDS and WALKER 1983, SCHMITT and EDWARDS 1983). Different opinions exist in the literature about the mechanism of this transport. Based on inhibition of glycine oxidation by mersalyl, DENCH et al. (1978) suggested that a glycine translocator was present in spinach leaf mitochondria. Also, CAVALIERI and HUANG (1980) suggested carrier-mediated transport of glycine and serine based on stereospecificity and inhibition by mersalyl of swelling. This study was, however, made on mitochondria from nongreen material and the validity for leaf mitochondria is uncertain. Mersalyl also inhibited swelling induced by glycine in mitochondria from *Arabidopsis* (SOMERVILLE and OGREN 1982). WALKER et al. (1982a) suggested that a specific glycine translocator is located in the mitochondrial inner membrane, based on the effect of inhibitors on swelling and glycine uptake measured by the silicone oil centrifugation technique. The glycine transport was dependent on a transmembrane pH gradient, as it was inhibited by the uncoupler carbonyl cyanide-p-trifluoromethoxy-phenylhydrazone (FCCP). DAY and WISKICH (1980) and PROUDLOVE and MOORE (1982), on the other hand, observed no effect of FCCP on glycine uptake measured by the silicone oil centrifugation technique. No saturation of the uptake at high glycine concentrations was observed either by the swelling technique (DAY and WISKICH 1980) or the silicone oil centrifugation technique (PROUDLOVE and MOORE 1982). Based on these results, these authors argued in favor of unspecific diffusion as the mechanism for glycine transport. It is worth noting that in the cases when mersalyl was reported to inhibit glycine uptake, the mitochondria were pre-incubated with the inhibitor (CAVALIERI and HUANG 1980, SOMMRVILLE and OGREN 1982) and when no inhibition of uptake was seen, no pre-incubation was used (DAY and WISKICH 1980, PROUDLOVE and MOORE 1982). DENCH et al. (1978) did not specify the method used. All the above-cited studies on uptake measured by silicone oil centrifugation were made on pea leaf mitochondria. Using purified mitochondria from sunflower cotyledons which had been induced to oxidize glycine by exposing plants to a 5 min day^{-1} light treatment, glycine was readily taken up via diffusion based on silicone oil centrifugation and shrinkage-swelling studies (ARRON and EDWARDS 1980b). Uptake measured by silicone oil centrifugation at low concentrations of glycine (0.77 mM), was not dependent on NADH or inhibited by mersalyl (preincubated 1 min at 0.1 mM) or aminoacetonitrile (ARRON and EDWARDS, unpublished).

In studies with pea leaf mitochondria, the reported State 3 rates of glycine oxidation were usually 65–100 nmol O_2 mg^{-1} protein min^{-1} (DAY and WISKICH 1980, WALKER et al. 1982b, PROUDLOVE and MOORE 1982). In the mitochondrial conversion of glycine to serine, 4 molecules of glycine are consumed for every O_2 consumed. Thus, in State 3 the uptake must be in the order of 250–400 nmol glycine mg^{-1} protein min^{-1}. The reported rates of glycine uptake were 4–40 nmol mg^{-1} protein min^{-1} (Day and Wiskich 1980, Walker et al. 1982a, Proudlove and Moore 1982). The measured transport rates are thus much lower than must be the case during glycine oxidation, and definite conclusions cannot be made from these studies. In addition, studies with pea leaf mitochondria have been with unpurified preparations containing thylakoids. Clearly, the kinetics of glycine transport must be studied in detail before the question is finally settled. The mechanism by which serine is exported from the mitochondria also remains to be elucidated.

5.3 Glycine Decarboxylase

Glycine decarboxylase (EC 2.1.2.10, glycine synthase) catalyzes the oxidative decarboxylation of glycine, yielding CO_2, NH_3, and the C_1 moiety of methylene-tetrahydrofolate; in the reaction NAD is also reduced to NADH. The enzyme complex is well characterized from animal tissue and bacteria, where it has been shown to consist of four different polypeptides (for review see Kikuchi and Hiraga 1982). In addition to oxidative decarboxylation of glycine, the enzyme catalyzes a ^{14}C-exchange between glycine and bicarbonate; in this reaction only two of the peptides in the enzyme are needed (Motokawa and Kikuchi 1974).

Plant glycine decarboxylase is located behind the mitochondrial inner membrane (Woo and Osmond 1977, Moore et al. 1977). At least one of its components is coded for by a nuclear gene (Somerville and Ogren 1982). It has been assumed that the enzyme is membrane-bound (Moore et al. 1977), analogous to the enzyme from rat liver mitochondria (Motokawa and Kikuchi 1971), but it has been difficult to study the intramitochondrial localization experimentally. This is because, for unknown reasons, the activity of the enzyme was lost when the inner membrane was ruptured (Woo and Osmond 1977). Woo and Osmond (1976) suggested that the enzyme system demands a membrane potential for activity. Alternatively, the components in the complex might dissociate upon the dilution of the matrix when the inner membrane is ruptured.

Recently Sarojini and Oliver (1983) obtained glycine decarboxylase activity, using an acetone powder from pea leaf mitochondria. The activity was dependent on pyridoxal phosphate and a thiol reagent for maximum activity. It was concluded that the enzyme was tightly bound to the mitochondrial membrane, as activity could not be obtained in a soluble form by sonicating the mitochondria or treating with detergents. However, glycine-bicarbonate exchange activity can also be detected in a concentrated matrix fraction from spinach leaf mitochondria (Ericson et al. 1984). This indicates that at least some of the components in the glycine decarboxylase are not firmly bound to the membrane.

After incubation of spinach leaf mitochondria with ^{14}C-glycine, some fractions after gel filtration were radioactively labeled. Interestingly, all of the labeled fractions contained at least one of the polypeptides specific for leaf mitochondria (see Sect. 4.2).

The purified 15,900 M.W. protein contained lipoic acid and had a low isoelectric point (4.2). It had no glycine-bicarbonate exchange activity on its own, but exchange activity was obtained after recombination with a fraction from gel filtration having very little activity alone (ERICSON et al. 1984). Thus, the 15,900 M.W. protein in leaf mitochondria appears to be equivalent to the H-protein in the glycine decarboxylase from animal mitochondria (KIKUCHI and HIRAGA 1982). Fractionation of the glycine decarboxylase complex from pea leaf mitochondria was also recently reported (SAROJINI and OLIVER 1984).

5.4 Serine Hydroxymethyltransferase

Serine hydroxymethyltransferase (EC 2.1.2.1) is present in both photosynthetic and nonphotosynthetic cells (COSSINS 1980). In spinach leaves the enzyme was mainly mitochondrial (WOO 1979). SOMERVILLE and OGREN (1981) characterized a mutant of *Arabidopsis thaliana* lacking mitochondrial serine hydroxymethyl-transferase and showed that the mitochondrial enzyme was coded for by a nuclear gene. The mutant retained 15% of the serine hydroxymethyltransferase activity compared to the wild type, but all of that activity was extramitochondrial.

In leaves, some activity has been reported to be associated with chloroplasts (SHAH and COSSINS 1970, KISAKI et al. 1971a). In etiolated mung beans, the enzyme was cytoplasmic and had allosteric properties (RAO and RAO 1982). The mitochondrial enzyme was liberated from the mitochondria upon rupture of the inner membrane, similar to malate dehydrogenase, suggesting that the enzyme is located in the mitochondrial matrix (WOO 1979). The pH optimum was 8.5 and the enzyme was absolutely dependent on tetrahydrofolate for activity, with pyridoxal phosphate having no effect (WOO 1979). The mitochondrial enzyme has not been purified from plants and its properties are thus not well known. The mitochondrial enzyme was suggested to be specifically associated with photorespiratory metabolism, as mutants deficient in the enzyme could survive under nonphotorespiratory conditions (SOMERVILLE and OGREN 1982).

5.5 Assay Methods for Glycine Decarboxylase

Several methods have been used to assay the glycine decarboxylase activity; the measurements have generally been made on intact organelles due to the difficulty in isolating glycine decarboxylase. The decarboxylation can be directly detected as $^{14}CO_2$ evolution from $1\text{-}^{14}C$-glycine (WOO and OSMOND 1976, MOORE et al. 1977), and the NH_3 released can also be measured (BERGMAN et al. 1981). The NADH formed in the reaction can be reoxidized via reduction of oxaloacetate to malate by malate dehydrogenase (WOO and OSMOND 1976) or via the respiratory chain. In the latter case, the oxygen consumption can be used as a measure of the activity (BIRD et al. 1972, WOO and OSMOND 1976, MOORE et al. 1977). Alternatively, the enzyme can be reoxidized by the artificial dye DCPIP and the change in absorbance detected (MOORE et al. 1980).

Functions other than glycine decarboxylase are involved in the different assays. First, the enzyme must be reoxidized, which involves the respiratory chain or malate dehydrogenase or DCPIP. Second, the mitochondrial pool of tetrahydrofolate must be recycled during the assay via serine hydroxymethyl-transferase activity in the matrix. Thus, the assay procedures depend on simultaneous activity of both glycine decarboxylase and serine hydroxymethyltransferase. As intact organelles have been used, transport limitations might also be involved. When ^{14}C-glycine-bicarbonate exchange (Woo and Osmond 1976) is used as the assay, no direct involvement of serine hydroxymethyltransferase is needed as no net reaction occurs.

The pH optimum for oxidation of glycine via the respiratory chain was 7.6, whereas the pH optimum for the $^{14}CO_2$ and NH_3 release in the presence of oxaloacetate was 8.1 (Bergman and Ericson 1983). This indicates that different steps might be rate-limiting in the different assays. The rate of glycine-bicarbonate exchange decreased with increasing pH in a manner similar to the concentration of dissolved CO_2 (Bergman and Ericson 1983). This indicates that CO_2 and not bicarbonate is the substrate for the glycine decarboxylase.

5.6 Inhibition of Glycine Metabolism

Isonicotinyl hydrazine has often been used to inhibit the conversion of glycine to serine (Bird et al. 1972, Oliver 1979, Berlyn 1980). The glycine analogs aminoacetonitrile and glycine hydroxamate also act as inhibitors (Usuda et al. 1980, Lawyer and Zelitch 1979). KCN has been shown to inhibit the reaction (Moore et al. 1980).

Most of the inhibitors affect both glycine decarboxylase and serine hydroxy-methyltransferase, but the cyanide inhibition was mainly on serine hydroxy-methyltransferase. A lower concentration of aminoacetonitrile was needed to inhibit glycine-bicarbonate exchange than serine hydroxymethyltransferase (Gardeström et al. 1981b). Studies by Walker et al. (1982a) indicate that aminoacetonitrile and glycine hydroxamate affect glycine transport into the mitochondria.

5.7 Reoxidation of NADH

When NADH is reoxidized via the respiratory chain, ATP can be produced in the mitochondria. Bird et al. (1972) obtained a ratio of 2 ATP per NADH oxidized, but more recent studies have shown that, in spinach mitochondria, 3 ATP can be produced per NADH oxidized (Douce et al. 1977, Moore et al. 1977). Of particular interest is recent evidence that the alternative pathway functions, in addition to the cytochrome pathway, in the light, where the level of free sugars and photorespiratory intermediates are high (Aczon-Bieto et al. 1983). After a dark period when the level of free sugars is low, there is little or no flow through the alternative pathway. This is consistent with the idea that the alternative pathway may serve as an "overflow" mechanism and could allow flexibility in glycine oxidation without tight coupling to the cytochrome pathway or shuttling of reductive power from the mitochondria. Oliver (1984), using pea leaf protoplasts, suggested that in vivo NADH produced from glycine oxidation was predominantly consumed by the mitochondrial electron transport chain with O_2 as the terminal electron acceptor.

It has been speculated that the NADH formed in the mitochondria from glycine decarboxylation can be shuttled out to the peroxisomes via a malate-oxaloacetate shuttle (Woo and Osmond 1976, Day and Wiskich 1981, Chen and Heldt 1984) or a malate-glutamate/aspartate-2-oxoglutarate shuttle (Journet et al. 1981). In the peroxisomes, NADH is required for hydroxypyruvate reduction to glycerate. Studies with isolated peroxisomes indicate that the malate dehydrogenase and glutamate aminotransferase activity, coupled in the direction required for generating NADH in the peroxisomes, is insufficient to accommodate the shuttle. Also intact peroxisomes were unable to reduce hydroxypyruvate to glycerate when supplied with malate and glutamate in the absence of exogenous pyridine nucleotides, suggesting that a shuttle is inoperative. The reductive power might be provided by direct uptake of reduced pyridine nucleotides by the peroxisomes (Schmitt and Edwards 1983). Whether the reductive power for peroxisomes is shuttled out of the mitochondria or the chloroplasts remains uncertain. It is also possible that the NADH from glycine decarboxylation may be shuttled out of the mitochondria and used for nitrate reduction in the cytoplasm (Woo et al. 1980).

5.8 Regulation of Glycine Oxidation

Not much is known about the regulation of the conversion of glycine to serine in leaf mitochondria, due to the fragility of the isolated glycine decarboxylase complex. Glyoxylate was reported to inhibit CO_2 release from glycine in tobacco leaf mitochondria (Peterson 1982). However, this is likely not significant in vivo since glyoxylate is not believed to accumulate other than under severe lack of amino donors (Oliver 1981, Somerville and Ogren 1981) and the concentrations of glyoxylate used inhibit glutamate: glyoxylate aminotransferase at low glutamate concentrations (Walton and Butt 1981).

The question also arises how the NADH produced in the glycine decarboxylation interferes with the oxidation of other substrates via the respiratory chain. Day and Wiskich (1981) showed that the rate of glycine oxidation in isolated pea leaf mitochondria was not decreased by the addition of other substrates (malate, 2-oxoglutarate or succinate). When pea leaf mitochondria oxidizing malate were supplied with glycine, the rate of malate oxidation in State 3 was decreased (Walker et al. 1982b). These observations taken together suggest that glycine is oxidized via the respiratory chain in preference to other substrates. Recent studies on spinach leaf mitochondria (Bergman and Ericson 1983) and pea leaf mitochondria (Dry et al. 1983) have confirmed that glycine oxidation indeed has "first pick" for the respiratory chain compared to malate, succinate, and NADH.

Glycine oxidation by spinach leaf mitochondria inhibited the oxidation of external NADH, whereas oxidation of malate alone did not (Bergman and Ericson 1983). Dry et al. (1983) found malate oxidation by pea leaf mitochondria, in the presence of glutamate + TPP to ensure maximum rates of TCA cycle turnover, also inhibited external oxidation of NADH. Thus, when internal rates of NADH oxidation are high, oxidation of NADH externally may be limited by an unidentified mechanism. The NADH produced in the mitochondrial matrix from malate oxidation (either malic enzyme or malate dehydrogenase, see NEUBURGER and DOUCE 1980) or glycine oxidation are probably in equilibrium in the mitochondrial matrix.

WALKER et al. (1982b) found that pea leaf mitochondria in State 4 were able to maintain a constant rate of malate oxidation, with additional capacity to oxidize glycine (respiration in State 4 with malate was stimulated by the addition of glycine). The increase in State 4 respiration with glycine was less sensitive to rotenone and antimycin A than respiration with malate. Thus, this increased capacity for glycine oxidation in State 4 may be through a bypass of Site I and some engagement of the alternative pathway. The preferred oxidation of NADH produced during the oxidation of glycine could be a means of regulating the tricarboxylic acid cycle and insuring that the glycine decarboxylase step is not limiting the glycolate pathway in the light. Much more work is needed in order to elucidate the mechanism that discriminates between glycine and other substrates for the respiratory chain.

5.9 Ammonia Refixation [1]

The ammonia released by glycine decarboxylase must be effectively reassimilated, since chloroplasts are known to be uncoupled by low concentrations of ammonia (KROGMANN et al. 1959). Furthermore, as the metabolic flux in photorespiration is very high, the plant would not survive due to deficiency of organic nitrogen if ammonia was not refixed.

Different mechanisms for the refixation of ammonia have been considered. TOLBERT (1979) suggested that ammonia could be fixed by mitochondrial glutamate dehydrogenase. The low affinity of the enzyme for ammonia (LEA and THURMAN 1972) makes it a less likely candidate for ammonia fixation. Direct studies have also shown that the glutamate dehydrogenase mediated ammonia fixation is very low in isolated mitochondria (HARTMAN and EHMKE 1980, BERGMAN et al. 1981). KEYS et al. (1978) proposed that glutamine synthetase (GS) was the primary ammonia-fixing enzyme, and that it functioned in sequence with chloroplastic glutamate synthase (GOGAT).

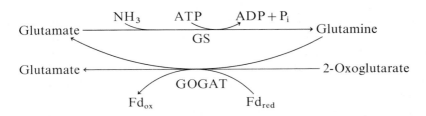

The importance of this system has been verified by SOMERVILLE and OGREN (1980) in studies using mutants deficient in GOGAT, and by ^{15}N-incorporation

---

1 See also Vol. 6 this Series, Chap. 5 (P.J. LEA and B.J. MIFLIN) pp. 445–456.

330   P. GARDESTRÖM and G.E. EDWARDS:

studies (Woo et al. 1982). Jackson et al. (1979b) proposed that sufficient activity of GS for reassimilation was associated with leaf mitochondria. Later studies have, however, found very low GS activity in leaf mitochondria (Keys et al. 1978, Wallsgrove et al. 1979, 1980, Bergman et al. 1981, Nishimura et al. 1982). The GS is rather located in the chloroplast and in the cytoplasm as two different isozymes (Mann et al. 1979, 1980, McNally et al. 1983). Depending on the relative distribution of the two isozymes, plants could be divided into four different groups (McNally et al. 1983). All plants except for achlorophyllous plant parasites had the chloroplast isozyme. In most C₃ plants, chloroplastic GS was the dominating isozyme. Some species (e.g., spinach) lacked the cytoplastic isozyme, whereas it was a minor component (less than 30%) in others (e.g., barley and pea). The fourth category, including several C₄ plants, had more than 30% of the GS activity in the cytoplasm. Woo and Osmond (1982) showed that isolated chloroplasts of spinach have the capacity to refix all the photorespiratory-released ammonia. This has also been directly shown using a recombinated system from spinach containing chloroplasts and mitochondria (Bergman et al. 1981).

NH₃ in the unprotonated form can pass through biological membranes by passive diffusion (Kleiner 1981). The transport of glutamine into chloroplasts is slow and inhibited by dicarboxylic acids (Barber and Thurman 1978). On the basis of the intracellular distribution of glutamine synthetase and permeability properties of NH₃ and glutamine, it appears that most of the NH₃ released in the mitochondria is reassimilated inside the chloroplasts.

6 Special Functions of Leaf Mitochondria – Role Relative to Decarboxylations in the C₄ Cycle

6.1 Introduction

As already noted, in C₄ and CAM plants[1] the level of photorespiration and the capacity of mitochondria from photosynthetic tissue to metabolize glycine is relatively low. However, mitochondria of certain C₄ and CAM species have important roles in photosynthetic carbon assimilation. In both C₄ and CAM plants, C₄ dicarboxylic acids are decarboxylated and CO₂ is donated to the photosynthetic carbon assimilation (PCA) cycle. In C₄ plants the CO₂ is fixed in the mesophyll cells through PEP carboxylase to oxaloacetate, which is then converted to aspartate and malate. The aspartate and malate are transported to bundle-sheath cells and decarboxylated to give CO₂, which is donated to the PCA cycle, and a 3-carbon compound which returns to the mesophyll cells. In CAM plants, the CO₂ is fixed at night through PEP carboxylase, with PEP being generated from breakdown of glucans. The oxaloacetate formed is converted to malate, which is stored in the vacuole. The following day the malate

---

[1] Refs. to C₄- and CAM metabolism in this Series: Vol. 12B, Chap. 15 (C.B. Osmond, K. Winter, H. Ziegler) pp. 479–547; Vol. 6, Chap. 6 (T.B. Ray and C.C. Black) pp. 77–10; Chap. 8 (M. Kluge) pp. 113–125; Chap. 9 (O. Queiroz) pp. 126–139; Vol. 3, Chap. 4 (M.D. Hatch and C.B. Osmond) pp. 144–184.

is decarboxylated to CO_2, which is donated to the PCA cycle, and a C_3 product which may be directly metabolized to glucans. There are three known decarboxylation mechanisms: NAD-malic enzyme, NADP-malic enzyme, and PEP carboxykinase. C_4 species are classified into three distinctive subgroups based on these differences. CAM species are currently divided into two biochemical subgroups; ME type (having NAD + NADP-malic enzyme as major decarboxylases), and PEP-CK type.

6.2 PEP Carboxykinase Types

In C_4 plants which have PEP carboxykinase as the primary decarboxylase, the enzyme is located in the cytoplasm in bundle-sheath cells, while in CAM plants which have PEP carboxykinase as the primary decarboxylase, the enzyme is located in the cytoplasm of the mesophyll cells (EDWARDS et al. 1982, EDWARDS and WALKER 1983). The enzyme requires 1 ATP per CO_2 released by decarboxylation of oxyloacetate. The released CO_2 is then refixed in the chloroplast through the PCA cycle. Therefore, ATP must be supplied to the enzyme at a rate equal to the rate of CO_2 assimilation. This ATP requirement can be met by mitochondrial respiration; alternatively the ATP may be shuttled out of the chloroplast through a triose-P/PGA exchange. If the ATP is generated by the mitochondria, then respiration would be required in the light to generate ATP at a rate equivalent to CO_2 fixation in the PCA cycle. Thus, in these species, it is very possible that a relatively high level of mitochondrial respiration is occurring in the light.

6.3 NADP-Malic Enzyme Types

In NADP-ME type species, NADP-malic enzyme is located in the bundle-sheath chloroplasts in C_4 plants and in the cytoplasm of mesophyll cells in CAM plants. There is no apparent role for the mitochondria in bundle-sheath cells of these C_4 species in relation to the C_4 cycle. In ME-type CAM plants, the NADPH generated in the cytoplasm via NADP-malic enzyme might either be oxidized by the mitochondria or utilized in gluconeogenesis in the cytoplasm during metabolism of pyruvate, the product of decarboxylation, to carbohydrate (EDWARDS et al. 1982). Mitochondria of Sedum praealtum, a ME-type species, readily oxidize NADPH with good respiratory control, similar to that with NADH as a substrate. Thus, it is possible that some of the NADPH produced by NADP-malic enzyme in the cytoplasm is oxidized by the mitochondria in vivo. Plant mitochondria from a number of tissues have been found to oxidize NADPH as well as NADH (ARRON and EDWARDS 1979).

6.4 NAD-Malic Enzyme Types

6.4.1 NAD-Malic Enzyme – A Mitochondrial Enzyme

All plant mitochondria examined have NAD-malic enzyme, and this is the basis for the complex malate oxidation in these mitochondria. In C_4 and CAM plants having NAD-malic enzyme as the major C_4 acid decarboxylase, the activity of the enzyme in leaf tissue is about 5- to 20-fold higher (on a chlorophyll or fresh weight basis) than the activity of the enzyme in leaf tissue of C_3 plants or other subgroups of C_4 and CAM plants. In ME type CAM plants, NAD-malic enzyme is located in the mitochondria, as demonstrated from fractionation of mesophyll protoplasts and organelle isolation (see Winter et al. 1982). In NAD-ME type C_4 plants, the enzyme is found in very low activity in mesophyll cells and in high activity in the bundle-sheath cells, where it is located in the mitochondria (see Edwards and Walker 1983, Gardeström and Edwards 1983).

Some enzymes, such as RuBP carboxylase, appear to be present exclusively in bundle-sheath chloroplasts and not in mesophyll chloroplasts of C_4 plants. However, this does not appear to be the case with respect to NAD-malic enzyme, since it is present in C_4 mesophyll preparations at levels, on a chlorophyll basis, similar to those of photosynthetic tissue in which the enzyme is not linked to C_4 photosynthesis (Gardeström and Edwards 1983). Thus, it appears that the synthesis of the enzyme in C_4 mesophyll cells is not repressed. Rather, the formation of the enzyme in bundle-sheath cells is greatly elevated (per unit cytochrome c oxidase activity) in comparison to its levels in other photosynthetic tissue (it is uncertain whether the mesophyll and bundle-sheath enzyme is the same isoenzyme). Therefore, the NAD-malic enzyme in the mitochondria of the C_4 mesophyll cell may play a role which is not linked to the C_4 cycle (i.e., allow respiration in the dark to proceed with malate as the major or only donor to the TCA cycle).

6.4.2 Properties of NAD-Malic Enzyme

NAD-malic enzyme is a complex enzyme which can display varying kinetic properties depending on the conditions of assay and source of the enzyme. For example, with the enzyme from the CAM plants *Crassula argentea* and *Kalanchoë daigremontiana*, a hysteretic lag is observed during assay (up to 30 to 40 min before reaching steady-state rate under some conditions; Wedding et al. 1981, Wedding 1982). The lag is reduced by NADH, Me^{2+}, sulfate ion, coenzyme A, and high malate levels, while chloride ion increases the lag. The lag which is seen under some conditions with the enzyme from the C_4 plant *Amaranthus retroflexus* was overcome by preincubating the enzyme with coenzyme A (Hatch et al. 1982). These lags may reflect a change in the oligomeric state of the enzyme during the assay period from a less active to a more active form. The enzyme from potato tubers has been found to exist as a dimer, a tetramer, and an octamer, with the dimer being least active (high K_m, low V_{max} for malate) and the tetramer and octomer being the most active forms (Grover and Wedding 1982). High ionic strength favors dissociation to the dimer, while high levels of malate, citrate, and NADH favor tetramer formation. Also, Me^{2+} and thiol reagents may be required to maintain the enzyme in the more aggregated state. In vivo, it is possible that conversions between oligomeric forms are a means of regulating the activity of the enzyme (e.g., high malate and/or Me^{2+} in vivo may favor a more active form of the enzyme).

Fig. 2. Mitochondrial metabolite transport and scheme for decarboxylation through NAD-malic enzyme in C_4 and CAM

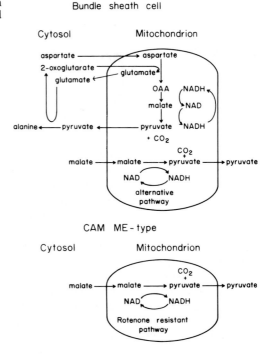

Whether coenzyme A, an activator of the enzyme which decreases the K_m malate, has its effect by changing the oligomeric state or by causing a conformational change is uncertain.

6.4.3 Function in NAD-ME Type C_4 Plants

In NAD-ME type C_4 plants, aspartate and malate are major initial products of CO_2 fixation, although aspartate is usually the predominant product. Both of these C_4 acids may be transported to bundle-sheath cells and metabolized through NAD-malic enzyme. The bundle-sheath mitochondria metabolize aspartate to pyruvate + CO_2 through the combined action of aspartate aminotransferase, NAD malate dehydrogenase and NAD-malic enzyme (Fig. 2). This results in recycling of the pyridine nucleotide, and is probably the major means through which decarboxylation occurs.

Malate may be transported to bundle-sheath cells and converted to aspartate in the cytoplasm, as illustrated in Fig. 3, and the aspartate then taken up by the mitochondria and decarboxylated. This would result in reducing power being generated in the cytoplasm which might be utilized in the reduction of PGA to glyceraldehyde 3-P. Part of the PGA formed in the bundle-sheath chloroplast through RuBP carboxylase may be exported and converted to

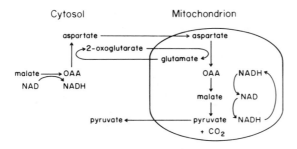

Fig. 3. Proposed pathway for malate as a donor through NAD-ME in C_4 or CAM with NADH being generated directly in the cytoplasm

triose-P either in the cytoplasm of the bundle-sheath cell or in the chloroplasts of the mesophyll cell.

Bundle-sheath mitochondria isolated from *Panicum miliaceum* oxidize malate via the alternative pathway, whereas the mesophyll cell mitochondria oxidize malate via the cytochrome linked pathway (Gardeström and Edwards 1983). Oxidation of external NADH by mitochondria of both cell types is linked to the cytochrome pathway. If malate is taken up directly by the mitochondria and oxidized through malic enzyme, the rate may be limited by the rate of oxidation of the NADH formed (Fig. 2). The oxidation of malate by the bundle-sheath mitochondria via the alternative pathway allows malate to be decarboxylated without being constrained by electron transport coupled to phosphorylation in the cytochrome pathway (Fig. 2). The maximum rate of malate oxidation by bundle-sheath cell mitochondria of NAD-ME type species is about 0.8 μmol mg^{-1} chl min^{-1}, whereas the bundle-sheath strands decarboxylate C_4 acids at a rate of 5 to 8 μmol mg^{-1} chl min^{-1} (Kagawa and Hatch 1975, Gardeström and Edwards 1983). Therefore, the major means of decarboxylation is via uptake and metabolism of aspartate by the mitochondria.

Bundle-sheath mitochondria of NAD-ME type C_4 species, like other plant mitochondria, can oxidize malate through malate dehydrogenase and/or malic enzyme, depending on conditions. At pH 7.2 in the presence of ADP (and arsenite to inhibit pyruvate dehydrogenase), respiration can be totally accounted for by pyruvate formation, indicating oxidation through malic enzyme (Table 2). However, with addition of glutamate and aspartate aminotransferase to remove the oxaloacetate as it is formed from malate via malate dehydrogenase, malic enzyme activity (as determined from pyruvate formation) accounted for only about 20% of the malate oxidation (Table 2). In vivo the bundle-sheath mitochondria have aspartate readily available from the carboxylation phase of the C_4 cycle. Therefore, the route of metabolism in the mitochondria in vivo will be from aspartate to oxaloacetate to malate, rather than in the opposite direction.

As shown in Table 2, there was no correlation between the degree of cyanide sensitivity and the oxidation of malate by NAD-malic enzyme (as indicated by pyruvate formation) or malate dehydrogenase by leaf mitochondria of *P. miliaceum*. This is in agreement with recent conclusions by Wiskich and Day (1982), using several mitochondrial sources and is not in accordance with the proposal by Rustin et al. (1980) that NADH produced by NAD-malic enzyme is preferably oxidized by the alternative pathway. In both bundle-sheath and

Table 2. Malate oxidation by mitochondria from mesophyll and bundle sheath cells of *Panicum miliaceum*. Oxygen uptake, pyruvate formation and KCN inhibition (GARDE-STRÖM and EDWARDS 1984). Numbers in parenthesis refer to % of O$_2$ uptake accounted for by pyruvate formation

Additions	O$_2$ uptake	Pyruvate formation	% Inhibition of O$_2$ uptake by 1 mM KCN
	nmol mg^{-1} protein min^{-1}		
Mesophyll mitochondria			
A	48	–	30
A + ADP	84	–	73
A + AAT + glutamate	35	18(26%)	59
A + ADP + AAT + glutamate	56	32(28%)	92
Bundle-sheath mitochondria			
A	76	106(70%)	18
A + ADP	70	138(99%)	38
A + AAT + glutamate	80	22(14%)	9
A + ADP + AAT + glutamate	75	32(21%)	32

A Mitochondria + respiration medium + 10 mM malate + 3 mM Na arsenite.
AAT Aspartate aminotransferase.
The concentration of ADP was 1 mM, glutamate 10 mM, and KCN 1 mM.

mesophyll mitochondria, the cyanide sensitivity was higher when ADP was added to the assay than in assay without ADP (Table 2). The addition of ADP allows coupled electron flow along the cytochrome chain and thus more activity through the cytochrome pathway. The results also indicate that with this NAD-ME type species the relative capacity of alternative pathway to cytochrome pathway is much higher in the bundle-sheath mitochondria compared to the mesophyll cell mitochondria.

During the decarboxylation phase in NAD-ME type C$_4$ species, the pyruvate formed in the mitochondria might either be metabolized in the TCA cycle or exported. Metabolism of pyruvate in the TCA cycle would be undesirable, as this would take carbon away from the C$_4$ cycle. As in CAM plant mitochondria, the metabolism of pyruvate may be limited by the capacity of pyruvate dehydrogenase or by a high adenylate energy charge. The rate of oxidation of externally supplied pyruvate by the bundle-sheath mitochondria of *P. miliaceum* is quite low (GARDESTRÖM and EDWARDS 1983), although this could be due to a transport limitation rather than limitation on pyruvate dehydrogenase. Therefore, how the export of pyruvate from the mitochondria occurs needs to be examined.

6.4.4 Function in ME Type CAM Plants

Among CAM plants, studies have been made on the respiratory properties of mitochondria of ME type species (Sect. 3.2.2) but not on those of PEP-CK type species. In mitochondria from ME type CAM plants, malate is decarboxylated at rates sufficient to accomodate the rate of metabolism of malate during deacidification in vivo (Day 1980, Spalding et al. 1980). Therefore, the capacity for mitochondrial decarboxylation of malate is high enough to contribute significantly to malate decarboxylation in vivo during crassulacean acid metabolism. The highest rates of malate decarboxylation by the isolated mitochondria of the NAD-ME type CAM plants occur in the presence of ADP, an uncoupler of oxidative phosphorylation (FCCP), in the presence of oxaloacetate, or with addition of aspartate + 2-oxoglutarate (Day 1980, Spalding et al. 1980). Each of these factors apparently increases malate decarboxylation by enhancing the oxidation of NADH, and thus preventing NAD from becoming limiting. Oxaloacetate increases the activity by allowing reoxidation of NADH through the mitochondrial malate dehydrogenase. Likewise, the effect of aspartate + 2-oxoglutarate is through uptake and conversion to oxaloacetate and glutamate via aspartate aminotransferase, and then the reoxidation of NADH via conversion of oxaloacetate to malate through malate dehydrogenase (Fig. 3).

There are two ways in which the NADH generated from NAD-malic enzyme might be utilized in the decarboxylation phase of CAM. When pyruvate, the product of malate decarboxylation, is metabolized through gluconeogenesis, the first step is conversion to PEP through pyruvate, P_i dikinase in the chloroplast. Based on enzyme compartmentation studies, the PEP is then proposed to be exported and converted to 3PGA in the cytoplasm, where it may be further converted to glyceraldehyde-3-P requiring ATP and NADH (Edwards et al. 1982). Therefore, it is conceivable that some of the NADH generated from NAD-malic enzyme in the mitochondria is shuttled to the cytoplasm by an aspartate + 2-oxoglutarate/glutamate + malate shuttle or an oxaloacetate/malate shuttle. Secondly, oxidation of part of the NADH by the cytochrome chain could provide ATP for the PGA kinase reaction. If conditions should exist where such demands for NADH are limited relative to the demand for CO_2 from decarboxylation, then alternative means of oxidizing NADH would be desirable. A distinctive feature of the mitochondria of ME type CAM plants is their strong resistance of malate oxidation to rotenone, a Site I electron transport inhibitor (Arron et al. 1979a, Day 1980). Thus, the rotenone bypass may be the major route of electron flow during malate oxidation, which would allow two, rather than three, coupling sites for each NADH oxidized (Fig. 2). Therefore, the oxidation of the NADH generated through NAD-malic enzyme in the mitochondria of CAM leaves would appear less tightly controlled than in mitochondria of C_3 plants. In addition, the mitochondria from ME type CAM leaves show some capacity for the alternative pathway, in that malate oxidation is only partially inhibited by antimycin A, a cytochrome pathway inhibitor, whereas antimycin A + SHAM (salicylhydroxamic acid) gives total inhibition. On the other hand, oxidation of reduced pyridine nucleotides added externally was completely inhibited by antimycin A, indicative of total dependence on the cytochrome pathway (Arron et al. 1979a).

In the inducible CAM plant *Mesembryanthemum crystallinum*, it has been reported that in the CAM mode the RC values and the ADP/O ratios for malate, succinate, and 2-oxoglutarate decreased when the plants were shifted from the C_3 to the CAM mode (Von Willert and Schwöbel 1978). Also, the mitochondria from leaves of plants in the CAM mode appeared to have a slightly higher level of alternative pathway (about 30% of respiration was cyanide insensitive) compared to the mitochondria from the plants in the C_3 mode (about 15% cyanide-insensitive). This suggests that the CAM mitochondria may have some differences in the mitochondrial electron transport systems which allow malate oxidation to proceed with a lower degree of coupling and dependence on phosphorylation. Again, this may remove some constraints on oxidation of NADH and be advantageous in CAM during malate decarboxylation. These differences were not so apparent in another study on these mitochondria from *M. crystallinum* (WINTER, ARRON, EDWARDS, unpublished data). In general, the results with mitochondria from ME-type-CAM plants suggest the NADH generated by malic enzyme may be reoxidized by the rotenone-resistant pathway or by the alternative pathway, allowing decarboxylation to proceed with a lower degree of coupling of electron transport to phosphorylation.

Whether the pyruvate generated from NAD-malic enzyme in the CAM leaf mitochondria is metabolized in the TCA cycle or exported and converted to starch through gluconeogenesis needs to be considered. DAY (1980) found that significant levels of the pyruvate formed from malate decarboxylation could be metabolized via pyruvate dehydrogenase if thiamine pyrophosphate is added as a cofactor. Therefore, in vivo the pyruvate dehydrogenase activity and the restrictions on electron transport by ATP/ADP levels might be the controlling factors. If all of the pyruvate is metabolized to starch through gluconeogenesis, then the cost of converting one malate to glucan is 6.7 ATP and 2 NADPH (3 ATP per pyruvate converted to triose P, 3 ATP and 2 NADPH per CO_2 fixed to the level of triose P, and 0.7 ATP per 1.3 triose P converted to glucan) (EDWARDS et al. 1982). Alternatively, if the malate were completely decarboxylated through the TCA cycle the equivalent of 5 NADH and 3 ATP would be generated (taking 1 FADH and 1 GTP generated in the cycle as equivalent to 3 ATP). The energy required to assimilate the 4 CO_2 to glucan is 12.7 ATP and 8 NADPH (3 ATP and 2 NADPH per CO_2 fixed to triose P and 0.5 ATP per triose P converted to glucan). The net requirement for converting the malate to glucan would then be 9.7 ATP and 3 NADPH. Therefore, the complete oxidation of malate and reassimilation of the CO_2 to glucan would cost about 50% more in energy than if the pyruvate, formed from malate decarboxylation, were directly converted to glucan. Obviously, if the plant utilizes the most energy-efficient route, pyruvate will be directly metabolized through gluconeogenesis to glucans. There is some evidence that this might be the case. HOLTUM and OSMOND (1981) found that ^{14}C pyruvate fed to leaf slices of ME type CAM plants was primarily converted to glucans, with only a minor portion entering the TCA cycle. This suggests the pyruvate generated from decarboxylation through NADP-ME in the cytoplasm would be largely metabolized through gluconeogenesis. However, it does not provide direct insight into the fate of pyruvate generated in the mitochondrial oxidation of malate through NAD-malic enzyme.

In CAM plants malate is the primary C_4 acid which donates carbon to the PCA cycle. Whether the malate is oxidized through NAD-malic enzyme or malate dehydrogenase may be controlled, in part, by pH. In mitochondria of *Kalanchoë daigremontiana*, a ME type CAM plant, malate oxidation occurs over a broad pH range (Day 1980). At low pH, oxidation of malate is primarily through malic enzyme. Above pH 7.2, the oxidation through malic enzyme drops dramatically and malate dehydrogenase is the primary means of malate oxidation. The reason why oxidation through malic enzyme is inhibited at high pH is not understood. However, bicarbonate is a known inhibitor of NAD-malic enzyme and of malate decarboxylation by mitochondria from both CAM and C_4 plants (Chapman and Hatch 1977, Spalding et al. 1980); and at high pH equilibration of CO_2 to bicarbonate is favored. Bicarbonate inhibits malate decarboxylation by isolated mitochondria of *Sedum praealtum* at pH 7.2 (Spalding et al. 1980). At pH 7.2, mitochondria isolated from *Mesembryanthemum crystallinum* in the CAM mode have high levels of NAD-malic enzyme and oxidize malate to pyruvate and CO_2, while mitochondria isolated from plants in the C_3 mode have low levels of malic enzyme and oxidize malate primarily to oxaloacetate. At higher pH, i.e., pH 8.0, NAD-malate dehydrogenase from both sources of mitochondria oxidizes malate to oxaloacetate (Winter, Arron, Edwards, unpublished). These data suggest that the pH of the cytoplasm may control malate decarboxylation through NAD-malic enzyme. When malic acid is released from the vacuole in the light, a lowering of cytoplasmic pH may occur which would favor malate decarboxylation through NAD-malic enzyme. Also, as the malate is released from the vacuole, a high concentration could become available to NAD-ME, favoring the aggregation to a higher molecular weight and more active form (see earlier discussion; Sect. 6.4.2).

6.5 Transport in Mitochondria Relative to C_4 and CAM Photosynthesis

In mitochondria from photosynthetic tissue of ME type CAM plants, malate is considered the major dicarboxylic acid transported. P_i is required for high rates of malate oxidation, and butylmalonate inhibits malate oxidation, which suggests that malate is taken up on the dicarboxylate carrier in exchange for P_i (Day 1980).

Whereas in CAM mitochondria the decarboxylation is probably linked to malate uptake as the major route, in C_4 bundle-sheath mitochondria having high NAD-malic enzyme, aspartate is considered the primary dicarboxylic acid which is transported. High rates of aspartate decarboxylation occur in these bundle-sheath mitochondria when, in addition to aspartate (i.e., ^{14}C-labeled to follow decarboxylation), malate, P_i, and 2-oxoglutarate are added (Kagawa and Hatch 1975). Metabolism of aspartate to pyruvate and CO_2 in the mitochondria requires uptake of aspartate and 2-oxoglutarate and export of glutamate and pyruvate. Malate and P_i serve as catalysts for the decarboxylation of aspartate. P_i, which could be taken up on the phosphate translocator in exchange for OH, may be required to exchange for malate entry on the dicarboxylate carrier. This would increase the malate pool in the mitochondria and may increase the NADH/NAD ratio through malic enzyme activity making the conversion of aspartate to malate, which requires NADH, more favorable. Alternatively, the malate taken up by the mitochondria could facilitate 2-oxoglutarate uptake via exchange on a malate/2-oxoglutarate carrier.

In C_4 plant mitochondria having high NAD-malic enzyme, pyruvate must be exported to the cytoplasm for the C_4 cycle to continue to function. In mitochondria of CAM plants having high NAD-malic enzyme, it is also likely that considerable export of pyruvate from the mitochondria occurs. Plant and animal mitochondria have a pyruvate translocator which catalyzes the uptake of pyruvate in exchange for OH. Evidence from such a carrier was also obtained by DAY (1980) in mitochondria for *K. daigremontiana* (CAM ME type) based on inhibition of the oxidation of externally supplied pyruvate by cyanocinnamic acid, a known inhibitor of the monocarboxylate carrier which facilitates pyruvate uptake. Whether a similar carrier catalyzes the export of pyruvate from the mitochondria following decarboxylation of malate in the CAM and C_4 ME types is not known.

In the mitochondria of both CAM and C_4 NAD-ME plants, high capacities of certain translocators are required to accommodate the rates of exchange of metabolites needed in photosynthesis. The mechanisms of transport of metabolites in these mitochondria remain to be studied.

6.6 Abundance and Ultrastructure of Mitochondria Relative to C_4 Photosynthesis

The mesophyll cells of C_4 plants and the mesophyll cells of C_3 plants are similar in the relative abundance of mitochondria versus chloroplasts (Table 3). In contrast, NAD-ME type C_4 species have a high ratio of mitochondria/chloroplast in bundle-sheath cells, the mitochondria are relatively large, and have well-developed cristae (HATCH et al. 1975, Table 3). Also, the ratio of area of mitochondria to area of chloroplasts in bundle-sheath cells, as determined from electron micrographs, is highest in NAD-ME type species, intermediate in PEP-

Table 3. Ratio of mitochondria to chloroplasts in mesophyll cells of C_3 plants and mesophyll and bundle-sheath cells of C_4 plants

Subgroup	Species	Mesophyll cell	Bundle-sheath cell
C_3	*Triticum aestivum*[a]	0.65	–
	Avena sativa[a]	0.54	–
	Panicum rivulare[b]	0.78	–
	Panicum hylaeicum[b]	0.68	–
	Panicum laxum[b]	0.75	–
C_4-NADP-ME	*Zea mays*[a]	0.57	1.3
	Sorghum sudanense[a]	0.68	1.3
	Panicum prionitis[b]	0.56	0.73
C_4PEP-CK	*Chloris gayana*[a]	0.44	3.2
C_4-NAD-ME	*Panicum miliaceum*[c]	–	4.2

[a] NEWCOMB and FREDERICK 1971
[b] BROWN et al. 1983
[c] Electron micrographs provided by V.E. GRACEN

CK-type species and lowest in NADP-ME type species (Hatch et al. 1975). In addition in the NAD-ME species, *P. miliaceum*, the amount of mitochondrial protein per unit Chl is about two-fold higher in isolated bundle-sheath mitochondria than in mesophyll mitochondria; and the activity of NAD-malic enzyme per unit cyt c oxidase was much higher in bundle-sheath than in mesophyll mitochondria (Gardeström and Edwards 1983). These results are consistent with the bundle-sheath mitochondria of NAD-ME type species being immanently involved in the C_4 cycle through NAD-malic enzyme. Metabolites, namely C_4 acids (aspartate + malate) and pyruvate, must be transported across the inner mitochondrial membrane at rates equal to the rate of CO_2 fixation in the PCA cycle in bundle-sheath cells. These rates should be at least an order of magnitude higher than mitochondrial metabolite transport during dark respiration (taking dark respiration to be about 10% of photosynthesis). These high rates of metabolite transport in the bundle-sheath mitochondria of NAD-ME types may be accommodated by the relatively large number of mitochondria in bundle-sheath cells, their large size, and their extensive development of cristae. Also bundle-sheath mitochondria isolated from *Panicum miliaceum* have a ratio of protein to cyt c oxidase about twice as high as in mesophyll mitochondria (Gardeström and Edwards 1983). This indicates that bundle-sheath mitochondria have more protein for other functions (i.e., metabolite transport) than respiration via the cytochrome chain. In NADP-ME species the size and number of mitochondria and degree of cristae development in bundle-sheath cells are more like those of C_3 plants and of C_4 mesophyll cells (Table 3; Hatch et al. 1975). The bundle-sheath mitochondria of NADP-ME species are not known to have any functions relative to the C_4 cycle. In the PEP-CK type C_4 plants examined, the numbers of mitochondria/chloroplast in bundle-sheath cells and the development of cristae are intermediate to those of the NAD-ME-type species and NADP-ME-type species (Table 3; Hatch et al. 1975). This suggests that the mitochondria in bundle-sheath cells have a role in C_4 photosynthesis, which, as previously noted, may be to provide ATP for PEP-carboxykinase. The rate of uptake of metabolites by the mitochondria would depend on whether the ATP was provided by the complete functioning of the TCA cycle or by decarboxylation of malate through NAD-malic enzyme with return of pyruvate to the mesophyll cells. These species do have NAD-malic enzyme in bundle-sheath cells at activities above those found in C_3 plants (Hatch et al. 1982). Therefore, malate oxidation through NAD-malic enzyme and oxidation of the NADH by the cytochrome chain could provide the ATP for PEP carboxykinase.

The relative abundance and ultrastructure of mitochondria in ME type and PEP-CK type CAM plants, in comparison to C_3 plants, have not been studied. This would be of considerable interest in evaluating the extent mitochondria of these CAM species may be associated with photosynthetic metabolism of malate in the light.

Acknowledgement. We are grateful to colleagues for making material available prior to publication. Support for the authors on work discussed in this review was provided by the Science and Educational Administration of the USDA through the Competitive Research Grant Office, the National Science Foundation, and The Swedish National Science Research Council.

References

Albertsson P-Å, Andersson B, Larsson C, Åkerlund H-E (1982) Phase partition – a method for purification and analysis of cell organelles and membrane vesicles. In: Glick D (ed) Methods of biochemical analysis, vol 28. John Wiley, New York, pp 115–150

Arron GP, Edwards GE (1979) Oxidation of reduced nicotinamide adenine dinucleotide phosphate by plant mitochondria. Can J Biochem 57:1392–1399

Arron GP, Edwards GE (1980a) Light-induced development of glycine oxidation by mitochondria from sunflower cotyledons. Plant Sci Lett 18:229–235

Arron GP, Edwards GE (1980) Uptake and oxidation of glycine by mitochondria isolated from sunflower cotyledons. Plant Physiol S 65:65

Arron GP, Spalding MH, Edwards GE (1979a) Isolation and oxidative properties of intact mitochondria from the leaves of Sedum praealtum. A crassulacean acid metabolism plant. Plant Physiol 64:182–186

Arron GP, Spalding MH, Edwards GE (1979b) Stoichiometry of carbon dioxide release and oxygen uptake during glycine oxidation in mitochondria isolated from spinach (Spinacia oleracea). Biochem J 184:457–460

Azcon-Bieto J, Osmond CB (1983) Relationship between photosynthesis and respiration. The effect of carbohydrate status on the rate of CO_2 production by respiration in darkened and illuminated wheat leaves. Plant Physiol 71:574–581

Azcon-Bieto J, Lambers H, Day DA (1983) The effect of photosynthesis and carbohydrate status on respiratory rates and the involvement of the alternative pathway in leaf respiration. Plant Physiol 72:598–603

Barber DJ, Thurman DA (1978) Transport of glutamine into isolated pea chloroplasts. Plant Cell Environ 1:297–303

Bergman A, Ericson I (1983) Effects of pH, NADH, succinate, and malate on the oxidation of glycine in spinach leaf mitochondria. Physiol Plant 59:421–427

Bergman A, Gardeström P, Ericson I (1980) Method to obtain a chlorophyll-free preparation of intact mitochondria from spinach leaves. Plant Physiol 66:442–445

Bergman A, Gardeström P, Ericson I (1981) Release and refixation of ammonia during photorespiration. Physiol Plant 53:528–532

Berlyn MB (1980) Isolation and characterization of isonicotinic acid hydrazide-resistant mutants of Nicotiana tabacum. Theor Appl Genet 58:19–26

Bird IF, Cornelius MJ, Keys AF, Whittingham CP (1972) Oxidation and phosphorylation associated with the conversion of glycine to serine. Phytochemistry 11:1587–1594

Brown RH, Bouton JH, Rigsby L, Rigler M (1983) Photosynthesis of grass species differing in carbon dioxide fixation pathways. VII. Ultrastructural characteristics of species in the Laxa group. Plant Physiol 71:425–431

Cavalieri AJ, Huang AHC (1980) Carrier protein-mediated transport of neutral amino acids into mung bean mitochondria. Plant Physiol 66:588–591

Chapman EA, Graham D (1974) The effect of light on the tricarboxylic acid cycle in green leaves. I. Relative rates of the cycle in the dark and the light. Plant Physiol 53:879–885

Chapman KSR, Hatch MD (1977) Regulation of mitochondrial NAD-malic enzyme involved in C_4 pathway photosynthesis. Arch Biochem Biophys 184:298–306

Chen J, Heldt HW (1984) Oxaloacetate translocator in plant mitochondria. Proc 6th Int Congr Photosynth, Brussels Vol III, pp 513–516

Cossins EA (1980) One-carbon metabolism. In: Davies DD (ed) The biochemistry of plants, vol II. Academic Press, London New York, pp 365–418

Day DA (1980) Malate decarboxylation by Kalanchoë daigremontiana mitochondria and its role in crassulacean acid metabolism. Plant Physiol 65:675–679

Day DA, Wiskich JT (1980) Glycine transport by pea leaf mitochondria. FEBS Lett 112:191–194

Day DA, Wiskich JT (1981) Glycine metabolism and oxaloacetate transport by pea leaf mitochondria. Plant Physiol 68:425–429

Day DA, Arron GP, Laties GG (1980) Nature and control of respiratory pathways in plants: The interaction of cyanide-resistant respiration with the cyanide-sensitive pathway. In: Davies DD (ed) The biochemistry of plants, a comprehensive treatise, vol II. Academic Press, London New York, pp 197–241

Dench JR, Briand Y, Jackson C, Hall DO, Moore AL (1978) Glycine oxidation in spinach leaf mitochondria. In: Ducet G, Lance C (eds) Plant mitochondria. Elsevier/North-Holland, Biomedical Press, Amsterdam New York, pp 133–140

Douce R, Christensen EL, Bonner WD (1972) Preparation of intact plant mitochondria. Biochim Biophys Acta 275:148–160

Douce R, Mannella CA, Bonner WD (1973) The external NADH dehydrogenases of intact plant mitochondria. Biochim Biophys Acta 292:105–116

Douce R, Moore AL, Neuburger M (1977) Isolation and oxidative properties of intact mitochondria isolated from spinach leaves. Plant Physiol 60:625–628

Dry IB, Wiskich JT (1982) Role of external adenosine triphosphate/adenosine diphosphate ratio in the control of plant mitochondrial respiration. Arch Biochem Biophys 217:72–79

Dry IB, Day DA, Wiskich JT (1983) Preferential oxidation of glycine by the respiratory chain of pea leaf mitochondria. FEBS Lett 158:154–158

Edman K-A, Ericson I (1984) Polar lipids in spinach leaf mitochondria. Proc 6th Int Congr Photosynth, Brussels, Vol III, pp 167–170

Edwards GE, Walker DA (1983) C_3, C_4: Mechanisms, cellular and environmental regulation of photosynthesis. Blackwell, Oxford

Edwards GE, Foster JG, Winter K (1982) Activity and intracellular compartmentation of enzymes of carbon metabolism in CAM plants. In: Ting IP, Gibbs M (eds) Crassulacean acid metabolism. Am Soc Plant Physiol. Waverly Press, Maryland, pp 92–111

Ericson I, Sahlström S, Bergman A, Gardeström P (1984) The glycine-decarboxylating system in spinach leaf mitochondria. Proc 6th Int Congr Photosynth, Brussels Vol III, pp 887–890

Fuchs R, Haas R, Wrage K, Heinz E (1981) Phospholipid composition of chlorophyll-free mitochondria isolated via protoplasts from oat mesophyll cells. Hoppe-Seylers Z Physiol Chem 362:1069–1078

Gardeström P, Edwards GE (1983) Isolation of mitochondria from leaf tissue of *Panicum miliaceum*, a NAD-malic enzyme type C_4 plant. Plant Physiol 71:24–29

Gardeström P, Edwards GE (1984) The regulation of electron flow between cytochrome oxidase and the alternative oxidase in mitochondria from *Panicum miliaceum*, a NAD-malic enzyme type C_4 plant. Proc 6th Int Congr Photosynth, Brussels Vol III, pp 641–644

Gardeström P, Ericson I, Larsson C (1978) Preparation of mitochondria from green leaves of spinach by differential centrifugation and phase partition. Plant Sci Lett 13:231–239

Gardeström P, Bergman A, Ericson I (1980) Oxidation of glycine via the respiratory chain in mitochondria prepared from different parts of spinach. Plant Physiol 65:389–391

Gardeström P, Bergman A, Ericson I, Sahlström S (1981a) Cytochromes of spinach mitochondria. In: Akoyunoglou G (ed) Photosynthesis, vol II. Balaban Int Sci Serv, Philadelphia, pp 633–640

Gardeström P, Bergman A, Ericson I (1981b) Inhibition of the conversion of glycine to serine in spinach leaf mitochondria. Physiol Plant 53:439–444

Gardeström P, Bergman A, Sahlström S, Edman K-A, Ericson I (1983) A comparison of the membrane composition of mitochondria isolated from spinach leaves and leaf petioles. Plant Sci Lett 31:173–180

Graham D (1980) Effects of light on "dark" respiration. In: Davies DD (ed) The biochemistry of plants, a comprehensive treatise, vol II. Academic Press, London New York, pp 525–579

Graham D, Chapman EA (1979) Interaction between photosynthesis and respiration in higher plants. In: Gibbs M, Latzko E (eds) The biochemistry of plants, a comprehensive treatise, vol VI. Academic Press, London New York, pp 150–162

Gronebaum-Turck K, Willenbrink J (1971) Isolierung und Eigenschaften von Mitochondrien aus Blättern von *Spineacia oleracea* und *Beta vulgaris*. Planta 100:337–346

Grover SD, Wedding RT (1982) Kinetic ramifications of the association–dissociation behavior of NAD malic enzyme. A possible regulatory mechanism. Plant Physiol 70:1169–1172

Hanson JB, Day DA (1980) Plant mitochondria. In: Tolbert NE (ed) The biochemistry of plants, a comprehensive treatise, vol I. Academic Press, London New York, pp 315–358

Hartman T, Ehmke A (1980) Role of mitochondrial glutamate dehydrogenase in the reassimilation of ammonia produced by glycine serine transformation. Planta 149:207–208

Hatch MD, Kagawa T, Craig S (1975) Subdivision of C_4-pathway species based on differing C_4 acid decarboxylating systems and ultrastructural features. Aust J Plant Physiol 2:111–128

Hatch MD, Tsuzuki M, Edwards GE (1982) Determination of NAD malic enzyme in leaves of C_4 plants: Effects of malate dehydrogenase and other factors. Plant Physiol 69:483–491

Holtum JAM, Osmond CB (1981) The gluconeogenic metabolism of pyruvate during deacidification in plants with crassulacean acid metabolism. Aust J Plant Physiol 8:31–44

Jackson C, Dench JE, Hall DO, Moore AL (1979a) Separation of mitochondria from contaminating subcellular structures using silica sol gradient centrifugation. Plant Physiol 64:150–153

Jackson C, Dench JE, Morris P, Lui SC, Hall DO, Moore AL (1979b) Photorespiratory nitrogen cycling: Evidence for a mitochondrial glutamine synthetase. Biochem Soc Trans 7:1122–1124

Journet E-P, Neuburger M, Douce R (1981) Role of glutamate-oxaloacetate transaminase and malate dehydrogenase in the regeneration of NAD^+ for glycine oxidation by spinach leaf mitochondria. Plant Physiol 67:467–469

Kagawa T, Hatch MD (1974) NAD malic enzyme in leaves with C_4-pathway photosynthesis. Its role in C_4 acid decarboxylation. Arch Biochem Biophys 160:346–349

Kagawa T, Hatch MD (1975) Mitochondria as a site of C_4 acid decarboxylation in C_4-pathway photosynthesis. Arch Biochem Biophys 167:687–696

Keys AJ, Bird IF, Cornelius MJ, Lea PJ, Wallsgrove RM, Miflin BJ (1978) Photorespiratory nitrogen cycle. Nature (London) 275:741–743

Kikuchi G, Hiraga K (1982) The mitochondrial glycine cleavage system. Mol Cell Biochem 45:137–149

Kisaki T, Imai A, Tolbert NE (1971a) Intracellular localization of enzymes related to photorespiration in green leaves. Plant Cell Physiol 12:267–273

Kisaki T, Yoshida N, Imai A (1971b) Glycine decarboxylase and serine formation in spinach leaf mitochondrial preparation with reference to photorespiration. Plant Cell Physiol 12:275–288

Kleiner D (1981) The transport of NH_3 and NH_4^+ across biological membranes. Biochim Biophys Acta 639:41–52

Krogmann DW, Jagendorf AT, Avron M (1959) Uncouplers of spinach chloroplast photosynthetic phosphorylation. Plant Physiol 34:272–277

Lawyer AL, Zelitch I (1979) Inhibition of glycine decarboxylation and serine formation in tobacco by glycine hydroxamate and its effect on photorespiratory carbon flow. Plant Physiol 64:706–711

Lea PJ, Thurman DA (1972) Intracellular location and properties of plant L-glutamate dehydrogenases. J Exp Bot 23:440–449

Lilley RMcC, Stitt M, Mader G, Heldt HW (1982) Rapid fractionation of wheat leaf protoplasts using membrane filtration. The determination of metabolite levels in the chloroplasts, cytosol, and mitochondria. Plant Physiol 70:965–970

Lorimer GH, Andrews TJ (1981) The C_2 chemo- and photorespiratory carbon oxidation cycle. In: Hatch MD, Boardman NK (eds) The biochemistry of plants, a comprehensive treatise, vol VIII. Photosynthesis. Academic Press London New York, pp 329–374

Mann AF, Fentem PA, Stewart GR (1979) Identification of two forms of glutamine synthetase in barley (*Hordeum vulgare*). Biochem Biophys Res Commun 88:515–521

Mann AF, Fentem PA, Stewart GR (1980) Tissue localization of barley (*Hordeum vulgare*) glutamine synthetase isoenzymes. FEBS Lett 110:265–267

McNally SF, Hirel B, Gadal P, Maun AF, Stewart GR (1983) Glutamine synthetase of higher plants. Evidence for a specific isoform content related to their possible physiological role and their compartmentation within the leaf. Plant Physiol 72:22–25

Moore AL, Jackson C, Halliwell B, Dench JE, Hall DO (1977) Intramitochondrial localization of glycine decarboxylase in spinach leaves. Biochem Biophys Res Commun 78:483–491

Moore AL, Dench JE, Jackson C, Hall DO (1980) Glycine decarboxylase activity in plant tissues meaasured by a rapid assay technique. FEBS Lett 115:54–58

Motokawa Y, Kikuchi G (1971) Glycine metabolism in rat liver mitochondria. V. Intramitochondrial localization of the reversible glycine cleavage system and serine hydroxymethyltransferase. Arch Biochem Biophys 146:461–466

Motokawa Y, Kikuchi G (1974) Glycine metabolism by rat liver mitochondria. Reconstitution of the reversible glycine cleavage system with partially purified protein components. Arch Biochem Biophys 164:624–633

Nash D, Wiskich JT (1983) Properties of substantially chlorophyll-free pea leaf mitochondria prepared by sucrose density gradient separation. Plant Physiol 71:627–634

Neuburger M, Douce R (1977) Oxydation du malate, du NADH et de la glycine par les mitochondries de plantes en C$_3$ et C$_4$. CR Acad Sci Ser D 285:881–884

Neuburger M, Douce R (1980) Effect of bicarbonate and oxaloacetate on malate oxidation by spinach leaf mitochondria. Biochim Biophys Acta 589:176–189

Neuburger M, Journet E-P, Bligny R, Carde J-P, Douce R (1982) Purification of plant mitochondria by isopynic centrifugation in density gradients of Percoll. Arch Biochem Biophys 217:312–323

Newcomb EH, Frederick SE (1971) Distribution and structure of plant microbodies (peroxisomes). In: Hatch MD, Osmond CB, Slatyer RO (eds) Photosynthesis and photorespiration. Wiley Interscience, New York, pp 442–457

Nishimura M, Graham D, Akazawa T (1976) Isolation of intact chloroplasts and other cell organelles from spinach leaf protoplasts. Plant Physiol 58:309–314

Nishimura M, Douce R, Akazawa T (1982) Isolation and characterization of metabolically competent mitochondria from spinach leaf protoplasts. Plant Physiol 69:916–920

Ohmura T (1955) Oxidative phosphorylation by a particulate fraction from green leaves. Arch Biochem Biophys 57:187–194

Oliver DJ (1979) Mechanism of decarboxylation of glycine and glyoxylate by isolated soybean cells. Plant Physiol 64:1048–1052

Oliver DJ (1981) Role of glycine and glyoxylate decarboxylation in photorespiratory CO$_2$ release. Plant Physiol 68:1031–1034

Oliver DJ (1984) Evidence for the involvement of the mitochondrial electron transport chain in photorespiratory glycine oxidation in vivo. Proc 6th Int Congr Photosynth, Brussels Vol III, pp 855–858

Palmer JM (1976) The organization and regulation of electron transport in plant mitochondria. Annu Rev Plant Physiol 27:133–157

Palmer JM (1979) The "uniqueness" of plant mitochondria. Biochem Soc Trans 7:246–252

Peterson RB (1982) Regulation of glycine decarboxylase and L-serine hydroxymethyltransferase activities by glyoxylate in tobacco leaf mitochondrial preparations. Plant Physiol 70:61–66

Pierpoint WS (1959) Mitochondrial preparations from the leaves of tobacco (*Nicotiana tabacum*). Biochem J 71:518–528

Pierpoint WS (1962) Mitochondrial preparations from the leaves of tobacco (*Nicotiana tabacum*). Biochem J 82:143–148

Proudlove MO, Moore AL (1982) Movement of amino acids into isolated plant mitochondria. FEBS Lett 147:26–30

Rao DN, Rao NA (1982) Purification and regulatory properties of mung bean (*Vigna radiata* L.) serine hydroxymethyltransferase. Plant Physiol 69:11–18

Rustin P, Moreau F, Lance C (1980) Malate oxidation in plant mitochondria via malic enzyme and the cyanide-insensitive electron transport pathway. Plant Physiol 66:457–462

Sahlström S, Ericson I (1984) Comparative electrophoretic studies of polypeptides in leaf, petiole and root mitochondria from spinach. Physiol Plant 61:45–50

Sarojini G, Oliver DJ (1983) The isolation and partial characterization of the glycine decarboxylase multienzyme complex from pea leaf mitochondria. Plant Physiol 72:194–199

Sarojini G, Oliver DJ (1984) Physical properties of glycine decarboxylase multienzyme complex from pea leaf mitochondria. Proc 6th Int Congr Photosynth, Brussels Vol III, pp 553

Schmitt MR, Edwards GE (1983) Provision of reductant for the hydroxypyruvate to glycerate conversion in leaf peroxisomes: A critical evaluation of the proposed malate/aspartate shuttle. Plant Physiol 72:728–734

Shah SPJ, Cossins EA (1970) The biosynthesis of glycine and serine by isolated chloroplasts. Phytochemistry 9:1545–1551

Somerville CR, Ogren WL (1980) Inhibition of photosynthesis in *Arabidopsis* mutants lacking leaf glutamate synthase activity. Nature (London) 286:257–259

Somerville CR, Ogren WL (1981) Photorespiration mutants of *Arabidopsis thaliana* lacking mitochondrial serine transhydroxymethylase activity. Plant Physiol 67:666–671

Somerville CR, Ogren WL (1982) Mutants of the cruciferous plant *Arabidopsis thaliana* lacking glycine decarboxylase activity. Biochem J 202:373–380

Spalding MH, Arron GP, Edwards GE (1980) Malate decarboxylation in isolated mitochondria from the crassulacean acid metabolism plant *Sedum praealtum*. Arch Biochem Biophys 199:448–456

Stitt M, Lilley RMcC, Heldt HW (1982) Adenine nucleotide levels in the cytosol, chloroplasts, and mitochondria of wheat leaf protoplasts. Plant Physiol 70:971–977

Tolbert NE (1979) Glycolate metabolism by higher plants and algae. In: Gibbs M, Latzko E (eds) Encyclopedia of plant physiology, vol VI. Photosynthesis II. Photosynthetic carbon metabolism and related processes. Springer, Berlin Heidelberg New York, pp 338–352

Usuda H, Arron GP, Edwards GE (1980) Inhibition of glycine decarboxylation by aminoacetonitrile and its effects on photosynthesis in wheat. J Exp Bot 31:1477–1483

Walker GH, Sarojini G, Oliver DJ (1982a) Identification of a glycine transporter from pea leaf mitochondria. Biochem Biophys Res Commun 107:856–861

Walker GH, Oliver DJ, Sarojini G (1982b) Simultaneous oxidation of glycine and malate by pea leaf mitochondria. Plant Physiol 70:1465–1469

Wallsgrove RM, Lea PJ, Miflin BJ (1979) Distribution of the enzymes of nitrogen assimilation within the pea leaf cell. Plant Physiol 63:232–236

Wallsgrove RM, Keys AJ, Bird IF, Cornelius MJ, Lea PJ, Miflin BJ (1980) The location of glutamine synthetase in leaf cells and its role in the reassimilation of ammonia released in photorespiration. J Exp Bot 31:1005–1017

Walton NJ, Butt VS (1981) Glutamate and serine as competing donors for amination of glyoxylate in leaf peroxisomes. Planta 153:232–237

Wedding RT (1982) Characteristics and regulation of the NAD malic enzyme from *Crassula*. In: Ting IP, Gibbs M (eds) Crassulacean acid metabolism. Am Soc Plant Physiol. Waverly Press, Maryland, pp 170–192

Wedding RT, Canellas PF, Black MK (1981) Slow transients in the activity of the NAD malic enzyme from *Crassula*. Plant Physiol 68:1416–1423

Willert Von DJ, Schwöbel H (1978) Change in mitochondria substrate oxidation during development of a crassulacean acid metabolism. In: Ducet G, Lance C (eds) Plant mitochondria. Elsevier/North Holland, Biomedical Press, Amsterdam New York, pp 403–410

Winter K, Foster J, Edwards GE, Holtum JAM (1982) Intracellular localization of enzymes of carbon metabolism in *Mesembryanthemum crystallinum* exhibiting C_3 photosynthetic characteristics or crassulacean acid metabolism. Plant Physiol 69:300–307

Wiskich JT, Day DA (1982) Malate oxidation, rotenone-resistance, and alternative path activity in plant mitochondria. Plant Physiol 70:959–964

Woo KC (1979) Properties and intramitochondrial localization of serine-hydroxymethyl-transferase in leaves of higher plants. Plant Physiol 63:783–787

Woo KC, Osmond CB (1976) Glycine decarboxylation in mitochondria isolated from spinach leaves. Aust J Plant Physiol 3:771–785

Woo KC, Osmond CB (1977) Participation of leaf mitochondria in the photorespiratory carbon oxidation cycle: glycine decarboxylation activity in leaf mitochondria from different species and its intramitochondrial location. In: Miyachi S, Katoh S, Fujita Y, Shibata K (eds) Special issue of plant cell physiology, No 3. Center Acad Publ, Tokyo, pp 315–323

Woo KC, Osmond CB (1982) Stimulation of ammonia and 2-oxoglutarate-dependent O_2 evolution in isolated chloroplasts by dicarboxylates and the role of the chloroplast in photorespiratory nitrogen recycling. Plant Physiol 69:591–596

Woo KC, Jokinen M, Canvin DT (1980) Nitrate reduction by dicarboxylate shuttle system in a reconstituted system from spinach leaves. Aust J Plant Physiol 7:123–130

Woo KC, Morot-Gaudry JF, Summons RE, Osmond CB (1982) Evidence for the gluta-mine synthetase/glutamate synthase pathway during the photorespiratory nitrogen cycle in spinach leaves. Plant Physiol 70:1514–1517

12 Starch and Sucrose Degradation

M. STITT and M. STEUP

1 Introduction

Mitochondrial respiration requires an oxidizable substrate which, in most cases, is provided by the degradation of carbohydrates (AP REES 1980a). Although in plants, as in animals and bacteria, lipids and proteins can provide the substrate for mitochondrial respiration, this occurs in plant cells less frequently. In order to maintain respiration, most plant cells contain a pool of carbohydrates which can be degraded via glycolysis or the oxidative pentosephosphate cycle and the tricarboxylic acid cycle, thus providing the substrate for the mitochondrial respiratory chain. During photosynthesis, the Calvin cycle can provide sugar phosphates as a substrate for the carbohydrate oxidation but, on the other hand, mitochondrial respiration may be suppressed to some extent by photosynthesis (cf. GRAHAM and CHAPMAN 1979), so the carbohydrate supply for mitochondrial respiration is usually derived from sugars synthesized during a preceding period of photosynthetic carbon dioxide fixation. Similarly, in carbon-heterotrophic cells the substrate for mitochondrial respiration is provided by an import of photosynthate which has been produced in a photoautotrophic part of the plant. In fact, the direct or indirect dependence of plant cells on photosynthetic carbon metabolism may be considered as the reason for the dominance of carbohydrate oxidation for respiration. In most higher plants there are two important carbohydrate pools which are consumed for respiration, starch and sucrose. Starch can be replaced to some extent (or in some rare cases even completely) by other polysaccharides such as mannans or fructans (cf. MEIER and REID 1982), but is usually predominant. Likewise, in addition to or instead of sucrose some sugar alcohols (cf. BIELESKI 1982) or other oligosaccharides (KANDLER and HOPF 1982) occur in higher plants, but doubtless starch and sucrose are of prime importance because of their wide distribution. In lower plants, besides starch and sucrose, a variety of other sugars or sugar derivatives are found (BIELESKI 1982, MANNERS and STURGEON 1982). Nevertheless, this chapter will concentrate on the two most important carbohydrate sources for respiration, i.e., starch and sucrose.

Plant cells usually contain starch and sucrose in a quantity which far exceeds the amount of carbohydrate needed for the observed rates of respiration. This

---

Abbreviations. CAM: Crassulacean acid metabolism; $F2,6P_2$: Fructose 2,6 bisphosphatase; F6P: Fructose 6 phosphate; F6P, 2-kinase: Fructose 6 phosphate, 2-kinase; FBP: fructose 1,6 bisphosphate; FBPase: fructose 1,6 bisphosphatase; PFK: phosphofructokinase; P_i: inorganic phosphate; P: organic phosphate; PP_i: inorganic pyrophosphate; PFP: pyrophosphate: fructose 6 phosphate phosphotransferase; UDPglc: UDP glucose.

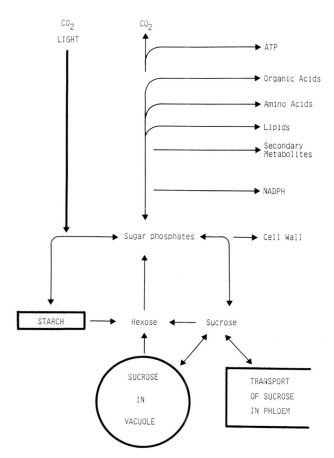

Fig. 1. Schematic diagram of the interaction between sucrose and starch metabolism, mitochondrial respiration, biosynthesis, photosynthesis, and sucrose transport

is because both starch and sucrose fulfill multiple functions in plant cell metabolism. Usually, carbohydrate breakdown provides substrates for respiration, but also for a variety of biosynthetic pathways (Fig. 1). This may lead to growth, or to storage of starch or sucrose, or to export of sucrose, depending on the tissue. Therefore, the relationship between the degradation of carbohydrate and respiration may be expected to be quite flexible. Considering the regulation of the plant cell metabolism two basic questions arise: (1) How are the substrate requirements of respiration sustained by sucrose or starch breakdown? (2) By which regulatory mechanisms are the fluxes of carbon toward respiration and other metabolic pathways balanced when the carbon supply by starch or sucrose degradation exceeds the requirements of mitochondrial respiration.

The aim of this chapter is not to describe the basic enzymology of starch or sucrose degradation; for this subject, the reader is referred to several recent articles (Preiss and Levi 1979, 1980, Whittingham et al. 1979, Avigad 1982, Preiss 1982a, b, Jenner 1982). Instead the intention is to discuss the present knowledge of how both starch and sucrose breakdown are coupled to respiration, and also how the intracellular carbon fluxes from starch or sucrose toward respiration and other metabolic pathways are adjusted under various physiological conditions.

2 Properties of Starch and Sucrose

2.1 Starch

Starch is composed of a mixture of amylose and amylopectin. Whilst the former consists of $(1 \rightarrow 4)$-α-linked D-glucosyl residues the latter contains, in addition, branching via $(1 \rightarrow 6)$-α-bindings. Both types of polymers together form the water insoluble starch particles (for details of the physical and chemical properties see FRENCH 1975). Most higher plants contain two types of starch granules which differ in morphology and physiology:
1. relatively small particles with a fast turnover (transitory starch which is common in photosynthetically active cells) and
2. usually larger particles which are formed over an extended period of time and are then degraded and disappear (reserve starch). The size and shape of at least the larger starch granules appear to be under genetic control. Furthermore, as indicated by several cereal mutants, the amylose amylopectin ratio in the reserve starch granule is genetically determined (SHANNON and CREECH 1973, LAVINTMAN 1966).

Despite these differences, the starch found in higher plants is rather homogeneous compared to the storage polyglucans occurring in lower plants, especially in algae. Storage glucans of lower plants show a considerable diversity both in structure and in intracellular localization. Whereas some algae, e.g., green algae, contain starch similar to that observed in higher plants, $(1 \rightarrow 3)$-β- and $(1 \rightarrow 6)$-β-linked polyglucans occur in other algae (for details see MANNERS and STURGEON 1982). The algal storage polyglucans can be located inside the chloroplast, in the perichloroplast space, or in the cytoplasm. Beside other chloroplast features (such as photosynthetic pigments, thylakoid pattern, chloroplast-surrounding membranes), the site and the nature of the storage polyglucans have been taken as an indication for different types of evolutionary pathways leading to a chloroplast-containing cell. According to this interpretation, in the original symbiotic relationship both host and symbiont possessed the capacity to accumulate a storage polyglucan. This ability has then been lost in one or the other during progressive integration (WHATLEY and WHATLEY 1981). These considerations deserve special interest in the light of the recently observed extrachloroplast location of some starch-metabolizing enzymes in higher plant cells (see below).

2.2 Sucrose

Sucrose plays a predominant role, in terms of quantity and distribution, among the plant oligosaccharides. The universal occurrence of sucrose has been attributed to physicochemical properties of the sucrose molecule (Avigad 1982), although such considerations are speculative. The metabolism of sucrose has been recently reviewed extensively (Avigad 1982, Whittingham et al. 1979) and will be summarized only shortly here. Despite a long controversy in the past, it is now generally accepted that the cytosol is the site of sucrose biosynthesis. However, significant sucrose concentrations inside the chloroplast are observed during frost-hardening (Santarius and Milde 1977), when sucrose acts as a cryoprotective. It is not known whether this implies an additional site of sucrose biosynthesis or an effective sucrose transport into the chloroplast, although the inner chloroplast envelope is impermeable to sucrose in short-term experiments (Heber and Heldt 1981).

In higher plants, the major intracellular storage site of sucrose is the vacuole (Willenbrink 1982), which accumulates sucrose up to concentrations of 200 mM. Thus, at least in higher plants, starch and sucrose as the two main carbohydrate sources for respiration are stored outside the cytosol and can be re-used for respiration only after being transferred into the cytosol.

3 Degradation of Starch and Sucrose

Starch- and sucrose-degrading enzymes are listed in Table 1. Degradation of particulate starch must be initiated by a reaction forming a soluble product from an insoluble substrate with a highly ordered structure. Such a reaction has some features which differ from the action of enzymes on soluble substrates like sucrose. In some cases the morphology of the starch particles during degradation has been studied. In vivo studies with germinating mung beans (Harris 1976) or barley seeds (Gram 1982) showed that the degradation of the reserve starch granules was initiated at very few areas of the particle surface. It is unknown whether these areas of the initial enzymatic attack possess some structural differentiation or are selected randomly. Starting from these areas, corrosion channels are formed into the central part of the granule, which then are extended to a central cavity (Gram 1982). This indicates an enhanced susceptibility of the interior part of the starch granule toward enzymatic degradation, whereas the outer part is more resistant. In leaf starch granules such corrosion channels or central cavities have never been observed; at least during in vitro degradation of transitory starch granules the enzymatic degradation is initiated at many sites of the granule surface (Steup et al. 1983). Therefore, an unequal supramolecular order within a starch particle and between different types of starch particles has to be taken into account.

It is likely that these differences are physiologically important because the rate of starch degradation is determined by both the amount of degradative enzyme activity and the accessibility of the particulate substrate to the enzymatic attack; in some cases the latter might be more important than the former. Starch–sucrose conversion in *Chlorella* was found to be dependent both on the temperature during starch degradation and during the preceding period of starch synthesis (Nakamura and Miyachi 1982), suggesting that the tempera-

Table 1. Glucan and sucrose-degrading enzymes

Enzyme	Reaction catalyzed
A) Glucan	
a) α-Amylase	Hydrolytic cleavage of α-1,4-bonds in a glucan
b) β-Amylase	Liberation of maltose from the nonreducing end of a glucan
c) α-Glucosidase	Hydrolytic cleavage of α-1,4- or α-1,6-bonds in oligosaccharides
d) Debranching enzyme	Hydrolysis of α-1,6-bonds
e) D-enzyme	$G_{donor} + G_{acceptor} \leftrightarrows G_{donor+acceptor-1} + glucose$
f) Glucosyl-glucan transferase	Transfer of glycosyl-, maltosyl- or maltotriosyl residues to glucose
g) Phosphorylase	Glucose 1-phosphate + α-glucan $\leftrightarrows$ α-glucosyl-glucan + orthophosphate
h) Maltose phosphorylase	Maltose + orthophosphate $\leftrightarrows$ glucose 1-phosphate + glucose
i) Maltose synthase	2 Glucose 1-phosphate $\leftrightarrows$ maltose + 2 orthophosphate
B) Sucrose	
a) Sucrose hydrolase (acid or alkaline invertase)	Sucrose + $H_2O \leftrightarrows$ glucose + fructose
b) Sucrose synthase	UDP glucose + fructose $\leftrightarrows$ sucrose + UDP
c) UDP glucose pyrophosphorylase	UDP glucose + pyrophosphate $\leftrightarrows$ UTP + glucose 1-phosphate
d) UDP glucose phosphorylase	UDP glucose + orthophosphate $\leftrightarrows$ UDP + glucose 1-phosphate
e) Sucrose phosphorylase	Sucrose + orthophosphate $\leftrightarrows$ glucose 1-phosphate + fructose

ture during starch formation also affects the accessibility of the granule to degradation. In cotton leaves (CHANG 1980) the ratio between a linear component which is readily mobilized and a branched component which is more resistant to breakdown varied. During aging of a leaf this branched component becomes increasingly dominant, leading to the starch turnover during the night being incomplete. Furthermore, the starch granule population within the same tissue may be heterogenous in size and metabolic state. This is indicated by the simultaneous degradation of large starch particles and the formation of small granules during germination of legumes (BRIARTY and PEARCE 1982).

Starch degradation is initiated by a depolymerization reaction. Cleavage of an $(1 \rightarrow 4)$-α-bond can be, in principle, hydrolytic or phosphorolytic, whereas $(1 \rightarrow 6)$-α-linkages removal is exclusively hydrolytic. Hydrolytic glucan degradation finally results in hexoses which must be converted into hexosephosphates for oxidative degradation by an ATP-consuming hexokinase reaction. This ATP input is avoided when glucan degradation occurs by phosphorolytic cleavage. Obviously, phosphorolysis is favorable for the net ATP yield obtained by glucan degradation.

However, unlike glycogen mobilization in animals, fungi, or bacteria, there is no evidence that the plant starch granule itself is degraded phosphorolytically.

Rather, most information available indicates that the starch particle initially is attacked by an amylase forming soluble glucans as the substrate for the phosphorylase. For transitory starch in higher plant leaf cells this degradative pathway has been postulated for several reasons: (1) The chloroplast phosphorylase from spinach leaves, like the maltodextrin phosphorylase of *Escherichia coli* (Schwartz and Hofnung 1967), has a high apparent affinity toward linear low molecular weight glucans but a poor affinity toward branched polyglucans with a high degree of polymerization (Steup and Schächtele 1981, Shimomura et al. 1982). (2) Isolated native starch granules were readily degraded by α-amylase but were not a substrate for the chloroplast phosphorylase, which was strictly dependent upon a preceding amylolytic attack on the starch granules (Chang 1982, Steup et al. 1983). Therefore, during in vivo granule degradation phosphorolysis is probably restricted to a pool of soluble glucan intermediates, although the presence of these polysaccharides has yet to be demonstrated. In pea chloroplasts, hydrolytic starch degradation has been questioned because earlier reports failed to detect amylase activity in the isolated chloroplasts. But the subsequent discovery of a non-Ca^{2+}-dependent heat-labile endoamylase in *Arum* spadix (Bulpin and ap Rees 1978), spinach (Okita et al. 1979, Okita and Preiss 1980) and pea (Beck, unpublished) chloroplasts supports the above-mentioned degradative pathway.

For reserve starch granules it is generally assumed that the initial attack at the particle occurs purely hydrolytically by an endoamylase (Preiss and Levi 1980, Banks and Muir 1980), whereas the involvement of exoamylase is questionable (Hildebrand and Hymowitz 1981). It is likely that even in those tissues which possess an excess of phosphorylase activity (like potato tubers) phosphorolysis is restricted to soluble intermediates; recent comparative studies on glucan specificity of various plant phosphorylases have provided evidence that potato phosphorylase is similar to the enzyme from spinach chloroplasts in having a poor affinity toward highly branched polyglucans such as glycogen, but a high apparent affinity toward unbranched glucans (Fukui et al. 1982).

Starch degradation also requires an attack on the $(1 \rightarrow 6)$-α-branchpoints. Starting at the nonreducing end of a glucan, phosphorolysis ceases near branching points. Therefore, the complete polyglucan degradation requires, in addition to phosphorylase, the cooperation of a debranching enzyme activity. Because cleavage of the 1,6-linkages results in free sugars, the relative glucose 1-phosphate yield is expected to decrease when the number of branching points increases. However, yeast cells or muscle tissue of mammals possess a more complicated debranching mechanism which is devised to increase the degradative glucose 1-phosphate formation. In this "indirect debranching", a two-component system is used, composed of an amylo-1,6-glucosidase and an oligo-1,4-glucan transferase (Lee and Whelan 1971). Debranching is initiated by the transfer of the outer glucose units from the A chain to the nonreducing end of the B chain thus forming a substrate for further phosphorolytic degradation (Fig. 2). At the branching point, a single glucose unit is left which is linked by a 1,6-bond to the rest of the molecule. In the second step, this glucosyl residue is then hydrolyzed by amylo-1,6-glycosidase. By this indirect debranch-

Fig. 2. Scheme of indirect debranching of a branched polyglucan. *G* glycosyl moiety

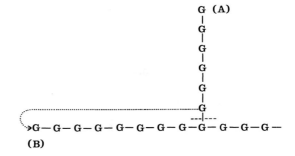

ing, glucose 1-phosphate formation is greatly enhanced over the release of free sugars. It is likely that in plants the same result is reached in a different way: by a combined action of a direct debranching enzyme (resulting in the liberation of the entire chain A) and of an endoamylase, soluble glucans are formed. In a second step the degree of polymerization in these degradation products is then increased by the action of D-enzyme or other transglycosidases, forming a substrate for phosphorolytic degradation. Likewise, maltotetraose, the end-product of phosphorolytic glucan degradation in leaves (STEUP and SCHÄCHTELE 1981) can be reconverted to a substrate for further phosphorolytic glucose 1-phosphate formation in a similar way.

Phosphorylase has sometimes been attributed a role during starch synthesis as well as starch degradation. Increases in the levels of glucan phosphorylase occur during accumulation of starch in many tissues. In mung bean seedlings, high activities of phosphorylase develop during the early stages of starch accumulation before starch synthetase activity rises (TSAY et al. 1983). Phosphorylase has been shown by immunofluorescence to be present in the vicinity of the starch grains in potato (SCHNEIDER et al. 1981), but the function of an enzyme of starch degradation during accumulation of starch is unclear. A massive flux of carbon toward starch via phosphorylase reaction is unlikely because of an unfavorable G1P/P_i ratio usually found in vivo (PREISS and LEVI 1979). However, present evidence does not totally exclude a temporary elongation of a glucan chain, even if the entire reaction is not phosphorolytic (STEUP and SCHÄCHTELE 1981) and a possible involvement of phosphorylase in the de novo synthesis of starch cannot be excluded (SIVAK et al. 1981).

As in the case of starch degradation, sucrose can be degraded hydrolytically, by invertase, forming free sugars (Reaction a in Table 1), or, alternatively, by retaining part of the energy of the glycosidic bound to form UDPglucose plus fructose by sucrose synthetase (Reaction b in Table 1). From UDPglucose the glucosyl moiety can be converted to glucose 1-phosphate either by UDPglucose pyrophosphorylase (resulting in the gain of 1 mol UTP at the expense of 1 mol pyrophosphate; Reaction c in Table 1) or by the recently described UDPglucose phosphorylase (GIBSON and SHINE 1983; reaction d in Table 1) which converts UDPglucose plus orthophosphate to UDP plus glucose 1-phosphate. In each case, the degradation of the fructosyl moiety of the sucrose molecule requires the conversion to hexosephosphate via a hexokinase reaction.

4 Relation Between Carbohydrate Mobilization and Respiration in Various Plant Tissues

Plant tissues vary in the way in which carbohydrate is utilized (Fig. 1). In a small number of plant tissues, carbohydrate is almost totally respired to CO_2. In others, it is used as a substrate for general growth; this requires respiration to provide energy, but simultaneously withdrawal of carbon from the respiratory pathways at many different points to allow synthesis of a wide range of end-products. In storage tissues, large quantities of a restricted range of storage products are synthesized during their development, and must be remobilized and exported later, without large-scale consumption in the respiratory pathways. Finally, in photosynthetic tissues, carbohydrate synthesis and mobilization alternate rapidly. In this section the mobilization of carbohydrates is considered in these different tissues, starting with those where respiration is predominant, and moving on to others where storage or export of carbohydrate is the predominant feature. Emphasis will be placed on how the relationship between carbohydrate mobilization and respiration is controlled in these differing plant tissues, i.e., which regulatory factors determine the proportion of the carbohydrate being converted to CO_2.

4.1 Tissues Having a High Respiratory Carbohydrate Consumption

There are comparatively few plant tissues where the majority of the carbohydrate is respired, and even fewer where this involves a near-quantitative conversion to CO_2. A high respiratory carbohydrate consumption is often found in growing and/or differentiating tissues, but in these cases respiration is tied closely to growth and biosynthesis so that there is a considerable withdrawal of carbon for use in biosynthesis. This results in a nonquantitative conversion of the degraded carbohydrates to CO_2. For example, in growing tomato, up to 50% of the incoming sucrose was respired (WALKER et al. 1978); in barley, more than 35% of photosynthate was released as CO_2 again within 24 h, while 40–60% was retained in protein and cell wall (GORDON et al. 1977). During rapid growth in rose cell callus (HUNT and FLETCHER 1976) about 20% of the carbon entering the Krebs cycle was retained for biosynthesis, and much more would have been previously removed from the respiratory pathway during glycolysis, to provide substrates for synthesis of cell walls, amino acids, and pigments.

4.1.1 Tissues Where Starch is Almost Completely Respired to CO_2

One of the few plant tissues where a massive conversion of carbohydrate to CO_2 occurs is the *Arum* spadix (AP REES 1977), where within 1–2 days the total starch, representing 40% of the dry weight (HALMER and BEWLEY 1982), is metabolized at rates approaching 7 μmol hexose $min^{-1} g^{-1}$ fr.w. (AP REES 1977). This is a highly specialized tissue with a largely CN-resistant respiration where the function of respiration is thermogenesis. During the development of the club there are 25–100-fold increases in the activities of enzymes of glycolysis (AP REES 1977), and of an endoamylase responsible for starch breakdown (BULPIN and AP REES 1978). However, maximal activities of these enzymes are attained before the rise in respiration (AP REES 1977), which can suddenly rise 70-fold within 3 h. Measurement of the levels of glycolytic intermediates did not reveal how these massive enzyme activities are regulated in situ (AP REES et al. 1977), so it remains an open question as to how the mobilization of starch for respiration is being controlled even in this comparatively simple system.

4.1.2 Cell Culture and Callus

The onset of growth or organogenesis in plant cell cultures is often preceded by a temporary accumulation of starch (THORPE and MURASHIGE 1968, LYNDON and ROBERTSON 1976), and subsequent growth and increased respiration is often accompanied by increased activities of acid invertase and sucrose synthetase (KLIS and MAK 1972, COPPING and STREET 1972, GRAHAM and JOHNSON 1978). Alterations in the levels of adenine nucleotides and pyridine nucleotides (SHIMIZU et al. 1977, BROWN and THORPE 1980) have also been reported, but their precise significance remains unclear. Thus it appears that onset of growth can be preceded by a pre-storage of carbohydrate and is not merely dependent on an increased import during the growth phase. The initiation of flowering in carnation and *Sinapis* is also accompanied by a temporary accumulation of starch followed by an increase in the activities of amylases, or invertase (BODSON 1977, FAWZI and EL FOULY 1979, PRYKE and BERNIER 1978); sucrose synthetase can also be present (HAWKER et al. 1976).

4.1.3 Root Differentiation

The distribution of enzymes during development of tissues of roots and storage tissues has received considerable attention (AP REES 1980b, AVIGAD 1982) and provides evidence that alterations in the amount and type of degradative enzyme play an important role in controlling the mobilization of imported sucrose in a growing tissue. In pea, alkali invertase is found throughout the root, but acid invertase is located in the cortex, and its measurable activity is associated with a decreased sucrose content and increased elongation (LYNN and AP REES 1972). Sucrose synthase is particularly high in the stele where cell-wall biosynthesis is occurring. The incoming sucrose is rapidly respired and used for growth in the cortex, but a substantial amount can be stored in the stele (DICK and AP REES 1976). A similar distribution is found in lupin for the invertases (ROBERTSON and TAYLOR 1973), and here the nodules also contain high alkali invertase and very little acid invertase.

Such studies, and similar results with developing storage tissue (see below), the ample documentation of a correlation between acid invertase and elongation, and sucrose synthase and cell-wall or starch biosynthesis (see below) led to the view (AP REES 1974, AVIGAD 1982) that alkaline invertase and sucrose synthase are located in the cytosol and mobilize sucrose for growth – sucrose synthase especially for transformation into polysaccharides. Acid invertase is in the vacuole (see below) and is associated with rapid growth and elongation (AP REES 1974, AVIGAD 1982) and (see below) the storage and mobilization of sucrose in the vacuole.

4.2 Tissues Where Sucrose is Metabolized but Diverted to an Increasing Extent into Storage Products

In tissues which are specializing for a particular storage or structural function, incoming sucrose is diverted to an increasing extent away from respiration into a restricted range of products such as lipid, starch, or even storage as sucrose. Depending upon the end-product, the incoming carbohydrate is mobilized and metabolized via pathways which lead to respiration, but is then diverted away from respiration at points which vary, depending upon which storage product is being synthesized. Simultaneously, however, some respiration is needed to provide the energy for synthesis of the storage product, and indeed in many cases the accumulation of a storage product, and the growth of the cells in which the storage product is accumulating, occur at the same time. For such tissues, a control of the mobilization and use of carbohydrate would be expected to be important, and to vary from tissue to tissue.

4.2.1 Development of Lipid-Storing Seeds

Much of the incoming sucrose is converted to lipid and must consequently be degraded to Ac-CoA but diverted away from the Krebs cycle. The cytosol of developing castor bean contains very low activities of sucrose synthase, while the activity of acid invertase (SIMCOX et al. 1977) is barely adequate to catalyze the rate of fat accumulation. The level of alkali invertase is unknown. Adequate activities of hexokinase are distributed between the cytosol (70%) and plastid (30%) and both compartments contain a complete glycolytic sequence with different isoenzymes in the two compartments (see Chap. 13, this Vol., MIERNYK and DENNIS 1982). The relative fluxes in the plastid and cytosol are not known. Recently, substantial activities of PFP (see Sect. 5.2.3.3) have been reported (KRUGER et al. 1983b).

   The significance of the compartmentation in controlling partitioning of carbon between fat accumulation and respiration is not known. Although the enzyme activities in the cytosol are higher (SIMCOX et al. 1977), fatty acid synthesis (ZILKEY and CANVIN 1972) is in the plastid and a major source of Ac-CoA is probably the pyruvate dehydrogenase isoenzyme in the plastid (MIERNYK and DENNIS, unpublished). At some point the majority of the carbon must enter the plastid, but nothing is known about the transport capacities of these plastids for sugars, phosphorylated intermediates, or pyruvate.

4.2.2 Starch-Storing Tubers and Seeds

Although starch storage involves a massive accumulation of polyglucan, this is not necessarily rigidly separated from respiration and cell growth. In potato tubers (PLAISTED 1957), cell division and growth and starch accumulation occur simultaneously, and cell number appears to be an important determinant of the amount of starch stored (REEVE et al. 1973). In tubers of *Dioscorea* (MARTIN and ORTIS 1962) and sweet potato (WILSON and LOWE 1973), cell division and starch accumulation also occur simultaneously. On the other hand, in pea seeds (PATE 1975) and wheat (BRIARTY et al. 1979), all or most starch accumulation occurs after cell division has ceased, although cell number is an important determinant of how much starch is stored (BROCKLEHURST 1977).

Sucrose synthase is the most important enzyme for degrading the imported sucrose in these tissues (AVIGAD 1982). An increase in sucrose synthase activity during starch accumulation is found in many storage tissues including potato, wheat, and barley (for references, see AVIGAD 1982), and the activity was higher than that of the invertases. Alterations in invertases may also influence carbohydrate accumulation, as in soybean (BHATIA et al. 1980) where a decrease in invertase later in development, as starch accumulation slowed, apparently allowed some sucrose to be stored as well. On the other hand, in wheat, invertase was present only initially and starch accumulation was clearly correlated with sucrose synthase (KUMAR and SINGH 1980). In maize (SHANNON and DOUGHERTY 1972, FELKER and SHANNON 1980), no sucrose synthase was found, and it was suggested that sucrose is inverted as it moves from the pedicel into the endosperm free space.

Sucrose synthase has a high K_m for sucrose (AVIGAD 1982), but high levels of sucrose are reported in tissues accumulating starch. In wheat, 30–50 mM sucrose was estimated and starch accumulation was proportional to the sucrose concentration (JENNER and RATHJEN 1978, JENNER 1980). In rice, floating panicles on increasing concentrations of sucrose produced an increasing accumulation of starch (SINGH et al. 1978), but in rice grains on the plant no such relation was found (SINGH and JULIANO 1977). The sucrose fell from 280 to 80 mM as starch accumulation commenced in pea seeds (TURNER et al. 1957). Moreover, the level of sucrose in mature tubers and seeds is often as high as that during starch accumulation (MURATA 1970, JENNER and RATHJEN 1978, AMUTI and POLLARD 1977), or even increases (BHATIA et al. 1980) in soybean. It seems that although enhanced sucrose may in some cases increase starch accumulation, the development of the ability to mobilize this sucrose via increased levels of enzymes (TSAY and KUO 1980) is probably decisive.

The diversion of sucrose into starch appears to be favored by an increase in the level of enzymes involved in starch synthesis in many tissues, although comparative studies of alterations in the respiratory capacity of the tissues have not been made. Increases in UDP glucose pyrophosphorylase, ADP glucose pyrophosphorylase, ADP glucose starch synthase, and the bound UDP glucose starch synthase have been found in many species including potato, barley, wheat, maize, and pea (see AVIGAD 1982, JENNER 1980, TURNER and TURNER 1975).

However, the increases are often moderate and the initial activity exceeds the maximal rate of starch synthesis later attained. The cessation of starch synthesis is accompanied by a decline in the levels of starch-synthesizing enzymes in some cases, such as wheat (JENNER 1980), but apparently not always, as in rice (PEREZ et al. 1975).

4.2.3 Storage of Sugar in Root Tubers

During the storage of sugar in tubers (WILLENBRINK 1982), an increasing proportion of the incoming sugar must be directed toward storage in the vacuole. However, growth often accompanies sucrose accumulation (GIAQUINTA 1979, STEIN and WILLENBRINK 1976), so that partitioning between storage and respiration is required. The importance of continued sucrose mobilization even during sucrose accumulation is indicated by reports that the energy charge of sugar beet tubers increases from 0.6 to 0.8 during the development of sugar storage capacity (STEIN and WILLENBRINK 1976). Studies in sugar beet show that the onset of sucrose accumulation is accompanied by a disappearance of acid invertase, but this is replaced by a comparable increase of sucrose synthase activity (GIAQUINTA 1979). Interestingly, immature pre-storage tubers metabolize 40% of supplied ^{14}C sucrose within 30–60 min to cell wall, amino acids, and organic acids, while mature tubers retain more than 90% in sucrose even over a long period, suggesting that the high activity of sucrose synthase is somehow being regulated in vivo so that the sucrose can be accumulated. Other studies (WYSE 1974, SILVIUS and SNYDER 1979a) confirm the increase in sucrose synthase and decline in acid invertase, but also report an alkali invertase activity, which declines during tuber development. The sucrose synthase in tap roots is specific for ADP, as opposed to the enzyme in fibrous roots which favor UDP (SILVIUS and SNYDER 1979a), and the ADP content in tap roots is high (SILVIUS et al. 1982). How the alteration in substrate affinity is involved in the development of sucrose storage is at present unknown. In sugar cane, storage of sugar is also associated with a loss of the intracellular acid invertase (HAWKER and HATCH 1965), which in this tissue is replaced by an alkali invertase, which is presumably responsible for continued growth and respiration. The activity of the insoluble cell-wall acid invertase, which is thought to be involved in sucrose transport, remains unchanged.

Sucrose is accumulated in the vacuoles in tubers (LEIGH et al. 1979) and the question as to the mechanism of this uptake will be discussed below. It might be noted that active transport of sucrose into the vacuole (WILLENBRINK 1982) has important implications, since by lowering the cytosolic sucrose it will influence the degradation of sucrose for respiration via sucrose synthase and alkali invertase, and promote transport of sucrose into the system. The ability of the vacuole to take up sucrose, compared to the rate at which sucrose is being degraded in the cytosol, will presumably be an important factor in controlling the distribution of the incoming carbohydrate between growth and storage of sucrose.

However, soluble acid invertase is located in the vacuole (LEIGH et al. 1979) and by hydrolyzing the sucrose there, it would interfere with storage. In fact it has been shown that the amount of acid invertase activity is controlled so that little is present during sucrose accumulation, but that it increases dramatically in conditions when sucrose is being mobilized. For example, slicing storage tissue leads to a mobilization of sucrose, and in red beet it has been demonstrated that the re-appearance of acid invertase in the vacuole accompanies remobilization of the sucrose from the vacuole (LEIGH et al. 1979). This mobilization of sucrose is accompanied by increased respiration, protein synthesis, and nucleic acid synthesis. In sweet potato after slicing there is a well-documented mobilization of sucrose for respiration and growth which is accompanied by de novo synthesis of acid invertase and decreased levels of acid invertase inhibitor (see URITANI and ASAHI 1980).

4.3 Tissues in Which Starch is Being Mobilized Primarily for Conversion to Sucrose

The mobilization of starch in seeds and tubers for export of sucrose is in many cases extracellular, having no direct connection with respiration, and will not be dealt with here. It might be noted, however, that the resynthesis of sucrose, for export to the growing part of the plant, requires a shielding of the intermediates in the starch–sugar conversion so that they are not respired "wastefully". However, in tubers, small-scale mobilization of starch can be induced by cold or slicing (HALMER and BEWLEY 1982), which is related to increased respiration as well as sucrose production. This mobilization occurs with little or no increase in activities of phosphorylase or amylase (HALMER and BEWLEY 1982), which are present throughout storage. The reason why starch is not normally degraded, but can start in the cold or after slicing is unclear, although it has been observed that cold storage leads to disintegration of the amyloplast membrane (OHAD et al. 1971). The observation that potato PFK is cold-labile (DIXON et al. 1981) may account for the diversion of carbon away from glycolysis into sucrose, but does not on its own explain why starch mobilization occurs in the cold.

In many fruits there is an accumulation of starch in the early stages, which is later remobilized to sugars, and may be accompanied by a respiratory burst. The starch mobilization may occur during the last stages of fruit growth before fruit ripening, or it may occur at the climacteric and be closely associated with the respiratory climacteric, as in banana and mango (HALMER and BEWLEY 1982). Although considerable attention has been paid to the control of glycolysis in fruit tissues, the cause of the stimulation of carbohydrate mobilization and of glycolysis remains unclarified (RUFFNER and HAWKER 1977). In grape berry (DOWNTON and HAWKER 1973) there are marked alterations in levels of acid invertase and sucrose synthase, as well as glucan phosphorylase during ripening, which are related to alterations in accumulation of sugar and malate.

4.4 Photosynthetic Tissues with a Rapid Alteration Between Synthesis and Mobilization of Carbohydrate

In the plant tissues discussed so far, there is a predominantly one-way flux, either net mobilization of carbohydrate for export or respiration, or accumulation of carbohydrate with some degree of associated respiration. In photosynthetic tissues, however, there is a rapid alternation between carbohydrate accumulation and breakdown, and alterations in the amounts of enzymes might be expected to play a less important role in control. Although photosynthetic tissues are a highly adapted system, their metabolism at night probably has similarities with other plant tissues respiring carbohydrate (Stitt and ap Rees 1978). The rapid alterations in fluxes in such tissues actually makes them particularly suitable for studies of metabolism, now that techniques exist for analyzing their complex compartmentation (Robinson and Walker 1980, Wirtz et al. 1980, Hampp et al. 1982, Lilley et al. 1982, Kaiser et al. 1982). In any case, the complexities of compartmention in nonphotosynthetic plant metabolism are only now being appreciated (Chap 13, this Vol.) and studies of photosynthetic metabolism are relevant to the problems posed in nongreen tissues.

4.4.1 Photosynthetic Metabolism

Both starch and sucrose play an important role in leaf metabolism and the fluxes occurring during their mobilization are very variable. During photosynthesis, triose P is converted in varying amounts into starch in the chloroplast and sucrose in the cytosol. Most of the sucrose is exported, but some can be retained in the vacuole (Kaiser et al. 1982), where it functions as a short-term storage product (Gordon et al. 1980a, b, Fondy and Geiger 1982, Stitt et al. 1983b). When the rate of CO_2 fixation exceeds the rate at which sucrose is being synthesized for export or storage in the leaf, a proportion of the photosynthate is diverted into starch (Preiss 1982a, 1982b, Stitt 1983). Starch is degraded later, usually overnight (but see below), when it can be converted to sucrose and exported (Sharkey and Pate 1976, Fondy and Geiger 1982). Starch and/or sucrose are presumably also removed by respiration at night.

In the long term, the amount of starch accumulated during the day does appear to be controlled (see Silvius and Snyder 1979b, Chatterton and Silvius 1980), so that alterations in the partitioning of photosynthate between sucrose and starch occur over a period of days in response to alterations in the conditions in which a plant is growing. This suggests some long-term control over the amount of carbohydrate being accumulated in a leaf, but the mechanisms involved are not yet understood. Moreover, the amount of starch depends on the species. In peanut, tobacco, soybean, sunflower, and pea substantial starch can accumulate, while wheat and barley accumulate little starch (Huber 1981a), and onion (Winkler 1898) has none.

4.4.2 Sucrose Degradation

Little is known about the enzymes of sucrose mobilization in leaves. Leaves contain ample activities of acid invertase and sucrose synthase (see Table 2) which decrease only slightly during the transition from a sink leaf (dependent upon imported sucrose for its growth) to a source leaf (which is a net exporter of sucrose). Significant levels of acid invertase have also been found in sugarcane leaves (Sampietro et al. 1980), *Ricinus communis* leaves (Vattuone et al. 1983), tobacco leaves (Herold 1978), wheat (Roberts 1973), and oats (Greenland

Table 2. Activity of sucrose-degrading enzymes in leaves. The results for sugar beet and *Lolium temulentum* leaves are from POLLOCK and LLOYD (1977) and GIAQUINTA (1978) respectively. In sugar beet it was demonstrated that sink leaves imported and metabolized sucrose, while source leaves did not

Leaf material		Activity			
		Sucrose synthase	Acid Invertase		
			Total	Soluble	Particulate
		(μmol mg^{-1} g fr. wt.)			
Lolium	young	0.5	47	35	–
	old	0.4	47	35	12
Sugar	sink	2.5	60	–	–
Beet	source	3.3	40	–	–

and LEWIS 1981), which decline only slightly during development. In *Lolium* (POLLOCK and LLOYD 1977, POLLOCK 1976) and sugar beet (GIAQUINTA 1979) no alkaline invertase was found but activity of about 2 μmol mg^{-1} chl h^{-1} occurs in spinach (STITT unpublished) which is comparable with the intracellular acid invertase activity.

Although mature leaves do not contain appreciably fewer enzymes for sucrose mobilization than young leaves, their ability to degrade the sucrose in situ seems to be restricted. In *Lolium* (POLLOCK and LLOYD 1977) the point when the lowest invertase activities (15 μmol g^{-1} fr. wt. h^{-1}) were found was also that at which the most sucrose (17 μmol g^{-1} fr. wt. h^{-1}) and the least glucose and fructose were present; nevertheless the presence of even this amount of sucrose implies that the acid invertase was either inactive or separated from the sucrose. In tobacco, the ability of leaf discs to convert exogenous sucrose to starch declined sharply with age, while starch synthesis from CO_2 remained constant (HEROLD 1978). The most convincing demonstration is in sugar beet (GIAQUINTA 1978), where young leaves metabolized ^{14}C sucrose rapidly to protein, cell-wall, starch amino acids, and hexoses, while in old leaves almost all the radioactivity remained as sucrose.

In older leaves, part of this slow mobilization may be due to alteration in the phloem unloading (FELLOWS and GEIGER 1974), as autoradiography revealed that much of the exogenously supplied ^{14}C sucrose remained in the veins, while it spread through all the cells of a young leaf. Nevertheless, the photosynthetically produced sucrose in source leaves must also be protected from potential sucrose-degrading enzymes if these tissues are to synthesize sucrose from CO_2 in the light, and from starch in the dark, for export without wasteful simultaneous degradation, or futile cycling. Moreover, in some leaves, including barley (GORDON et al. 1980a, b), oat, and rye (STITT unpublished), *Lolium* (POLLOCK and LLOYD 1977), and spinach (WIRTZ et al. 1980, STITT et al. 1983b), considerable amounts of sucrose (10–30 μmol mg^{-1} Chl) can be accumulated during the day and remobilized at night. Curiously, other leaves such as sugar beet (FONDY and GEIGER 1982) and tomato (HO 1976) contain very little sugar. The reason for these differences are not known. In *Ricinus communis* leaves the simultaneous presence of acid invertase and sucrose in the vacuole has been shown; it remains unclear how the invertase activity is being controlled in such conditions (Vattuone et al. 1983), although preliminary evidence has been presented for the presence of a proteinaceous activator of invertase in *Ricinus communis*, which disappears in conditions when growth is slow (Sampie-tro 1983).

4.4.3 Starch Degradation

The route of starch degradation in leaves has been amply reviewed recently (Preiss 1982a, b, Preiss and Levi 1980, Stitt 1983) and will only be summarized here (Fig. 3). The initial hydrolytic attack is on the starch granule, which (see Sect. 2) probably leads to release of soluble glucans. These oligo- or polysaccharides are thought to serve as a substrate for phosphorylase as well as for further hydrolytic attack, possible via D-enzyme (Okita et al. 1979, Stitt 1983) leading to maltose and glucose (Peavey et al. 1977, Stitt and Heldt 1981). The glucose 1-P produced by the phosphorylase is metabolized further within the chloroplast to trioseP and 3-phosphoglycerate (Peavey et al. 1977, Stitt and Heldt 1981 b) (for details see Chap. 13, this Vol.). Most of the information available on chloroplast starch breakdown comes from spinach, but other work on pea chloroplasts

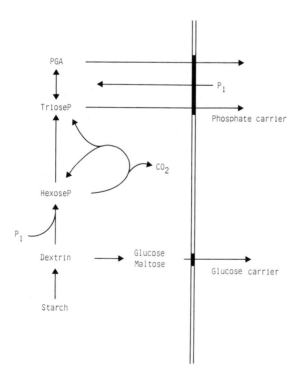

Fig. 3. Degradation of starch in chloroplasts from C_3 leaves

indicates that there may be differences in the details of starch degradation in other species (KRUGER and AP REES 1983). In contrast to spinach, which lacks maltose phosphorylase, pea chloroplasts contain a maltose phosphorylase which is responsible for synthesis of at least some of the maltose produced during starch mobilization (KRUGER and AP REES 1983). Investigations with pea chloroplasts have confirmed that maltose is a major degradation product rather than glucose (STITT and AP REES 1980, KRUGER and AP REES 1983), in contrast to spinach where glucose is the major product of hydrolytic breakdown (STITT and HELDT 1981a). It is not known what fate these different products have in the cytosol. It is attractive to speculate that the triose P and PGA are respired, rather than converted back to sucrose, and that glucose and maltose represent the precursors for production of sucrose in the dark; but further studies are needed to establish under what conditions glycolysis and gluconeogenesis function in the cytosol of leaf cells in the dark.

The question arises how starch mobilization is controlled with respect to the respiratory needs of the plant, and the amount of sucrose available in the leaf. Two studies suggest a restriction of starch mobilization at the start of the night until the amount of sucrose in leaves of barley (GORDON et al. 1980a, b) or sugar beet has been lowered (FONDY and GEIGER 1982). The cause of the inhibition remains unclear, but it appears in these tissues that sucrose may be preferentially degraded under conditions when it is present at high levels in the dark. Later starch mobilization occurs, leading in some cases to net synthesis and export of sucrose (see above). Under these conditions it is not known how far simultaneous remobilization of sucrose may in turn be restricted.

Considered in terms of the regulatory properties of the enzymes of starch breakdown, very little is known about the control of starch mobilization in chloroplasts. The conversion of starch to glucose and maltose is not fully understood (see above) and as yet no regulatory properties have been found for the endoamylase (OKITA and PREISS 1980) and the D-enzyme has not been studied. The rate of glucose and maltose production during rapid starch breakdown in isolated spinach chloroplasts (but not in pea chloroplasts, see above) does decrease when phosphorolysis increases (STITT and HELDT 1981a), but it is unclear whether this involves a regulation of the enzymes, or competition with phosphorylase for a common substrate such as dextrins. Studies on the isolated α-glucan phosporylase have also failed to find regulatory properties (PREISS et al. 1980, STEUP and SCHACHTELE 1981) except a competitive inhibition by maltotetraose.

However, experiments with intact isolated chloroplasts suggest that the rate of starch phosphorolysis is related to the requirement of the cell for carbon substrate. The observation that the rate of phosphorolysis and glycolysis, can be stimulated severalfold by P_i in isolated chloroplasts from pea (STITT and AP REES 1980) and spinach (PEAVEY et al. 1977, HELDT et al. 1977, STITT and HELDT 1981a) points to these being controlled in vivo. Recently, it has also been observed that oxidation of the pyridine nucleotides in chloroplasts also leads to a stimulation of phosphorolysis and glycolysis when P_i is limiting (STITT and HELDT, unpublished). In these treatments, where phosphorolysis and glycolysis were stimulated, there was a marked decrease in stromal triose P and hexose P, and an increase in P_i (Stitt and Heldt unpublished). The decline in stromal hexose P points to regulation of phosphofructokinase playing an important role. Chloroplast phosphofructokinase is stimulated by increasing P_i/PGA ratios (Kelly and Latzko 1977) and has recently also been shown to be inhibited by NADPH (Cseke et al. 1982a), and these properties could well contribute to the observed stimulation of glycolysis in spinach chloroplasts by high P_i or oxidized NADP(H). To what extent the resulting decrease of the hexose P/P_i ratio is responsible in turn for increased polyglucan mobilization via phosphorylase remains unclear. However, these results imply that in situ phosphorolysis might be stimulated in the dark by a requirement for carbon skeletons or reducing equivalents, suggesting a link to respiration.

In addition to regulation in the dark, control of starch breakdown might be expected in the light when starch is being accumulated. In fact, there are several reports that the turnover of starch in the light is restricted (Fondy and Geiger 1982, Kruger et al. 1983a), but turnover does occur (Jones et al. 1959, Dickson and Larwood 1973), and in any case inhibition of starch degradation does not seem to necessarily occur in the light, as net starch mobilization in the light has been found in leaves from pea (Kruger et al. 1983a), spinach (Pongratz and Beck 1978), and barley (Stitt unpublished) as well as in algae (Hirokawa et al. 1982). Experiments with isolated spinach chloroplasts (Stitt and Heldt 1981b) suggest that starch degradation in the light is not completely restricted, but that up to 70% inhibition does occur under conditions when the available P_i is limiting, and it is also in these conditions that starch accumulates (Steup et al. 1976, Heldt et al. 1977). The inhibition was ascribed to an accumulation of metabolites and low levels of P_i, possibly analogous to that occurring in the dark under limiting P_i when phosphorolysis is restricted.

4.4.4 Carbohydrate Mobilization in CAM Plants

In CAM plants in the dark, starch is degraded primarily to generate CO_2 acceptor for dark fixation, leading to malate which is stored in the vacuole (Osmond 1978). The slow growth of CAM plants implies a tight energy budget, so that wasteful use of ATP and the consequent need for high respiration can be minimized in this system. In the absence of CO_2 (Kluge 1969) as a substrate for dark fixation, the breakdown of starch is, in fact, very restricted. On the other hand, some dark respiration does occur (Kaplan et al. 1976). At 10° C this is only 17% of the gross dark fixation, rising to 72% at 24° C, and this is associated with a sharp decrease in the growth rate, showing how delicate the balance between product accumulation and respiration can be in plant tissues. The route, compartmentation, and control of starch mobilization in CAM plants is not known. A phosphorolytic mobilization of starch would require less energy and initial reports found phosphorylase but no amylase in *Kalachnoë diagremontiana* (Sutton 1975), but later studies (Vieweg and de Fekete 1977, Schilling and Dittrich 1979) also found amylases. The extent to which phosphorylase or amylase are located in the chloroplast is unknown.

5 General Features of the Control
of Carbohydrate Respiration

There are at least two important points of control in the respiration of carbohydrates in plants. On one hand, the accessibility of starch or sucrose to metabolism – namely their compartmentation and the availability of the necessary
enzymes to mobilize them – and, on the other hand, the distribution of their
immediate breakdown products between the various respiratory and biosynthetic pathways.

As illustrated in Fig. 1, the ultimate sources of carbohydrate will be starch
granules in the plastid, sucrose in the vacuole, or sucrose entering the tissue
via the phloem. In gluconeogenetic tissues, carbohydrates are derived by de
novo synthesis – from photosynthesis or from other storage products as in
lipid-storing seeds. The use of these carbohydrates will depend, in the first
place, upon the ability of the degradative enzymes to gain access to them,
as well as on the properties and activity of these enzymes. The extent to which
access to imported sucrose may be controlled in part by factors affecting phloem
unloading will not be discussed here further (but see GIAQUINTA 1983).

After hydrolysis of carbohydrates, and phosphorylation of the resulting hexoses, or after the phosphorolytic attack, the immediate products are hexose
P and UDP glucose. These, in turn, are the substrate for a large number of
reactions leading to biosynthesis of carbohydrates, cell walls, and other end-
products (see Fig. 1) as well as respiration. Moreover, the hexose P are also
compartmented between the plastid and cytosol, so that factors like transport
and enzyme distribution must be considered, as well as the activity and properties of enzymes.

For reference, a schematic representation of the metabolism in different
compartments is given in Fig. 4. As this scheme is based almost exclusively
on studies on spinach and pea leaves, its application for other tissues must
be seen as extremely tentative. Nevertheless, as discussed elsewhere in this review, and in Chap. 13, this Vol., reactions involved in carbohydrate synthesis
and degradation are now known to double up in the cytosol and plastid in
at least three different tissues – green leaves, amyloplasts, and plastids of fat-
storing seeds.

5.1 Control of Mobilization

5.1.1 Coarse Control of Enzymes

The previous discussion makes it clear that alteration in the enzyme activities
in a tissue often plays an important role in control of carbohydrate mobilization
(see Sect. 4). In some cases it has already been established that the increases
are indeed de novo synthesis such as starch degradation in cereals (HALMER
and BEWLEY 1982), acid invertase in sweet potatoes (MATSUHITA and URITANI

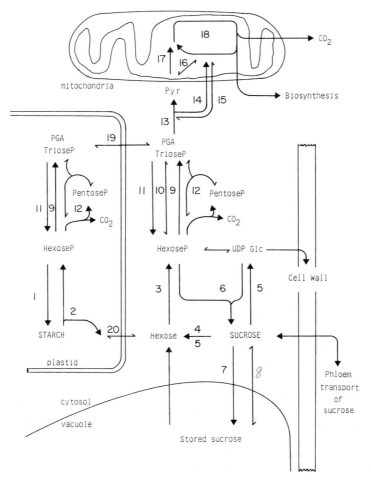

Fig. 4. Compartmentation of carbohydrate metabolism between cytosol, plastid and vacuole in an idealized plant cell. The scheme is based on available evidence and it relies heavily on results from C_3 leaves. Reversible reaction are shown as $\leftrightharpoons$; irreversible reaction $\rightarrow$. Reactions are numbered; *1* ADPglc pyrophosphorylase, starch synthase; *2* amylase, phosphorylase, D-enzyme; *3* hexokinase; *4* invertase; *5* sucrose synthase; *6* sucrose P synthase, sucrose 6 phosphatase; *7* active sucrose uptake in vacuole; *8* passive sucrose uptake in vacuole; *9* PFK; *10* PFP; *11* FBPase; *12* OPP cycle; *13* pyruvate kinase; *14* phosphoenol pyruvate carboxylase; *15* phosphoenolpyruvate carboxykinase; *16* malic enzyme; *17* pyruvate dehydrogenase; *18* Krebs cycle; *19* phosphate carrier; *20* glucose carrier

1977) as well as acid invertase inhibitor (MATSUSHITA and URITANI 1977, URITANI and ASAHI 1980). However, the analysis of the significance of protein turnover and synthesis remains an understudied area in this respect. Although levels of sucrose and hormones have been implicated in control of the level of acid invertase (SEITZ and LANG 1968, GLASZIOU et al. 1966, KAUFMANN et al.

1972, KLIS et al. 1974), the extent to which this is a causal relationship remains unestablished. More recently, the synthesis and intracellular distribution of acid invertase in radish cotyledons has been shown to be modified by phytochrome (ZOUAGHI et al. 1979).

While alterations in the level of enzymes make an important contribution to the control of carbohydrate mobilization, they do not provide a full explanation even in those tissues where synthesis and degradation are clearly separated in time. Enzyme activities are often in excess of fluxes; indeed in many cases the enzyme activities initially present before they rise are anyway ample to catalyze the higher fluxes reached during growth and accumulation. Also, the increase in enzyme activity often occurs before the rise in carbohydrate mobilization, as in *Arum* spadices (AP REES 1977), or, as in potato tubers, unchanging amounts of degradative enzymes are available throughout a long storage period, and mobilization can even commence without any increase in the amount of enzyme (see Sect. 4).

5.1.2 Multiple Forms of Enzymes

Enzymological studies on starch and sucrose metabolism are severely complicated by the fact that many of the enzymes occur as multiple forms which are usually separable by ion-exchange chromatography or by electrophoretic techniques (see PREISS 1982a, b). The relation between the various forms of a single enzyme activity are often unclear. A conversion of a particular potato phosphorylase form into another has been attributed to limited proteolysis occurring both in vivo and in vitro (IWATA and FUKUI 1973, GERBRANDY et al. 1975, SCHNEIDER et al. 1981), indicating that enzyme multiplicity can be brought about by an artifact. However, in many cases the enzyme forms differ in kinetic properties such as substrate affinity of specificity (phosphorylase: MATHESON and RICHARDSON 1978, PREISS et al. 1980, STEUP and SCHÄCHTELE 1981, SHIMOMURA et al. 1982, starch synthetase: NELSON et al. 1978, PREISS 1982a, amylase: OKITA and PREISS 1980), intracellular localization (phosphorylase: OKITA et al. 1979, STEUP and LATZKO 1979; amylase, debranching enzyme, and D-enzyme: OKITA et al. 1979), immunological properties (SCHÄCHTELE and STEUP, unpublished results for phosphorylase), and amino acid sequence at the amino terminal (HAMMOND and PREISS 1983). In some cases, e.g., invertase, the enzyme forms may differ in their carbohydrate content (see AVIGAD 1982). These observations suggest an enzyme diversity in vivo. However, according to the I.U.B. definition the term "isozyme" should be avoided as long as the primary structure of the different enzyme forms is unknown.

For enzymological studies on starch or sucrose metabolizing enzymes, the occurrence of multiple forms has two implications. First, a correlation between the velocity of a biosynthetic or degradative pathway observed in vivo and the in vitro activity of an enzyme may be masked by other enzyme forms which have a different physiological function but contribute to the total measurable activity. Likewise, an observed correlation may be misleading. Second, kinetic properties of purified enzymes are difficult to interpret as long as the enzyme preparation contains several forms of activity. It is particularly puzzling that distinct forms of amylases, phosphorylases, and debranching enzymes are found outside the chloroplast in leaves (see above). Recently (Hammond and Preiss 1983), it has been found that the activities of the chloroplast and extra-chloroplast phosphorylases vary independently during leaf growth, and that an increase of the extrachloroplast enzyme corresponded with a decline in levels of starch which is in the chloroplast. The function of these extrachloroplast enzymes, which should have no direct access to starch in the plastid, remains unclear.

5.1.3 Compartmentation of Starch

In higher plants, the synthesis of starch occurs within plastids (Dennis and Miernyck 1982), either amyloplasts or chloroplasts. The extraplastidic location of starch occurring in algae has been mentioned (Sect. 2.1), but very little information is available on the glucan metabolism in these organisms. In the higher plants starch degradation also occurs inside the chloroplast, while storage starch is sometimes surrounded by a membrane (Dennis and Miernyk 1982, Jenner 1982, Fishwick and Wright 1980), but this may disintegrate (Ohad et al. 1971). At least in leaves, the cytosol and plastid stroma have differing levels of ions, cofactors, metabolites, and pH (see below), and the significance of these differences for the separate control of starch and sucrose turnover has still to be evaluated, but it may be crucial in providing conditions where degradation of one can be accompanied by synthesis of the other (see below).

In tissues with sucrose–starch conversion, the separation of sucrose synthase in the cytosol (Avigad 1982) from starch synthesis in the plastid raises unanswered questions about how the UDP glucose produced in the cytosol from sucrose can be transferred to the plastid to form UDP glucose or ADP glucose (see below) for synthesis of starch, as is proposed. In photosynthetic tissues the major fluxes of carbon between the stroma and the cytosol occur as triose P or 3P 3-phosphoglycerate GA during photosynthesis, but even in this tissue it is possible that during starch metabolism in the dark, a significant portion of the transfer may occur by other processes, for example transfer of glucose or maltose via the glucose carrier (Schäfer et al. 1977).

In developing seeds and tubers, insufficient is known about the contents (Liu and Shannon 1981) or the transport properties of amyloplasts to show whether they are actually similar to chloroplasts, as has been suggested (Jenner 1980, 1982). Recently, it has been shown that potato amyloplasts contain the enzymes necessary for the interconversion of triose P and hexose P, as well as the metabolites found in glycolysis or gluconeogenesis (MacDonald and ap Rees 1983 a, b). This is consistent with the presence of carbon transfer into amyloplasts by a system analogous to the phosphate translocator of chloroplasts, but direct studies of the transport properties of the amyloplast envelope still have to be carried out. In fact, available evidence suggests that the transport properties in different sorts of plastid can be variable (Heber and Heldt 1981), and the enzyme complement of plastids from leaves, developing and germinating castor beans are also quantitatively different (Dennis and Niernyk 1982, Stitt and ap Rees 1979, Mieryk and Dennis 1982, Nishimura and Beevers 1981).

5.1.4 Compartmentation of Sucrose

There is now evidence from numerous sources that a large proportion of the sucrose in the cell can be stored in the vacuole (WILLENBRINK 1982), both in long term storage in beet (LEIGH et al. 1979, DOLL et al. 1979, WAGNER 1979), the short-term accumulation in vacuoles of the germinating castor bean (NISHI-MURA and BEEVERS 1978) and during the diurnal accumulation and depletion of sucrose in the leaf (FISHER and OUTLAW 1979, GERHARDT unpublished). At present little is known about what controls the uptake into the vacuole, or the level to which sucrose can be accumulated. Recent studies point to transport into the vacuole involving cotransport with H^+ (GUY et al. 1979, WILLEN-BRINK and DOLL 1979, KOMOR et al. 1982), possibly utilizing a gradient generated by a Mg^{2+}-dependent ATPase on the tonoplast (ADMAN et al. 1981), but these investigations are complicated by the isolated vacuoles apparently differing from those in situ (WILLENBRINK and DOLL 1979, DOLL and HAUER 1981, KO-MOR et al. 1982). Other studies with vacuoles from barley leaf protoplasts, however, do not provide support for active uptake of sucrose always being involved in transfer into the vacuole (KAISER, unpublished).

   Factors involved in remobilization of the sucrose are also not clear. During the mobilization of long-term-stored sucrose, the synthesis and turnover of soluble acid invertase (see Sect. 4.2.3) and of the invertase inhibitor protein play a role, but it is unclear whether the more rapid turnover of sucrose in vacuoles of tissues like leaves (GORDON et al. 1980a, b, GERHARDT unpublished) occurs by the same mechanism, or whether sucrose moves back into the cytosol without prior hydrolysis. Recent studies (KAISER personal communication) suggest that sucrose may be transported passively out of barley mesophyll protoplasts.

5.1.5 Fine Control of Enzymes

Very little is known about the extent to which the enzymes of carbohydrate mobilization are under fine control although such a control seems necessary. The enzyme activities are often substantial, the K_m values low (AVIGAD 1982) compared to the likely levels of sucrose in the cell which range from 20–200 mM (TURNER et al. 1957, Jenner and RATHJEN 1978, GERHARDT unpublished). Moreover, compartmentation does not remove the need for regulation of enzyme activity; for example the sucrose being stored in vacuoles in the sugar beet tap root must still pass through the cytosol, where substantial sucrose synthase is present (WYSE 1979). In many cases, growth and respiration, as well as accumulation of storage material, occur simultaneously – as in tomato and potato – so that a continuous control of partitioning of carbohydrate between respiration, growth, and storage is needed rather than an abrupt switch from respiration and growth to storage. In tissues where rapid turnover of starch or sucrose occur, fine control is also required, and our understanding of such control in photosynthetic tissues has already been discussed (see Sect. 3.6). In general, enzymes involved in sucrose or starch breakdown seem to lack strong regulatory properties, in contrast to those in glycolysis (TURNER and TURNER 1975, 1980) and synthesis of starch (PREISS and LEVI 1979, 1980, PREISS 1982a, b) and sucrose (AKAZAWA and OKAMOTO 1980, AVIGAD 1982).

5.1.5.1 Starch

There is a need for a study of the enzymes of starch degradation using their natural substrates – starch grains and defined dextrins. With somewhat idealized substrates such as solubilized starch, no regulation of the various amylases has yet been found. For phosphorylase, inhibition by ADP-glucose has been found with a K_i of 1.3 mM in pea seed (MATHESON and RICHARDSON 1978). This may inhibit phosphorylase during starch accumulation (TURNER and TURNER 1980) either in chloroplasts (KANAZAWA et al. 1972) or in seeds when the ADP glucose levels increase (JENNER 1968). In general, phosphorylases from photosynthetic tissues seem less sensitive to inhibition than those from nonphotosythetic tissue (see also Sect. 4.4.3), but recently the phosphorylase from *Chlorella vulgaris* has been found to be sensitive to regulation by ATP, ADP, ADP glucose, and UDP glucose (NAKAMURA and IMAMURA 1983). For higher plant phosphorylases, so far no regulation by allosteric transitions or by covalent modification has been reported, although both are present in the enzyme from animals, e.g., in the extensively studied rabbit muscle phosphorylase. This difference in regulatory properties has been attributed to dissimilarities in the amino acid sequence at the NH_2-terminal region of the enzyme molecules (NAKANO et al. 1980), which contrasts with a high conservation of amino acid sequence in other regions of these molecules (FUKUI et al. 1982, TAGAYA et al. 1982). An unanswered question for the starch phosphorylase is the extent to which its activity can be restricted by accumulation of the product glucose 1P, or low levels of P_i. As discussed above, results with isolated chloroplasts are consistent with such an effect, but the mechanism remains unclear.

5.1.5.2 Sucrose

The properties of the enzymes degrading sucrose have been reviewed thoroughly (AVIGAD 1982) and in general show a remarkable lack of potential regulatory properties. Acid invertase has a K_m of 2–13 mM. Recently, the enzyme from cane sheaths has been analyzed (SAMPIETRO et al. 1980) and a competitive inhibition by fructose (32 mM K_i) and noncompetitive inhibition by glucose ($K_i = 27$ mM) reported. The levels of fructose in the tissue were high enough to regulate the enzyme. Alkali invertase is even less studied. It has a K_m for sucrose of 9–25 mM (AVIGAD 1982), and a weak inhibition of the enzyme from sweet potato by glucose and glucose 6P (MATSUSHITA and URITANI 1974) is probably of little physiological interest. The enzyme in nodules of lupin was stimulated fourfold by 10–24 mM P_i (KIDBY 1966). It might be noted that many studies finding no inhibition of enzymes degrading sucrose have been carried out at 100 mM sucrose levels, and there is a need for another survey of the effects of a wide range of metabolites on these enzymes at truly limiting sucrose concentrations.

Following inversion, the reducing sugar must be phosphorylated by hexokinase or fructokinase. The properties of these enzymes have been reviewed (TURNER and TURNER 1975, 1980, TURNER and COPELAND 1981). Different isoenzymes occur, which are specific for glucose (I and II) or fructose (III and IV) with K_m values of 50–100 μM. Inhibition by mM levels of ADP occurs and up to 25% inhibition were achieved by 5–10 mM hexose P. Since glucose 6P may reach values of 15 mM in the cytosol (STITT et al. 1983a, MARTIN et al. 1982) a reinvestigation of these properties might be worthwhile. Hexokinases from mammalian tissues, with the exception of the low affinity liver glucokinase, are all inhibited by glucose 6P, and this effect is alleviated by P_i, so that the phosphorylation of glucose responds to the glycolytic requirement of the cell (see TURNER and TURNER 1980).

Sucrose synthase has been purified from numerous tissues (for details, see AVIGAD 1982, TURNER and TURNER 1980, PREISS 1982a, b, AKAZAWA and OKAMOTO 1980, WHITTINGHAM et al. 1979). An oligomer with four subunits, the enzyme often displays weak substrate specificity, sometimes even favoring ADP instead of UDP (see Sect. 4.2.3). The K_m for sucrose in the cleaving direction ranges form 10–400 mM, in contrast to the K_m for fructose (1.6–7 mM) and UDP glucose (0.1–8 mM) in the reverse direction, which is lower. The K_m (UDP) is 0.1–1.7 mM. The relatively poor K_m for sucrose and the reversibility of the reaction [K_{eq} for (sucrose) (UDP)/(UDP-glucose) (fructose) between 1.3 and 2.0 at pH 7.5] suggests that it would cleave sucrose only at high levels, such as those found in storage tissue, and if the fructose is being efficiently removed by fructokinase. The level of sucrose in storage tissue can reach 0.5 M (AVIGAD 1982), but can be much lower, for example 200 mM decreasing to 40 mM in potato during starch accumulation (TURNER et al. 1957). In leaves, levels of 10–60 mM in the cytosol have been proposed (AVIGAD 1982, GERHARDT unpublished), although these figures are uncertain. Sigmoidal substrate dependence has sometimes been found for sucrose (SU et al. 1977, MURATA 1971), but the significance of these results is unclear. Inhibition of sucrose synthase by high glucose or fructose, as well as by high UDP, as been reported (see AVIGAD 1982 for references). However, many of these studies have been carried out in the direction of sucrose synthesis rather than degradation.

In view of the lack of alternatives, it has been proposed that sucrose synthase is regulated by the availability of sucrose (AVIGAD 1982). However, this alone would not readily account for the decline in sucrose during starch accumulation in potato (TURNER et al. 1957), nor the ability of many sinks, in which sucrose synthase is the main enzyme of sucrose degradation, to divert sucrose toward them (see Sect. 4.2.2, also CLAUSSEN 1983). Even less is known about the ability of the cell to remove the products of the sucrose synthase reaction, fructose, and UDP glucose. The removal of fructose by fructokinase has been discussed above. The interconversion of hexosemono P and UDP glucose in plant tissues is required for both synthesis and degradation of sucrose, but little is known about this process in plants. Plants contain high activities of UDP glucose pyrophosphorylase, especially in tissues synthesizing starch (see above), but the direction of this reaction will depend on the amount of PP_i and UTP available, and the cytosolic UTP or PP$_i$ levels in plants are unknown (see below). The interconversion of hexosemono P and UDP glucose is made more complex by a recent report that potatoes possess a UDP glucose phosphorylase (Gibson and Shine 1983), catalyzing the reaction.

UDP glucose + P$_i$ ↔ UDP + Glucose 1P.

It is not known how widely this enzyme is distributed in plants and its contribution to sucrose synthesis and degradation remains to be elucidated but it is known that the enzyme is dependent upon fructose 2,6-bisphosphate (see Sect. 5.2.3.2).

5.2 Control of the Utilization of Hexose P

The hexose P made available from starch or sucrose can be utilized further in glycolysis or the oxidative pentose phosphate pathway, for biosynthesis of amino acid, pigments, fats, or conversion to CO_2 in the Krebs cycle for the supply of ATP for growth. Alternatively, it can be metabolized via nonrespiratory pathways, particularly for cell-wall biosynthesis, synthesis of starch for storage, and of sucrose for storage or export (Fig. 4). The fate of hexose P will depend upon a coordinated control of phosphofructokinase, glucose-6-phosphate dehydrogenase, and enzymes of cell wall synthesis, sucrose synthesis, and starch synthesis.

5.2.1 Coarse Control Hexose Phosphate Metabolism

The role of coarse control in altering the distribution of hexose P between the respiratory pathways (Turner and Turner 1975, 1980, ap Rees 1974, 1980b) has been amply documented, as has the role of changes in the levels of enzymes of sucrose and starch metabolism in favoring accumulation and mobilization of these end-products (see above). However, a factor that has received little attention is the integration of these processes. There are few studies (Bulpin and ap Rees 1978) where alterations of respiratory enzymes have been compared with alteration in the capacity of a tissue to mobilize and synthesize different carbohydrates. Since effective growth or storage often requires a balancing of the use of carbohydrates for respiration, or accumulation for future use and biosynthesis, more integrated studies are needed.

5.2.2 Hexose Phosphate Metabolism and Compartmentation

As discussed in Chapter 13, this Volume, this control is not essentially achieved by compartmentation because the plastids and cytosol both contain degradative and gluconeogenetic pathways; however, it remains possible that the different compartments provide varying conditions which are important in permitting different processes to occur simultaneously. This could be of importance in allowing degradation of starch and synthesis of sucrose (or vice versa) to occur simultaneously in the same tissue, as occurs during every sucrose–starch transformation. Thus, the allocation of starch to the chloroplast and sucrose to the cytosol might be a fundamental aspect in plant metabolism. In leaves, the cytosol and chloroplast stroma can vary in pH (HEBER and HELDT 1981) and concentrations of cations and metabolites (STITT et al. 1981, GIERSCH et al. 1980). For example, it appears that the chloroplast ATP/ADP ratio undergoes marked fluctuations between the light and dark, while the cytosolic ratio remains basically unaltered, at a higher level (HAMPP et al. 1982, STITT et al. 1982a). This may be of importance during conversion of sucrose to starch in the dark. Starch synthesis in the dark is probably inhibited by increased ADP and higher P_i (KAISER and BASSHAM 1979, COPELAND and PREISS 1981), which inhibits ATP glucose pyrophosphorylase, while in the cytosol or the dark, sucrose synthesis may continue using carbon derived from breakdown of starch. On the other hand, during photosynthesis, the onset of rapid starch synthesis in the stroma implies a withdrawal of carbon from sucrose synthesis in the cytosol. For this, a re-adjustment of the relative activities of the FBPases in the cytosol and plastid would be needed. Recent work suggests that an increase in fructose 2,6-bisphosphate (see below), which is only in the cytosol and so specifically restricts the activity of the cytosolic enzyme, may be involved in a redirection of carbon from one carbohydrate product to the other (STITT et al. 1983b).

5.2.3 Fine Control of Hexose Phosphate Metabolism

This topic has been recently reviewed (TURNER and TURNER 1975, 1980) and the present discussion will be restricted to problems raised by recent developments. An integrated system of regulation was proposed previously by TURNER and TURNER (1975) in which metabolites have opposite effects on key enzymes competing for a common substrate, thus allowing a direction of metabolism in one direction or another. For example, they considered the balance between phosphofructokinase and ADPglucose pyrophosphorylase which use hexose P for respiration and starch synthesis respectively. 3-Phosphoglycerate and phosphoenolpyruvate inhibit PFK and activate ADP glucose pyrophosphorylase, while P_i stimulates PFK and activates ADP glucose pyrophosphorylase. Alternatively, citrate inhibits PFK and pyruvate kinase, and activates sucrose P synthase in wheat germ. As a method for redirecting carbon between different pathways this proposal remains valuable, although the details for individual tissues still await clarification, and the implications of compartmentation will complicate the working of any such regulation. In this section some factors are discussed affecting the activities of key enzymes which have become clearer in the recent past.

5.2.3.1 Phosphofructokinase

The quantitatively most important route (AP REES 1980a) for consumption of hexose P in plant tissues, apart from those with massive starch or sucrose synthesis, is phosphofructokinase (PFK). An integrated control of glycolysis has been proposed, in which citrate and ATP inhibit pyruvate kinase (see TURNER and TURNER 1980) and accumulation of phosphoenolpyruvate and

3-phosphoglycerate then inhibits PFK, which is also affected by citrate and, more weakly, ATP. This scheme is supported by numerous in situ studies showing an interaction of pyruvate kinase and PFK in regulating glycolysis, but requires a series of qualifications due to recent findings.

The level of phosphoenolpyruvate in the cytosol (estimated at 0.3–0.5 mM) should totally inhibit PFK; however, it has been shown that salts, including chloride and phosphate, will relieve the inhibition (TURNER et al. 1980) and the question remains as to the significance of these results. The regulation of pyruvate kinase and PFK by the ATP/ADP ratio is problematic, as very few data are available on the cytosolic levels of these compounds, and there is no evidence available that it alters under conditions when the rate of glycolysis alters. Moreover, the widespread occurrence of CN-insensitive respiration raises the question as to how far plant respiration may be controlled solely by the availability of ATP and ADP (DAY and LAMBERS 1983). In leaves, for example, the rate of respiration varies greatly during the night (AZCÓN-BIETO and OSMOND 1983), and it may be questioned how such observations are consistent with a primary control of glycolysis occurring via adenylates. It seems possible that respiration varies in leaves, depending on amount of substrate available (AZCÓN-BIETO and OSMOND 1983), and that the CN-resistant alternative pathway becomes more important during rapid respiration in the presence of high levels of substrates such as sucrose and starch (AZCÓN-BIETO et al. 1983). Respiration via the alternative pathway is thought to decrease the yield of ATP by up to two thirds. In fruits at climacteric the rise in respiration, possibly linked to a CN-insensitive electron transport, does not lead to a decrease in the ATP/ADP (YOUNG and BIALE 1967, RHODES 1980), presumably because the increased flux compensates for the decrease in the number of phosphorylation sites. In *Kalanchoë* (SMITH et al. 1982) the ATP/ADP actually rises in the dark during starch mobilization and glycolysis. It must be stressed that these measurements are of the total adenine nucleotides in tissues and take no account of the extremely heterogenous intracellular distribution. However, recent studies (HAMPP et al. 1982, STITT et al. 1982a) have investigated the cytosolic ATP/ADP ratios in light and dark in cereal leaf protoplasts. Again, contrary to expectation, the cytosolic ATP/ADP does not decrease in the dark, raising the question as to whether alterations in adenine nucleotides are a central feature of the control of cytosolic carbohydrate metabolism in transition from light to dark.

Some theoretical considerations also suggest that the control of glycolysis may be more complex in plants than previously thought. It seems likely that considerable fluxes may occur via phosphoenolpyruvate carboxylase and malic enzyme rather than via pyruvate kinase; this would allow the Krebs cycle to function anapleurotically to provide carbon for biosynthesis as well as for oxidation to provide ATP (Fig. 4). The presence of the alternative (nonphosphorylating) pathway might speculatively be seen as allowing a flexible relationship between these two different products of fluxes through the Krebs cycle. However, such a view would also require a flexible control of enzymes converting hexose P through to phosphoenolpyruvate, rather than linking their activity strictly to the energy household in the cytosol. Three recent areas of research provide possibilities for how such a flexible control might be feasible.

5.2.3.2 Fructose 2,6-Bisphosphate

From investigations first carried out with liver, it has become clear that fructose 2,6-bisphosphate ($F2,6P_2$) plays an important role in the regulation of glycolysis and gluconeogenesis. Evidence is accumulating for a function in plants. $F2,6P_2$ is present in a mung bean (SABULARSE and ANDERSON 1981), spinach leaves (CSEKE et al. 1982b, STITT et al. 1982b) as well as maize leaves, marrow cotyledons and pea roots (STITT, unpublished). In spinach leaves, $F2,6P_2$ is located in the cytosol at concentrations of 2–15 µM (STITT et al. 1983b). $F2,6P_2$ inhibits the cytosolic FBPase (CSEKE et al. 1982b, STITT et al. 1982b), but no effect has been found on the cytosolic ATP-dependent PFK, in contrast to liver, where PFK is stimulated. Stimulation of the plastid ATP-PFK has been found in castor bean cotyledons (MIERNYK and DENNIS 1981) but not in spinach leaf chloroplasts (CSEKE et al. 1982b). In animal tissues, no other enzymes are affected by µM concentrations of $F2,6P_2$, but in plants three other enzymes, PFP (see below), UDP glucose phosphorylase (GIBSON and SHINE 1983), and ATP pyrophosphohydrolase (ECHEVERIA and HUMPHRIES 1983) have already been found to be activated by $F2,6P_2$ (for reactions see Table 3).

Synthesis and degradation of $F2,6P_2$ in plants may occur by routes similar to those in animals. In spinach leaves, an enzyme synthesing $F2,6P_2$ from F6P and ATP has been found (CSEKE and BUCHANAN 1983), the F6P, 2-kinase, which is analogous to the enzyme found in liver (Table 4). This same preparation contained a phosphatase with a high affinity for $F2,6P_2$, probably representing

Table 3. Reactions using pyrophosphate or being regulated by fructose 2,6-bisphosphate in plants

Enzyme	Reaction catalyzed			Effect of $F2,6P_2$
PFP	$PP_i + F6P$	$\rightleftharpoons P_i$	$+ F1,6P_2$	Activate
PFK	$ATP + F6P$	$\rightarrow ADP$	$+ F1,6P_2$	–
FBPase	$P_i + F6P$	$\longleftarrow$	$F1,6P_2$	Inhibit
UDPGlc pyrophosphorylase	$PP_i + UDPGlc \rightleftharpoons UTP + GlP$			–
UDPGlc phosphorylase	$P_i + UDPGlc \rightleftharpoons UDP + GlP$			Activate
ATP pyrophosphohydrolase	$ATP \longrightarrow AMP + PP_i$			Activate
Pyrophosphatase	$PP_i \longrightarrow P_i + P_i$			–

Table 4. Properties of enzymes which synthesize and degrade $F2,6P_2$ in spinach leaves. (Data from CSEKE and BUCHANAN 1983, CSEKE et al. 1983, STITT et al. 1984)

Enzyme	F6P, 2-kinase	F2,6P$_2$ phosphatase
Reaction catalyzed	$F6P \xrightarrow{\text{ATP} \quad \text{ADP}} F2,6P_2$	$F2,6P_2 \xrightarrow{P_i} F6P$
Stimulators	F6P, P_i	–
Inhibitors	PGA, DHAP	F6P, P_i

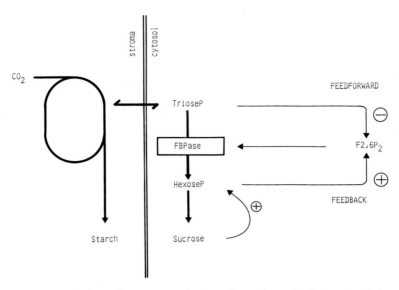

Fig. 5. Regulation of sucrose synthesis and starch synthesis in spinach leaves via fructose 2,6-bisphosphate

the $F2,6P_2$ phosphatase (Cseke et al. 1983) responsible for degrading $F2,6P_2$. These enzyme activities are regulated by various metabolites.

The precise role of $F2,6P_2$ in plants needs more study, but recent experiments on spinach leaves indicate a complicated control of the $F2,6P_2$ concentration, which allows a re-adjustment of the rate of sucrose synthesis to the requirement for sucrose and the rate of photosynthesis. Increases in the rate of photosynthesis, due to provision of light or CO_2, are accompanied by a decreased level of $F2,6P_2$ (Stitt et al. 1983b, Herzog and Stitt, unpublished). Studies show that F6P, 2-kinase is strongly inhibited by triose P (Stitt and Cseke, unpublished) and PGA (Cseke and Buchanan 1983) both of which increase during rapid photosynthesis (Stitt et al. 1983a), and this could account for the observed decline in $F2,6P_2$ levels, which in turn would allow enhanced activity of the cytosolic FBPase and stimulate sucrose synthesis (Fig. 5).

However, the concentration of F6P probably also plays an important role in controlling $F2,6P_2$ in situ. F6P, 2-kinase is stimulated by F6P, while $F2,6P_2$ phosphatase is inhibited by F6P (Cseke et al. 1983), so that an increase in F6P concentration in situ may be expected to be accompanied by an raised concentration of $F2,6P_2$. This is observed when the sucrose or glucose levels of leaves is increased by adding sugars exogenously, or by allowing them to accumulate endogenously (Stitt et al. 1983b) during prolonged photosynthesis. These results suggest that the level of $F2,6P_2$ may vary in response to the requirement for further production of hexose P, so that in conditions when the generation of hexose P by cytosolic FBPase exceeds the rate of use of hexose P for sucrose synthesis and other reactions, then an accumulation of hexose P leads to an increased $F2,6P_2$ concentration, which in turn restricts the activity of the cytosolic FBPase (Fig. 5). It is conceivable that the increase in F2,6P$_2$ plays a part in rediverting photosynthate from sucrose into starch under conditions when sucrose accumulates (STITT et al. 1983 b).

The experiments with leaves suggest that F2,6P$_2$ plays a role in coordinating metabolism by adjusting the flux between the C$_3$ and C$_6$ P-esters so that they take into account the availability of carbon in these different cell pools. The extent to which this is also true of nonphotosynthetic tissues remains unclear, as is the contribution of F2,6P$_2$ to regulation of enzymes other than fructose bisphosphatase. This provides the possibility of switching between glycolysis and gluconeogenesis in response to the carbohydrate status of a cell and without requiring changes in the adenine nucleotide levels, as has been observed in wheat protoplasts, which can change between synthesis and degradation of sucrose without any alteration of the ATP/ADP ratio in the cytosol (STITT et al. 1983 a).

It is intriguing that one of the effects of F2,6P$_2$ on the FBPase is to greatly enhance its sensitivity to inhibition by AMP (CSEKE et al. 1982 b, STITT et al. 1982 b). In animals, the inhibition of the PFK by ATP is relieved by F2,6P$_2$ (AVIGAD 1982 for references). These observations suggest a second aspect in the role of F2,6P$_2$; that alterations of F2,6P$_2$ may change the sensitivity of enzymes to regulation by adenine nucleotides. This speculation, that carbohydrate metabolism may be "locked" to a varying extent to the adenine nucleotide levels depending on the conditions, including the F2,6P$_2$ level, is also suggested by the occurrence of an unusual enzyme in plants which could potentially operate as a phosphofructokinase using PP$_i$ instead of ATP.

5.2.3.3 Pyrophosphate: Fructose 6-phosphate Phosphotransferase

The discovery of F2,6P$_2$ has led to an appreciation of the significance of the pyrophosphate:fructose 6-phosphate phosphotransferase (PFP) (see Table 3 for reaction). The enzyme is F2,6P$_2$-dependent (SABULARSE and ANDERSON 1981, CSEKE et al. 1982 b, STITT et al. 1982 b, VAN SCHAFTINGEN et al. 1983). This enzyme is present in a wide range of plant tissues including mung bean seedlings (SABULARSE and ANDERSON 1981), leaves of C$_3$ plants (CSEKE et al. 1982 b, STITT et al. 1982 b), CAM plants (CARNAL and BLACK 1979, 1983), and maize leaves (STITT unpublished) and potato tubers (VAN SCHAFTINGEN et al. 1982). Recent studies on castor bean seedlings (KRUGER et al. 1983 b) show that PFP is present at activities comparable with those of the ATP dependent PFK and of FBPase in all tissues. In spinach and wheat leaves, PFP was located exclusively in the cytosol (CSEKE et al. 1982 b, STITT et al. 1982 b).

It remains unclear what contribution PFP makes to metabolism. The reaction catalyzed is readily reversible, and a K$_{eq}$ (in the PP$_i$-utilizing direction) of 2–4 is found with potato tuber PFP (STITT unpublished). This means that the direction of the reaction will be determined by the relative concentrations of the four substrates. In leaves, the P$_i$ concentration is likely to be several times higher than F6P; this would imply that the enzyme will only operate in the glycolytic direction when the PP$_i$ concentration is equal to, or higher than, the FBP level. The cytosolic FBP is estimated at 0.5 mM in the light

(Giersch et al. 1980, Gerhardt unpublished) and is probably five- to ten-fold lower in the dark (Stitt et al. 1983a, 1983b). It is at present unknown if the PP_i concentration in the cytosol reaches levels of 100 µM or more (see Sect. 5.2.3.4). It might be noted that since PFP catalyzes a readily reversible reaction, the net flux catalyzed by this enzyme in situ is likely to be considerably lower than the available enzyme activity, and this might be borne in mind when comparing activities of PFP with those of PFK and FBPase. However, simultaneous activity of PFP and either PFK or FBPase could lead to a futile cycle (see Table 3) and some control of PFP might be expected.

To date, the only regulator reported for PFP is $F2,6P_2$ (Sabularse and Anderson 1981, Cseke et al. 1982b, Stitt et al. 1982b, Van Schaftingen et al. 1982). Since much lower levels of $F2,6P_2$ are required to fully activate the potato tuber (Van Schaftingen et al. 1982) or spinach leaf (Cseke et al. 1982b) PFP when it operates in the glycolytic direction than when it operates in the gluconeogenic direction, it has been suggested (Van Schaftingen et al. 1982) that PFP is primarily active in glycolysis. However, the levels of $F2,6P_2$ estimated in leaf material (2–15 µM) (Stitt et al. 1983b) would be adequate to keep the PFP fully active in both directions. Also, significant activity of the gluconeogenic reaction is attained even in the absence of $F2,6P_2$ especially at higher FBP concentrations (Cseke et al. 1982b, Van Schaftingen et al. 1982). Whether PFP activity is also effected by other metabolites, or the sensitivity to $F2,6P_2$ is modified by other factors in situ, requires further attention. To date it is only known that PFP is inhibited by a wide range of anions (Van Schaftingen et al. 1982), some of which modify the affinity for $F2,6P_2$ as well as the affinity for substrates. For example, the K_m for PP_i is 10 µM (Van Schaftingen et al. 1982) but this is increased four- to six-fold in the presence of 10–40 mM levels of anions (Stitt, unpublished), again raising the question as to whether adequate PP_i is available in the cytosol.

5.2.3.4 Pyrophosphate

The cytosol contains a variety of enzymes generating or using PP_i (Table 3), but very little is known at present about PP_i concentration, or which direction these various reactions actually function in situ. For PFK, FBPase and PFP, as for UDP glucose pyrophosphorylase, and UDP glucose phosphorylase, a cycle exists in which PP_i could be degraded to P_i or generated from ATP or UTP, depending on the direction in which these reactions function. Recently an enzyme producing PP_i and AMP from ATP has also been reported in corn scutellum (Echeveria and Humphries 1983), the ATP-pyrophosphohydrolase. This enzyme is activated by $F2,6P_2$. The pyrophosphatase in plants has itself attracted little attention. Reported K_m values for PP_i range between 750 µM in sugarcane leaf (Bucke 1970), 76 µM in *Typha latifolia* pollen (Hara et al. 1980) 10, 30, and 70 µM in spinach leaves (Klemme and Jacobi 1974) and 6 µM in maize leaves (Simmons and Butler 1969). Of these only an isoenzyme from spinach (K_m 70 µM) was identified as cytosolic, the remainder probably being stromal, where most of the activity in leaves is found (Bucke 1970).

Considerable work is needed before the significance of PP_i in plant metabolism, or the exact route of sucrose mobilization and synthesis is clear. It might be noted that an efficient removal of PP_i has been considered necessary to drive several synthetic reactions, including ADP-glucose and UDP-glucose pyrophosphorylase in starch, sucrose, and cell-wall synthesis. The presence of PP_i in the cytosol, especially if subject to alterations, might have profound implications for the relative activities of enzymes competing for the hexose P for respiratory and nonrespiratory purposes.

5.2.3.5 Nonrespiratory Use of Hexose Phosphate

A variety of enzymes use hexose P or related compounds as the starting point for biosynthesis of structural or storage carbohydrates. UDP glucose dehydrogenase represents the first stage to cell-wall synthesis. In contrast to animal tissues, where an efficient regulation by UDP-gluconate occurs, the enzyme in plant has been studied from several sources and shows weak or little inhibition by products (TURNER and TURNER 1975). In view of the quantitative importance of cell-wall synthesis in plants, this absence is somewhat puzzling.

ADP glucose pyrophosphorylase is the first irreversible enzyme leading to starch and its regulation has recently been reviewed (PREISS and LEVI 1980, PREISS 1982 a, b) and varies in tissues, but the enzyme appears to be activated under conditions where ample amounts of substrate are available for conversion into starch. Normally P_i operates as inhibitor, opposed by a metabolite acting as an indicator of the supply of carbon available. Thus in photosynthetic tissues the carbon comes from CO_2 fixation and the regulation is via 3-phosphoglycerate/P_i quotient, while in other plant tissues the enzyme responds to alterations in 3-phosphoglycerate/P_i ratios with varying sensitivity (see PREISS 1982 a, b for details) having a weak response in maize, and a sensitive response in potato (SOWOKINOS and PREISS 1982) and stomata (ROBINSON et al. 1983). The significance of this regulation remains unclear until more is known about transport into the plastid and plastid metabolism during starch accumulation nonphotosynthetic tissues.

Sucrose phosphate synthase catalyzes the key regulatory reaction in the synthesis of sucrose, and our understanding of the regulation of this enzyme is still inadequate (AKAZAWA and OKAMOTO 1980, PREISS 1982). In leaves, the enzyme is activated by increasing UDP glucose/P_i and F6P/P_i quotients (AMIR and PREISS 1982). In wheat germ, citrate is a positive effector, and the F6P dependence is sigmoid in both tissues (PREISS 1982 b). Recently it has been shown that glucose 6P is an effective activator of sucrose P synthetase (DOEHLERT and HUBER 1983). This suggests an activation of sucrose synthesis in conditions where ample substrate is available, analogous to control of ADP-glucose pyrophosphorylase. There are also indications of feedback control when the products accumulate, as UDP inhibits in wheat germ (SALERNO and PONTIS 1976) and spinach leaf (HABRON et al. 1981), and sucrose inhibits the enzyme from some but not all sources (SALERNO and PONTIS 1978, HUBER 1981 b). In sugarcane, sucrose phosphate phosphatase is inhibited by sucrose (HAWKER

1967) and sucrose-6-phosphate is itself an efficient inhibitor of the enzyme from spinach leaf (AMIR and PREISS 1982). Although this data is suggestive of a feedback control of sucrose synthesis as sucrose accumulates, there is still no case where clear evidence for such control has been obtained from combined enzymological and physiological studies of a specific tissue.

5.2.4 A Possible Integration of Pathways

The limited information available at present on control of hexose P metabolism indicates that the synthetic pathways are subject to a positive feed-forward control so that synthesis of starch or sucrose occurs when adequate substrate is available. The possibility also exists that accumulation of sucrose may sometimes restrict its further synthesis; to what extent and how the growth of a starch grain affects further deposition is unknown. Consumption of hexose P in the respiratory pathways is affected by the availability of substrates such as sucrose and starch; in addition, the accumulation of end-products such as ATP, phosphoenolpyruvate, citrate, and depletion of free P_i may also influence the rate at which hexose P is diverted into respiration. Although more work is needed, it appears that $F2,6P_2$ might play a role both in allowing carbohydrate turnover to respond to the availability of carbon in different metabolic pools in the cytosol, as well as, speculatively, by modifying the "linkage" between carbohydrate breakdown and adenine nucleotide levels. More work is needed to clarify under what conditions the level of $F2,6P_2$ varies, what the role of PFP is in plant respiration, and how far PP_i plays a role in plant metabolism. However, it might be imagined that the differentiated control of glycolysis allows it to respond to the requirement of the cell for growth or energy, in part independently of shifts in substrates which seem to control starch and sucrose turnover. Finally, it is to be expected that the relation between starch and sucrose turnover might be decisively influenced by their compartmentation between the plastid and cytosol, respectively, but more work is needed on the transport properties of plastid membranes in different tissues and the conditions in the plastid and cytosol during metabolism of starch and sucrose in various tissues, before it will be clear how independent and sometimes simultaneous degradation and synthesis of these carbohydrates is achieved. In addition, selective alterations in the level of the enzymes of starch and sucrose metabolism, and respiration, play a crucial role in generating the framework within which the control via compartmentation, transport of metabolites and regulators might then produce the finer tuning of metabolism.

References

Adman A, Jacoby B, Goldschmidt EE (1981) Some characteristics of the Mg^{2+}-ATPase of isolated red beet vacuoles. Plant Sci Lett 22:89–96
Akazawa T, Okamoto K (1980) Biosynthesis and metabolism of sucrose. In: Preiss J (ed) The biochemistry of plants, Vol III. Academic Press, London New York, pp 199–220

Amir J, Preiss J (1982) Kinetic characterisation of spinach leaf sucrose-phosphate synthetase. Plant Physiol 69:1027–1030

Amuti KS, Pollard CJ (1977) Soluble carbohydrates of dry and developing seeds. Phytochemistry 16:529–532

ap Rees T (1974) Pathways of carbohydrate breakdown in higher plants. In: Northcote
DH (ed) MTP International review of science. Biochem Ser One Vol 11. Butterworths,
London, pp 129–158

ap Rees T (1977) Conservation of carbohydrate by the nonphotosynthetic cells of higher
plants. Soc Exp Biol Symp XXXI:7–32

ap Rees T (1980a) Assessment of the contributions of metabolic pathways to plant
respiration. In: Davies DD (ed) The biochemistry of plants, vol II. Academic Press,
London New York, pp 1–30

ap Rees T (1980b) Integration of pathways of synthesis and degradation of hexose phosphates. In: Preiss J (ed) The Biochemistry of plants, vol III. Academic Press, London
New York, pp 1–42

ap Rees T, Fuller WA, Wright BW (1976) Pathways of carbohydrate oxidation during
thermogenesis by the spadix of *Arum maculatum*. Biochim Biophys Acta 437:22–35

ap Rees T, Fuller WA, Wright BW (1977) Measurement of glycolytic intermediates during
the onset of thermogenesis in the spadix of *Arum maculatum*. Biochim Biophys Acta
461:274–282

Avigad G (1982) Sucrose and other disaccharides. In: Loewus FA, Tanner W (eds)
Encyclopedia of plant physiology, vol XIII/A. Springer, Berlin Heidelberg New York,
pp 217–347

Azcón-Bieto J, Osmond CB (1983) Relationship between photosynthesis and respiration.
The effect of carbohydrate status on the rate of CO_2 production by respiration in
darkened and illuminated wheat leaves. Plant Physiol 71:574–581

Azcón-Bieto J, Lambers H, Day DA (1983) Effect of photosynthesis and carbohydrate
status on respiratory rates and involvement of the alternative pathway in leaf respiration. Plant Physiol 72:598–603

Banks W, Muir DD (1980) Structure and chemistry of the starch granule. In: Preiss
J (ed) The biochemistry of plants, vol III. Carbohydrates: structure and function.
Academic Press, London New York, pp 321–369

Bhatia IS, Gumber SC, Singh R (1980) Metabolism of free sugars in relation to starch
synthesis in the developing *Sorghum vulgare* grain. Physiol Plant 49:248–254

Bieleski (1982) Sugar alcohols. In: Loewus FA, Tanner W (eds) Encyclopedia of plant
physiology, vol XIII/A. Springer, Berlin Heidelberg New York, pp 158–192

Bodson M (1977) Changes in the carbohydrate content of the leaf and apical bud of
Sinapis during transition to flowering. Planta 135:19–23

Briarty LG, Pearce NM (1982) Starch granule production during germination on legumes.
J Exp Bot 33:506–510

Briarty LG, Hughes CE, Evers AD (1979) The developing endosperm of wheat, a stereo-
logical analysis. Ann Bot (London) 44:641–658

Brocklehurst PA (1977) Factors controlling grain weight in wheat. Nature (London)
266:348–349

Brown DCW, Thorpe TA (1980) Adenosine phosphate and nicotinamide adenine dinucleotide pool sizes during shoot initiation in tobacco callus. Plant Physiol 65:587–
590

Bucke C (1970) The distribution and properties of alkaline pyrophosphatase from higher
plants. Phytochemistry 9:1303–1309

Bulpin PV, ap Rees T (1978) Starch breakdown in the spadix of *Arum maculatum*. Phytochemistry 17:391–396

Carnal NW, Black CC (1979) Pyrophosphate-dependent 6-phosphofructokinase, a new
glycolytic enzyme in pineapple leaves. Biochem Biophys Res Commun 86:20–26

Carnal NW, Black CC (1983) Phosphofructokinase activities in photosynthetic organisms.
The occurrence of pyrophosphate-dependent 6-phosphofructokinase in plants and
algae. Plant Physiol 71:150–155

Chang C (1980) Starch depletion and sugars in developing cotton leaves. Plant Physiol
65:844–847

Chang CW (1982) Enzymic degradation of starch in cotton leaves. Phytochemistry 21:1263–1269

Chatterton NJ, Silvius JE (1980) Acclimation of photosynthetic partitioning and photosynthesis rates to changes in length of the daily photosynthetic period. Ann Bot (London) 46:739–745

Claussen W (1983) Investigations on the relationship between distribution of assimilates and sucrose synthetase activity in *Solanum melongena* L. II Distribution of assimilates and sucrose synthetase activity. Z Pflanzenphysiol 110:175–182

Copeland L, Preiss J (1981) Purification of spinach leaf ADP glucose pyrophosphorylase. Plant Physiol 68:996–1001

Copping LG, Street HE (1972) Properties of the invertases of cultured sycamore cells and changes in the activities during culture growth. Physiol Plant 26:346–354

Cséke C, Buchanan B (1983) An enzyme-synthesizing fructose 2,6 bisphosphate occurs in leaves and is regulated by metabolite effectors. FEBS Lett 155:139–142

Cséke C, Nishizawa AN, Buchanan BB (1982a) Modulation of chloroplast phosphofructokinase by NADPH. Plant Physiol 70:658–661

Cséke C, Weeden NF, Buchanan BB, Uyeda K (1982b) A special fructose bisphosphate functions as a cytoplasmic regulatory metabolite in green leaves. Proc Natl Acad Sci USA 79:4322–4326

Cséke C, Stitt M, Balogh A, Buchanan B (1983) A product regulated fructose-2,6 bisphosphatase occurs in green leaves. FEBS Lett 162:103–106

Day DA, Lambers H (1983) The regulation of glycolysis and electron transport in roots. Physiol Plant 58:155–160

Dennis DT, Miernyk J (1982) Compartmentation of nonphotosynthetic metabolism. Annu Rev Plant Physiol 33:27–50

Dick PS, ap Rees T (1976) Sucrose metabolism by roots of *Pisum sativum*. Phytochemistry 15:255–259

Dickson RE, Larwood PR (1973) Incorporation of ^{14}C photosynthate into major chemical fractions of source and sink leaves in cottonwood. Plant Physiol 56:185–193

Dixon WL, Franks F, ap Rees T (1981) Cold lability of phosphofructokinase from potato tubers. Phytochemistry 20:969–972

Doehlert DC, Huber SC (1983) Spinach leaf sucrose phosphate synthase: activation by glucose 6-phosphate and inactivation with inorganic phosphate FEBS Lett 153:293–298

Doll S, Hauer R (1981) Determination of the membrane potential of vacuoles isolated from red beet storage tissue. Planta 152:153–158

Doll S, Rodier F, Willenbrink J (1979) Accumulation of sucrose in vacuoles isolated from red beet tissue. Planta 144:407–411

Downton WJS, Hawker JS (1973) Enzymes of starch metabolism in leaves and berries of *Vitis vinifera*. Phytochemistry 12:1557–1563

Echeveria E, Humphries T (1983) Evidence for sucrose synthetase as a mechanism for sucrose breakdown. Plant Physiol (Suppl) 72:66

Fawzi AFA, El Fouly MH (1979) Amylase and invertase activities and carbohydrate contents in relation to physiological sink in carnation. Physiol Plant 47:245–249

Felker FC, Shannon JC (1980) Movement of ^{14}C-labelled assimilates into kernels of *Zea mays* L. III. An anatomical and microautoradiographical study of assimilate transfer. Plant Physiol 65:864–870

Fellows RJ, Geiger DR (1974) Structural and physiological changes in sugar beet leaves during sink to source conversion. Plant Physiol 54:877–885

Fisher DB, Outlaw WH (1979) Sucrose compartmentation in the palisade parenchyma of *Vicia faba* L. Plant Physiol 64:481–483

Fishwick MJ, Wright AJ (1980) Isolation and characterization of amyloplast envelope membranes from *Solanum tuberosum*. Phytochemistry 19:55–61

Fondy BR, Geiger DR (1982) Diurnal pattern of translocation and carbohydrate metabolism in source leaves of *Beta vulgaris* L. Plant Physiol 70:671–676

French D (1975) Chemistry and biochemistry of starch. In: Whelan WJ (ed) MTP international review of science. Biochem Ser 1, vol V: Biochemistry of carbohydrates. Butterworths/Univ Park Press, London Baltimore, pp 267–335

Fukui T, Shimomura S, Nakano K (1982) Potato and rabbit muscle phosphorylases: compartive studies on the structure, function and regulation of regulatory and nonregulatory enzymes. Mol Cell Biochem 42:129–144

Gerbrandy SJJ, Shankar V, Shivaram KN, Stegemann H (1975) Conversion of potato phosphorylase isozymes. Phytochemistry 14:2231–2333

Giaquinta RT (1978) Sucrose translocation and storage in the sugar beet. Plant Physiol 63:828–832

Giaquinta RT (1979) Source and sink leaf metabolism in relation to phloem translocation. Plant Physiol 61:380–385

Giaquinta RT (1983) Phloem loading. Annu Rev Plant Physiol 34:347–388

Gibson DM, Shine WE (1983) UDP-glucose breakdown is mediated by a unique enzyme activated by fructose 2,6-bisphosphate in *Solanum tuberosum*. Proc Natl Acad Sci USA 80:2491–2494

Giersch Chr, Heber U, Kaiser G, Walker DA, Robinson SP (1980) Intracellular metabolite gradients and flow of carbon during photosynthesis of leaf protoplasts. Arch Biochem Biophys 205:246–259

Glasziou KT, Waldron JB, Bull BT (1966) Control of invertase synthesis in sugar cane. Loci of auxin and glucose effects. Plant Physiol 41:282–288

Gordon AG, Ryle GJA, Powell CE (1977) The strategy of carbon utilization in uniculm barley I. The chemical fate of the photosynthetically accumulated ^{14}C. J Exp Bot 28:1258–1269

Gordon AG, Ryle GJA, Powell CE, Mitchell DE (1980a) Export, mobilization and respiration of assimilates in uniculm barley during light and darkness. J Exp Bot 31:461–473

Gordon AG, Ryle GJA, Webb G (1980b) The relationship between sucrose and starch during "dark" export from leaves of uniculm barley. J Exp Bot 31:845–850

Graham D, Chapman EA (1979) Interaction between photosynthesis and respiration in higher plants. In: Gibbs M, Latzko E (eds) Encyclopedia of plant physiology, vol VI. Springer, Berlin Heidelberg New York, pp 150–162

Graham LL, Johnson MA (1978) Sucrose synthetase from triploid quaking aspen callus. Phytochemistry 17:1231–1233

Gram NH (1982) The ultrastructure of germinating barley seeds. II. Breakdown of starch granules and cell walls of the endosperm in three barley varieties. Carlsberg Res Commun 47:173–185

Greenland AJ, Lewis DH (1981) The acid invertases of the developing third leaf of oat. I. Changes in the activity of invertase and concentration of ethanol soluble carbohydrates. New Phytol 88:265–277

Guy M, Reinhold L, Michaeli D (1979) Direct evidence for a sugar transport mechanism in isolated vacuoles. Plant Physiol 64:61–64

Habron S, Foyer C, Walker DA (1981) The purification and properties of sucrose-phosphate synthetase from spinach leaves: the involvement of this enzyme and fructose bisphosphatase in the regulation of sucrose biosynthesis. Arch Biochem Biophys 212:237–246

Halmer P, Bewley JD (1982) Control by external and internal factors over the mobilization of reserve carbohydrates in high plants. In: Loewus FA, Tanner W (eds) Encyclopedia of plant physiology, vol XIII/A. Springer, Berlin Heidelberg New York, pp 748–794

Hammond JBW, Preiss J (1983) Spinach leaf intra and extra chloroplast phosphorylase activities during growth. Plant Physiol 73:709–712

Hampp R, Goller M, Ziegler H (1982) Adenylate levels, energy charge and phoshphorylation potential during dark–light and light–dark transition in chloroplasts, mitochondria and cytosol of mesophyll protoplasts from *Avena sativa* L. Plant Physiol 69:448–455

Hara A, Kawamoto K, Funagama T (1980) Inorganic pyrophosphatase from pollen of *Typha latifolia*. Plant Cell Physiol 21:1475–1482

Harris N (1976) Starch grain breakdown in cotyledon cells of germinating mung bean. Planta 129:271–272

Hawker JS (1967) Inhibition of sucrose phosphatase by sucrose. Biochem J 102:401–406

Hawker JS, Hatch MD (1965) Mechanism of sugar storage by mature stem tissue of sugar cane. Physiol Plant 18:444–453

Hawker JS, Walker RR, Ruffner HP (1976) Invertase and sucrose synthase in flowers. Phytochemistry 15:1441–1443

Heber U, Heldt HW (1981) The chloroplast envelope: structure, function and role in leaf metabolism. Annu Rev Plant Physiol 32:139–168

Heldt HW, Chon CJ, Maronde D, Herold A, Stankovic ZS, Walker DA, Kraminer A, Kirk MR, Heber U (1977) Role of orthophosphate and other factors in the regulation of starch formation in leaves and isolated chloroplasts. Plant Physiol 59:1146–1155

Herold A (1978) Starch synthesis from exogenous sugars in tobacco leaf discs. J Exp Bot 29:1391–1401

Hildebrand DF, Hymowitz T (1981) Role of β-amylase in starch metabolism during soybean seed development and germination. Physiol Plant 53:429–434

Hirokawa T, Hata M, Takeda H (1982) Correlation between the starch level and the rate of starch synthesis during the developmental cycle of *Chlorella elipsoidea*. Plant Cell Physiol 23:813–820

Ho LC (1976) The relationship between rates of carbon transport and of photosynthesis in tomato leaves. J Exp Bot 27:87–97

Huber SC (1981a) Inter- and intraspecific variation in photosynthetic formation of starch and sucrose. Z Pflanzenphysiol 101:49–54

Huber SC (1981b) Interspecific variation in activities and regulation of leaf sucrose phosphate synthetase. Z Pflanzenphysiol 102:443–450

Hunt L, Fletcher JS (1976) Estimated drainage of carbon from the tricarboxylic acid cycle for protein synthesis in suspension cultures of Paul's scarlet Rose cells. Plant Physiol 57:304–307

Iwata S, Fukui T (1973) The subunit structure of α-glucan phosphorylase from potato. FEBS Lett 36:222–226

Jenner CF (1968) The composition of soluble nucleotides in the developing wheat grain. Plant Physiol 43:41–49

Jenner CF (1980) The conversion of sucrose to starch in developing fruits. Ber Dtsch Bot Ges 93:289–298

Jenner CF (1982) Storage of starch. In: Loewus FA, Tanner W (eds) Encyclopedia of plant physiology, vol XIII/A. Springer, Berlin Heidelberg New York, pp 700–747

Jenner CF, Rathjen AJ (1978) Physiological basis of genetic differences in the growth of grains of six varieties of wheat. Aust J Plant Physiol 5:249–262

Jones H, Martin RV, Porter HK (1959) Translocation of ^{14}C carbon in tobacco following assimilation of ^{14}C carbon dioxide by a single leaf. Ann Bot (London) 23:493–508

Kaiser G, Martinoisa E, Wiemken A (1982) Rapid appearance of photosynthetic products in the vacuoles isolated from barley mesophyll protoplasts by a new, fast method. Z Pflanzenphysiol 107:103–113

Kaiser WM, Bassham JA (1979) Light-dark regulation of starch metabolism in chloroplasts. II. Effect of chloroplastic metabolite levels on the formation of ADP-glucose by chloroplast extracts. Plant Physiol 63:109–113

Kanazawa T, Kanazawa K, Kirk MR, Bassham JA (1972) Regulatory effects of ammonia on carbon metabolism in *Chlorella pyrenoidosa* during photosynthesis and respiration. Biochim Biophys Acta 256:656–669

Kandler O, Hopf H (1982) Oligosaccharides based on sucrose (sucrosyl oligosaccharides). In: Loewus FA, Tanner W (eds) Encyclopedia of plant physiology, vol XIII/A. Springer, Berlin Heidelberg New York, pp 348–383

Kaplan A, Gale J, Poljakoff-Mayber A (1976) Resolution of net dark fixation of carbon dioxide into its respiration and gross dark fixation components in *Bryophyllum daigremontianum*. J Exp Bot 27:220–230

Kaufmann PB, Ghoshel NP, Croiv La JD, Soni SL, Ikuma M (1972) Regulation of invertase levels in *Avena* stems sections by gibberellin, sucrose, glucose and fructose. Plant Physiol 52:221–228

Kelly GJ, Latzko E (1977) Chloroplast phosphofructokinase. II. Partial purifications, kinetic and regulatory properties. Plant Physiol 60:295–299

Kidby DK (1966) Activation of plant invertase by inorganic phosphate. Plant Physiol 41:1139–1144

Klemme B, Jacobi G (1974) Separation and characterization of two inorganic pyrophosphatases from spinach leaves. Planta 120:147–153

Klis FM, Mak A (1972) Wallbound invertase activity in *Convolvulus* callus: increase after subculturing, and paradoxical effects of actinomycin D, cycloheximide and phenylalanine. Physiol Plant 26:364–368

Klis FM, Groot de C, Veriver R (1974) Effect of carbon source, giberellins, and ethylene on wall-bound invertase activity in callus of *Convolvulus arvensis*. Physiol Plant 30:334–336

Kluge M (1969) Zur Analyse des CO_2-Austausches von *Bryophyllum*. II Messung des nächtlichen Stärkeabbaues in CO_2-verarmter Atmosphäre. Planta 86:142–150

Komor E, Thom M, Maretzki A (1982) Vacuoles from sugarcane suspension cultures. III. Proton motive potential difference. Plant Physiol 69:1326–1330

Kruger NJ, ap Rees T (1983) Maltose metabolism by pea chloroplasts. Planta 158:179–184

Kruger NJ, Bulpin PV, ap Rees T (1983a) The extent of starch degradation in the light in pea leaves. Planta 157:271–273

Kruger NJ, Kombrink E, Beevers H (1983b) Pyrophosphate: fructose-6-phosphate phosphotransferase in germinating castor bean seedlings. FEBS Lett 153:409–412

Kumar R, Singh R (1980) The relationship of starch metabolism to grain size in wheat. Phytochemistry 19:2299–2303

Lavintman N (1966) The formation of branched glucans in sweet corn. Arch Biochem Biophys 116:1–8

Lee EYC, Whelan WJ (1971) Glycogen and starch debranching enzymes. In: Boyer PD (ed) The enzymes, vol V, 3rd edn. Academic Press, London New York, pp 191–234

Leigh RA, ap Rees T, Fuller WA, Banfield J (1979) The location of acid invertase activity and sucrose in the vacuoles of storage roots of beetroot (*Beta vulgaris*). Biochem J 178:539–547

Lilley R McC, Stitt M, Mader G, Heldt HW (1982) Rapid fractionation of wheat leaf protoplasts using membrane filtration. Plant Physiol 70:965–970

Liu T-TY, Shannon JC (1981) Measurement of metabolites associated with nonaqueously isolated starch granules from immature. *Zea mays* L. endosperm. Plant Physiol 67:525–529

Lyndon RF, Robertson EF (1976) The quantitative ultrastructure of the pea shoot apex in relation to leaf initiation. Protoplasma 87:387–402

Lynn RL, ap Rees T (1972) Sucrose metabolism in stele and cortex isolated from roots of *Pisum sativum*. Phytochemistry 11:2171–2174

MacDonald F, ap Rees T (1983a) Enzymic properties of amyloplasts from suspension cultures of soybean. Biochim Biophys Acta 755:81–89

MacDonald F, ap Rees T (1983b) The labelling of carbohydrate by (^{14}C)-glycerol supplied to suspension cultures of soybean. Phytochemistry 22:1141–1144

Manners DJ, Sturgeon RJ (1982) Reserve carbohydrates of algae, fungi, and lichens. In: Loewus FA, Tanner W (eds) Encyclopedia of plant physiology, vol XIII/A. Springer, Berlin Heidelberg New York, pp 472–514

Martin FW, Ortis SB (1962) Oritin and anatomy of tubers of *Dioscorea floribunda* and *D. spiculoflora*. Bot Gaz 124:416–421

Martin J-B, Bligny R, Rebeille F, Douce R, Leguay J-J, Mathieu Y, Guern T (1982) A ^{31}P nuclear magnetic reasonance study of intracellular pH of plant cells cultured in liquid medium. Plant Physiol 70:1156–1161

Matsushita K, Uritani K (1974) Change in invertase activity of sweet potato in response to wounding and purification and properties of its invertase. Plant Physiol 54:60–66

Matsushita K, Uritani I (1977) Synthesis and apparent turnover of acid invertase in relation to invertase inhibition in wounded sweet potato root tissue. Plant Physiol 59:879–883

Matheson NK, Richardson RM (1978) Kinetic properties of two starch phosphorylases from pea seeds. Phytochemistry 17:195–200

Meier H, Reid JSG (1982) Reserve polysaccharides other than starch in higher plants. In: Loewus FA, Tanner W (eds) Encyclopedia of plant physiology, vol XIII/A. Springer, Berlin Heidelberg New York, pp 418–471

Miernyk JA, Dennis DT (1981) Activation of the plastic isoenzyme of phosphofructokinase from developing endosperm of *Ricinus communis* by fructose 2,6-bisphosphate. Biochem Biophys Res Commun 105:793–798

Miernyk JA, Dennis DT (1982) Isoenzymes of glycolytic enzymes in endosperm from developing castor oil seeds. Plant Physiol 69:825–828

Murata T (1970) Enzymic mechanism of starch synthesis in sweet potato roots III. The composition of carbohydrates and soluble nucleotides in the developing sweet potate root. J Agric Chem Soc Jpn 44:412–421

Murata T (1971) Sucrose synthase of rice grains and potato tubers. Agric Biol Chem 36:1815–1818

Nakamura Y, Imamura M (1983) Characeristics of α-glucan phosphorylase from *Chlorella vulgaris*. Phytochemistry 22:835–840

Nakamura Y, Miyachi S (1982) Effect of temperature on starch degradation in *Chlorella vulgaris* 11h cells. Plant Cell Physiol 23:333–341

Nakano K, Fukui T, Matsubara H (1980) Structural basis for the difference of the regulatory properties between potato and rabbit phosphorylases. The NH_2-terminal sequence of the potato enzyme. J Biol Chem 255:9255–9261

Nelson OE, Chourey PS, Chang MT (1978) Nucleoside diphosphate sugar–starch glucosyl transferase activity of wx starch granules. Plant Physiol 62:383–386

Nishimura M, Beevers H (1978) Hydrolases in vacuoles from castor bean endosperm. Plant Physiol 62:44–48

Nishimura M, Beevers H (1981) Isoenzymes of sugar phosphate metabolism in endosperm of germinating castor beans. Plant Physiol 67:1255–1258

Ohad I, Friedberg N, Ne'eman Z, Schramm M (1971) Biogenesis and degradation of starch I The fate of the amyloplast membrane during maturation and storage of potato tubers. Plant Physiol 47:465–477

Okita TW, Preiss J (1980) Starch degradation in spinach leaves: isolation and characterization of the amylases and R-enzyme of spinach leaves. Plant Physiol 66:870–876

Okita TW, Greenberg E, Kuhn DN, Preiss J (1979) Subcellular localization of the starch degradative and biosynthetic enzymes of spinach leaves. Plant Physiol 64:187–192

Osmond CB (1978) Crassulacean acid metabolism: a curiosity in context. Annu Rev Plant Physiol 29:379–414

Pate JS (1975) Pea. In: Evans LT (ed) Crop physiology. Some case histories. Cambridge Univ Press, London, pp 191–224

Peavey DG, Steup M, Gibbs M (1977) Characterization of starch breakdown in the intact spinach chloroplast. Plant Physiol 60:305–308

Perez CM, Perdon AA, Resurreccion AP, Villareal RM, Juliano BO (1975) Enzymes of carbohydrate metabolism in the developing rice grain. Plant Physiol 56:579–583

Plaisted PM (1957) Growth of the potato tuber. Plant Physiol 32:445–453

Pollock CJ (1976) Changes in the activity of sucrose-synthesizing enzymes in developing leaves of *Lolium temulentum*. Plant Sci Lett 7:27–31

Pollock CJ, Lloyd EJ (1977) The distribution of acid invertase in developing leaves of *Lolium temulentum* L. Planta 133:197–200

Pongratz P, Beck E (1978) Diurnal oscillation of amylolytic activity in spinach chloroplasts. Plant Physiol 62:687–689

Preiss J (1982a) Biosynthesis of starch and its regulation. In: Loewus FA, Tanner W (eds) Encyclopedia of plant Physiology, vol XIII/A. Springer, Berlin Heidelberg New York, pp 397–417

Preiss J (1982b) Regulation of the biosynthesis and degradation of starch. Annu Rev Plant Physiol 33:431–454

Preiss J, Levi C (1979) Metabolism of starch in leaves. In: Gibbs M, Latzko E (eds) Encyclopedia of plant physiology, vol VI. Springer, Berlin Heidelberg New York, pp 282–312

Preiss J, Levi C (1980) Starch synthesis and degradation. In Preiss J (ed) The biochemistry of plants, vol III. Academic Press, London New York, pp 371–423

Preiss J, Okita TW, Greenberg E (1980) Characterization of the spinach leaf phosphorylases. Plant Physiol 66:864–869

Pryke JA, Bernier G (1978) Levels of acid invertase in the apex of *Sinapis alba* during transition and flowering. Ann Bot (London) 42:747–749

Reeve RM, Timm H, Weaver ML (1973) Parenchyma cell growth in potato tubers. Am Potato J 47:148–162

Rhodes MJC (1980) Respiration and senescence of plant organ. In: Stumpf PK, Conn EE (eds) The biochemistry of plants, vol II. Academic Press, London New York, pp 419–462

Roberts DWA (1973) A survey of the multiple forms of invertase in the leaves of winter wheat *Triticium aestivum* L. emend Thell ssp. *vulgare*. Biochim Biophys Acta 321:220–227

Robertson JG, Taylor MP (1973) Acid and alkali invertases in roots and nodules of *Lupinus augustifolium* infected with *Rhizobium lupini*. Planta 112:1–6

Robinson NL, Zeiger E, Preiss J (1983) Regulation of ADPglucose synthesis in guard cells of *Commelina communis*. Plant Physiol 73:862–865

Robinson SP, Walker DA (1980) Rapid separation of the chloroplast and cytoplasmic fractions from intact leaf protoplasts. Arch Biochem Biophys 196:319–323

Ruffner JP, Hawker JS (1977) Control of glycolysis in ripening berries of *Vitis vinifera*. Phytochemistry 16:1171–1175

Sabularse DC, Anderson RL (1981) D-fructose 2,6-bisphosphate: a naturally occurring activator for inorganic pyrophosphate: D-fructose-6-phosphate I-phosphotransferase in plants. Biochem Biophys Res Commun 103:845–855

Salerno GL, Pontis HG (1976) Studies on sucrose phosphate synthetase: reversal of UDP inhibition by divalent ions. FEBS Lett 64:415–418

Salerno GL, Pontis HG (1978) Studies on sucrose phosphate synthetase. The inhibitory action of sucrose. FEBS Lett 86:263–267

Sampietro AR (1983) Regulation of *Ricinus communis* invertase. Plant Physiol (Suppl) 72:36

Sampietro AR, Vattuone HA, Prado FE (1980) A regulatory invertase from sugar cane leaf sheaths. Phytochemistry 19:1637–1642

Santarius KA, Milde H (1977) Sugar compartmentation in frost-hardened cabbage leaf cells. Planta 136:163–166

Schäfer G, Heber U, Heldt HW (1977) Glucose transport into spinach chloroplasts. Plant Physiol 60:286–289

Schaftingen Van E, Lederer B, Bartrons R, Hers H-G (1982) A kinetic study of pyrophosphate: fructose-phosphate phosphotransferase from potata tubers. Eur J Biochem 129:191–195

Schilling N, Dittrich P (1979) Interaction of hydrolytic and phosphorolytic enzymes of starch metabolism in *Kalanchoë daigremontiana*. Planta 147:210–215

Schneider EM, Becker JV, Volkman D (1981) Biochemical properties of potato phosphorylase change with its intracellular localization revealed by immunological methods. Planta 151:124–134

Schwartz M, Hofnung M (1967) La maltodextrin phosphorylase d'*Escherichia coli*. Eur J Biochem 2:132–145

Seitz K, Lang A (1968) Invertase activity and cell growth in lentil epicotyls. Plant Physiol 43:1075–1082

Shannon JC, Creech RG (1973) Genetics of storage polyglucosides in *Zea mays* L. Ann NY Acad Sci 210:279–289

Shannon JC, Dougherty CT (1972) Movement of ^{14}C labelled assimilates into kernels of *Zea mays* L. II Invertase activities of the pedicel and placentochalazal tissues. Plant Physiol 49:203–206

Sharkey PJ, Pate JS (1976) Translocation from leaves to fruits of legume, studied by a phloem-bleeding technique: diurnal changes and effect of continual darkness. Planta 128:63–72

Shimizu T, Clifton A, Kommamine A, Fowler HW (1977) Changes in the metabolite levels during growth of *Acer pseudoplatanus* cells in batch suspension culture. Physiol Plant 40:125–129

Shimomura S, Nagai M, Fukui T (1982) Comparative glucan specificities of two types of spinach leaf phosphorylase. J Biochem 91:701–717

Silvius JE, Snyder FW (1979a) Comparative enzymic studies of sucrose metabolism in the tap roots and fibrous roots of *Beta vulgaris* L. Plant Physiol 64:1070–1073

Silvius JE, Snyder FW (1979b) Photosynthate partitioning and the enzymes of sucrose metabolism in sugar beet roots. Physiol Plant 46:169–173

Silvius JE, Snyder FW, Kremer DF (1982) Nucleoside diphosphate levels in tap roots and fibrous roots of *Beta vulgaris* L. Plant Physiol 70:316–317

Simcox PD, Reid EE, Canvin DT, Dennis DT (1977) Enzymes of the glycolytic and pentose phosphate pathways in proplastid from the developing endosperm of *Ricinus communis* L. Plant Physiol 59:1128–1132

Simmons S, Butler LG (1969) Alkaline inorganic pyrophosphatase of maize leaves. Biochim Biophys Acta 172:150–157

Singh R, Juliano BO (1977) Free sugars in relation to starch accumulation in developing rice grains. Plant Physiol 59:417–421

Singh R, Perez CM, Pascual CG, Juliano BG (1978) Grain size, sucrose level and starch accumulation in developing rice grain. Phytochemistry 17:1869–1874

Sivak MN, Tandecarz JS, Cardini CE (1981) Studies on potato tuber phosphorylase-catalayzed reaction in the absence of an exogeneous acceptor I. Characterization and properties of the enzyme. Arch Biochem Biophys 212:525–536

Smith JAC, Marigo G, Lüttge U, Ball E (1982) Adenine nucleotide levels during crassulacean acid metabolism and the energetics of malate accumulation in *Kalanchoe tubiflora*. Plant Sci Lett 26:13–21

Sowokinos JR, Preiss J (1982) Pyrophosphorylases in *Solanum tuberosum* III Purification, physical, and catalytic properties of ADP glucose pyrophosphorylase in potato. Plant Physiol 69:1459–1466

Stein M, Willenbrink J (1976) On the accumulation of sucrose in the growing sugar beet. Z Pflanzenphysiol 79:310–322

Steup M, Latzko E (1979) Intracellular localization of phosphorylases in spinach and pea leaves. Planta 145:69–75

Steup M, Schächtele Chr (1981) Mode of glucan degradation of purified phosphorylase forms from spinach leaves. Planta 153:351–361

Steup M, Peavey DG, Gibbs M (1976) The regulation of starch metabolism by inorganic phosphate. Biochem Biophys Res Commun 72:1554–1562

Steup M, Robenik H, Melkonian M (1983) In-vitro degradation of starch granules isolated from spinach chloroplast. Planta 158:428–436

Stitt M (1983) Degradation of starch chloroplasts: a buffer to sucrose metabolism. In: Lewis DH (ed) Plant storage carbohydrates. Cambridge Univ Press, Cambridge (in press)

Stitt M, ap Rees T (1978) Pathways of carbohydrate oxidation in leaves of *Pisum sativum* and *Triticum aestivum*. Phytochemistry 17:1251–1256

Stitt M, ap Rees T (1979) Capacities of pea chloroplasts to catalyse the oxidative pentose phosphate pathway and glycolysis. Phytochemistry 18:1905–1911

Stitt M, ap Rees T (1980) Carbohydrate breakdown by chloroplasts of *Pisum sativum*. Biochim Biophys Acta 627:131–143

Stitt M, Heldt HW (1981a) Physiological rates of starch breakdown in isolated intact spinach chloroplast. Plant Physiol 68:755–761

Stitt M, Heldt HW (1981b) Simultaneous synthesis and degradation of starch in spinach chloroplasts in the light. Biochim Biophys Acta 638:1–11

Stitt M, Bulpin PV, ap Rees T (1978) Pathway of starch breakdown in photosynthetic tissues of *Pisum sativum*. Biochim Biophys Acta 544:200–214

Stitt M, Wirtz W, Heldt HW (1980) Metabolite levels during induction in the chloroplast and extrachloroplast compartments of spinach protoplasts. Biochim Biophys Acta 593:85–102

Stitt M, Lilley R McC, Heldt HW (1982a) Adenine nucleotide levels in the cytosol, chloroplasts and mitochondria of wheat leaf protoplasts. Plant Physiol 70:971–977

Stitt M, Mieskes G, Söling H-D, Heldt HW (1982b) On a possible role of fructose 2,6-bisphosphate in regulating photosynthetic metabolism in leaves. FEBS Lett 145:217–222

Stitt M, Wirtz W, Heldt HW (1983a) Regulation of sucrose synthesis by cytoplasmic fructosebisphosphatase and sucrose phosphate synthetase during photosynthesis in varying light and carbon dioxide. Plant Physiol 72:767–774

Stitt M, Gerhardt R, Kürzel B, Heldt HW (1983b) A role for fructose 2,6 bisphosphate in the regulation of sucrose synthesis in spinach leaves. Plant Physiol 72:1139–1141

Stitt M, Cseke C, Buchanan BB (1984) Regulation of fructose-2,6-bisphosphate concentration in spinach leaves. Eur J Biochem 143:89–93

Su J-C, Wu J-L, Yang C-L (1977) Purification and properties of sucrose synthase from shoots of bamboo Leleba oldhami. Plant Physiol 60:17–21

Sutton BG (1975) Kinetic properties of phosphorylase and phosphofructokinase of Kalanchoe daigremontiana and Atriplex spongiosa. Aust J Plant Physiol 2:403–411

Tagaya M, Nakano K, Shimumora S, Fukui T (1982) Structural similarities in the active-site region between potato and rabbit muscle phosphorylases: a lysyl residue located close to the pyridoxal 5'-phosphate. J Biochem 91:599–606

Thorpe JA, Murashige T (1968) Starch accumulation in shoot forming tobacco callus cultures. Science 160:421–422

Tsay CS, Kuo CG (1980) Enzymatic activities of starch synthesis in potato tubers of different sizes. Physiol Plant 48:460–462

Tsay CS, Kuo WL, Kuo CG (1983) Enzymes involved in starch synthesis in the developing mung bean seed. Phytochemistry 22:1573–1576

Turner JF, Copeland L (1981) Hexokinase II of pea seeds. Plant Physiol 68:1123–1127

Turner JF, Turner DH (1975) The regulation of carbohydrate metabolism. Annu Rev Plant Physiol 26:159–186

Turner JF, Turner DH (1980) The regulation of glycolysis and the pentose phosphate pathway. In: Stumpf PK, Conn EE (eds) The biochemistry of plants, vol II. Academic Press, London New York, pp 279–316

Turner JF, Turner DH, Lee JB (1957) Physiology of pea fruits, IV. Changes in sugars in the developing seed. Aust J Biol Sci 10:407–413

Turner JF, Tomlinson JD, Caldwell RA (1980) Effects of salts on the activity of carrot phosphofructokinase. Plant Physiol 66:973–977

Uritani I, Asahi T (1980) Respiration and related metabolic activity in wounded and infected tissues. In: Davies DD (ed) The biochemistry of plants, vol II. Academic Press, London New York, pp 463–487

Vattuone MA, Fleischmacher OL, Prado FE, Vinals AL, San Pietro AR (1983) Localization of invertase activities in Ricinus communis leaves. Phytochemistry 22:1361–1365

Vieweg GH, Fekete de MAR (1977) Tagesgang der Amylasenaktivität im Blatt von Kalanchoë diagremontiana. Z Pflanzenphysiol 81:74–79

Wagner CJ (1979) Content and vacuole-extravacuole distribution of neutral sugars, free amino acids and anthrocyanin in protoplasts. Plant Physiol 64:88–93

Walker AJ, Ho LC, Baker DA (1978) Carbon translocation in tomato: pathways of carbon metabolism in the fruit. Ann Bot (London) 42:901–909

Whatley JM, Whatley FR (1981) Chloroplast evolution. New Phytol 87:233–247

Whittingham CP, Keys AJ, Bird IF (1979) The enzymology of sucrose synthesis in leaves. In: Gibbs M, Latzko E (eds) Encyclopedia of plant physiology, vol VI. Springer, Berlin Heidelberg New York, pp 313–326

Willenbrink J (1982) Storage of sugars in higher plants. In: Loewus FA, Tanner W (eds) Encyclopedia of plant physiology, vol XIII/A. Springer, Berlin Heidelberg New York, pp 684–699

Willenbrink J, Doll S (1979) Characterization of the sucrose uptake system of vacuoles isolated from red beet tissue. Kinetics and specificity of the sucrose uptake system. Planta 147:159–162

Wilson LA, Lowe SB (1973) The anatomy of the root system in west indian sweet potato (*Ipomoea batatas* L. Lan) cultivars. Ann Bot (London) 37:633–643

Winkler H (1898) Untersuchungen über die Stärkebildung in den verschiedenartigen Chromatophoren. Jahrb Wiss Bot 32:525–556

Wirtz W, Stitt M, Heldt HW (1980) Enzymic determination of metabolites in the subcellular compartments of spinach protoplasts. Plant Physiol 66:187–193

Wyse R (1974) Enzymes involved in the post-harvest degradation of sucrose in *Beta vulgaris* L root tissue. Plant Physiol 53:507–508

Wyse R (1979) Sucrose uptake by sugar beet tap root tissue. Plant Physiol 64:837–841

Young RE, Biale JB (1967) Phosporylation in advocado fruit slices in relation to the respiratory climacteric. Plant Physiol 42:1357–1362

Zilkey DF, Canvin DT (1972) Localization of oleic acid biosynthesis enzymes in the proplastids of developing castor bean endosperm. Can J Bot 50:323–326

Zouaghi M, Klein-Eude D, Rollin P (1979) Phytochrome-regulated transfer of fructosidase from cytoplasm to cell wall in *Raphanus sativus* L. hypocotyls. Planta 147:7–13

13 The Organization of Glycolysis and the Oxidative Pentose Phosphate Pathway in Plants

T. AP REES

1 Introduction

The first requirement for the understanding of a metabolic pathway is a knowledge of the sequence of the reactions involved and the way in which they are organized in the cell. The aim of this chapter is to consider the extent to which we have such knowledge for the only two pathways of carbohydrate oxidation known to operate in plants, glycolysis and the oxidative pentose phosphate pathway. I have said little about regulation because we cannot study it effectively until we know the pathways and their organization, and it is better to learn to walk before we try to run, and also because consideration of organization and control would make too large a chapter. Various aspects of the two pathways have been reviewed recently (TURNER and TURNER 1980, AP REES 1980a, b, DENNIS and MIERNYK 1982). By and large I have taken the picture presented by these articles as my starting point, and in general I have not dwelt on matters established in them, but have concentrated on recent developments. References have been reduced to those needed to establish the argument in an attempt to produce a review, not a catalogue.

The oxidative pentose phosphate pathway starts at glucose 6-phosphate. The beginning of glycolysis in plants is more difficult to define. To take glucose as the starting point is adequate for an organism that uses glucose as its respiratory substrate. It is inadequate for plants because sucrose is the key substrate (AP REES 1977). Breakdown of sucrose gives glucose and fructose via invertase, and fructose and UDPglucose via sucrose synthase. Breakdown of starch gives glucose or glucose 1-phosphate. Thus, although glucose is an important substrate in plants, it is by no means the only one, and may not be the dominant one in all tissues. I have taken glucose, fructose, and glucose 6-phosphate formed by phosphorolysis of starch as the starting points of glycolysis. This is arbitrary, and possibly misleading, as the chosen sequences have been removed from context because it is convenient to do so as the metabolism of sucrose and starch are discussed in Chap. 12, this Vol.

2 Reactions of Glycolysis

There is overwhelming evidence that plants convert the immediate products of sucrose and starch breakdown to pyruvate via the classic intermediates of glycolysis. It is clear that glycolysis is not only universal in plants, but is also the dominant pathway of carbohydrate oxidation in plants (AP REES 1980a, b).

2.1 Enzymes of Glycolysis

Although we lack the wealth of information available for glycolytic enzymes of animals, a number of the plant enzymes have been studied in detail and a general picture of their properties has emerged (TURNER and TURNER 1980). Further work on pyruvate kinase is needed. First, because we still do not know the relationship between pyruvate kinase and the phosphatase activity that extracts of most plants show towards phosphoenolpyruvate. The latter makes it difficult to study the properties of the kinase. Second, there remains some contradiction between the lack of sophisticated regulatory properties of plant pyruvate kinase in vitro and the evidence that the enzyme makes a major contribution to the control of glycolytic flux in vivo (AP REES 1980b).

It is becoming clear that plants contain a variety of kinases for hexose. TURNER and his colleagues have demonstrated four such enzymes in extracts of pea seeds: two hexokinases whose affinity for glucose so exceeds that for fructose that only glucose is likely to be used in vivo, and two fructokinases (see TURNER and COPELAND 1981). DENNIS and MIERNYK (1982) report that there are three hexokinases in the developing endosperm of castor bean. In addition, spinach leaves have been shown to contain a fructokinase, and at least one hexokinase with a preference for glucose (BALDUS et al. 1981). The presence of a range of enzymes to phosphorylate hexoses in plants is not surprising if we consider the multiple sources of hexose, the compartmentation of plant cells, and the fact that both fructose and glucose are formed in substantial but variable amounts. The precise functions of these kinases are not yet known, and the links between sucrose and starch, and hexose 6-phosphates still remain one of the least well-characterized aspects of glycolysis in plants.

Evidence continues to accumulate that in many plant cells many of the glycolytic enzymes are present as two isoenzymes. ANDERSON and her colleagues have shown that pea leaves contain two isoenzymes of fructose-bisphosphate aldolase, triosephosphate isomerase and phosphoglycerate kinase. Leaves also contain isoenzymes of phosphoglucomutase and glucosephosphate isomerase (DENNIS and MIERNYK 1982). The temptation to ascribe one of each of the above pairs of isoenzymes to the reductive pentose phosphate pathway and one to glycolysis should be resisted because leaves have been shown to contain isoenzymes for steps that are unique to glycolysis: phosphofructokinase (KELLY and LATZKO 1977a) and pyruvate kinase (IRELAND et al. 1979). No information is available as to whether leaves contain isoenzymes of enolase and phosphoglyceromutase, two more enzymes that are involved in glycolysis but not in the reductive pentose phosphate pathway. Isoenzymes for glycolysis are not confined to photosynthetic cells of plants. The developing endosperm of castor bean contains isoenzymes for hexokinase and for every enzyme needed to convert glucose 6-phosphate to pyruvate via glycolysis (DENNIS and MIERNYK 1982). The endosperm of germinating castor beans has been shown to contain isoenzymes of phosphoglucomutase, glucose-phosphate isomerase, phosphofructoki-nase and fructose-bisphosphate aldolase (NISHIMURA and BEEVERS 1981). The situation in other nonphotosynthetic cells of plants is not known and urgently needs investigation. For the enzymes likely to catalyse equilibrium reactions in vivo, the properties of the glycolytic isoenzymes appear to be very similar. In contrast, for the isoenzymes of phosphofructokinase and pyruvate kinase, enzymes likely to play a dominant role in control (AP REES 1980b), distinct kinetic and regulatory differences have been demonstrated (DENNIS and MIER-NYK 1982).

2.2 Pyrophosphate: Fructose 6-Phosphate 1-Phosphotransferase

One of the most important recent discoveries that affects our understanding and investigation of glycolysis in plants is the demonstration that a wide range of plant tissues contain substantial activities of pyrophosphate: fructose 6-phos-phate 1-phosphotransferase (EC 2.7.1.90). This enzyme catalyses the readily re-versible reaction

$$\text{Fru-6-P} + \text{PP}_i \rightleftharpoons \text{Fru-1,6-P}_2 + \text{P}_i \qquad \Delta G = -2.93 \text{ kJ mol}^{-1}$$

and was formerly regarded as being characteristic of a small number of rather specialized microorganisms, e.g., *Entamoeba histolytica, Propionibacterium sher-manii*. CARNAL and BLACK (1979) found PP_i: fructose 6-phosphate 1-phospho-transferase in extracts of pineapple leaves; SABULARSE and ANDERSON (1981a) reported it to be present in mung bean shoots. More recently, CARNAL and BLACK (1983) have demonstrated significant activities of the enzyme in extracts of a very wide range of photosynthetic cells. The phosphotransferase has also been found in nonphotosynthetic cells of higher plants, e.g., potato tuber (VAN SCHAFTINGEN et al. 1982), endosperm and roots of castor bean (KRUGER et al. 1983), spadix of *Arum maculatum* and roots of peas (AP REES et al. 1985). The enzyme is clearly widespread amongst plants but we do not know if it is univer-sal. Activity was not detected in extracts of some of the species examined by CARNAL and BLACK (1983), but this may well have been due to inhibition of the enzyme during the preparation of the extracts.

The properties of PP_i: fructose 6-phosphate 1-phosphotransferase have been investigated with partially purified preparations from a number of plant tissues: potato tubers and spinach leaves (VAN SCHAFTINGEN et al. 1982), mung bean sprouts (SABULARSE and ANDERSON 1981a, b), castor bean endosperm (KOM-BRINK et al. 1983). In these studies no substitute was found for PP_i, although ATP, GTP, CTP, ITP, UTP, ADP, AMP, phosphoenolpyruvate and sodium tripolyphosphate were tried. A requirement for a divalent metal ion such as Mg^{2+} has been demonstrated for the mung bean and castor bean enzymes. The enzyme is inhibited by anions, particularly phosphate. A variety of kinetic constants has been reported. Estimates of K_m^{app} for PP_i have given low values, 10–30 µM, as have those for fructose-1,6-bisphosphate, 20–90 µM; higher values have been reported for fructose 6-phosphate, 0.3–2.2 mM, and for P_i, 0.7–0.8 mM.

Perhaps the most noteworthy property of plant PP_i : fructose 6-phosphate 1-phosphotransferase is that it is markedly stimulated by fructose 2,6-bisphosphate. The latter has been regarded as the most potent regulator of mammalian glycolysis (Hers and van Schaftingen 1982). The precise effects of fructose 2,6-bisphosphate on plant PP_1 : fructose 6-phosphate 1-phosphotransferase that have been reported vary a little. The mung bean enzyme was almost completely dependent upon fructose 2,6-bisphosphate, which was shown to lower K_m^{app} for fructose 6-phosphate to increase V^{app} and to cause a 500-fold activation. Hyperbolic kinetics towards fructose 6-phosphate were found in the presence and absence of fructose 2,6-bisphosphate (Sabularse and Anderson 1981 b). A similar, but more detailed, picture is available for the castor bean enzyme (Kombrink et al. 1983). The latter enzyme was dependent upon fructose 2,6-bis-phosphate. In the direction of fructose 1,6-bisphosphate formation, fructose 2,6-bisphosphate decreased K_m^{app} for fructose 6-phosphate and PP_i and increased V^{app}. The activation by fructose 2,6-bisphosphate was hyperbolic, enhanced by fructose 6-phosphate and diminished by P_i. K_a for fructose 2,6-bisphosphate was low but was markedly affected by P_i. At 5 mM fructose 6-phosphate the addition of 5 mM P_i increased K_a from 12 to 123 nM. In the other direction, fructose 2,6-bisphosphate decreased K_m^{app} for fructose 1,6-bisphosphate, in-creased that for P_i, and had no detectable effect on V^{app}. K_a for fructose 2,6-bis-phosphate was higher than in the other direction and was increased by raising the concentration of P_i. Hyperbolic kinetics were found with fructose 6-phos-phate, PP_i, fructose 1,6-bisphosphate and P_i in the absence and in the presence of fructose 2,6-bisphosphate.

The known properties of PP_i : fructose 6-phosphate 1-phosphotransferase from potato tubers and spinach leaves are comparable to those described above except in two respects (van Schaftingen 1982, Cséke et al. 1982). First, the potato and spinach enzymes were reported to show significant activity in the absence of fructose 2,6-bisphosphate. Second, sigmoid kinetics for fructose 6-phosphate in the absence of fructose 2,6-bisphosphate were reported for the potato and spinach enzymes. These may represent differences between species, or they may be artefacts caused by the unsuspected presence of fructose 2,6-bis-phosphate in the commercial preparations of the fructose 6-phosphate used to assay PP_i : fructose 6-phosphate 1-phosphotransferase (Kombrink et al. 1983).

It is unlikely that plant PP_i : fructose 6-phosphate 1-phosphotransferase has been completely characterized. However, it is clear that it is capable of being regulated. It is dependent upon, and sensitive to, low concentrations of fructose 2,6-bisphosphate. The enzyme's response to this compound can be affected by fructose 6-phosphate and P_i, substances involved in glycolysis. For example, at a limiting concentration of fructose 2,6-bisphosphate an increase in the ratio fructose 6-phosphate : P_i would increase the enzyme's activity.

The properties of PP_i : fructose 6-phosphate 1-phosphotransferase imply that plants contain fructose 2,6-bisphosphate. Although we lack the rigorous chemi-cal identification made for mammalian cells, the work of Sabularse and Ander-son (1981 b), Stitt et al. (1982a) and Cséke et al. (1982) leave little doubt that this compound occurs in plants. Methods for measuring fructose 2,6-bisphosphate in plants are still in their infancy. The more convenient depend upon measurement of the extent to which extracts of plants stimulate either mammalian phosphofructokinase (VAN SCHAFTINGEN et al. 1980) or plant PP_i: fructose 6-phosphate 1-phosphotransferase (VAN SCHAFTINGEN et al. 1982). The hysteretic behaviour of the former and our incomplete knowledge of the latter enzyme suggest that data obtained by these methods should be confirmed by the more defined assay of HUE et al. (1982). With the enzymic assay STITT et al. (1982a) estimated that spinach leaves contained 0.05–0.2 nmol fructose 2,6-bisphosphate mg^{-1} chlorophyll. Assuming that the compound was confined to either chloroplast or cytosol, they calculated that the above would give a concentration of fructose 2,6-bisphosphate in vivo of 1–5 µM. CSÉKE et al. (1982), in contrast, estimate that the cytosolic concentration in leaves of spinach and peas could be as high as 300 µM. STITT et al. provided evidence that their assays were reliable, CSÉKE et al. did not. The differences between the above estimates may reflect differences in the metabolic condition of the leaves or difficulties in using the current assays. It is important to note that even the lowest estimate is appreciably greater than published values of K_a for activation of plant PP_i: fructose 6-phosphate 1-phosphotransferase by fructose 2,6-bisphosphate. In order to assess the role of the latter in plant metabolism we need authenticated measurements of its concentration in vivo when the rate and direction of carbohydrate metabolism is being changed. To date all that we know is that fructose 2,6-bisphosphate is present in spinach leaves in the light and the dark (STITT et al. 1982a) and that it increases as sucrose accumulates (STITT et al. 1983).

CSÉKE and BUCHANAN (1983) have shown that spinach leaves contain an enzyme, 6-phosphofructo-2-kinase, that catalyses the synthesis of fructose 2,6-bisphosphate from ATP and fructose 6-phosphate. This kinase is specific for fructose 6-phosphate and ATP, is activated by P_i and inhibited by 3-phosphoglycerate. Fractionation of spinach leaves has led to the view that in this tissue both the kinase (CSÉKE and BUCHANAN 1983) and fructose 2,6-bisphosphate (CSÉKE et al. 1982) are located in the cytoplasm. This may be so, but the above experiments do not prove it. The authors do not show that all of the enzyme and the fructose 2,6-bisphosphate in the unfractionated homogenate were recovered in the cytosolic fraction under conditions in which known organelles were maintained intact. There is indirect evidence that fructose 2,6-bisphosphate is present in the cytosol but not the chloroplast. CSÉKE et al. (1982) could not detect fructose 2,6-bisphosphate in blue-green algae, and have provided evidence that it affects fructose bisphosphatase from the cytosol but not the chloroplasts of spinach. Further, PP_i: fructose 6-phosphate 1-phosphotransferase, the enzyme that shows the greatest known response to fructose 2,6-bisphosphate, is located in the cytosol in leaves (STITT et al. 1982a) and in the endosperm of germinating castor beans (KRUGER et al. 1983). However, it is unwise to conclude that fructose 2,6-bisphosphate is always confined to the cytosol in plants. First, much of the above evidence for its absence from chloroplasts is negative. Second, and more important, MIERNYK and DENNIS (1982) have shown that in developing endosperm of castor bean the plastid phosphofructokinase, but not that in the cytosol, is stimulated by fructose 2,6-bisphosphate.

The role of PP_i : fructose 6-phosphate 1-phosphotransferase in carbohydrate metabolism in plants is not known. It has been implied that the enzyme forms fructose 1,6-bisphosphate in vivo and is a component step of plant glycolysis. However, the enzyme is readily reversible, and could catalyse the conversion of fructose 1,6-bisphosphate to fructose 6-phosphate during photosynthesis and gluconeogenesis. The available evidence does not allow us to decide whether it works in one direction or the other, or both, in vivo. The evidence in favour of it working in the glycolytic direction, i.e., catalysing the synthesis of fructose 1,6-bisphosphate, is as follows. First is that from comparative biology: there is evidence that PP_i : fructose 6-phosphate 1-phosphotransferase mediates glycolysis in *Entamoeba histolytica* and *Propionibacterium shermanii* (Wood et al. 1977). In reply it may be argued that the metabolisms of the above organisms are too highly specialized to serve as a basis for any generalization that may be extended to plants. Second, K_a of the plant enzyme for fructose 2,6-bisphosphate is lower in the glycolytic direction than in the other direction. The significance of this observation is difficult to assess, particularly as the K_a can be affected so markedly by the amounts of fructose 6-phosphate and P_i present. Third, the effects of fructose 2,6-bisphosphate on PP_i : fructose 6-phosphate 1-phosphotransferase are opposite to those it has on cytosolic fructose bisphosphatase from leaves (Stitt et al. 1982a, Cséke et al. 1982). This suggests that the two enzymes may form a cycle that amplifies the effect of fructose 2,6-bisphosphate and controls the flow of carbon between sucrose and triose phosphate in leaves. Finally, PP_i : fructose 6-phosphate 1-phosphotransferase appears to be sufficiently widespread and active to be involved in glycolysis. It is present in tissues that are neither photosynthetic nor gluconeogenic. Glycolysis operates in such tissues, but we would not expect these tissues to be converting fructose 1,6-bisphosphate to fructose 6-phosphate in the cytosol.

Possibly the best way to discover the role of PP_i : fructose 6-phosphate 1-phosphotransferase is to measure its maximum catalytic activity, and those of phosphofructokinase and fructose bisphosphatase, and compare these values with the rates of glycolysis and gluconeogenesis in tissues with differing carbohydrate metabolism. Few data of this sort are available. Carnal and Black (1983) reported estimates of PP_i : fructose 6-phosphate 1-phosphotransferase and phosphofructokinase from a large number of species but did not show that their values reflect the maximum catalytic activities of the tissues. Thus, no comparison may be made either between enzymes or between tissues, because the extent to which the reported values underestimate the true ones is not known. Measurements made by Kruger et al. (1983) show that, in a range of tissues of the castor bean, the maximum catalytic activity of the phosphotransferase is appreciably greater than that of the phosphofructokinase, and in the endosperm of germinating seeds is high enough to mediate either gluconeogenesis or glycolysis. We (ap Rees et al. 1985) found a different picture in the club of the spadix of *Arum maculatum*. In the young, rapidly growing club the activity of the phosphotransferase exceeded that of phosphofructokinase and the rate of glycolysis. During development the activity of phosphofructokinase rose about 100-fold, whereas that of the phosphotransferase increased only two- to threefold. Thus, in the mature club the activity of phosphofructokinase substantially exceeded that of the phosphotransferase and was high enough to catalyse glycolysis during the rapid respiration at thermogenesis. More significantly the maximum catalytic activity of the phosphotransferase was substantially less than the observed rate of glycolysis.

It is conceivable that there are two entries into glycolysis in plants, via phosphofructokinase, and via PP_i: fructose 6-phosphate 1-phosphotransferase. The former may predominate when the major product of carbohydrate oxidation is ATP, and the latter during periods of marked biosynthesis when there is extensive movement of respiratory carbon into new cellular material and when polymer synthesis would provide a supply of pyrophosphate. The latter would be needed for any glycolytic role for the phosphotransferase. It is often assumed that cellular concentrations of pyrophosphate are maintained at a very low level by a ubiquitous pyrophosphatase, and that this plays a vital role in making key biosynthetic reactions irreversible, e.g., those catalysed by DNA polymerase and RNA polymerase. We know too little about the activity and location of pyrophosphatase in plant cells to know if this assumption is true for the cytosol. The level of pyrophosphate could be kept low by PP_i: fructose 6-phosphate 1-phosphotransferase or by pyrophosphatase. Clearly, if the former is a glycolytic enzyme, the interaction between it and pyrophosphatase would be crucial.

Whatever the role of PP_i: fructose 6-phosphate 1-phosphotransferase in plants, its presence must affect our investigation of carbohydrate metabolism in plants. It will complicate comparison between glycolytic capacity and glycolytic flux. It will also complicate experiments designed to elucidate and measure pathways of carbohydrate metabolism, because it provides a means whereby carbon supplied as labelled isotopes can be reversibly exchanged between fructose 6-phosphate and fructose 1,6-bisphosphate.

3 Reactions of the Oxidative Pentose Phosphate Pathway

3.1 Enzymes of the Pathway

The enzymes of this pathway have not been studied in as much detail as those of glycolysis: our knowledge of them has been reviewed by TURNER and TURNER (1980) and DENNIS and MIERNYK (1982). As with the enzymes of glycolysis, there is clear proof that at least some of the enzymes of the oxidative pentose phosphate pathway are present in plants as two isoenzymes. For photosynthetic cells we know there are two isoenzymes each of glucose 6-phosphate dehydrogenase, 6-phosphogluconate dehydrogenase and ribosephosphate isomerase (DENNIS and MIERNYK 1982). Recent investigations of transketolase from leaves failed to provide conclusive evidence of isoenzymes (MURPHY and WALKER 1982a, FEIERABEND and GRINGEL 1983). Neither transaldolase nor ribulosephosphate 3-epimerase from leaves appears to have been examined for isoenzymes.

Data from nonphotosynthetic cells of plants are even more sparse. Isoenzymes of 6-phosphogluconate dehydrogenase have been demonstrated in extracts of the endosperm of developing castor beans. However, determined attempts to demonstrate isoenzymes of glucose 6-phosphate dehydrogenase in the same extracts failed, and provide substantial evidence that this enzyme is present in only one form in this tissue (Dennis and Miernyk 1982). There is also evidence that tissues that are neither photosynthetic nor involved in massive synthesis of fat contain two isoenzymes of 6-phosphogluconate dehydrogenase (Ashihara and Fowler 1979). Other tissues and the other enzymes do not seem to have been investigated. They should be, so that we can obtain a clear picture of which enzymes of the pathway are duplicated in which cells.

Whether fructose bisphosphatase should be regarded as a constituent of the oxidative pentose phosphate pathway is a moot point. If it is present and active it would permit recycling through the pathway of fructose 1,6-bisphosphate formed from the glyceraldehyde 3-phosphate produced in the nonoxidative section of the pathway. Fructose bisphosphatase is clearly present in photosynthetic and gluconeogenic tissues. Recently it has been reported from tissues that lack these properties, viz. roots of pea and maize (Harbron et al. 1981) and suspension cultures of soybean (Macdonald and ap Rees 1983). Purification of the enzyme from the latter tissues will be needed to eliminate the possibility that the activity observed was not due to PP_i : fructose 6-phosphate 1-phosphotransferase. Whatever the outcome, it is now evident that the presence of the phosphotransferase might permit more extensive recycling through the oxidative pentose phosphate pathway.

3.2 The Nonoxidative Reactions of the Pathway

It is still useful to distinguish between the oxidative steps of the pathway, that convert glucose 6-phosphate to CO_2 and ribulose 5-phosphate, and the nonoxidative steps that convert ribulose 5-phosphate to fructose 6-phosphate and glyceraldehyde 3-phosphate. The sequence of the former needs no further comment, but that of the latter is increasingly a matter for dispute. The ready reversibility of the reactions and the broad specificities of transketolase and transaldolase have prevented the precise definition of the nonoxidative steps in the pathway. Even if we restrict ourselves to the conventionally accepted enzymes and labeling patterns, it is possible to formulate the nonoxidative steps in more than one way (Davies et al. 1964). Relatively recently, J.F. Williams and his colleagues have proposed a more radical deviation from the conventional sequence. Williams (1980) has argued that rat epididymal fat pad is the only tissue among 17 different animal and plant preparations examined that uses the conventional pathway. The other 16 tissues were held to use a different sequence, called by Williams the L-pathway. This view has not been accepted by Katz and Rognstad and their colleagues (Rognstad et al. 1982). The issue is of considerable importance. Unless we know the sequence of the pathway we cannot study its control and function effectively, and we cannot measure flux through it.

The case against the conventional representation of the pathway (Fig. 1) is made by Williams et al. (1978a) and by Williams (1980, 1981), and, in the main, comprises the following arguments.

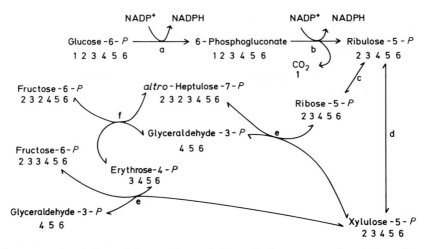

Fig. 1. The conventional view of the reactions of the oxidative pentose phosphate pathway. *Numbers* show the way the pathway redistributes the six carbons of the incoming glucose 6-phosphate. *Letters* denote the enzymes: *a* glucose-6-phosphate dehydrogenase; *b* 6-phosphogluconate dehydrogenase; *c* ribosephosphate isomerase; *d* ribulosephosphate 3-epimerase; *e* transketolase; *f* transaldolase

In liver, the maximum catalytic activities of the enzymes of the pentose phosphate pathway, and the C_6/C_1 ratios, suggest a much higher activity of the pathway than is apparent from assessments based on the assumption that the conventional pathway operates. C_6/C_1 ratios are not a reliable estimate of the activity of the pathway (AP REES 1980a). The presence of a given amount of enzyme in an extract does not prove that it is all used in vivo. This is particularly true of the oxidative pentose phosphate pathway, in which regulation is almost certainly achieved by de-inhibition (EGGLESTON and KREBS 1974).

When crude extracts of liver were supplied with ribose 5-phosphate, not all of the carbon metabolized could be accounted for as intermediates and products of the conventional pathway. The discrepancies, though referred to as up to 20% (WILLIAMS 1980), were not great: 19.2, 17.3, 13.0, 10.2, 9.8 and 4.2% (WILLIAMS et al. 1978a). These are not surprising if we bear in mind that an uncharacterized cell-free preparation was used, and that our knowledge of carbohydrate metabolism is incomplete.

When 1-^{14}C-ribose 5-phosphate was incubated with crude extracts of liver, the distribution of label in ^{14}C-hexose 6-phosphates was not, in general, that predicted by the conventional pathway (WILLIAMS et al. 1978a). The latter should lead to the almost exclusive labeling of C-1 and C-3 of hexose 6-phosphate in the ratio 1:2 (Fig. 1). Only at 17 h did the observed pattern approach that predicted. In briefer incubations there was little label in C-1 or C-3, and heavy labeling of C-2 and C-6 (WILLIAMS et al. 1978a). This type of experiment played a vital role in the conception of the conventional scheme for the nonoxidative steps of the pathway. However, as evidence for the scheme, such experiments have been largely superseded by data from experiments carried out with whole cells or tissues. The preparation used by Williams et al. (1978a) was that fraction of an acetone powder of liver that sedimented between 40 and 65% saturation with ammonium sulphate. This treatment would have disrupted cellular organization and control. Thus the preparation may well have contained enzymes, known and unknown, that metabolized carbohydrates in an uncontrolled and artefactual manner. This could have produced the observed pattern at the expense of the predicted one. Such experiments in vitro are of limited use in revealing what does happen in vivo, and almost no use in determining what does not happen.

The distribution of label in ^{14}C-glucose 6-phosphate isolated from liver by Williams et al. (1971) after supplying 2-^{14}C-glucose or 1-^{14}C-ribose, and by Longenecker and Williams (1980) from hepatocytes supplied with specifically labeled glucose does not accord with the accepted pathway. This is certainly so, but the reason for the discrepancy is not established. Further, the above results must be set against data from four separate studies, presented, or referred to, by Rognstad et al. (1982), that show that the labeling of glucose by hepatocytes supplied with a range of labeled substrates is very close to that predicted by the conventional pathway. Williams (1981) has argued that the restricted amount of data that he has reported from experiments in vivo is more definitive than the type of data reported by Rognstad et al. (1982). Williams queries, unspecifically, the use by Rognstad, and others, of compounds such as ^{14}C-xylitol, ^{14}C-xylose and ^{14}C-glycerol to study the pathway, because these compounds are not normally considered to be precursors of the pathway. However, the above compounds are metabolized to intermediates of the pathway, and any proposal as to the sequence of the pathway should be able to survive tests using these substrates.

Williams has argued that there is little value in using the labeling pattern in compounds other than phosphorylated intermediates or products of the pathway to determine its sequence. In studies with liver cells, most people have relied upon the distribution of label in glucose. Williams (1981) states that "there is no equity or indeed similarity in the distribution of ^{14}C isotope from 2-^{14}C-glucose or 1-^{14}C-ribose in liver glucose, glycogen units, fructose 6-phosphate or glucose 6-phosphate", but quotes only one example from experiments in vivo that are described in sufficient detail to be assessed (Williams et al. 1971). Here, after supplying 2-^{14}C-glucose and 1-^{14}C-ribose to the livers of rats in which the hepatic veins had been clamped, the labeling patterns of glucose, glucose 6-phosphate and fructose 6-phosphate were found to be quite different. In contrast, Rognstad et al. (1982) found good agreement between the labeling of glucose and glucose 6-phosphate when hepatocytes were supplied with 3-^{14}C-xylitol. Further, H.R. Williams and Landau (1972) found a similar distribution in glycogen, glucose and glucose 6-phosphate when 6-^{14}C-fructose was supplied to rat liver. The cause of the differences between the above sets of results is not known, nor is it known if inequality of labeling between glucose and glucose 6-phosphate is widespread.

In general, the argument that compounds close to the pathway should be used for determination of labeling pattern is sound. However, the pattern in a product is likely to be more reliable than that in an intermediate. The former should reflect the mean metabolic pattern over the whole time of the experiment. The labeling in an intermediate reflects the pattern at only one point in the experiment. Such a pattern is particularly susceptible to interference from pools of the intermediate involved in other pathways or cells in the sample of tissue. In future, both glucose 6-phosphate and its immediate derivatives should be examined.

I think that, at present, the only significant evidence against the operation of the conventional pathway is the labeling of glucose 6-phosphate in vivo described by WILLIAMS et al. (1971). As this pattern is not always found, it seems to me too early to abandon the conventional scheme, even for liver. For higher plants there are too few data for us to decide. There are no in vivo studies of the labeling of glucose 6-phosphate or any other intermediate. However, the labeling of the monomers of cellulose and xylan after feeding ^{14}C-glucose, ^{14}C-xylose, and ^{14}C-sedoheptulose has been determined (NEISH 1955, ALTERMATT and NEISH 1956). Considering that these experiments were done in the light and that some respired $^{14}CO_2$ could have been refixed in photosynthesis, the patterns obtained are consistent with the conventional scheme. Similar agreement was found when ^{14}C-xylose was supplied and the labeling of the glucosyl moiety of sucrose was determined (GINSBURG and HASSID 1956).

As a result of their investigations, which were mainly with liver, WILLIAMS et al. (1978b) proposed an alternative sequence for the nonoxidative steps of the pathway (Fig. 2). This involves the following additional intermediates: D-arabinose 5-phosphate, D-altro-heptulose 1,7-bisphosphate, D-glycero-D-ido-octulose 8-phosphate and D-glycero-D-ido-octulose 1,8-bisphosphate. Two additional enzymes are included: one, tentatively called arabinose phosphate 2-epimerase, is said to catalyse

$$\text{D-arabinose 5-P} \rightleftharpoons \text{D-ribose 5-P}$$

The other enzyme is a phosphotransferase said to catalyse

$$\text{D-glycero-D-ido-octulose 1,8-P}_2 + \text{altro-heptulose 7-P} \rightleftharpoons$$
$$\text{D-glycero-D-ido-octulose 8-P} + \text{altro-heptulose 1,7-P}_2$$

The main evidence for this proposal is as follows. First, it accounts for all of the carbon metabolized, and the labeling patterns obtained in the experiments in vitro described by WILLIAMS et al. (1978a). This does not prove that the sequence operates in vivo. Second, it explains the labeling pattern of glucose 6-phosphate obtained in vivo by WILLIAMS et al. (1971), and LONGENECKER and WILLIAMS (1980). However, the new formulation does not explain the more extensive labeling data obtained, in vivo, by KATZ, ROGNSTAD and their colleagues and discussed above. Until the doubts about using hexose 6-phosphates for the determination of the labeling patterns are resolved, and the discrepancies between the different sets of experiments in vivo are explained, it seems premature to accept the alternative hypothesis.

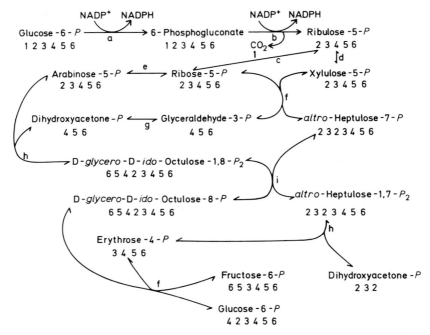

Fig. 2. The reactions of the oxidative pentose phosphate pathway proposed by Williams (1980). *Numbers* show the way the pathway redistributes the six carbons of the incoming glucose 6-phosphate. *Letters* denote the enzymes: *a* glucose-6-phosphate dehydrogenase; *b* 6-phosphogluconate dehydrogenase; *c* ribosephosphate isomerase; *d* ribulosephosphate 3-epimerase; *e* arabinose phosphate 2-epimerase; *f* transketolase; *g* triosephosphate isomerase; *h* aldolase; *i* a phosphotransferase

    The assertion that the alternative pathway operates in photosynthetic tissues (Williams 1980) seems to me to be, as yet, without any published direct experimental support. It is said that extracts of *Chlorella* contain phosphotransferase, but no evidence is presented. Such activity in an extract would not be too surprising, but would hardly demonstrate the operation of the alternative pathway. The other evidence is that the alternative pathway would explain the Gibbs Effect in photosynthetic carbon metabolism (Clark et al. 1974). There is already an adequate explanation of the Gibbs Effect (Bassham 1964).

    In conclusion, I suggest that the experiments of Williams and his colleagues have demonstrated the presence of additional enzymes and substrates of carbohydrate metabolism, but are not yet definitive enough to establish the alternative pathway or to cause us to abandon the scheme in Fig. 1 as the best working hypothesis. What is clear is that for plants we have insufficient information to decide on the reactions of the nonoxidative section of the oxidative pentose phosphate pathway. We need to determine, at different times, the manner in which label, from precursors supplied to cells or tissues, is distributed amongst the carbons of glucose 6-phosphate, fructose 6-phosphate and compounds immediately derived from these intermediates. As will become clear from the next section, in carrying out the above experiments we will have to distinguish between the pathway in the cytosol and that in the plastid.

4 Location and Inter-relationship of Glycolysis and the Oxidative Pentose Phosphate Pathway

The presence of isoenzymes for many of the steps of the two pathways raises the question of their intra-cellular location. It is now known that plant cells are distinguished from other eukaryotic cells in having all, or substantial fractions, of both pathways of carbohydrate oxidation duplicated in the cytosol and the plastids. This is directly related to the fact that in plants a great deal of biosynthesis, in addition to the reductive pentose phosphate pathway, is located in the plastid. This crucial role of plastids in biosynthesis is not confined to photosynthetic cells, and appears to be a general feature of plant cells as a whole. The dual location of the pathways of carbohydrate oxidation greatly complicates the study of their inter-relationship.

4.1 Carbohydrate Oxidation in the Cytosol

It is difficult to show that an enzyme is cytosolic. The evidence is usually negative in that, on fractionation of an extract, the enzyme is not found to be associated with a known structure. Such evidence is unconvincing unless accompanied by proof that the fractionation procedure preserved intact a reasonable proportion of the recognized organelles. There is evidence of the latter sort that the cytosol of nonphotosynthetic and photosynthetic cells of plants contains each of the enzymes of glycolysis. NISHIMURA and BEEVERS (1979) fractionated the endosperm of germinating castor beans so carefully that essentially all of the plastids were recovered intact. Nonetheless, activity of each glycolytic enzyme was found in the cytosolic fraction. DENNIS and MIERNYK (1982) have summarized the convincing evidence that the cytosol of developing castor beans contains the glycolytic enzymes needed to convert glucose to pyruvate. For leaves there is sound evidence for the presence in the cytosol of: phosphoglucomutase, glucosephosphate isomerase (HERBERT et al. 1979), phosphofructokinase, phosphoglyceromutase, enolase, pyruvate kinase (STITT and AP REES 1979), aldolase, triosephosphate isomerase, phosphoglycerate kinase (ANDERSON and ADVANI 1970), and glyceraldehyde-phosphate dehydrogenase (LATZKO and KELLY 1979).

It is not clear which of the hexose kinases are present in the cytosol, and further studies of their location are needed. In spinach leaves (BALDUS et al. 1981) and pea stems (TANNER et al. 1983) fructokinase (EC 2.7.1.4) appears to be entirely cytosolic. Activity with glucose is generally found in the soluble and in the particulate fractions of plant extracts (STITT et al. 1978, BALDUS et al. 1981). In extracts of pea leaves (DRY et al. 1983) and stems (TANNER et al. 1983) much and all, respectively, of the hexokinase activity was mitochondrial and associated with the outer membrane. Thus, proof of a cytosolic location of this enzyme will require evidence that its presence in the soluble fraction is not due to the rupture of the outer mitochondrial membrane. A location of hexokinase on the outer mitochondrial membrane would place the enzyme next to a source of ATP but keep it accessible to the cytosol. There seems little doubt that the cytosol of plant cells can catalyse the complete glycolytic sequence.

For the oxidative pentose phosphate pathway, the cytosol of the endosperm of developing (Dennis and Miernyk 1982) and germinating (Nishimura and Beevers 1979) castor beans has been shown to contain both the dehydrogenases, transaldolase and transketolase. The same four enzymes, plus ribulosephosphate 3-epimerase have been shown to be present in the cytosol of pea roots (Emes and Fowler 1979b). For leaves, there is adequate proof that the cytosol contains both dehydrogenases (Stitt and ap Rees 1979), transketolase (Murphy and Walker 1982a, Feierabend and Gringel 1983), and ribose phosphate isomerase (Park and Anderson 1973). The last reference also demonstrates a cytosolic isomerase in nonphotosynthetic cells. Although we need more information on transaldolase, the epimerase and the isomerase, it is probably safe to conclude that the cytosol of plant cells can catalyse the complete oxidative pentose phosphate pathway.

As glycolysis and the oxidative pentose phosphate pathway share intermediates, a close relationship between the two pathways in the cytosol would be expected. Previously I have argued for the relationship shown in Fig. 3 (ap Rees 1974, 1980a, b). The bulk of the products of the oxidative pentose phosphate pathway were regarded as being converted to pyruvate by the reactions of glycolysis. Limited, and probably variable, recycling of fructose 6-phosphate through the pentose phosphate pathway was envisaged. In nonphotosynthetic and nongluconeogenic tissues recycling of glyceraldehyde 3-phosphate was thought unlikely.

Recent studies affect the above conclusions. First, as will be shown later, the scheme in Fig. 3 must now be regarded as incomplete because it ignores the appreciable carbohydrate oxidation that occurs in plastids. Second, the presence of PP_1 : fructose 6-phosphate 1-phosphotransferase provides the potential for a second point of entry into glycolysis, and also a means whereby glyceraldehyde 3-phosphate, after conversion to fructose 1,6-bisphosphate, could be recycled through the pentose phosphate pathway in nonphotosynthetic and nongluconeogenic cells. However, the latter is still unlikely, as the major evidence against it, the low yield of $^{14}CO_2$ from 6-^{14}C-glucose, remains (ap Rees 1974). Finally is the substantial point that the labeling experiments that gave rise to Fig. 3 were designed and interpreted on the assumption that both pathways were confined to the cytosol. It is reasonable to assume that in these experiments the added isotopes did label the intermediates of the cytosolic pathways. Thus the results may still be said to support Fig. 3 as a representation of what happens in the cytosol. We do not know the extent to which the plastid pathways became labeled. Any such labeling is unlikely to affect the basic conclusion that most of the products of the oxidative pentose phosphate pathway in the cytosol

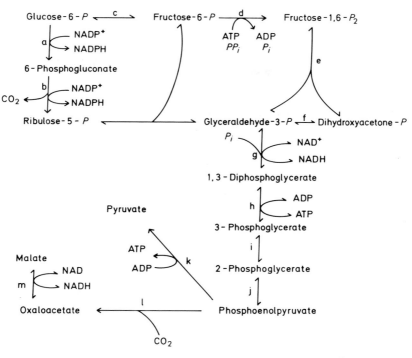

Fig. 3. Relationship between glycolysis and the oxidative pentose phosphate pathway in the cytosol of plant cells. *Letters* denote: *a* glucose-6-phosphate dehydrogenase; *b* 6-phosphogluconate dehydrogenase; *c* glucosephosphate isomerase; *d* phosphofructokinase and or PP_i:fructose 6-phosphate 1-transferase; *e* fructose-bisphosphate aldolase; *f* triosephosphate isomerase; *g* glyceraldehydephosphate dehydrogenase; *h* phosphoglycerate kinase; *i* phosphoglyceromutase; *j* enolase; *k* pyruvate kinase; *l* phosphoenolpyruvate carboxylase; *m* malate dehydrogenase. For the reactions that convert ribulose 5-phosphate into fructose 6-phosphate and glyceraldehyde 3-phosphate see Fig. 1

are converted to pyruvate via the reactions of glycolysis. Plastid metabolism may have contributed differentially to release of $^{14}CO_2$ from ^{14}C-glucose and ^{14}C-fructose in studies of the recycling of fructose 6-phosphate (AP REES et al. 1965). However, the data still show that in a number of tissues fructose 6-phosphate enters the pentose phosphate pathway much less readily than does glucose 6-phosphate. We cannot tell from these results if this is a property of the cytosolic pathway, the plastid pathway, or both.

Some of the glycolytic phosphoenolpyruvate is converted to oxaloacetate by phosphoenolpyruvate carboxylase (EC 4.1.1.31) (AP REES 1980a). This is a property of cytosolic glycolysis as the carboxylase is confined to the cytosol (QUAIL 1979). Studies with the spadices of Araceae have revealed an additional feature of this carboxylation (AP REES et al. 1981, 1983). In these tissues the oxaloacetate produced is reduced to malate in the cytosol. The malate then enters the mitochondrial matrix where it is converted to pyruvate by NAD malic enzyme (EC 1.1.1.39): the pyruvate is subsequently metabolized by the tricarboxylic acid cycle. The significance of this sequence is not known. It may be a way of re-oxidizing cytosolic NADH, or an additional means of transporting the products of carbohydrate oxidation into mitochondria.

4.2 Carbohydrate Oxidation in Plastids

In photosynthetic and non-photosynthetic cells plastids play a key role in biosynthesis, e.g., synthesis of starch (MACDONALD and AP REES 1983), fat (STUMPF 1980), isoprenoid compounds (PORTER and SPURGEON 1981), amino acids (ROTHE et al. 1983), and the assimilation of nitrate and ammonia (MIFLIN and LEA 1982). Although plastids in certain cells become specialized for particular syntheses, in almost all living cells of plants significant synthesis is likely to be occurring in plastids. Thus, plastids need the three basic products of carbohydrate oxidation: intermediates, ATP, and reducing power. The relative needs will vary with the plastid and the cell.

Two mechanisms may meet the above needs. One is import from the cytosol, either directly or via a shuttle. A range of such systems is known for chloroplasts (HEBER and HELDT 1981) but we do not know whether they operate in other plastids. The unitary view, that all plastids use the same transport systems, has the virtue of simplicity but is, as yet, without much experimental support. A priori there appears to be no reason why differentiation of plastids should not involve the development of transport systems characteristic of each type of plastid. The other way to meet the plastids' needs for intermediates, ATP and reducing power is to make them in situ by carbohydrate oxidation within the plastids. The relative contributions of the latter and of direct import are not known, but are likely to vary with the plastid, as are the pathways of plastid carbohydrate oxidation. Not enough is known to generalize further, and the major types that have been studied are considered separately.

4.2.1 Chloroplasts

In addition to the enzymes of the reductive pentose phosphate pathway, chloroplasts contain one of the isoenzymes of each dehydrogenase of the oxidative pathway (DENNIS and MIERNYK 1982) and transaldolase (LATZKO and GIBBS 1968). Thus they can catalyse the complete oxidative pentose phosphate pathway with recycling of the glyceraldehyde 3-phosphate as well as that of the fructose 6-phosphate. Chloroplasts also contain one of the isoenzymes of phosphofructokinase (KELLY and LATZKO 1977b), phosphoglucomutase and glucosephosphate isomerase (HERBERT et al. 1979). α-Glucan phosphorylase is also present (STITT et al. 1978). There is adequate evidence to show that chloroplasts can convert $(1 \rightarrow 4)$-α-glucans to 3-phosphoglycerate via glycolysis. The extent to which chloroplasts can catalyse the remaining reactions of glycolysis is not resolved. The point is of considerable importance because of the chloroplasts' need for precursors for the synthesis of lipids, isoprenoids and amino acids. Pea chloroplasts contain relatively modest amounts of enolase and pyruvate kinase (STITT and AP REES 1979). A chloroplast isoenzyme of pyruvate kinase has been demonstrated in leaves of castor bean (IRELAND et al. 1979). It seems likely that these two enzymes are present in chloroplasts. The key to this question is phosphoglyceromutase. There is considerable evidence that chloroplasts from pea shoots do not contain significant activities of this enzyme (STITT and AP REES 1979). The enzyme was readily detected in crude and purified preparations of chloroplasts, but it always behaved as a cytosolic contaminant rather than as a chloroplast enzyme. When the isolated chloroplasts metabolized starch or glucose in the dark triose phosphates and 3-phosphoglycerate were exported and there was no evidence of complete glycolysis within the chloroplast (STITT and AP REES 1980a).

In contrast to the above, there is evidence, reviewed by DENNIS and MIERNYK (1982) and GIVAN (1983), and added to by MURPHY and WALKER (1982b), that spinach chloroplasts contain the complete glycolytic sequence. It may well be that chloroplasts from different species, or even from leaves of different ages from the same species, differ in the extent to which they contain the complete glycolytic sequence. However, as yet no one has published a direct demonstration that spinach chloroplasts do contain phosphoglyceromutase. This requires measurement of the enzyme itself and proof that the activity is within the chloroplast. The evidence for complete glycolysis in spinach chloroplasts is indirect in that it consists of demonstrations that purified chloroplasts can convert products of photosynthesis to fatty acids. Even the best of this evidence (MURPHY and WALKER 1982b) fails to show that the chloroplasts were free of cytosolic phosphoglyceromutase. The absence of detectable phosphoenolpyruvate carboxylase suggests that the chloroplasts were not contaminated by cytosol, but the absolute activity of the carboxylase in the unfractionated extracts is likely to have been very much lower than that of phosphoglyceromutase. In peas the ratio of the former to the latter is 3.3:181 (STITT and AP REES 1979). Thus absence of detectable carboxylase does not prove that there was not enough cytosolic phosphoglyceromutase present to convert some 3-phosphoglycerate, exported from the chloroplast, to 2-phosphoglycerate that re-entered the chloroplast. It is pertinent to note that there has been no report that photosynthetic cells contain isoenzymes of phosphoglyceromutase. A great deal more direct work on a wider range of species is required to resolve the question of whether chloroplasts can catalyse the whole of glycolysis.

The inter-relationships between the pathways of carbohydrate oxidation in chloroplasts have been studied by following the metabolism of starch and glucose by chloroplasts from peas (STITT and AP REES 1980a) and spinach (STITT and HELDT 1981). The results are consistent with the scheme in Fig. 4. This shows that glucose 6-phosphate may enter glycolysis and be converted directly to triose phosphates or 3-phosphoglycerate which are then exported to the cytosol. Glucose 6-phosphate may also enter the oxidative pentose phosphate pathway and both the products of this pathway, fructose 6-phosphate and glyceraldehyde 3-phosphate, may be recycled. Alternatively, the fructose 6-phosphate could enter glycolysis, and the glyceraldehyde 3-phosphate be exported either as triose phosphate or after conversion to 3-phosphoglycerate.

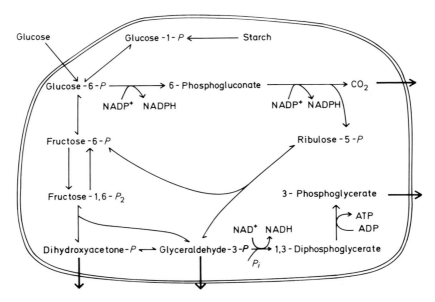

Fig. 4. Tentative scheme for pathways of carbohydrate oxidation in chloroplasts

The crucial evidence for the scheme in Fig. 4 comes from the enzymic composition of the chloroplasts, the observation that during the breakdown of starch in the dark triose phosphates and 3-phosphoglycerate are exported, and the patterns of $^{14}CO_2$ production when isolated chloroplasts are supplied with specifically labeled glucose in the dark. The fact that glucose carbons 2 and 6 are both converted to CO_2 by chloroplasts provides the main evidence for recycling of fructose 6-phosphate and 3-phosphoglyceraldehyde. Appreciable operation of the scheme in Fig. 4 is almost certainly restricted to the dark. Evidence for this, and possible mechanisms of control, are presented by Bassham (1979) and Bassham and Buchanan (1982). Perhaps the greatest drawback of the scheme in Fig. 4 is the need for fructose bisphosphatase to show sufficient activity in the dark to permit the recycling of glyceraldehyde 3-phosphate. We know that the phosphatase is severely inhibited in the dark (Laing et al. 1981) but we do not know whether enough activity remains for the recycling.

The capacity of chloroplasts to catalyse carbohydrate oxidation is substantial when compared to that of the cytosol. In young leaves of peas, about 40% of the activity of the dehydrogenases of the oxidative pentose phosphate pathway and 25% of that of phosphofructokinase is in the chloroplast (Stitt and ap Rees 1979). This shows that the ratio of the capacities of the two pathways in the chloroplast differs from that in the cytosol. There is evidence that suggests that these differences in capacity are reflected by the relative activities of the pathways in vivo. In pea chloroplasts, the fact that the oxidative pentose phosphate pathway is the only means whereby CO_2 can be released from glucose removes many of the factors that complicate assessment of the pathway (ap

Rees 1980a, Stitt and ap Rees 1980b). In pea chloroplasts we calculated that for every 100 molecules of glucose 6-phosphate metabolized in the dark 35–56 entered the oxidative pentose phosphate pathway. The comparable figures for spinach chloroplasts were 37–47 (Stitt and Heldt 1981). These estimates are significantly higher than those made for pea leaves as a whole (Stitt and ap Rees 1978). Thus the activity of the oxidative pentose phosphate pathway relative to that of glycolysis appears to be greater in the chloroplast than in the cytosol.

Carbohydrate oxidation in chloroplasts requires a substrate. For most chloroplasts, at least in the initial period of darkness, starch is almost certainly the principal substrate. In peas the rate of starch breakdown is rapid during the first 3–6 h of darkness, thereafter the rate is low (Stitt et al. 1978). Under the latter conditions the chloroplast may import substrate, as will chloroplasts that do not make starch. Both pea (Stitt and ap Rees 1980a) and spinach (Stitt and Heldt 1981) chloroplasts metabolize exogenous glucose at rates that might maintain at least a basal rate of metabolism in the chloroplast in the dark. The manner in which this glucose is phosphorylated is not established. Although pea chloroplast preparations showed appreciable hexokinase activity it was not within the chloroplast (Stitt et al. 1978). As glucose 6-phosphate does not readily cross the chloroplast envelope it is unlikely that the above hexokinase is responsible for the provision of glucose 6-phosphate in the chloroplast. The latter should be examined to see if they contain their own hexokinase.

Too small a range of chloroplasts has been studied for us to be sure of the complete significance of carbohydrate oxidation in chloroplasts. The following picture emerges from the available data. The evidence that chloroplasts lack, or have very little, phosphoglyceromutase suggests that provision of acetyl-CoA for chloroplast biosynthesis is not a quantitatively important aspect of chloroplast glycolysis. The data of Murphy and Walker (1982b) suggest that this is true even of spinach chloroplasts that may contain the complete glycolytic sequence. Support for the above view is provided by the evidence of alternative sources of acetyl-CoA for the chloroplast. For example, Stumpf (1980) points out that acetate is an excellent precursor for fatty acids when supplied to chloroplasts. Subsequently, Kuhn et al. (1981) have shown that leaves contain acetate, and that chloroplasts possess acetyl-CoA synthetase (EC 6.2.1.1). In addition, mitochondrial preparations have been shown to contain a short-chain acyl-CoA hydrolase that can convert acetyl-CoA to acetate (Liedvogel and Stumpf 1982). Thus acetyl-CoA, formed in the mitochondria from pyruvate produced by cytosolic glycolysis, could be metabolized therein to acetate, which could then pass from the mitochondria to the chloroplast and there be converted to acetyl-CoA. Other possibilities are discussed by Givan (1983). It is conceivable that each of the above means of forming acetyl-CoA for the chloroplast operates to a different extent in chloroplasts from leaves of different ages and different species.

The main significance of carbohydrate oxidation in chloroplasts is likely to be that it not only provides a means for exporting the products of starch breakdown via the phosphate translocator, but also permits the synthesis of ATP and NADPH in the darkened chloroplast. Although the amounts of ATP (Stitt et al. 1982b) and NADPH (Krause and Heber 1976) in chloroplasts fall on darkening, significant levels are maintained in the dark. The presence of the two routes to triose phosphate, and the option of exporting either triose phosphate or 3-phosphoglycerate means that the scheme in Fig. 4 is sufficiently versatile to permit ATP to be made in the chloroplast independently of NADPH formation, and for both processes to be independent of the mobilization and export of chloroplast starch.

4.2.2 Plastids Involved in Massive Synthesis of Fat

Dennis and his colleagues have made a very comprehensive study of carbohydrate oxidation in the fatty acid-synthesizing plastids from the endosperm of developing castor bean (Dennis and Miernyk 1982). They have shown that each glycolytic enzyme and pyruvate dehydrogenase is represented in the plastid by a distinct isoenzyme. The activities in the plastids are substantial in relation to the rate of fatty acid synthesis in vivo. Further, these plastids have been shown to contain transketolase, transaldolase, and their own isoenzyme of 6-phosphogluconate dehydrogenase. Ribosephosphate isomerase and ribulose-phosphate 3-epimerase do not appear to have been studied. Extensive attempts to discover glucose 6-phosphate dehydrogenase in these plastids have given negative results, and led to the conclusion that this enzyme is not in these plastids. This conclusion is supported by the observation that only one form of this enzyme can be detected in the endosperm. As with leaves, the proportions of the total activities of the enzymes in the endosperm that are present in the plastid are high, 20–70%. These plastids are distinguished from chloroplasts in that they can catalyse the whole of glycolysis but cannot catalyse the complete oxidative pentose phosphate pathway.

As yet there are no data from experiments with labeled substrates on the precise relationships between the pathways in plastids from developing endosperm of castor bean. The enzymic complement of these plastids suggests complete glycolysis plus oxidation of 6-phosphogluconate via the pentose phosphate pathway to fructose 6-phosphate and glyceraldehyde 3-phosphate which are then metabolized via glycolysis. This could provide the acetyl-CoA required for fatty acid synthesis plus ATP, NADH and NADPH. The question of substrates to supply these plastid pathways has been examined by Miernyk and Dennis (1983), who compared the ability of different compounds to provide carbon for lipid synthesis by isolated plastids. Sucrose and glucose 6-phosphate were ineffective precursors when supplied to intact plastids; fructose, glucose, 3-phosphoglycerate, acetate and pyruvate, in increasing order of effectiveness, did label lipids. These results, plus the fact that the only enzymes known to break down sucrose, invertase and sucrose synthase, are cytosolic (Dennis and Miernyk 1982), make it most unlikely that sucrose enters these plastids. However, there appears to be considerable flexibility in respect of precisely which compounds do enter. Hexose and phosphoglycerate could enter to support glycolysis and supply carbon for fatty acid synthesis; pyruvate and acetate could also do the latter. The relative contributions of these possibilities are not known, but it is of interest to note that, even where the plastid contains the complete glycolytic sequence, carbon for fatty acid synthesis can still be imported as either pyruvate or acetate formed during cytosolic glycolysis. In order to supply the oxidative pentose phosphate pathway in these plastids, 6-phosphogluconate, or conceivably a precursor other than glucose 6-phosphate, must enter.

In the endosperm of germinating castor beans, gluconeogenesis is the main metabolic activity, but there is some lipid synthesis. Preparations of plastids from this tissue showed no significant activity of glucose 6-phosphate dehydrogenase, but did contain 6-phosphogluconate dehydrogenase, transketolase, transaldolase and each enzyme of glycolysis (NISHIMURA and BEEVERS 1979). One of the isoenzymes of each of the following: phosphoglucomutase, glucose-phosphate isomerase, phosphofructokinase, aldolase and 6-phosphogluconate dehydrogenase (NISHIMURA and BEEVERS 1981) is present in the plastids. The proportions of enolase and phosphoglyceromutase recovered in the plastid preparations are low. In view of the arguments over the location of phosphoglycero-mutase in chloroplasts, we need further evidence that this enzyme and enolase are within the plastids of endosperm from germinating seeds. Nonetheless, the above data suggest a similar picture to that presented for developing endosperm except that for most of the enzymes the proportion present in the plastid was different.

Whether plastids from the endosperm of germinating castor beans contain pyruvate dehydrogenase is not established. NISHIMURA and BEEVERS (1979), and FRITSCH and BEEVERS (1979) detected this enzyme in plastid preparations, but RAPP and RANDALL (1980) did not. Again it will be important to determine if the activity in the preparation is actually within the plastid. Absence of pyruvate dehydrogenase from these plastids would raise two significant points. First, it would show qualitative differences in carbohydrate oxidation in plastids performing the same function in different nonphotosynthetic tissues. Second, it would show that the significance of carbohydrate oxidation in plastids from germinating castor bean was not provision of acetyl-CoA for fatty acid synthesis. In this instance the latter could be formed as described earlier or via ATP citrate-lyase (EC 4.1.3.8) activity in the plastid (FRITSCH and BEEVERS 1979).

4.2.3 Other Plastids

Studies of other types of plastid have been less common and intensive. EMES and FOWLER (1979 a, b) have investigated carbohydrate metabolism in the plastids involved in nitrate assimilation in the apices of pea roots. Preparations of plastids from homogenates of pea roots contained significant proportions (6–40%) of the total activities of each of the enzymes of the oxidative pentose phosphate pathway. There is additional evidence that plastids from roots contain both dehydrogenases of the pentose phosphate pathway (MIFLIN and BEEVERS 1974). These results strongly suggest that these plastids contain the complete sequence of the oxidative pentose phosphate pathway. The extent to which glycolysis is present in these plastids is not known. EMES and FOWLER (1979 a, b) present evidence that triosephosphate isomerase is in pea root plastids. In a recent paper (EMES and FOWLER 1983), they report that these plastids also contain pyruvate kinase, but they provide no data on this point. It is important to discover which of the remaining enzymes of glycolysis are present, and thus complete our knowledge of the capacity of these plastids to oxidize carbohydrate. However, the data that are available are sufficient to demonstrate the important point that plastids from nonphotosynthetic tissues differ significantly in their complement of enzymes of carbohydrate oxidation. Glucose-6-phosphate dehydrogenase is present in plastids involved in nitrate assimilation in roots but not in plastids involved in massive synthesis of fat.

EMES and FOWLER (1983) supplied 1-^{14}C- and 6-^{14}C-glucose 6-phosphate to prepararations of plastids from pea roots and found that C-1 was more readily converted to $^{14}CO_2$ than was C-6. These results are consistent with the operation in the plastids of the complete oxidative pentose phosphate pathway with re-cycling of the glyceraldehyde 3-phosphate. However, proof of this will require a more complete characterization of the enzymic complement of the plastids and, above all, proof that the metabolism of the labeled glucose 6-phosphate was due to intact plastids and was not affected by cytosolic contamination or damaged plastids. The above results also imply that glucose 6-phosphate entered the plastids. Again, this conclusion must be resisted until we have proof that the metabolism was due to intact plastids. The presence of contaminating phosphatases could have converted the ^{14}C-glucose 6-phosphate to ^{14}C-glucose that entered the plastid. Alternatively, the metabolism may have been due to damaged plastids. Indeed, the range of intermediates that was found to support nitrite reduction by the plastid preparation suggests either that these plastids have very different permeability properties from types of plastid that have been studied, or that the plastids were damaged. The key to these questions is the development of techniques that will permit the isolation of uncontaminated fully functional plastids from nonphotosynthetic cells.

There have been very few studies of amyloplasts, probably because of the difficulty of isolating them. Besides a source of substrate for starch synthesis, amyloplasts are likely to need ATP for the synthesis of ADPglucose, and a means of converting the stored starch into a soluble compound(s) for export to the cytosol. If the latter occurs via the phosphate translocator, then one or both of the pathways of carbohydrate oxidation will be needed to convert hexose to triose phosphate in the amyloplast. This need would not arise if hexose or hexose phosphates were exported to the cytosol during starch breakdown. At present we have too little information to answer these questions. Amyloplasts from suspension cultures of soybean have been shown to contain phosphofructokinase (MACDONALD 1982), α-glucan phosphorylase, phosphoglucomutase, glucosephosphate isomerase, aldolase, glyceraldehyde-phosphate dehydrogenase and triosephosphate isomerase (MACDONALD and AP REES 1983). They thus have the capacity to convert $(1 \rightarrow 4)$-α-glucans to triose phosphate via glycolysis.

Carbohydrate oxidation in chromoplasts has been neglected almost as badly as that in amyloplasts. Recent studies have led to the suggestion that carbon to support isoprenoid synthesis in chromoplasts is imported from the cytosol as isopentenyl pyrophosphate (KREUZ and KLEINIG 1981). Mevalonate kinase (EC 2.7.1.36) is a cytosolic enzyme (GRAY and KEKWICK 1973), and there is appreciable evidence that this is true of the remaining enzymes needed for the synthesis of isopentenyl pyrophosphate (KREUZ and KLEINIG 1981). Further, isolated chromoplasts have been shown to convert isopentenyl pyrophosphate to isoprenoid compounds (BEYER et al. 1980). If this hypothesis is correct and applicable to plastids as a whole, then the provision of intermediates and energy for isoprenoid synthesis must be regarded mainly as a function of carbohydrate oxidation in the cytosol, not the plastid. An attractive feature of the hypothesis is that it offers an explanation of how the plastid might control the flow of carbon to fatty acids and isoprenoid compounds. Acetyl-CoA in the plastid, whether made by plastid glycolysis, or by any other route, would not be metabolized to isoprenoids, as the enzymes needed to convert it to isopentenyl pyrophosphate would be absent. Conversely, the confinement of fatty acid synthesis to plastids would allow cytosolic acetyl-CoA to be channelled to isopentenyl pyrophosphate, which could enter the plastid as a unique precursor for plastid isoprenoid compounds. These suggestions raise an additional problem in that the scheme requires there to be acetyl-CoA in the cytosol.

Our knowledge of carbohydrate oxidation in plastids is still very fragmentary. Nonetheless, the following very important points emerge. First, in many, if not all, plant cells all or part of glycolysis and the oxidative pentose phosphate pathway are duplicated to a significant extent in the plastid. Second, there is considerable variation between cells in respect of which sections of the pathways are duplicated, and in the quantitative distribution of the pathways between the two compartments. Third, the relative activities of glycolysis and the oxidative pentose phosphate pathway in the plastid may be quite different from those in the cytosol.

The proportion of total carbohydrate oxidation, and the relative activities of the pathways, found in the plastid almost certainly vary with the demands of the particular plastid. The form of the pathways in the plastid, i.e., whether a pathway is complete or not, is presumably related to the particular specialist form of metabolism of the plastid. For example, the lack of phosphoglyceromutase in pea chloroplasts may be a means of preventing interaction between pyruvate kinase and the reductive pentose phosphate pathway. Much more information is required before we can be sure of the significance of the different arrangements of carbohydrate oxidation found in different plastids.

The fact that carbohydrate oxidation in plants is compartmented should be taken into account in future investigations of glycolysis and the oxidative pentose phosphate pathway in plants. In future it will be necessary either to show that in one's chosen experimental material the plastid pathways are too minor to affect the data, a difficult task and an unlikely situation, or to devise an experimental procedure that takes compartmentation into account. Thus studies and measurements of enzymes must distinguish between plastid and cytosolic forms.· Measurements of flux and of changes in substrate levels must make a similar distinction, as must studies based on the metabolism of labeled substrates. The difficulties imposed by these restrictions are to some extent offset by the development of more sophisticated methods for fractionation and analysis of plant cells, e.g., that described by LILLEY et al. (1982).

Acknowledgements. I am grateful to Professor D.T. DENNIS for his helpful suggestions and to Dr. S. MORRELL for her criticism of the manuscript.

References

Altermatt HA, Neish AC (1956) The biosynthesis of cell wall carbohydrates III. Further studies on formation of cellulose and xylan from labelled monosaccharides in wheat plants. Can J Biochem Physiol 34:405–413

Anderson LE, Advani VR (1970) Chloroplast and cytoplasmic enzymes. Three distinct isoenzymes associated with the reductive pentose phosphate cycle. Plant Physiol 45:583–585

ap Rees T (1974) Pathways of carbohydrate breakdown in higher plants. In: Northcote DH (ed) Int Rev Biochem, vol XI. Butterworths, London, pp 89–127

ap Rees T (1977) Conservation of carbohydrate by the non-photosynthetic cells of higher plants. Symp Soc Exp Biol 31:7–32

ap Rees T (1980a) Assessment of the contributions of metabolic pathways to plant respiration. In: Davies DD (ed) The biochemistry of plants, vol II. Academic Press, London New York, pp 1–29

ap Rees T (1980b) Integration of pathways of synthesis and degradation of hexose phosphates. In: Preiss J (ed) The biochemistry of plants, vol III. Academic Press, London New York, pp 1–42

ap Rees T, Blanch E, Graham D, Davies DD (1965) Recycling in the pentose phosphate pathway: comparison of C_6/C_1 ratios measured with glucose-C^{14} and fructose-C^{14}. Plant Physiol 40:910–914

ap Rees T, Fuller WA, Green JH (1981) Extremely high activities of phosphoenolpyruvate carboxylase in thermogenic tissues of Araceae. Planta 152:79–86

ap Rees T, Bryce JH, Wilson PM, Green JH (1983) Role and location of NAD malic enzyme in thermogenic tissues of Araceae. Arch Biochem Biophys 227:511–521

Ashihara HA, Fowler MW (1979) 6-Phosphogluconate dehydrogenase species from roots of *Pisum sativum*. Int J Biochem 10:675–681

Baldus B, Kelly GJ, Latzko E (1981) Hexokinases of spinach leaves. Phytochemistry 20:1811–1814

Bassham JA (1964) Kinetic studies of the photosynthetic carbon reduction cycle. Annu Rev Plant Physiol 15:101–120

Bassham JA (1979) The reductive pentose phosphate cycle. In: Gibbs M, Latzko E (eds) Photosynthesis II. Encyclopedia of plant physiology, vol VI, New Ser. Springer, Berlin Heidelberg New York, pp 9–30

Bassham JA, Buchanan BB (1982) Carbon dioxide fixation pathways in plants and bacteria. In: Govindjee (ed) Photosynthesis, vol II. Academic Press, London New York, pp 141–189

Beyer P, Kreuz K, Kleinig H (1980) β-Carotene synthesis in isolated chromoplasts from *Narcissus pseudonarcissus*. Planta 150:435–438

Carnal NW, Black CC (1979) Pyrophosphate-dependent 6-phosphofructokinase, a new glycolytic enzyme in pineapple leaves. Biochem Biophys Res Commun 86:20–26

Carnal NW, Black CC (1983) Phosphofructokinase activities in photosynthetic organisms. Plant Physiol 71:150–155

Clark MG, Williams JF, Blackmore PF (1974) Exchange reactions in metabolism. Catal Rev Sci Eng 9:35–77

Cséke C, Buchanan BB (1983) An enzyme-synthesizing fructose 2,6-bisphosphate occurs in leaves and is regulated by metabolic effectors. FEBS Lett 155:139–142

Cséke C, Weeden NF, Buchanan BB, Uyeda K (1982) A special fructose bisphosphate functions as a cytoplasmic regulatory metabolite in green leaves. Proc Natl Acad Sci USA 79:4322–4326

Davies DD, Gionvanelli J, ap Rees T (1964) Plant biochemistry. Blackwell, Oxford

Dennis DT, Miernyk JA (1982) Compartmentation of nonphotosynthetic carbohydrate metabolism. Annu Rev Plant Physiol 33:27–50

Dry IB, Nash D, Wiskich JT (1983) The mitochondrial localization of hexokinase in pea leaves. Planta 158:152–156

Eggleston LV, Krebs HA (1974) Regulation of the pentose phosphate cycle. Biochem J 138:425–435

Emes MJ, Fowler MW (1979a) The intracellular location of the enzymes of nitrate assimilation in the apices of seedling pea roots. Planta 144:249–253

Emes MJ, Fowler MW (1979b) Intracellular interactions between the pathways of carbohydrate oxidation and nitrate assimilation in plant roots. Planta 145:287–292

Emes MJ, Fowler MW (1983) The supply of reducing power for nitrite reduction in plastids of seedling pea roots (*Pisum sativum* L.). Planta 158:97–102

Feierabend J, Gringel G (1983) Plant transketolase: subcellular distribution, search for multiple forms, site of synthesis. Z Pflanzenphysiol 110:247–258

Fritsch H, Beevers H (1979) ATP citrate lyase from germinating castor bean endosperm. Plant Physiol 63:687–691

Ginsburg V, Hassid WZ (1956) Pentose metabolism in wheat seedlings. J Biol Chem 223:277–284

Givan CV (1983) The source of acetyl coenzyme A in chloroplasts of higher plants. Physiol Plant 57:311–316

Gray JC, Kekwick RGO (1973) Mevalonate kinase in green leaves and etiolated cotyledons of the French bean *Phaseolus vulgaris*. Biochem J 133:335–347

Harbron S, Foyer C, Walker DA (1981) The purification and properties of sucrose-phosphate synthetase from spinach leaves: the involvement of this enzyme and fructose bisphosphatase in the regulation of sucrose biosynthesis. Arch Biochem Biophys 212:237–246

Heber U, Heldt HW (1981) The chloroplast envelope: structure, function and role in leaf metabolism. Annu Rev Plant Physiol 32:139–168

Herbert M, Burkhard Ch, Schnarrenberger C (1979) A survey for isoenzymes of glucose-phosphate isomerase, phosphoglucomutase, glucose-6-phosphate dehydrogenase and 6-phosphogluconate dehydrogenase in C_3-, C_4- and Crassulacean-acid-metabolism plants, and green algae. Planta 145:95–104

Hers H-G, Schaftingen Van E (1982) Fructose 2,6-bisphosphate 2 years after its discovery. Biochem J 206:1–12

Hue L, Blackmore PF, Shikama H, Robinson-Steiner A, Exton JH (1982) Regulation of fructose 2,6-bisphosphate content in rat hepatocytes perfused hearts and perfused hindlimbs. J Biol Chem 257:4308–4313

Ireland RJ, DeLuca V, Dennis DT (1979) Isoenzymes of pyruvate kinase in etioplasts and chloroplasts. Plant Physiol 63:903–907

Kelly GJ, Latzko E (1977a) Chloroplast phosphofructokinase. I Proof of phosphofructokinase activity in chloroplasts. Plant Physiol 60:290–294

Kelly GJ, Latzko E (1977b) Chloroplast phosphofructokinase. II Partial purification, kinetic and regulatory properties. Plant Physiol 60:295–299

Kombrink E, Kruger NJ, Beevers H (1983) Kinetic properties of pyrophosphate:fructose-6-phosphate phosphotransferase from germinating castor bean endosperm. Plant Physiol 74:395–401

Krause GH, Heber U (1976) Energetics of intact chloroplasts. In: Barber J (ed) The intact chloroplast. Elsevier, Amsterdam, pp 171–214

Kreuz K, Kleinig H (1981) On the compartmentation of isopentenyl diphosphate synthesis and utilization in plant cells. Planta 153:578–581

Kruger NJ, Kombrink E, Beevers H (1983) Pyrophosphate:fructose 6-phosphate phosphotransferase in germinating castor bean seedlings. FEBS Lett 153:409–412

Kuhn DN, Knauf M, Stumpf PK (1981) Subcellular localization of acetyl-CoA synthetase in leaf protoplasts of *Spinacia oleracea*. Arch Biochem Biophys 209:441–450

Laing WA, Stitt M, Heldt HW (1981) Control of CO_2 fixation. Biochim Biophys Acta 637:348–359

Latzko E, Gibbs M (1968) Distribution and activity of enzymes of the reductive pentose phosphate cycle in spinach leaves and in chloroplasts isolated by different methods. Z Pflanzenphysiol 59:184–194

Latzko E, Kelly GJ (1979) Enzymes of the reductive pentose phosphate cycle. In: Gibbs M, Latzko E (eds) Photosynthesis II. Encyclopedia of plant physiology, vol VI, New Ser. Springer, Berlin Heidelberg New York, pp 239–250

Liedvogel B, Stumpf PK (1982) Origin of acetate in spinach leaf cell. Plant Physiol 69:897–903

Lilley R McC, Stitt M, Mader G, Heldt HW (1982) Rapid fractionation of wheat leaf protoplasts using membrane filtration. Plant Physiol 70:965–970

Longenecker JP, Williams JF (1980) Quantitative measurement of the L-type pentose phosphate cycle with [2-^{14}C]glucose and [5-^{14}C]glucose in isolated hepatocytes. Biochem J 188:859–865

Macdonald FD (1982) The synthesis of starch from sucrose in nonphotosynthetic cells of higher plants. PhD Thesis, Univ Cambridge

Macdonald FD, ap Rees T (1983) Enzymic properties of amyloplasts from suspension cultures of soybean. Biochim Biophys Acta 755:81–89

Miernyk JA, Dennis DT (1982) Activation of the plastid isozyme of phosphofructokinase from developing endosperm of *Ricinus communis* by fructose 2,6-bisphosphate. Biochem Biophys Res Commun 105:793–798

Miernyk JA, Dennis DT (1983) The incorporation of glycolytic intermediates into lipids by plastids isolated from the developing endosperm of castor oil seeds (*Ricinus communis* L.). J Exp Bot 34:712–718

Miflin BJ, Beevers H (1974) Isolation of intact plastids from a range of plant tissues. Plant Physiol 53:870–874

Miflin BJ, Lea PJ (1982) Ammonia assimilation and amino acid metabolism. In: Boulter D, Parthier B (eds) Nucleic acids and proteins in plants 1. Encyclopedia of plant physiology, vol XIV/A, New Ser. Springer, Berlin Heidelberg New York, pp 5–64

Murphy DJ, Walker DA (1982a) The properties of transketolase from photosynthetic tissue. Planta 155:316–320

Murphy DJ, Walker DA (1982b) Acetyl coenzyme A biosynthesis in the chloroplast. Planta 156:84–88

Neish AC (1955) The biosynthesis of cell wall carbohydrates II. Formation of cellulose and xylan from labelled monosaccharides in wheat plants. Can J Biochem Physiol 33:658–666

Nishimura M, Beevers H (1979) Subcellular distribution of gluconeogenetic enzymes in germinating castor bean endosperm. Plant Physiol 64:31–37

Nishimura M, Beevers H (1981) Isoenzymes of sugar phosphate metabolism in endosperm of germinating castor beans. Plant Physiol 67:1255–1258

Park KEY, Anderson LE (1973) Appearance of 3 chloroplast isoenzymes in dark-grown pea plants and pea seeds. Plant Physiol 51:259–262

Porter JW, Spurgeon SL (1981) Biosynthesis of isoprenoid compounds, vol I. John Wiley, New York

Quail PH (1979) Plant cell fractionation. Annu Rev Plant Physiol 30:425–484

Rapp BJ, Randall DD (1980) Pyruvate dehydrogenase complex from germinating castor bean endosperm. Plant Physiol 65:314–318

Rognstad R, Wals P, Katz J (1982) Further evidence for the classical pentose phosphate cycle in the liver. Biochem J 208:851–855

Rothe GM, Hengst G, Mildenberger I, Scharer H, Utesch D (1983) Evidence for an intra- and extraplastidic pre-chorismate pathway. Planta 157:358–366

Sabularse DC, Anderson RL (1981a) Inorganic pyrophosphate: D-fructose-6-phosphate 1-phosphotransferase in mung beans and its activation by D-fructose 1,6-bisphosphate and D-glucose 1,6-bisphosphate. Biochem Biophys Res Commun 100:1423–1429

Sabularse DC, Anderson RL (1981b) D-fructose 2,6-bisphosphate: a naturally occurring activator for inorganic pyrophosphate: D-fructose-6-phosphate 1-phosphotransferase. Biochem Biophys Res Commun 103:848–855

Schaftingen E van, Hue L, Hers H-G (1980) Control of the fructose 6-phosphate/fructose 1,6-bisphosphate cycle in isolated hepatocytes by glucose and glucagon. Biochem J 192:887–895

Schaftingen E van, Lederer B, Bartrons R, Hers H-G (1982) A kinetic study of pyrophosphate: fructose-6-phosphate phosphotransferase from potato tubers. Eur J Biochem 129:191–195

Stitt M, ap Rees T (1978) Pathways of carbohydrate oxidation in leaves of *Pisum sativum* and *Triticum aestivum*. Phytochemistry 17:1251–1256

Stitt M, ap Rees T (1979) Capacities of pea chloroplasts to catalyse the oxidative pentose phosphate pathway and glycolysis. Phytochemistry 18:1905–1911

Stitt M, ap Rees T (1980a) Carbohydrate breakdown by chloroplasts of *Pisum sativum*. Biochim Biophys Acta 627:131–143

Stitt M, ap Rees T (1980b) Estimation of the activity of the oxidative pentose phosphate pathway in pea chloroplasts. Phytochemistry 19:1583–1585

Stitt M, Heldt H (1981) Physiological rates of starch breakdown in isolated intact spinach chloroplasts. Plant Physiol 68:755–761

Stitt M, Bulpin PV, ap Rees T (1978) Pathway of starch breakdown in photosynthetic tissues of *Pisum sativum*. Biochim Biophys Acta 544:200–214

Stitt M, Mieskes G, Söling H-D, Heldt HW (1982a) On a possible role of fructose 2,6-bisphosphate in regulating photosynthetic metabolism in leaves. FEBS Lett 145:217–222

Stitt M, Lilley R McC, Heldt HW (1982b) Adenine nucleotide levels in the cytosol, chloroplasts, and mitochondria of wheat leaf protoplasts. Plant Physiol 70:971–977

Stitt M, Wirtz W, Heldt HW (1983) Regulation of sucrose synthesis by cytoplasmic fructosebisphosphatase and sucrose phosphate synthase during photosynthesis in varying light and carbon dioxide. Plant Physiol 72:767–774

Stumpf PK (1980) Biosynthesis of saturated and unsaturated fatty acids. In: Stumpf PK (ed) The biochemistry of plants, Vol IV. Academic Press, London New York, pp 177–204

Tanner GJ, Copeland L, Turner JF (1983) Subcellular localization of hexose kinases in pea stems: mitochondrial hexokinase. Plant Physiol 72:659–663

Turner JF, Copeland L (1981) Hexokinase II of pea seeds. Plant Physiol 68:1123–1127

Turner JF, Turner DH (1980) The regulation of glycolysis and the pentose phosphate pathway. In: Davies DD (ed) The biochemistry of plants, vol II. Academic Press, London New York, pp 279–316

Williams HR, Landau BR (1972) Pathways of fructose conversion to glucose and glycogen in liver. Arch Biochem Biophys 150:708–713

Williams JF (1980) A critical examination of the evidence for the reactions of the pentose pathway in animal tissues. Trends Biochem Sci 5:315–320

Williams JF (1981) The pentose cycle in liver. Trends Biochem Sci 6:XVI–XVII

Williams JF, Rienits KG, Schofield PJ, Clark MG (1971) The pentose phosphate pathway in rabbit liver. Biochem J 123:923–943

Williams JF, Clark MG, Blackmore PF (1978a) The fate of ^{14}C in glucose 6-phosphate synthesized from [1-^{14}C]ribose 5-phosphate by enzymes of rat liver. Biochem J 176:241–256

Williams JF, Blackmore PF, Clark MG (1978b) New reaction sequences for the non-oxidative pentose phosphate pathway. Biochem J 176:257–282

Wood HG, O'Brien WE, Michaels G (1977) Properties of carboxytransphosphorylase; pyruvate, phosphate dikinase; pyrophosphate-phosphofructokinase and pyrophosphate-acetate kinase and their roles in the metabolism of inorganic pyrophosphate. Adv Enzymol 45:85–155

14 Respiration in Intact Plants and Tissues: Its Regulation and Dependence on Environmental Factors, Metabolism and Invaded Organisms

H. Lambers

1 Introduction

In this chapter respiration as an integrated part of the metabolism of the intact plant will be considered; its aim is to discuss the qualitative and, wherever possible, quantitative links of respiration with various other aspects of metabolism.

About 50 years ago, respiration was widely considered as a largely wasteful process; an imperfection in the machinery involved in the conversion of carbohydrates and other nutrients into structural dry matter (ALGERA 1932). This concept was based on meticulous measurements of the production of CO_2 and heat and the consumption of substrates during the production of mycelium by *Aspergillus niger*. ALGERA observed that essentially all the energy in the carbon source of *A. niger* which was not recovered in the mycelium itself was lost as heat. The conclusion, therefore, appeared unavoidable, that no respiratory energy was required for the synthesis of mycelium. Admittedly, a slight discrepancy was found between the heat produced during growth of *A. niger* and that which was calculated to be produced during substrate oxidation if all the energy in the substrate had been lost as heat. However, this discrepancy was a mere 6 to 7% and was ascribed to experimental error. Fifty years later we only have to consult a biochemical handbook to explain this 6 or 7% "discrepancy": the synthesis of the peptide bond in proteins requires approximately 30 kcal when occurring in a cell, although its hydrolysis releases only 0.5 kcal. Thus, what ALGERA ascribed to an experimental error is in fact in very good agreement with theoretical values, based on known energy costs of the cell's biosynthetic reactions. It was not until the late 1950's, when the role of such compounds as ATP, trait d'unions between respiration and biosynthesis, was elucidated, that the controversy about the significance of respiration was finally resolved.

Once the role of respiration in energy conservation had been established, it became of interest to study the quantitative relationship between respiration and such processes as growth, maintenance of cellular structures and ion uptake (Sect. 2). It also became relevant to investigate to what extent, and how, the rate of respiration depends on the demand for respiratory energy (Sect. 4.1). One aspect of the regulation of plant respiration involves the alternative pathway, a nonphosphorylating mitochondrial electron transport chain (Sect. 3). In more recent years the role of respiration in adaptation of a plant to a specific environment (Sect. 6) and its relation to growth rate and yield (Sect. 7) has attracted considerable interest.

2 Respiration Associated with Growth, Maintenance and Ion Uptake

Once the qualitative link between energy conservation in respiratory processes and energy utilization in biosynthetic reactions became clear, physiologists tried to establish a more quantitative relationship [cf. KANDLER's (1953) Synthetischer Wirkungsgrad, which is the ratio of assimilated and respired glucose]. Upon the realization that respiratory energy is not only required for the net synthesis of plant matter, but also for its maintenance, methods were developed to separate a growth component and a maintenance component of respiration (MCCREE 1974, THORNLEY 1970). More recently, a method has been described to quantify a third component of respiration, associated with ion uptake (VEEN 1980). Clearly, the relationship between respiration and ion uptake has been the subject of many studies [cf. LUNDEGÅRDH's (1955) anion respiration], but the reinterpretation of these early data is such that they need treatment in a separate section (5.1) and need not be discussed here.

2.1 Is There a Justification for the Concept of Growth Respiration?

The justification for the concept of growth respiration versus maintenance respiration has to be sought in whole plant physiology and comparative physiology. Instead of simply measuring the rate of respiration, physiologists have tried to relate this rate to the growth rate, and to make a comparison of a ratio of the respiration rate and the growth rate as measured in plant organs or whole plants of different species; but when it is recognized that nongrowing tissues also require respiratory energy (maintenance energy), this simple ratio of respiration and growth rate becomes meaningless (LAMBERS et al. 1983 b).

2.1.1 Definitions and Basic Assumptions

A list of definitions as used in this field and the dimensions of the various parameters and constants are included in Table 1. MCCREE (1970) observed that the rate of respiration of whole white clover plants could be described by the following equation:

$$R = aP + bW,\tag{1}$$

where R is the rate of respiration of the whole plant, W is the dry weight and P is the rate of gross photosynthesis; a and b are constants. THORNLEY (1970) analyzed this empirical equation, making assumptions analogous to those in similar work on microorganisms (PIRT 1975). *First assumption:* To produce 1 g of biomass from sugars and mineral nutrients a certain amount of sugars will have to be respired to generate the energy required for biomass production.

Table 1. List of definitions and symbols

Item	Definition	Symbol or abbre- viation	Further remarks	Reference
Respiration rate per plant		R		
Respiration rate per g plant		r	Equals $\dfrac{R}{W}$	
Synthetischer Wirkungsgrad = synthetic efficiency	Ratio between assimilated and respired C	S.W.		Kandler (1950)
Yield (Conversion efficiency)	Ratio between dry weight increment and (dry weight increment + substrate utilized in respiration)	Y_G		Thornley (1970)
Maintenance coefficient	Substrate consumption for the production of energy which is *not* utilized for growth	m		Thornley (1970)
Production value	Ratio between weight of the end product and the weight of substrate required for C-skeletons and energy production	pv	Calculated for individual compounds or dry matter of known composition	Penning de Vries et al. (1974)
True growth yield	Ratio between dry weight increment and substrate respired to produce this increment	Y_{EG}		Pirt (1975)
Growth respiration	The amount of oxygen consumed to generate energy required in the synthesis of dry matter	$\dfrac{1}{Y_{EG}}$	This is the inverse of Pirt's true growth yield	Lambers et al. (1978a)
Growth efficiency	Ratio between dry weight increment and total amount of substrate utilized, or: the ratio of dry weight increment and (dry weight increment + substrate utilized in respiration)	GE	Equals Thornley's Y_G	Yamaguchi (1978)

Thornley defined the yield:

$$Y_G = \frac{\varDelta W}{\varDelta W + \varDelta S_r},$$

(2)

where $\varDelta W$ is the increment in dry matter during an interval of time, and $\varDelta S_r$

is the amount of substrate respired to produce ΔW units of plant dry matter [ΔW and ΔS_r have to be expressed in the same units, e.g., CO_2 equivalents (THORNLEY 1970) to make the obtained value meaningful.] Theoretically, the value for Y_G must depend on the chemical composition of the produced biomass; e.g., the synthesis of 1 g of protein requires a higher input of respiratory energy than that of 1 g of starch (PENNING DE VRIES et al. 1974). Also, the value for Y_G depends on the predominant respiratory pathway (see Sect. 3). *Second assumption:* The amount of sugars respired to maintain biomass is proportional to the weight of biomass to be maintained. THORNLEY defined the maintenance coefficient, i.e., the rate of substrate utilization for maintenance per unit plant weight:

$$m = \frac{1}{W} \cdot \frac{\Delta S_m}{\Delta t}, \tag{3}$$

where W is the dry weight of the investigated plant or plant organ, ΔS_m is the amount of substrate required to generate energy for maintenance processes, and Δt represents an interval of time during which the substrate is consumed to generate the maintenance energy. Theoretically, the value for m must depend on the biochemical composition of the biomass to be maintained. Environmental factors which affect the rate of maintenance processes should also affect the maintenance coefficient. Like Y_G, m will depend on the efficiency of energy conservation in respiratory pathways.

Based on Eqs. (2) and (3), the rate of respiration, R, is described:

$$R = \frac{\Delta S_r}{\Delta t} + \frac{\Delta S_m}{\Delta t}. \tag{4}$$

Combination of Eqs. (2), (3) and (4) yields, after some rearrangement:

$$R = \frac{(1-Y_G)}{Y_G} \cdot \frac{\Delta W}{\Delta t} + m.W. \tag{5}$$

Combination of Eq. (5) with the following:

$$\Delta S = \Delta W + \Delta S_r + \Delta S_m, \tag{6}$$

where ΔS is the total amount of substrate consumed during an interval of time (expressed in the same units as used to calculate Y_G), yields:

$$R = \frac{1-Y_G}{\Delta t}(\Delta S - \Delta S_m) + \frac{\Delta S_m}{\Delta t} = (1-Y_G)\frac{\Delta S}{\Delta t} + m\,Y_G W, \tag{7}$$

which is of the form of McCree's Eq. (1), since $\frac{\Delta S}{\Delta t}$ equals P.

If we want to express respiration per gram, r, instead of per plant, R, then Eq. (5) develops into:

$$r = \frac{(1-Y_G)}{Y_G} \cdot \frac{1}{W} \cdot \frac{\Delta W}{\Delta t} + m. \tag{8}$$

This equation shows that as long as $\frac{\Delta W}{W \Delta t}$ (= the relative growth rate) is constant, i.e., during exponential growth, the respiration rate expressed per gram will also be constant, provided that neither the "yield", Y_G, nor the maintenance coefficient, m, change. This points to the main experimental problem in separating the two components: how can the system be manipulated in such a way that the rate of maintenance respiration is no longer a constant proportion of the total respiration rate, as it is during exponential growth, without affecting the conversion efficiency or the maintenance coefficient? This problem will be further discussed in the next section.

If the aim of the study is to investigate the respiratory efficiency, rather than the C-economy of the plant, it is more appropriate to work with:

$$Y_{EG} = \frac{\Delta W}{\Delta S_r} \tag{9}$$

This term is generally designated as the "true growth yield" in the microbiological literature (cf. PIRT 1975).

Combining Eqs. (8) and (9) yields:

$$r = \frac{1 - \dfrac{\Delta W}{\Delta W + \Delta S_r}}{\dfrac{\Delta W}{\Delta W + \Delta S_r}} = \frac{\dfrac{\Delta S_r}{\Delta W + \Delta S_r}}{\dfrac{\Delta W}{\Delta W + \Delta S_r}} = \frac{1}{Y_{EG}} \cdot \frac{1}{W} \cdot \frac{\Delta W}{\Delta t} + m \tag{10}$$

2.1.2 Experimental Approaches

Two principally different approaches have been followed to disrupt the constant relationship between growth rate and respiration rate that may exist during exponential growth [cf. Eq. (8) and comments]. MCCREE (1970) subjected plants of white clover to different light intensities for a short period. In this way he obtained a range of values for R and for $\frac{\Delta S}{\Delta t}$ [cf. Eq. (7)], so that he could find both the conversion efficiency and the maintenance coefficient (Method I A). HANSEN and JENSEN (1977) followed a similar approach. They grew plants for 3 days at low light intensity, followed by 3 days at high light intensity, and so on. Thus, they also obtained a range of values for both R and $\frac{\Delta S}{\Delta t}$ (Method I B). A major problem with this approach, if used for (partly) autotrophic organs, is that it is assumed without any evidence that the respiration during photosynthesis is the same as that in the dark. As discussed further in Section 6.1, this is a doubtful assumption.

The second approach used, for example, by KIMURA et al. (1978), is to investigate respiration and growth of plants during a nonexponential growth stage. Plotting the values obtained for r against the values for the relative growth rate, $\frac{1}{W} \cdot \frac{\Delta W}{\Delta t}$ [cf. Eq. (8)], will then yield the maintenance coefficient as the intercept with the y-axis, and Y_{EG} as the slope of the obtained line (Method II A). A modification of this method was used by LAMBERS and STEINGRÖVER (1978a), who worked with plants in a stage of linear growth (Method II B). Nonexponential growth in Method II is associated with mutual shading.

In a series of elegant experiments, in which growth, volume, ion uptake, and the respiration of maize roots were measured, VEEN (1980) tried to discriminate three components of root respiration. Variation in the four measured parameters was obtained both by varying the light intensity and measuring what happened during the recovery phase after excision of part of the root system. Multiple regression analysis was subsequently used to separate respiration for growth, maintenance, and ion uptake (Method III).

A different approach (Method IV) to estimate the maintenance coefficient has been to place plants in the dark for up to 2 days (MCCREE 1974). It is assumed that the rate of respiration after this period in the dark represents the rate of maintenance respiration. Subtracting this value from the rate of respiration before the start of this treatment, allows Y_G to be derived. However, when CHALLA (1976) used this method to estimate the rate of maintenance respiration in cucumber roots, he found that after only 12 h in the dark, structural compounds formed part of the respiratory substrate, indicating that the tissue was not "maintained" in its original state. Therefore, at least in this one instance where the method has been investigated, it was found that it could *not* be used to estimate maintenance respiration.

The merits and problems of the various methods have been evaluated in a recent review (LAMBERS et al. 1983b).

2.1.3 Experimentally Derived Values for Y_G and Y_{EG} Compared with Theoretical Values

PENNING DE VRIES et al. (1974) made extensive calculations on theoretical values for Y_G in whole plants, calculated on the basis of known energy costs of biochemical reactions and the conservation of energy in respiratory pathways. A major conclusion from these calculations, which has subsequently been ignored by many other workers in this field, e.g., HANSEN and JENSEN (1977), is that a change in the ADP:O ratio of respiration from 3 to 2 affects the Y_G only by ca. 5% [Fig. 6 in PENNING DE VRIES et al. (1974)]. The implication of this is that an experimentally obtained Y_G value cannot give information on the efficiency of respiratory processes, unless the differences in respiratory efficiency are very large, e.g. ADP:O ratio of 1 vs. 3, which affect Y_G by about 20%.

A further complicating factor, which has become clear from the work of Hansen and Jensen (1977), Lambers et al. (1979), Szaniawski (1981), and Szaniawsi and Kielkiewicz (1982), is that the conversion efficiencies of roots and shoots are considerably different. If subsequently a value for a whole plant is calculated, this value may be very close to that calculated by Penning de Vries et al. (1974) for whole plants.

However, if the conversion efficiency for the shoots is so high that it cannot be explained on the basis of the calculations made by Penning de Vries et al. (1974), and if that of the roots is lower than predicted by these calculations, then a combination of the two Y_G values is not meaningful. If this combined value agrees well with the theoretical values, as given by Penning de Vries et al. (1974), this must be fortuitous and cannot possibly justify the conclusion that plants work with biochemically maximal efficiency (see also Sect. 3).

Unlike the conversion efficiency (Y_G), the respiratory efficiency (Y_{EG}, cf. Eq. (9)] is strongly affected by the ADP:O ratio (Penning de Vries et al. 1974). Theoretical calculations, based on energy costs of biosynthetic reactions and transport processes, have been made for roots of two *Senecio* species (Lambers and Steingröver 1978b). The theoretical values for growth respiration ($1/Y_{EG}$ in mg O_2 g^{-1} dry wt. of roots) were significantly lower than the experimental values, which was explained in part by the large contribution of the alternative respiratory pathway in root respiration.

It is concluded that a comparison of theoretical and experimental Y_G values does not provide useful information about the efficiency of energy utilization and that Y_{EG} values are a more sensitive tool for this aim. A comparison of theoretical values with experimental values for Y_{EG} (or growth respiration) in roots shows that experimental values tend to be higher than theoretical ones. For leaves or shoots, a comparison of experimental with theoretical values is not yet possible. This is due to the absence of information on the chemical composition of dry matter (content of protein, starch, ions, etc.) of the shoots which were used to determine Y_G. Moreover, there is no clear information available on the engagement of respiratory processes during photosynthesis.

2.1.4 Some Experimentally Derived Values and Their Significance

Only a selection of values for shoots or leaves, roots, and generative organs will be discussed here. For further information on experimental values see Ruget (1981).

Hansen and Jensen (1977) used *Lolium multiflorum* in an investigation of the conversion efficiency and the maintenance coefficient (Method I B). A considerably higher conversion efficiency was found for the shoots than for the roots, whilst the maintenance coefficient was largest for the roots. Yamaguchi (1978) observed the highest values for the growth efficiency (which will approximate the conversion efficiency, if the proportion of maintenance respiration remains low; Table 1) in leaf blades of vegetative rice plants. Intermediate values were observed for the stem, while the lowest values were observed for the roots.

The growth efficiency of the panicle was approximately the same as that of the leaf blades (0.65–0.75), similar to that of the ear of maize, but higher than for the pods of soybean. YAMAGUCHI et al. (1975, cited by YAMAGUCHI 1978) observed a ca. 10% lower growth efficiency in the roots as compared with the shoots of maize.

LAMBERS (1979), in an investigation of the respiratory efficiency of roots of 14 species in a stage of linear growth (Method II B), found conversion efficiencies ranging from 0.40 to 0.75. Using the same method for shoots, a value close to 1.00 was found for shoots of two *Senecio* species (LAMBERS et al. 1979). Accordingly, considerably higher rates of maintenance respiration were found in the roots than in the shoots. Similarly, SZANIAWSKI (1981) found values for the conversion efficiency for shoots and roots of Scots pine of 0.86 and 0.66, respectively, whilst the rate of maintenance respiration was about 3.5 times higher in the roots than in the shoots (Method II A). In sunflower the conversion efficiency was also highest in the shoots, while the rate of maintenance respiration was highest in the roots (SZANIAWSKI and KIELKIEWICZ 1982).

Despite the use of different methods, the wide range of experimental values and the use of different species, one predominant conclusion arises. The conversion efficiency of the roots is quite often rather low, while that of shoots and other aboveground organs tends to be higher. A number of explanations for the difference in conversion efficiency between shoots and roots have been discussed by LAMBERS et al. (1983 b). They include (1) the role of the roots in ion uptake, to satisfy not only their own demand but also that of the shoot. (2) The respiratory efficiency (ADP:O ratios of mitochondrial electron transport). (3) The possibility that photosynthetically derived energy is used for biosynthetic reactions in autotrophic plant parts.

It is concluded that both the conversion efficiency and the growth respiration differ between roots and leaves. Values for growth respiration of autotrophic organs have to be considered with reservation, in view of the possibility that photosynthesis may directly supply ATP for growth. Energy costs of ion uptake are likely to affect the conversion efficiency significantly, and more experimental data on this subject are needed.

2.1.5 Summarizing Remarks

The determination of respiration for either growth or maintenance is only useful when roots and shoots are considered separately and not when whole plants are studied. Even so, large differences are expected if roots are compared which differ largely in their net rates of ion export to the shoots. For this reason it is important to quantify a third component of respiration, related to ion uptake. Thanks to the concept of respiration as related to growth, maintenance, and ion uptake, we now have a better understanding of the quantitative significance of the various sinks for respiratory energy and of the efficiency of respiratory processes in plants. We need more information on the energy costs of ion transport, on the respiratory efficiency of reproductive organs, and on the regulatory aspects of the coordination between growth and respiration.

2.2 Respiration as an Aspect of the C-Economy of a Plant

The preceding section contains information on the C-requirements for growth, maintenance, and ion uptake. In this section respiration will be discussed as an aspect of the plant's C-economy. How much of the recently acquired carbon is subsequently lost in respiratory processes?

In many species 30–60% of the photosynthates produced each day are lost in respiration; these values tend to decline with increasing age (Table 2). A significant portion of these losses occurs in the roots; their contribution to respiration is greater than would be expected from their weight as a portion of total plant weight. On an annual basis the respiratory losses were as high as 70% of the annual photosynthesis in the evergreen perennial herb *Pteridophyllum racemosum* (Table 2).

In a review on respiration and growth, Ryle (1984) concluded that ca. 30% of the carbohydrates assimilated in the youngest fully expanded leaves of a range of gramineous species is lost in respiration within 1–2 days. Subsequently, respiratory loss of carbon continues, so that a total of ca. 50% of the C in the original photosynthate molecules is lost in respiration.

Obviously respiration is a major sink for photosynthates. In Sect. 7, differences between genotypes will be discussed in the context of the yield and growth rate of the plant.

Table 2. Loss of C in respiratory processes, expressed as a percentage of the C gained in photosynthesis

Species	Growth conditions	Special remarks	Age days	C-loss in		Reference
				Total respiration	Root respiration	
Hordeum distichum	Nutrient solution	–	7 24	60 38	– –	Farrar (1980)
Lupinus albus	Sand + nutrients	N_2-fixing NO_3-fed	ca. 60 ca. 60	43 31	36 23	Pate et al. (1979)
Nicotiana tabacum	Field-grown		22–57	41–47	–	Peterson and Zelitch (1982)
Plantago lanceolata	Nutrient solution		28 44	49 40	29 12	Lambers et al. (1981 b)
Plantago major	Nutrient solution		28 44	50 47	28 15	Lambers et al. (1981 b)
Pteridophyllum racemosum	Growth in natural habitat (central Japan)	Data on an annual basis		77	–	Kimura (1970)
Zea mays	Nutrient solution		56	39	18	Massimino et al. (1980)

3 Cyanide-Resistant Respiration:
Its Distribution and Physiological Significance

The first observation on cyanide-resistant respiration in plants was made over 50 years ago, when GENEVOIS (1929) found that the respiration of seedlings of *Lathyrus odorata* (sweet pea) was resistant to cyanide. It is now well documented that cyanide-resistant respiration is a widespread phenomenon in higher plants (both monocotyledons and dicotyledons); it also occurs in a few animal species and in fungi, bacteria, and algae (HENRY and NYNS 1975). The biochemistry of the cyanide-resistant, or "alternative", respiratory pathway will be discussed in detail in LANCE et al. Chap. 8, this Vol. Suffice it here to repeat that the alternative path is located in the inner mitochondrial membrane and that no energy is conserved during transport of electrons via the alternative pathway, between the branching region with the cytochrome path (ubiquinone) and the terminal electron acceptor (oxygen). In this section the various hypotheses to explain the physiological significance of the cyanide-resistant pathway will be discussed.

3.1 Cyanide-Resistant Respiration in Vivo: Some Methodological Aspects

3.1.1 Cyanide-Resistant Oxygen Uptake

Respiration of roots, leaves and storage organs is more often than not resistant to cyanide. Since the alternative oxidase is by no means the only cyanide-resistant oxidase in plants, it is important to know whether cyanide-resistant respiration in vivo can be ascribed to the alternative path. In slices of white potato (*Solanum tuberosum*) the development of cyanide-resistant respiration in vivo (upon aging) is reflected in a similar development in isolated mitochondria (DIZENGREMEL 1975, THEOLOGIS and LATIES 1978a). Also in the roots of five other plant species and in the leaves of two, cyanide resistance in vivo was reflected in the mitochondria (LAMBERS et al. 1983a).

In these roots there was not always a good quantitative correlation between cyanide resistance of intact roots and that of mitochondria isolated from these roots. This has been explained by the extent to which the cytochrome path is used; in isolated mitochondria under State 3 conditions and with an adequate supply of substrate, the full capacity of the cytochrome path is used, whereas in vivo the flow of electrons through the cytochromes is often restricted by the availability of ADP (DAY and LAMBERS 1983), as will be further discussed in Sect. 4.1.

In slices of storage tissues (THEOLOGIS and LATIES 1978a) and soybean callus cells (DE KLERK-KIEBERT et al. 1981) two cyanide-resistant components of respiration have been described. One is sensitive to hydroxamic acids and has been ascribed to the cyanide-resistant path, whereas the other was designated a "residual" component, the nature of which remains obscure (DAY et al. 1980).

3.1.2 Cyanide-Sensitive Oxygen Uptake

Cyanide inhibits cytochrome oxidase, since the undissociated HCN combines with the ferri form of cytochrome a_3; 10^{-4} M HCN fully inhibits cytochrome a_3. Other enzymes, e.g., catalase and peroxidase, are inhibited in the same fashion. Copper-containing oxidases are also sensitive to cyanide, although at somewhat higher concentrations. As well as being an inhibitor of these oxidases, cyanide can combine with a variety of intermediates, e.g., pyruvate and oxalo-acetate, and inhibit a number of other enzymes (HACKETT 1960). It is, therefore, not surprising that cyanide inhibition in vivo cannot always fully be ascribed to inhibition of the cytochrome path. KANO and KAGEYAMA (1977) concluded that two cyanide-sensitive oxidases participate in the respiration of *Cucumis melo* roots. One is cytochrome oxidase, which is sensitive to 0.1 mM cyanide. The other is an unknown oxidase, with a lower affinity for oxygen and insensitive to 0.1 mM but sensitive to 1 mM cyanide. This unidentified oxidase was estimated to contribute 6–16% in root respiration of *C. melo*. There is no information on the contribution of such an oxidase to respiration in other higher plant tissues.

A further complicating factor is that the inhibitory effects of cyanide, azide, and antimycin, all inhibitors of the cytochrome path, are not always exactly the same. Oxygen uptake of wheat root tips was 30% inhibited by 0.5 mM cyanide, whereas azide inhibited 42%; in older segments (30–40 mm from the tip) cyanide and azide inhibited oxygen uptake by 46 and 78% respectively, (ELIASSON and MATHIESEN 1956). It is unlikely that 0.5 mM cyanide was too low to inhibit the cytochrome path fully, since the K_i of KCN for respiration of intact *Triticum aestivum* roots is 0.01 mM (LAMBERS et al. 1983a). Therefore, either azide must have inhibited respiration of these root segments in another fashion in addition to its inhibition of cytochrome oxidase, or cyanide must have stimulated oxygen uptake in addition to inhibiting cytochrome oxidase. Side effects of azide have indeed been described, e.g., inhibition of alcoholic fermentation in yeast (FALES 1953). In many other tissues cyanide and azide have very similar effects [see BEEVERS (1961) for a compilation of data]. There is also a disparity between inhibition of respiration by cells of *Nicotiana glutinosa* by antimycin, azide, and cyanide (HORN and MERZ 1982). In these experiments it could be excluded that the antimycin concentration was too low, since the combination of antimycin and 2 mM SHAM gave full inhibition.

Severe side effects of cyanide are illustrated in Fig. 1, showing data on root respiration of Fe-deficient *Zea mays* plants. The capacity of the alternative cyanide-resistant path in these roots, as determined by measuring respiration in the presence of 0.1 mM KCN, was ca. 45% of root respiration. Nonetheless, 0.5 mM cyanide inhibited respiration by ca. 85%. Thus in the Fe-deficient maize roots, inhibition by cyanide concentrations above 0.1 mM should be interpreted as a side effect.

Apart from these pathological cases (Fig. 1), the information on cyanide inhibition of respiration indicates that inhibition by cyanide is largely due to inhibition of the cytochrome path and only to a minor extent to inhibition of other oxidases. The above data should be taken as a warning that so-called

Fig. 1. Root respiration of Fe-deficient *Zea mays* plants as dependent on the concentration of KCN. Maximum inhibition of root respiration of healthy plants was achieved with a KCN concentration of 0.1 mM (*broken line*). SHAM (10 mM) did not inhibit root respiration of Fe-deficient plants in the absence of KCN. Mitochondria, isolated from the roots of the Fe-deficient plants, were equally as cyanide-resistant (ca. 40%) as mitochondria from healthy roots. Thus, the inhibition of root respiration in Fe-deficient plants by KCN concentrations above 0.1 mM must be due to "side effects", i.e., it is not an effect on cytochrome oxidase. Unpublished data from H. LAMBERS and D.A. DAY. The control rate of respiration was 385 μmol O_2 h^{-1} $(g^{-1}$ dry wt., in the Fe-deficient plants)

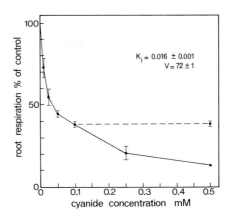

specific inhibitors of the cytochrome path *can* have other effects on respiration as well. After all, in the words of M. GIBBS, specific inhibitors are like beautiful virgins: rare!

3.1.3 SHAM-Sensitive Oxygen Uptake

A range of inhibitors of the alternative path is available for in vitro studies (SCHONBAUM et al. 1971, GROVER and LATIES 1978, SIEDOW and BICKET 1981), but unfortunately some of these, e.g., disulfiram (GROVER and LATIES 1981), propyl gallate (D. KUIPER, personal communication), cannot be used in vivo, due to their sequestration, lack of penetration, or interference with other aspects of metabolism. Thus far, only the substituted hydroxamic acids have been successfully used in vivo (LATIES 1982, LAMBERS 1982), and even these have to be used with caution, since they are known to have some side effects, such as inhibition of chloride uptake, decreasing the level of ATP in the cell and the plasmalemma membrane potential, and stimulating other oxidases (Table 3). They are also known to inhibit other oxidases, such as lipoxygenase, tyrosinase and horseradish peroxidase (Table 3). Benzhydroxamic acid, as well as being an inhibitor of peroxidase, can serve as a substrate for the same enzyme (AVIRAM 1981).

FRITZ et al. (1958) used $^{18}O_2$ to estimate the participation of lipoxygenase in the oxygen uptake of 2-day-old etiolated *Zea mays* seedlings. Chloroform-soluble substances in these plants incorporated ^{18}O at a rate which demonstrated that lipoxygenase accounted for less than 0.5% of oxygen uptake in these seedlings. ^{18}O incorporation into chloroform-insoluble residues was much less than into the lipid fraction.

LATIES (1982) stated that "when SHAM is inhibitory in the presence of CN but not in its absence, its effect can be taken to be on the alternative oxidase". In view of the observation that SHAM can also stimulate respiration (Table 3), a lack of inhibition in the absence of cyanide might point to the

Table 3. The effects of substituted hydroxamic acids on the in vitro activity of some enzymes and a number of physiological parameters, other than the alternative path. Note that the side effects on physiological parameters refer to relatively *low* concentrations of inhibitors

Tyrosinase	SHAM, m-ClAM	1 mM	>90% inhibition	RICH et al. (1978)
	BHAM	2 mM	>90%inhibition	RICH et al. (1978)
Horseradish peroxidase	SHAM	1 mM	>90% inhibition	RICH et al. (1978)
Buttermilk xanthine oxidase	SHAM	"high"	Some inhibition	RICH et al. (1978)
Lipoxygenase	SHAM	1 mM	>90% inhibition	GOLDSTEIN et al. (1980)
Chloride uptake in potato and fresh preclimacteric banana slices	m-ClAM	0.7 mM	ca. 40% inhibition	THEOLOGIS and LATIES (1981)
Transmembrane potential in root slices of *Beta vulgaris*	SHAM	5 mM	Decrease from −216 to −188 mV after 40–90 min incubation	MERCIER and POOLE (1980)
ATP content in root discs of *Beta vulgaris*	SHAM	5 mM	ca. 20%, decrease after 40–90 min incubation	MERCIER and POOLE (1980)
Respiration in cells of *Nicotiana glutinosa*	SHAM	2 mM	Stimulation of 30% at the end of the growth cycle	HORN and MERTZ (1982)
Respiration in roots of *Pisum sativum*	SHAM	1–5 mM	15–25% stimulation	DE VISSER and BLACQUIÈRE (1984)
Respiration in slices of preclimacteric *Cucumis melo* fruits	SHAM	0.5 mM	ca. 30% stimulation	PASSAM and BIRD (1978)

compensation of an inhibitory by a stimulatory effect. This is precisely what DE VISSER and BLACQUIÈRE (1984) observed in roots of *Pisum sativum* and *Plantago* species. It is, therefore, recommended that the rate of respiration in the presence of varying concentrations of substituted hydroxamic acids, be determined both in the presence and absence of an inhibitor of the cytochrome path, and that a "*p*-plot" (BAHR and BONNER 1973, THEOLOGIS and LATIES 1978a) be constructed.

3.1.4 The Determination of the Activity of the Alternative Path

Both in isolated mitochondria and in vivo the alternative path is only engaged upon saturation or inhibition of the cytochrome path (see DAY et al. 1980, LATIES 1982). Upon saturation or inhibition of the cytochrome path, electrons spill over into the alternative path. Thus, the respiration in the presence of cyanide can provide information on the *capacity* of the alternative path only. Cyanide resistance does not give insight into the *activity* of the alternative path,

Fig. 2. Respiration of *Triticum aesti-*
vum leaves as dependent on the concentration of SHAM, measured in the
absence (●) and presence (o) of
0.3 mM KCN (**A**). The data obtained
in the absence of KCN were plotted
against a similar set of titration
values, obtained in the presence of
KCN [g(i)alt; **B**]. The slope of this line
(p) designates the fraction of the *ca-*
pacity of the alternative path which
is *engaged*. (After LAMBERS et al.
1983a)

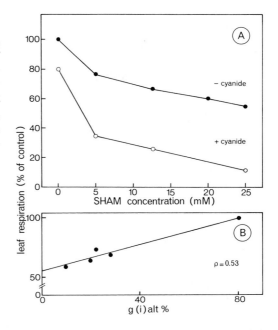

in the absence of an inhibitor of the cytochrome path, since addition of KCN
may "re-route" electrons to an alternative path not fully engaged.

The *capacity* of the alternative path cannot simply be determined by adding cyanide,
if a residual component contributes to the cyanide resistance. Also, when cyanide stimulates respiration (as it may when respiration rates are limited by adenylate control of
glycolysis, and the alternative path capacity is greater than control oxygen uptake), the
resistant component does not necessarily reflect the full capacity of the alternative path.

Specific inhibitors of the alternative path are required to determine the *activi-*
ty of the alternative path. Respiration can be titrated with these inhibitors.
The titration values obtained in the presence of an inhibitor of the cytochrome
path (KCN in Fig. 2a) give the maximum possible flux through the alternative
path in the presence of a given concentration of an inhibitor of the alternative
path (SHAM, in Fig. 2a). The values obtained in the absence of cyanide give
the total flux of electrons at each SHAM concentration. The set of values
obtained in the absence of KCN is plotted against the first set (Fig. 2b), and
this will yield a straight line if altering the activity of the alternative path does
not alter the flux through the cytochrome path, which is generally the case
(DAY et al. 1980). The slope of the line (p) gives the extent to which the capacity
of the alternative path is used, and can vary from 0 to 1.

SHAM is commonly used in concentrations up to 1 mM in vitro (LATIES
1982). In *Solanum tuberosum* slices this concentration is also sufficient to give
full inhibition (THEOLOGIS and LATIES 1978a). In slices of fruits of *Persea ameri-*
cana (avocado; THEOLOGIS and LATIES 1978c, cultured cells of *Glycine max*
(DE KLERK-KIEBERT et al. 1981) and slices of *Triticum aestivum* leaves (AZCÓN-

Bieto et al. 1983b), slightly higher concentrations of SHAM are required. In intact leaves of wheat and also in intact roots of a range of species, a concentration of 10 or 25 mM SHAM is needed to give full and immediate inhibition of the alternative path (Lambers et al. 1983a). Titration experiments ("p plots") and a comparison of inhibitor sensitivity between intact tissues and isolated mitochondria have shown that even these high concentrations of SHAM can safely be used to give an estimate of the engagement of the alternative path in vivo (Lambers et al. 1983a). However, these high concentrations should only be used in short-term experiments (< 20 min).

However, similar titration experiments with roots of *Pisum sativum* and *Plantago* species revealed a stimulatory effect of SHAM at low concentrations and inhibition at 25 mM (De Visser and Blacquière 1984). In these roots, the stimulation by SHAM was suggested to be due to a third oxidase, whereas SHAM inhibition (at 25 mM) was concluded to be associated with both the alternative path and the hypothetical third oxidase.

It is concluded that the capacity of the alternative path can be determined by measuring the respiration in the presence of cyanide, provided that the cyanide-resistant component is sensitive to substituted hydroxamic acids. The activity of the alternative path *can* be estimated using suitable concentrations of substituted hydroxamic acids, but care has to be taken to make sure that SHAM does indeed inhibit only the alternative path and does not interfere with other oxidative systems.

3.2 Cyanide-Resistant Respiration and Heat Production

Thermogenesis in *Arum* was discovered by Lamarck in 1778 (Meeuse 1975). Garreau (1851, cited by Meeuse 1975) showed a close relationship between heat production and respiration, and Van Herk and Badenhuizen (1934) demonstrated the cyanide insensitivity of the *Sauromatum* spadix. Until quite recently it has been popular to ascribe the heat production to the engagement of the nonphosphorylating pathway (Meeuse 1975). However, this has not been done with much justification, since, as explained in Section 3.1, the presence of the cyanide-resistant path does not imply its engagement. Moreover, in the spadix of *Arum maculatum*, respiration is not only cyanide-resistant during thermogenesis, but also before (Wedding et al. 1973). Up to now the engagement of the alternative path during thermogenesis has only been demonstrated in *Arum* spadix slices (Wedding et al. 1973); no information is available on the engagement of the alternative path before or after thermogenesis. The respiration at climax in *Philodendron selloum*, where the capacity of the alternative path is close to that of the cytochrome path, is via the cytochrome path only ($p = 0$) (Laties 1982). Presumably these experiments have been done with slices. Since the spadix (of *Symplocarpus* at least, Knutson 1974) does not have starch reserves and begins to cool immediately on being severed from the parent plant, there is no convincing evidence that in situ the alternative path is not engaged in *Philodendron*. This gives some pause before we must conclude that there is no close correlation between heat production and the engagement of the alternative path and that the high respiration rates of climax spadices must be attributed in addition to a high activity of the cytochrome path. Nonetheless, as demonstrated by PEARSON (1948) for shrews (*Sorex* spp.) and other small mammals, organisms and tissues need a minimal size of more than 2.5 cm³, like that of the smallest spadices of Araceae, before heat production can significantly raise their temperature. Thus we are still left with the task of finding an explanation for the cyanide resistance of respiration in smaller organs, like leaves and fibrous roots.

3.3 Cyanide Resistance and Ion Uptake: "Anion Respiration"

LUNDEGÅRDH (1954) has stressed the relation between anion uptake and a component of respiration, which is sensitive to cyanide: "anion respiration". LUNDE-GÅRDH believed that the cyanide-resistant component of respiration ("ground respiration") was always operative, even in the absence of absorbable ions, but could not support ion uptake. Since respiration of roots in distilled water was highly cyanide-resistant and became more sensitive to cyanide upon addition of absorbable ions, it was believed that the ground respiration was operative in distilled water, whereas the anion respiration became operative in the presence of ions (LUNDEGÅRDH 1954, 1955). However, since the cyanide resistance per se does not give any information on the activity of the alternative path, LUNDE-GÅRDH's conclusion is no longer justified. This matter will be further discussed in Section 5.1.

3.4 The Alternative Path, Fruit Ripening, Ethylene Production and the Synthesis of Stress Metabolites

The ripening of many fruits is attended by a burst of respiratory activity, the well-known climacteric. Respiration during climacteric is cyanide-resistant and this cyanide resistance can be induced by ethylene (SOLOMOS and LATIES 1976). This, in combination with the belief that the alternative path would produce H_2O_2 and thus be involved in the biosynthesis of ethylene, led THEOLOGIS and LATIES (1978c) to investigate whether the alternative path was engaged during this phase. In slices of avocado (*Persea americana*) and banana (*Musa cavendishii*) fruits, the alternative path is engaged neither before nor during the climacteric. THEOLOGIS and LATIES (1978c) concluded that the respiratory climacteric in intact avocado and banana fruits is cytochrome path-mediated. A role of the alternative path in ripening of avocado and banana was, therefore, considered nonexistent. This conclusion hinges on the belief that the respiration in slices of fruits in various stages is a true reflection of that in intact fruits. BAUER and WORKMAN (1964) demonstrated that respiration of intact banana fruits shows a pattern during ripening different from that of slices made at various stages during fruit development. In *Cucumis melo* (honeydew melon), the pattern of CO_2 production by the intact fruit is different from that in discs from various parts of the fruit. Preclimacteric discs respire at a rate similar to that of climacteric intact fruits, while the climacteric peak is probably higher in slices than in the intact fruit (PASSAM and BIRD 1978). Knowledge on the activity of both the cytochrome path and the alternative path is still lacking. However, even if the alternative path were involved in fruit ripening, this cannot be associated with the synthesis of H_2O_2, as originally thought. H_2O is the only product of the alternative oxidase and the H_2O_2, originally supposed to be produced, was later found to originate from other sources (HUQ and PALMER 1978, RICH 1978).

ALVES et al. (1979) found that the synthesis of a range of sesquiterpenoid stress metabolites in tuber slices of *Solanum tuberosum* upon infection with *Phytophtora infestans* was inhibited by SHAM, an inhibitor of the cyanide-resistant path. The authors suggested, therefore, that the alternative path is involved in the production of stress metabolites. However, SHAM, as pointed out in Sect. 3.2, has a number of side effects, and in long-term experiments, effects of SHAM cannot simply be ascribed to its effect on the alternative path. Moreover, it remains uncertain if the alternative path was engaged at all in the investigated tuber slices.

3.5 Cyanide Resistance as a Mechanism to Tolerate Cyanide in the Environment

PASSAM (1976) investigated the respiration of root tubers of *Manihot esculenta*, which contain linamarin and lotostraulin, cyanogenic glycosides which release HCN upon hydrolysis. The tuber respiration (O_2 uptake) of six investigated cultivars was stimulated by 1 mM KCN. Succinate oxidation by mitochondria isolated from these tubers which contained less than 50 mg CN kg^{-1} fr. wt. was ca. 70% inhibited by 1 mM KCN, whereas only ca. 30% inhibition was found in mitochondria isolated from tubers containing more than 50 mg CN kg^{-1} fr. wt. PASSAM (1976) concluded that the cyanide resistance of mitochondria isolated from tubers containing more than 50 mg CN kg^{-1} fr.wt. was considerably more than that of mitochondria from other root crops, and suggested that the electron transport capacity of the alternative pathway relates to the cyanogen content of the tissue.

HALL et al. (1971) studied two millipedes, which secrete HCN in a defensive mechanism in sufficient amounts to kill other animals in the immediate neighborhood. Respiration of the millipedes was cyanide-resistant and the oxidation of succinate by mitochondria isolated from one of the two millipedes (*Euryurus leachii*) was only 30% inhibited by 1 mM KCN.

A third example of an association between cyanide-resistant respiration and HCN production was found by CHALKEY and MILLAR (1982). In *Coprinus psychromorbidus* (a cyanogenic low-temperature basidiomycete), both growth and respiration can proceed when the alternative path has been induced by 0.05 mM cyanide. Unlike other cyanide-tolerant fungi, e.g., *Stemphylium loti* (RISSLER and MILLAR 1977), *C. psychromorbidus* does not detoxify cyanide via formamide hydro-lyase, and thus appears to depend on the alternative path for life in an environment which has been enriched by HCN from endogenous origin (CHALKEY and MILLAR 1982).

The three examples above suggest that the alternative path can be of significance when the organism is exposed to an environment containing cyanide or other inhibitors of the cytochrome path. However, it is unlikely that coping with such inhibitors in the environment is the prime role of the alternative path.

COLLINGE and HUGHES (1982) investigated cyanogenesis in *Trifolium repens*, whose leaves may contain up to 1% of the fresh weight as cyanogenic glucosides, depending on the genotype. Two genes which are involved in the synthesis of the cyanogenic compounds, linamarin and R-lotaustralin, exhibit polymorphism in most natural populations and in commercial varieties (DADAY 1954). To explain this polymorphism it has been suggested that grazing favors cyanogenesis, whereas low temperature and frost favor acyanogenesis (DADAY 1954, FOULDS and YOUNG 1977). These abiotic factors would tend to cause some damage to the plant and disturb the compartmentation inside the cell. This would enable the glycosides to meet linamarase, the enzyme catalyzing their hydrolysis, and lead to the release of HCN. If this explanation were correct, it would appear that the alternative path has very little survival value in white clover with respect to the functioning in an environment containing HCN.

3.6 The Alternative Path in Relation to Anaplerotic Functions of Mitochondria

BAHR and BONNER (1973) suggested that there may be an advantage in having a nonphosphorylating electron transport chain, if the role of the mitochondria were not simply to supply ATP but to play a role in the metabolism of carbon compounds. For example, in cells which synthesize large quantities of organic acids which subsequently may be stored or used in biosynthesis (Fig. 3) and in which the requirement for ATP is small, oxidation of NADH without concomitant production of ATP can be important. MØLLER and PALMER (1982) speculated that not only the alternative path but also the rotenone-insensitive dehydrogenase, which bypasses phosphorylation site 1 (BRUNTON and PALMER 1973, MARX and BRINKMANN 1978), is significant to keep anaplerotic routes going. This is an attractive hypothesis, and it is to be regretted that no physiological experiments have been performed to corroborate it. It would seem that a high rate of production of intermediates for biosynthesis (i.e., a major functioning of the Krebs cycle in anaplerotic reactions) coincides with a high requirement for ATP in the cell. If this were so, there would be no need for nonphosphorylating pathways to allow production of organic acids.

However, under conditions where more of the supplied cation than the anion is absorbed, i.e., when a mobile cation, such as K^+ of Na^+, is supplied with a less mobile anion, such as SO_4^{2-}, organic acids balance the "excess cation absorption" (ULRICH 1941). Under such conditions the R.Q. of respiration decreases and accumulation of organic acids is found. Reverse changes occur during "excess anion absorption", e.g., from $CaCl_2$ solution. Whether an increased engagement of the alternative path occurs or is to be expected during synthesis of organic acids associated with K^+ uptake from a K_2SO_4 solution must also depend on the energy costs of ion uptake. Only if these costs are relatively small can an increased engagement of the alternative path be expected. No data on in vivo experiments are available so far. However, it is noteworthy that MØLLER (1978) found differences in mitochondria isolated from wheat

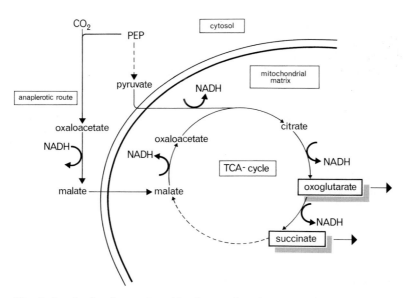

Fig. 3. Synthesis of organic acids via anaplerotic routes. Production of such compounds as α-oxoglutarate and succinate may involve the net production of NAD(P)H. If ATP is not concomitantly required in the cell, nonphosphorylating electron pathways may be significant to allow the synthesis of these intermediates

roots, grown in concentrated versus diluted nutrient solution. The maximal capacity of the alternative path in mitochondria isolated from the plants grown in diluted nutrient solution was highest, while the pathway was only fully engaged in mitochondria isolated from roots of plants grown in a concentrated solution. MØLLER (1978) speculated that the larger engagement of the alternative path might be related to a higher rate of organic acid production in roots of plants grown in a concentrated solution; these roots contained more organic acids. If it were mainly malic acid which accumulates, there would still be no need for an alternative path; malic acid is synthesized from OAA (via MDH) which in turn is derived from PEP in a reaction catalyzed by PEP-carboxylase and/or PEP-carboxykinase (POPP et al. 1982) and production of malate from glucose in this way does not involve the *net* production of NAD(P)H. On the other hand, rapid breakdown of malic acid, e.g., upon addition of $CaCl_2$, does produce NADH and might involve the engagement of the alternative path.

A second example of rapid synthesis of organic acids is that of CAM plants. KINRAIDE and MAREK (1980) measured the cyanide resistance of respiration in sectioned leaves of *Bryophyllum tubiflorum*. Thick (8-mm) leaf sections accumulated organic acids at the same rate as intact leaves (KINRAIDE and BEHAN 1975). However, KINRAIDE and MAREK (1980) concluded from the lack of inhibition by SHAM in the absence of cyanide that the alternative path in 8-mm sections was not engaged. THOMAS (1951) measured the oxygen uptake in leaves of *Kalanchoë* both during rapid acid accumulation and when the maximum amount of acids had already been produced, and found the same rate in both situations. One is thus left with the observation that extensive synthesis of organic acids can proceed without an increase in respiration or the operation of the alternative path. In fact, the synthesis of malate from glucose does not involve a net production of NADH, as stated above.

Increased respiration via the alternative path in *Nicotiana tabacum* callus cells after addition of sugars to the medium has been described in terms of the role of the alternative path in anaplerotic reactions (DE KLERK-KIEBERT et al. 1981). This explanation is an unlikely one, since nitrate was the predominant N-source for these cells and no excess of NADH from metabolism is to be expected as the reduction of one molecule of nitrate to ammonia already requires the equivalent of four molecules of NADH.

3.7 The "Energy Overflow" Model

As a modification of BAHR and BONNER's (1973) suggestion, it has been proposed that the alternative path functions as an "energy overflow" (LAMBERS and STEINGRÖVER 1978b). It was hypothesized that the alternative pathway is involved in the oxidation of sugars in excess of those required for the production of carbon skeletons for growth, ATP production, osmoregulation, and storage as carbohydrate reserves. Supporting evidence for this model has been presented by STEINGRÖVER (1981), who observed that the contribution of the alternative path in root respiration decreased upon the commencement of storage in the tap-root of *Daucus carota*. LAMBERS et al. (1981a) observed a decrease in activity of the alternative path in the roots of *Plantago coronopus* upon transfer of the plants to saline conditions which triggered the synthesis of sorbitol, a compatible solute in *Plantago*. The carbohydrates saved in respiration equalled those utilized in sorbitol synthesis. In leaves of *Triticum aestivum* harvested at the end of the night, respiration via the alternative path could be stimulated by the addition of sugars (sucrose, glucose, or fructose) to the level in leaves, harvested at the end of the day (AZCÓN-BIETO et al. 1983a). Similarly, VAN DER PLAS and WAGNER (1983) observed a positive correlation between the sucrose concentration in the medium and the engagement of the alternative path in callus-forming potato tuber tissue discs. A number of other examples are included in a recent review (LAMBERS 1982); they indicate an association between the concentration of metabolically available sugars and the engagement of the alternative path. Taken together, these data support the energy overflow function of the alternative path. However, this model has been criticized by LATIES (1982), who considered that the concentration of SHAM used in the experiments that provided the supporting evidence might have been too high and have produced aspecific effects. In subsequent experiments (BLACQUIÈRE and DE VISSER 1984, DE VISSER and BLACQUIÈRE 1984, LAMBERS et al. 1983a) it was convincingly demonstrated that this criticism was *not* justified. A second point of LATIES' criticism, that control points in glycolysis had to be bypassed to explain activity of the alternative path, will be discussed in Section 4. Evidently, this is a point of criticism that applies equally to all hypotheses on the significance of the alternative path.

3.8 The Alternative Path and an Increased Demand for Metabolic Energy

In a number of species, saturation of the cytochrome path and concomitant engagement of the alternative path has been demonstrated to be due to a restriction of the electron flux via the cytochromes by oxidative phosphorylation (Sect. 4.1). When synthesis of cytochrome components was inhibited in cultured *Glycine max* cells by chloramphenicol, total respiration and the engagement of the alternative path increased (DE KLERK-KIEBERT et al. 1982). In these cells, saturation of the cytochrome path may have been due to complete saturation of the capacity (as distinct from saturation under restriction from adenylates). This leads to the hypothesis that the alternative path might be of significance in such cells as those of cultured *Glycine max* upon an increased demand for metabolic energy (cf. BLACQUIÈRE and DE VISSER 1984). This increased demand would lead to an increased flux of carbon through glycolysis. In the presence of the alternative path, electrons could be transferred to O_2, be it with only little ATP production (via Site 1 only). This situation would last until sufficient components of the phosphorylating cytochrome path have been synthesized, to match the increased demand for metabolic energy.

4 Regulatory Aspects of Respiration in Vivo

Various points of control of respiration can be envisaged: (1) the supply and/or availability of substrate for glycolysis, (2) glycolysis per se, (3) transport of organic acids across the mitochondrial membrane, (4) the TCA cycle, (5) the electron transport pathways. In this section information concerning regulatory aspects will be discussed in the context of the respiration of intact tissues. For reasons of convenience, and *not* because this would be the predominant mechanism, the regulation of the electron transport pathway(s) will be presented first. Figure 4 illustrates some of the topics discussed in this section.

4.1 The Regulation of the Activity of the Cytochrome and the Alternative Pathways

PASSAM and BIRD (1978) observed significant stimulation of respiration in *Cucumis melo* (honeydew melon) fruit slices by exogenous succinate (0.1 M), and a further increase by the subsequent addition of ADP (3 mM); addition of ADP by itself gave no significant stimulation. This indicated that in these slices, before the addition of substrates, the cytochrome path was restricted by the supply of reducing power to the mitochondria.

These honeydew melon fruit slices were used during their "climacteric" peak of respiration. Climacteric fruits show a typical rise in respiration at the onset of ripening: the "climacterium" (i.e., a critical phase in life). Nonclimacteric fruits do not have a climacteric peak, but rather a slow downward drift in respiration following detachment (RHODES 1980).

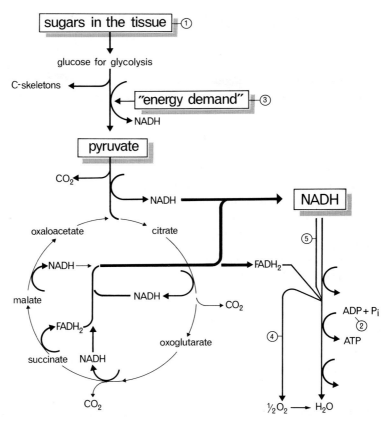

Fig. 4. A simplified scheme of respiration and its points of control. Controlling factors include the concentration of substrate, e.g., glucose (*1*) and adenylates (*2, 3*). Adenylates may exert their control on electron transport via a constraint by oxidative phosphorylation (*2*), or on glycolysis, via inhibition of key enzymes ("energy demand", *3*). When input of electrons into the respiratory chain exceeds the capacity of the cytochrome path, the alternative path accepts this excess (*4*). There is some evidence that also the rotenone-insensitive bypass (*5*) only operates when the rate of production of endogenous NADH is high

THEOLOGIS and LATIES (1978a) concluded that in aged slices of *Solanum tuberosum* (white potato) the substrate supply to the mitochondria limited the flux of electrons through the cytochrome path. In the absence of uncoupler, the activity of the cytochrome path was only 80 μl O_2 g^{-1} fr. wt. h^{-1} and the alternative path was not engaged ($\rho=0$). In the presence of uncoupler the activity of the cytochrome path rose to 116 μl O_2 g^{-1} fr. wt. h^{-1} and the alternative path became fully operative ($\rho=1$). The increase in activity of both pathways by uncouplers was attributed to an increase in the flux of carbon through glycolysis (via a decreased inhibition through the control by energy demand in Fig. 4) and the Krebs cycle, with the result that the electron transport capacity of the cytochrome path was exceeded, thus causing spill-over of electrons into the alternative path. Therefore, in aged potato slices and likewise in fresh slices of *Ipomoea batatas* (sweet potato), *Persea americana* (avocado) and *Musa cavendishii* (banana) fruits (Theologis and Laties 1978b, c), the cytochrome path did not operate at its full capacity. Whether the activity of the cytochrome path in these tissues was restricted by the supply of substrate to the mitochondria only, or perhaps simultaneously controlled by adenylates in such a way that the flux of electrons through the cytochrome path matched the import of organic acids into the mitochondria cannot be concluded with certainty.

In *Glycine max* axes (18 h after the start of inhibition), 1 µM CCP stimulated respiration and this increased respiration was fully inhibited by SHAM (1 mM) (Leopold and Mushgrave 1980). This indicates that the mitochondria in these axes (in the absence of uncoupler) received just enough substrate to saturate the cytochrome path. An increased supply of substrate to the mitochondria by addition of CCP caused a spill-over of electrons into the alternative path.

In intact roots of *Spinacea oleracea* and *Zea mays*, Day and Lambers (1983) found ca. 65% engagement of the alternative path in the absence of uncoupler ($\rho = 0.65$). FCCP (0.2 µM) stimulated both the cytochrome path (by 30%) and the alternative path (the engagement became 75% in *S. oleracea* and 100% in *Z. mays*). Therefore, in these intact roots the cytochrome path was restricted by adenylates, and the substrates imported into the mitochondria in excess of those to be oxidized by the cytochrome path were oxidized by the alternative path. This was also the case in leaves of *Phaseolus vulgaris* (Azcón-Bieto et al. 1983c). In other tissues, e.g., slices of wheat leaves (*Triticum aestivum*) harvested at the end of the day, 5 mM SHAM inhibited ca. 20% in the absence of 0.1 µM FCCP, but had no effect in the presence of uncoupler (Azcón-Bieto et al. 1983a). In these leaf slices, and to a lesser extent in the intact roots of the same species (Day and Lambers 1983), FCCP released the adenylate control of the cytochrome path without affecting the total rate of respiration; i.e., the increased flux of electrons through the cytochromes was at the expense of that through the alternative path. In all of these intact tissues, therefore, the cytochrome path was not restricted by the input of electrons, but by adenylates.

An intermediate situation was found in intact roots of *Phaseolus vulgaris* (Day and Lambers 1983). In these roots uncoupler did *not* stimulate the cytochrome path, but the engagement of the alternative path increased from 10 to 40%. Therefore, the electron transport capacity of the cytochrome path per se limited the flow of electrons through this chain. In *P. vulgaris*, the activity of the cytochrome path apparently almost matched the rate of input of electrons to the mitochondria, so that the alternative path was essentially not engaged. In roots of two *Plantago* species and those of *Pisum sativum*, alternative path engagement was found when the cytochrome path had reached its maximum capacity (Blacquière and De Visser 1984).

The conclusion that the cytochrome path in *Pisum sativum* and *Plantago* species operates at its maximum capacity hinges on the assumption that SHAM only inhibits the alternative path and does not have significant side effects. De Visser and Blacquière (1984) found that exactly in the above-mentioned species SHAM did have severe side effects. In *Pisum* roots uncoupler did stimulate CN-sensitive respiration at low concentrations of SHAM (but sufficiently high to inhibit the alternative path). Therefore, it is not yet absolutely certain that the cytochrome path operates at its maximum capacity in vivo in such tissues as pea roots.

It is concluded that electron transport through the cytochromes in vivo can be restricted by oxidative phosphorylation, and in some tissues possibly also by the electron transport capacity of one or more of the components of the path. In aged slices of some fruits and storage organs the full capacity of the cytochrome path is not realized, due to a limited supply of mitochondrial substrates, possibly in combination with a constraint by oxidative phosphorylation.

The engagement of the alternative path, when the cytochrome path is restricted by adenylates, raises a further question. How do electrons pass the first energy transducing site (complex 1) at a higher rate than they pass the subsequent sites? LATIES (1982) has pointed out the possibility that the proton motive force (pmf) involves localized energy domains. If these localized domains exist, they offer an explanation why site 1 is less restricted than the next two sites and thus allows a high flux of electrons through complex 1.

In fact, due to the relative redox-potential "drops" across certain sections of the chain, electron flow is likely to be more severely restricted at site 1 than at the other sites, and thus far, there is no evidence that the pmf is localized so severely.

The nonphosphorylating rotenone-insensitive bypass (see PALMER and WARD, Chap. 7. this Vol.) offers a more attractive explanation for a bypass of the first site, if this dehydrogenase acts in series with the alternative path. Kinetic properties of the two dehydrogenases (high K_m for NADH of the rotenone-insensitive dehydrogenase versus a low K_m of the normal dehydrogenase (MØLLER and PALMER 1982), and the observation that both the alternative path and the rotenone-insensitive bypass develop simultaneously during aging of potato slices (CAMMACK and PALMER 1973) provide supporting evidence for this attractive model.

4.2 The Regulation of Glycolysis

From the preceding section it transpires that the rate of supply of mitochondrial substrates can exceed the rate at which the cytochrome path oxidizes these substrates. Assuming that the bulk of the mitochondrial substrates is produced via glycolysis, the conclusion is inevitable that the rate of glycolysis is *not* regulated in such a fashion that it matches the activity of the cytochrome path; accordingly, an excess of electrons spills over into the alternative path. This conclusion disagrees with the general belief of a tight control of glycolysis at the level of pyruvate kinase and phosphofructokinase (by the "energy demand" in Fig. 4; TURNER and TURNER 1980). The question then arises whether the rate of glycolysis is controlled in a different fashion in tissues where the alternative path is engaged in comparison with those tissues in which this path is not engaged or in tissues which do not show cyanide resistance at all.

There are two lines of evidence for the theory that glycolysis is controlled at the level of phosphofructokinase and of pyruvate kinase (TURNER and TURNER 1980). First, these two key enzymes are controlled by a number of metabolites in such a fashion that the activity of these enzymes tends to decrease when the demand for metabolic energy and for intermediates is low and to increase when this demand is high. Secondly, it can be calculated that the concentrations of the substrates and products of these key enzymes is such that these chemical reactions are far from equilibrium; the concentrations of fructose phosphate and phosphoenolpyruvate, the substrates of phosphofructokinase and pyruvate kinase, respectively, are three to four orders of magnitude higher (in comparison with the concentrations of the products of these enzymes) than can be estimated from thermodynamic characteristics of the two reactions. The other steps in glycolysis, with the exception of hexokinase, are essentially in thermodynamic equilibrium. The latter line of evidence will be followed in this section.

AP REES et al. (1977) measured the concentrations of substrates and products of phosphofructokinase and pyruvate kinase in the spadix of *Arum maculatum*. These authors concluded that in the club (the thermogenic terminal portion of the spadix) the reactions catalyzed by both enzymes were considerably displayed from equilibrium. This was the case for both pre-thermogenic and thermogenic clubs. Presumably the alternative path is only engaged during thermogenesis (LATIES 1982, but see also Sect. 3.2), so that the results of AP REES et al. indicate that glycolysis is controlled in the same fashion both when the alternative path is not operative and also when it is engaged.

DAY and LAMBERS (1983) measured the concentration of a number of glycolytic intermediates in the roots of three species. In *Pisum sativum* the alternative path did not participate in root respiration ($\rho = 0$). In roots of *Spinacia oleracea* and *Triticum aestivum* the alternative path was 65% and 92% engaged, respectively. Despite the wide variation in engagement of the alternative path in the roots of these three species, no significant variation was observed in the ratios of concentrations of substrates and products of phosphofructokinase and pyruvate kinase.

Obviously, the engagement of the alternative path in the *Arum* spadix or roots of higher plants does not point to a lack of control of glycolysis. It is, therefore, concluded that this control is not so tight that it prevents an "overproduction" of substrates for the mitochondria. Alternatively, the pentose phosphate pathway might play a major role when the alternative path is to a large extent engaged, provided that the NADPH produced in this pathway has access to the alternative path. Whenever the input of organic acids into the mitochondria exceeds the capacity of the cytochrome path, as determined by one of its components or by the rate of oxidative phosphorylation, the excess of organic acids is oxidized via the alternative path.

4.3 Regulation by the Concentration or Supply of Respiratory Substrates

YEMM and SOMERS (cited by YEMM 1965) observed a close correlation between the concentration of hexoses (up to 20 mg hexose g^{-1} fr. wt.) and the rate of CO_2 production in leaves of *Hordeum vulgare*. At higher hexose concentrations, little effect of sugar content on the rate of respiration was observed. Similar results were found for roots of *Cucumis sativus* (CHALLA 1976) and

Table 4. Respiration (in mg O_2 h^{-1} g^{-1} dry wt.) and soluble carbohydrate content (mg g^{-1} dry wt.) in roots of *Plantago major* grown at a high (=optimal) and a low (= limiting) supply of nutrients in solution. Note that the total carbohydrate content does not equal the sum of the cytoplasmic and vacuolar contents, because some of the sugars are located in the free space. (Unpublished data from I. STULEN, C. WOLFF and R. HOFSTRA)

Nutrient supply	Root respiration		Soluble carbohydrate content		
	Total	Alternative path	Total	Cytoplasm	Vacuole
High	5.8	2.4	73	4	65
Low	5.0	2.6	117	4	111

leaves of *Triticum aestivum* (AZCÓN-BIETO and OSMOND 1983). Also the observations of McCREE (1970) with *Trifolium repens* and PENNING DE VRIES and VAN LAAR (1975) with *Zea mays* and *Helianthus annuus* that there is a correlation between respiration and photosynthesis can be interpreted in terms of substrate supply (cf. AZCÓN-BIETO and OSMOND 1983). Excised *Zea mays* root tips rapidly depleted endogenous sugars and the rate of respiration decreased in parallel with the level of carbohydrates (SAGLIO and PRADET 1980). At a low level of endogenous glucose, DNP (10^{-5} M) failed to stimulate respiration, whereas in the presence of 0.2 M exogenous glucose respiration rates were higher and could be ca. 50% increased by DNP. Root respiration in *Plantago media* decreased after excision from the shoot over a period of 24 h. This decrease could be prevented by exogenous sucrose (25 mM) and was fully due to a decreased engagement of the alternative path (A.H. DE BOER and D. DE BEER, personal communication). More examples of the effects of sugars on respiration and the engagement of the alternative path are included in Sect. 3.7.

The evidence presented in this section shows a general trend of an increased rate of respiration and an increased engagement of the alternative path at a high level of sugars in the tissue. However, the absolute rates of respiration at a given concentration of soluble sugars in the tissue vary widely between leaves of different species and between different leaves of the same plants. These differences even persist when plants are grown under exactly the same conditions (AZCÓN-BIETO et al. 1983a, b). They may reflect the biochemistry of the regulation of respiration by sugars. However, compartmentation of sugars in the cell is a likely factor contributing to variation in respiration. STULEN et al. (personal communication) investigated the variation of respiration in roots of *Plantago major*, due to variation in the supply of nutrients (Table 4). At a low supply of nutrients the soluble carbohydrate content was ca. 60% higher than at an optimal supply. However, the rate of root respiration was lowest in plants grown at a low supply of nutrients. This apparent discrepancy was explained by the observation that the sugars which accumulated at a low nutrient supply were all compartmented in the vacuole, and as such *not* available for respiration. (Note that the *decrease* in respiration rate at a low nutrient supply is largely due to expressing the data g^{-1} dry weight. After correction for the accumulation of both soluble and insoluble carbohydrates at a low nutrient supply, respiration rates are approximately the same).

4.4 Toward a Model of the Regulation of Respiration by Substrates and Adenylates

There can be no doubt that adenylates can regulate the rate of glycolysis and/or the activity of the cytochrome path. Neither is there doubt about the significance of carbohydrates in regulation of respiration. SAGLIO and PRADET (1980) have shown that at very low substrate levels adenylates do not control the rate of respiration; under these conditions respiration is obviously limited by respirable substrates (Fig. 5). These authors have also demonstrated that in the presence of high glucose levels adenylates control the rate of respiration. They may do so via their effect on key enzymes in glycolysis and/or via a constraint of electron transport via the cytochrome path (Fig. 4).

At even higher levels of substrates, that is when the alternative path is engaged, adenylates are also important in regulation of the flux through glycolysis in roots (DAY and LAMBERS 1983). At the same time the respirable substrates themselves regulate respiratory rates, but not because they are limiting the respiratory processes, as at low substrate levels (Fig. 5). AZCÓN-BIETO et al. (1983a) have also demonstrated the significance of the constraint by oxidative phosphorylation under conditions of high substrate levels in leaves. Respiration in flag

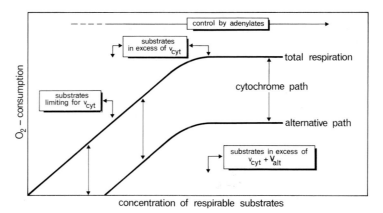

Fig. 5. The dependence of respiration and the alternative path on the sugar concentration in the tissue. Adenylates control the rate of respiration only marginally when the sugar concentration is so low that the alternative path is not engaged. The control of adenylates is stronger at higher sugar concentrations, including those at which both the cytochrome path and the alternative path are saturated. Note that despite the alternative path, which acts as an "energy overflow", the sugar concentration in the tissue varies over a wide range. V_{alt} capacity of the alternative path; v_{cyt} activity of the cytochrome path

leaf slices of *Triticum aestivum*, harvested after 6 h of photosynthesis, was insensitive to addition of sucrose, glycine, or FCCP alone. However, when sucrose or glycine was added in combination with FCCP, respiration was stimulated by ca. 30%. Since the oxidation of glycine, unlike that of sucrose, is not subject to adenylate control of glycolysis, the FCCP effects clearly point to a constraint of electron transport by oxidative phosphorylation. In leaf slices of the same wheat plants, harvested at the end of a 14 h night, respiration was ca. 30% less (AZCÓN-BIETO et al. 1983a). Respiration of these leaves was ca. 20% stimulated by both sucrose and glycine and only marginally by FCCP, when added alone. Moreover, the increased respiration upon addition of sucrose or glycine was partly due to a ca. 10% stimulation of the cytochrome path. This shows that when the input of respiratory substrates into the mitochondria restricts the cytochrome path, adenylates only marginally control respiration.

Only by the combined application of exogenous substrate and uncouplers and inhibitors of the alternative path can information be obtained on the regulation of respiration in each specific case. The regulation of respiration by both substrates and adenylates is illustrated in Fig. 5, which indicates that at a wide range of sugar concentrations in the cell the respiration rate is the same. This illustrates that the alternative path, although engaged to a larger extent at high than at low sugar concentration, cannot be considered a perfect mechanism to maintain a fixed sugar concentration in the cell (cf. VAN DER PLAS and WAGNER 1983). At best, the alternative path is a mechanism to damp otherwise severe oscillations in sugar concentrations. This mechanism is akin to that used to regulate the water table in a river during the threat of a severe flood: an increased water flow prevents the occurrence of a flood, but the water table nevertheless rises to a high level.

The engagement of the alternative path upon saturation of the cytochrome path can now be more fully appreciated. If *saturation* of the cytochrome path occurs because of restriction by oxidative phosphorylation, as in the roots of some crop species (DAY and LAMBERS 1983), the activity of the alternative path can truly be seen as an energy overflow. The cytochrome path could have accepted the electrons transport via the alternative path if the demand for ATP in the cell had been higher. However, the case is different in roots of some legumes and *Plantago* species (BLACQUIÈRE and DE VISSER 1984), where the cytochrome path reached its full capacity (presumably because one of the components reached its V_{max}). In these roots the alternative path also accepted electrons which could not be transported via the cytochrome path, and as such it also acted as an overflow. Two alternative hypotheses might be proposed to explain engagement of the alternative path in these roots (see also Sect. 3.8). (1) Similar to the situation in the roots of the crop plants mentioned, the substrate level might be responsible for a high flux of carbon through glycolysis and the TCA cycle, and thus produce large quantities of NADH. (2) If the energy requirement actually exceeded that which could be met by the cytochrome path, glycolysis and subsequently also the TCA cycle might be stimulated and generate so much NADH that the capacity of the cytochrome path was exceeded. In the latter case, the alternative path cannot be considered as a mechanism to remove excess substrates, but rather as a means to prevent accumulation of fermentation products. It is not known if the situation indicated here is purely hypothetical, and more information on the concentration of glycolytic key intermediates, for example, would be very welcome.

5 Respiration and Its Relation to Other Aspects of Metabolism

The substrates of respiration are produced in photosynthesis and transported across membranes and via the phloem. Products and intermediates of respiratory metabolism are utilized in many biosynthetic reactions. It is, therefore, not at all surprising that respiration depends on so many other aspects of plant metabolism, e.g., ion uptake, N-metabolism, photosynthesis, etc. In this section examples are chosen which illustrate the specific links between respiration and the rest of the plant's metabolism, or which reveal our lack of knowledge in certain areas.

5.1 Ion Uptake

The effect of O_2 on the rate of salt uptake implies that respiration is required in this process. Presumably respiration is required, since ion uptake requires metabolic energy, but it is also possible that the respiration merely maintains the integrity of the cell and is as such associated with salt uptake. LUNDEGÅRDH (1960) took the view that there is a direct association between respiration and ion uptake and introduced the concept of "anion respiration". Since 1960, when LUNDEGÅRDH reviewed this subject, very little information has become available.

The basic observation is that the respiration of tissues, kept in distilled water, increases upon addition of NaCl, KCl or KNO_3 (JACOBI 1899). LUNDE-GÅRDH (1960) stated that there should be a quantitative relationship of 4 between transported anions and consumed oxygen. The concept of anion respiration has been criticized, since respiration can also be stimulated by cations bound to a resin which prohibits anion absorption and other workers were unable to find the quantitative relationship of 4, as postulated by LUNDEGÅRDH (cf. BOWLING 1976).

LUNDEGÅRDH (1960) concluded from experiments in which he used KCN or CO that the ion-stimulated respiration was due to increased activity of the cyanide-sensitive path. However, now more is known on the biochemistry of cyanide-resistant respiration, it is clear that the use of inhibitors of the cyto-chrome path cannot provide information on the *activity* of either the cytochrome path or the alternative path (cf. Sect. 3.1.4). Specific inhibitors of the alternative path should be used before this aspect of LUNDEGÅRDH's hypothesis can be corroborated.

PITMAN et al. (1971) presented a model on the exchange of ions against sugars across the tonoplast. Based on this model, LAMBERS (1980) proposed that the stimulation of respiration by ions might be due to the increased availability of respirable substrates, originally stored in the vacuole (sugars and organic acids). This model is supported by the observation that upon growth in a $CaSO_4$ solution, the activity of the alternative path in *Zea mays* roots decreased to a low level, whereas the cytochrome path was much less affected. However, further experiments need to be done to explain the phenomenon of anion respiration.

5.2 The "Movement" of Plants

Leaflets in *Samanea saman* open and close according to a circadian rhythm. These movements are driven by the redistribution of K^+ and Cl^- between motor cells in the extensor and flexor regions of the pulvinus (the joint; SATTER and GALSTON 1973). The respiration rate of the pulvini oscillated in phase with the pulvinar movement (SATTER et al. 1979). It seems likely that the increase in respiration during pulvinar opening is associated with the transport of ions into the extensor cells and out of the flexor cells. Similarly, coiling of *Pisum sativum* tendrils was shown to correlate with respiration and to depend on the operation of the cytochrome path (RIEHL and JAFFE 1982). SHAM, an inhibitor of the alternative path, did not block the motor response of mechanically perturbed tendrils.

5.3 Flowering

HEW et al. (1978) observed a circadian rhythm and an increased rate of CO_2 production in some tropical orchid flowers upon the opening of the flower. The rhythmicity in CO_2 production correlated with the production of fragrance. This phenomenon appears akin to that in Araceae, where heat production, due to a burst in respiratory activity, is supposed to be significant in the volatilization of odoriferous compounds, which attract the pollinators (cf. MEEUSE 1975). However, considering the size of the orchid flowers thermogenesis is an unlikely phenomenon in these organs (cf. Sect. 3.2). Flowering in *Pisum sativum* did not increase root respiration, but the engagement of the alternative path increased (DE VISSER and LAMBERS 1983). This observation could be fully explained by the "energy overflow hypothesis" (Sect. 3.7), since the start of flowering was associated with reduced root growth, while the carbon flux to the roots was less affected. The increased activity of the alternative path was only transitory. The subsequent decrease in root respiration was associated with decreased transport of carbohydrates to the roots.

6 Respiration and Its Dependence on Environmental Factors

As is to be expected of a process which is linked to so many other aspects of metabolism, respiration is affected by many environmental factors. In this section no attempt has been made to make an inventory of all environmental factors and their effects on respiration. The reader is referred to Volume XII of the Encyclopedia of Plant Physiology (1960) for such information. In this section, only those effects are indicated which have led or may lead to a further understanding of the role of respiration in adaptive processes and of the functioning of respiration as an integrated part of metabolism.

6.1 Effects of Light

It is still in debate whether and to what extent dark respiration in green leaves continues during photosynthesis (see GARDESTRÖM and EDWARDS, Chap 11, this Vol.). In this section only the respiratory processes occurring in the dark, apart from photorespiration in a strict sense, will be discussed.

6.1.1 Leaf Respiration After a Period of Photosynthesis

It is well established (e.g., ROSENSTOCK and RIED 1960) that respiration following a period of photosynthesis is increased in comparison with that after a long dark period. This effect may disappear in about 1 h, e.g., in leaves of *Vallisneria spiralis* (PRINS and WALSARIE WOLFF 1974). The phenomenon is generally ascribed to the availability of substrate, and some blue light-dependent effects have been described which ultimately also indicate this mechanism (e.g., KOWALLIK and SCHÄTZLE 1980). HEICHEL (1970) investigated the CO_2 production in *Zea mays* leaves after a period of light and found that the increased respiration, as compared with that after several hours in the dark, was correlated with the light intensity prior to the measurement of respiration. The increase was not dependent on net CO_2 fixation, since it was also found when the plants were kept in CO_2 free air or CO_2 free N_2, containing 400 µl O_2 per l N_2 during the preceding light period. This light effect lasted 2 to 3 h and was dependent on O_2 in the dark. These effects are not readily explained by the light dependence of photosynthesis, which provides substrates for respiration. In *Larrea tridentata* (creosote bush) the rate of dark CO_2 release was found to correlate with the concentration of total nonstructural carbohydrates, but not with the integrated CO_2 fixation in the preceding light period (CUNNINGHAM and SYVERTSEN 1977). AZCÓN-BIETO and OSMOND (1983) found in *Triticum aestivum* (wheat) leaves that the CO_2 efflux 30 min after the beginning of the dark period was correlated with accumulated net CO_2 fixation in the preceding light period and with the carbohydrate level in the leaves. In these wheat leaves, both the extent and the duration of the increased respiration rate after photosynthesis were dependent on the temperature during the respiration measurements. At 30° C leaf respiration declined to the level found before several hours of photosynthesis in ca. 2 h, whereas at 20° C it took up to 5 h to reach this control level. The increase in respiration rates following a period of photosynthesis is biphasic (AZCÓN-BIETO and OSMOND 1983, AZCÓN-BIETO et al. 1983b), consisting of an initial fast phase of 15–20 min and a slower phase of much longer duration (it is the latter phase which is referred to above). Lowering the oxygen concentration from 21 to 3% for 20 min during photosynthesis, just before the start of the respiration measurements, affected the initial rate of subsequent leaf respiration, but did not affect the longer phase of increased respiration in wheat leaves. In leaves of *Pisum sativum*, the respiration rate immediately after photosynthesis was negatively correlated with the carbon dioxide concentration during photosynthesis and the extra respiration was sensitive to SHAM, suggesting the involvement of the alternative pathway (AZCÓN-BIETO et al. 1983b). The dependence of at least this initial part of respiration after photosynthesis on the concentrations of CO_2 and O_2 during photosynthesis, suggests the involvement of photorespiratory intermediates (and implies that decarboxylation of accumulated glycine in the dark in vivo occurs at least in part via the respiratory chain). However, the effects after 30 min in darkness do not seem related to prior photorespiration and may reflect an increased availability of respirable substrates and/or an increased demand for metabolic energy. The latter suggestion is not attractive, since this "extra" respiration occurs predominantly via the alternative path (AZCÓN-BIETO et al. 1983b). The former suggestion agrees with the finding that the dark CO_2 efflux in leaves of *Triticum aestivum* correlated with the concentration of carbohydrates which was varied by varying the time of, and CO_2 concentration during, photosynthesis (AZCÓN-BIETO and OSMOND 1983). Moreover, leaf slice respiration after several hours of darkness could be increased to the level after the light period by the addition of sugars; again exogenous sugars predominantly stimulated the alternative path (AZCÓN-BIETO et al. 1983a). The rate of respiration in wheat leaves was also correlated with accumulated net CO_2 fixation. However, in this context it is perhaps worth mentioning that in leaves of *Lycopersicon esculentum* the rate of respiration was more sensitive to changes in the light level during the preceding light period than to changes in the CO_2 concentration, even when similar amounts of carbon dioxide were assimilated during the light period (LUDWIG et al. 1975).

Thus, one is left with two major effects of the light intensity in the period preceding the dark. One effect is related to photosynthesis, presumably via the substrate supply. The other is a specific effect of light, the nature of which remains in the dark, as befits the conditions under which it occurs.

KOLKWITZ (1899), and many workers after him (for a review, see ROSENSTOCK and RIED 1960), have described effects of light on the respiration of chlorophyll-free tissue, including roots, storage tissue slices, seeds, and bacteria. Upon reviewing the early literature, ROSENSTOCK and RIED (1960) concluded that it could not fully be excluded that these light effects might, in fact, be effects of temperature, which would rise upon illumination. It has to be borne in mind that it cannot be excluded that some of the above light effects might also involve the temperature rise of leaves.

6.1.2 Respiration as Affected by Light Intensity During Growth

Respiration in sun leaves of *Cassia fistula* and *Stelechocarpus burahol* (both tropical trees) was higher than in shade leaves of the same species, both when expressed on an area and on a dry weight basis (STOCKER 1935). GRIME (1966) reported that the rate of mature leaf respiration (g^{-1} dry wt. of lamina) was consistently higher in shade-intolerant species than in shade-tolerant ones. Since fully expanded leaves were used, this can hardly be a reflection of differences in growth rate; these were also consistently higher in shade-intolerant species.

The rate of root respiration also depends on the light intensity. SZANIAWSKI and ADAMS (1974) exposed seedlings of *Tsuga canadensis* (eastern hemlock), grown at 340 μE m^{-1} s^{-1}, for 5 h to light intensities in the range of 100 to 850 μE m^{-2} s^{-1}, and found that root respiration increased with increasing rates of photosynthesis, up to a saturation level. MASSIMINO et al. (1981) observed that root respiration of *Zea mays* was reduced by ca. 50% during a day at low irradiance (i.e., about 1/7 of the irradiance during growth). CHALLA (1976) found that the rate of root respiration in *Cucumis sativus* grown at high light intensity and long days was high and showed no diurnal rhythm. During growth at low light intensity and short days, the rate of root respiration was less and declined at the end of the night. These results may seem obvious, in so far as root growth is also reduced after a decrease of the light intensity (RICHARDSON 1953). However, the explanation for this phenomenon is more complicated. The oscillations in root respiration of cucumber (cf. CHALLA 1976) are largely due to a variation in engagement of the alternative path (LAMBERS 1980). Also the lower rate of root respiration during the night as compared with that during the day in *Lupinus albus* (NO$_3$-fed) was largely due to a lower activity of the alternative path during the night (LAMBERS et al. 1980). Changes in the rate of root respiration can apparently be largely ascribed to the operation of an alternative path, and the conclusion is by no means justified that a decrease in the rate of respiration reflects a limitation of structural growth by carbohydrates.

6.1.3 Respiration as Affected by the Integrated Level of Radiation

In many experiments on root respiration in which the photoperiod was varied, the light intensity was simultaneously altered, so that it is not always clear if the effects are due to the photoperiod, the light intensity, or a combination of the two parameters. HANSEN and JENSEN (1977) found for both shoots and roots of *Lolium multiflorum* that with a photoperiod of 16 h and a high light intensity, the efficiency of converting assimilate into structural dry matter (Y$_G$, cf. Sect. 2) was higher than with a photoperiod of 8 h and lower light intensity. Intermediate values were found at a photoperiod of 12 h and an intermediate light intensity. Maintenance respiration rates (expressed g^{-1} dry wt.) increased for both shoots and roots with decreasing radiation. This suggests that the overall respiratory efficiency decreased with decreasing levels of radiation.

The results of HANSEN and JENSEN (1977) do not agree with those on *Trifolium repens* by McCREE and KRESOVICH (1978), who found no effect of day length on either the conversion efficiency (Y$_G$) or the maintenance coefficient (m). However, these authors presented data for *whole plants* only. Since both m and Y$_G$ are generally different between shoots and roots (Sect. 2.1), and since day length may affect the relative amount of shoot per plant, it is possible that effects of day length on respiration of white clover exist but have not been detected.

6.1.4 Further Remarks on Effects of Light

Upon reviewing the literature on effects of light on respiration, one is left with a number of observations on whole plants which may have been easy to understand before recent more detailed information became available on the effect on roots versus those on shoots, and on the involvement of the different respiratory pathways. It should be stressed that the effects of light intensity are not really fully understood. The energy overflow model offers at best a framework to interpret the results, but also leads to further questions as regards the rate of carbohydrate utilization for growth, which cannot yet be answered.

6.2 Effects of Temperature

6.2.1 The Q_{10} of Respiration

The Q_{10} of respiration is generally found to be around 2.0, i.e., the apparent activation energy (E_a) is ca. 12 kcal mol^{-1}. However, there are frequent instances in both the early and the more recent literature, where the Q_{10} declines steadily, and more than to be expected from the Arrhenius equation, as temperature rises (for a review of the early literature, see FORWARD 1960). The explanation for this phenomenon, which does not occur in all species (EARNSHAW 1981, SIMON et al. 1976, YOSHIDA and TAGAWA 1979), is still uncertain. The possibilities that O_2 would fail to diffuse to its place of reduction rapidly enough to meet the increased demand at high temperature, or that sugars become limiting at high temperatures have been excluded (FORWARD 1960).

LYONS and RAISON (1970) observed an E_a of 5 kcal mol^{-1} for succinate oxidation by mitochondria from chilling-sensitive tissues (tomato and cucumber fruit and sweet potato roots) within the temperature range of 11°–25 °C. In the range 1°–10 °C, E_a of the same tissue was 35 kcal mol^{-1}. Such a "break" was not found in mitochondria from chilling-resistant tissues (cauliflower buds, potato tubers, and beet roots). POMEROY and ANDREWS (1975) observed a breakpoint in the Arrhenius plot of oxygen uptake by leaf segments of *Triticum aestivum* and *Secale cereale*. A similar breakpoint was found for the State 3 rate of respiration by mitochondria isolated from the same tissue, although this breakpoint was not always at the same temperature as that for leaf segments. AZCÓN-BIETO and OSMOND (1983) compared the activation energy of respiration in leaves of *Triticum aestivum* harvested after several hours of photosynthesis (but measured after 30 min in the dark to avoid photorespiratory complications) with that in leaves harvested at the end of a night period. The E_a of leaf respiration at the end of the night was 12.9 kcal mol^{-1} in the whole range from 8° to 42 °C. Leaf respiration after a period of photosynthesis showed a value for E_a of 17.9 kcal mol^{-1} below 20 °C and a lower value (8.6 kcal mol^{-1}) at temperatures above 20 °C. Thus below 20 °C, E_a was lowest for leaves with low carbohydrate levels (i.e., after a period of darkness), whereas the opposite was found above 20 °C.

A similar anomaly in the Arrhenius plot was found by BREEZE and ELSTON (1978) for leaves of *Phaseolus vulgaris*, grown in summer. When the carbohydrate level in the leaves was low, the Q_{10} of respiration was 2.1, whereas at high carbohydrate content the Q_{10} was 1.7 below 20 °C and 1.5 above 20 °C. In winter-grown plants, Breeze and Elston (1978) found breakpoints in the Arrhenius plots at both high and low levels of soluble carbohydrates. The biochemical explanation of this phenomenon remains the subject of speculation, but could be the following.

McCaig and Hill (1977) and Kiener and Bramlage (1981) observed a break in the Arrhenius plot of cyanide-resistant respiration in mitochondria isolated from *Triticum aestivum* coleoptiles and in *Cucumis sativus* hypocotyls, respectively. Knutson (1974) found a negative temperature coefficient for the respiration of the spadix of *Symplocarpus foetidus*, in which the cyanide-resistant path is presumably fully engaged (cf. Sect. 3.2). In the wheat mitochondria, the capacity of the alternative path actually decreased with increasing temperature, above 17.5 °C. Between 2° and 30 °C, the overall rate of respiration increased smoothly with temperature, both in State 3 and State 4. Yoshida and Tagawa (1979) found up to 60% inhibition by SHAM of respiration by *Cornus stolonifera* callus cells at low temperatures and less than 20% inhibition above 15 °C. In many plant tissues (cf. Sect. 3.7), including wheat leaves (Azcón-Bieto et al. 1983a), the alternative path is only operative at a high carbohydrate level. The break in the Arrhenius plot of leaf respiration, which is apparently associated with a high carbohydrate level, might, therefore, reflect a decreasing activity of the alternative path, with increasing temperature.

Wolfe and Bagnall (1980) have argued that the relationship of a biochemical process and temperature should not produce "breaks" but rather be curvilinear. The biochemical basis of a breakpoint in the Arrhenius plot, or, rather, a curvilinear relationship, is often associated with phase transitions in membrane lipids, from a liquid-crystalline structure to a solid state (e.g., Lyons and Raison 1970). Waring and Laties (1977a) have shown that the development of the cyanide-resistant path in aging *Solanum tuberosum* slices is preceded by the synthesis of phospholipids. Inhibition of fatty acid synthesis in vivo by cerulenin curtailed the development of the cyanide insensitivity (Waring and Laties 1977b). These results indicate that phospholipids are essential for the functioning of the alternative path, and a breakpoint in the Arrhenius plots might involve these lipids.

6.2.2 Transient Effects of Temperature on Respiration

Respiration of plants in a natural environment varies in a complicated fashion. Strain (1969) observed that the rate of leaf respiration in four desert species, measured at 25 °C, was higher in winter than in summer. Even after correction for the lower temperature in winter (assuming a Q_{10} of ca. 2), the respiration was still higher than in summer.

Smakman and Hofstra (1982) observed a decreased activity of the cytochrome path and a similar increase in the activity of the alternative path in roots of *Plantago lanceolata*, 2 h after decreasing the root temperature from 21° to 13 °C. In the subsequent 2 h, the activity of the cytochrome path increased to its original level; the activity of the alternative path, on the other hand, decreased to a very low level, but returned to its original level after another 24 h. Upon an increase in the root temperature from 13° to 21 °C, the activity of both pathways initially increased by ca. 60%, but the activity of the alternative path declined again after ca. 2 h. After 24 h it was only 30% higher than in control plants. After several days of growth at both 13° or 21 °C, the rate of respiration via both respiratory pathways was the same for both temperatures. A similar tendency was observed in *Atriplex polycarpa* (CHATTERTON et al. 1970).

The interpretation of the quoted observations cannot yet be fully given. Taken at face value they contain the warning that a comparison of the rate and Q_{10} of respiration of different species cannot be made without taking these transient effects into account.

6.2.3 Effects of Chilling

After chilling-sensitive tissues have been kept at low temperature (ca. 2 °C), they respond with a respiration burst upon exposure to normal (25 °C) temperatures (KIENER and BRAMLAGE 1981, TÀNCZOS 1977, YOSHIDA and NIKI 1979). In leaves of *Cucumis sativus*, exposed to 2 °C for 24 h, not only the coupled rate of oxygen uptake but also that in the presence of DNP was increased (TÀNCZOS 1977). This shows that the increase cannot be ascribed to uncoupling of respiration and it is more probably due to the accumulation of soluble sugars (TÀNCZOS 1974). KIENER and BRAMLAGE (1981) investigated the oxygen uptake of *Cucumis sativus* hypocotyls during recovery at 25 °C from 48 h chilling at 2 °C. They found that the increased respiration immediately after chilling was partly, and ca. 12 h after chilling completely, due to an increased engagement of the alternative path. The immediate effect is likely to be due to either an increased demand for ATP or to partial uncoupling (stimulating glycolysis beyond the level that can be accommodated by the cytochrome path). The effects after 6 h and more might be connected with TÀNCZOS' (1974) observation that sugars accumulate under chilling conditions (cf. Sect. 3.7 on the "energy overflow hypothesis").

WOODSTOCK and POLLOCK (1965) observed that when the temperature during imbibition of *Phaseolus lunatus* (lima bean) seeds was below 15 °C, respiration measured at 25 °C in the subsequent 2 h was severely reduced, and growth was reduced accordingly. When soaking of *Glycine max* occurred at 4 °C, solutes leaked from the cotyledons and the respiratory rate was reduced. This reduced respiratory rate involved a decreased capacity of the cytochrome path. In cotyledons, but not in axes, chilling during imbibition led to an increased engagement of the alternative path in the post-chilling period (p increased from 0 to 1). Little or no chilling damage occurred if this treatment was given 5 to 15 min after the start of soaking (LEOPOLD and MUSHGRAVE 1979). The authors suggested that the decreased capacity of the cytochrome path might be due to a loss of respiratory components of the cytochrome path. Equally well, the decreased capacity of the cytochrome path could be due to a more severe restriction of electron transport by the proton motive force (cf. Sect. 4.1).

Special attention has been given to the effect of temperature on respiration in chilling-sensitive versus chilling-resistant tissues, especially with regard to the occurrence of a breakpoint in the Arrhenius plot (GRAHAM and PATTERSON

1982, LATIES 1982). KIENER and BRAMLAGE (1981) found a breakpoint in the Arrhenius plot of the capacity of the alternative path in chilling-sensitive hypocotyls of *Cucumis sativus* which was very similar to that in mitochondria of the chilling-tolerant *Triticum aestivum* seedlings (McCAIG and HILL 1977). In mitochondria from chilling-sensitive callus cells of *Cornus stolonifera* and from chilling-tolerant cells of *Sambucus sieboldiana*, no breaks in either the cyanide-resistant or the total State 3 rate of respiration were found (YOSHIDA and TAGAWA 1979). Root respiration of the cold-tolerant *Ligusticum scoticum* showed a breakpoint in the Arrhenius plot around 18 °C (CRAWFORD and PALIN 1981). Thus, there is no consistent information indicating a difference in temperature sensitivity of the alternative path between chilling-sensitive and chilling-resistant tissues. However, in the presence of SHAM a break occurred in the Arrhenius plot of succinate oxidation by mitochondria from *Cornus stolonifera*, but not by those from *Sambucus sieboldiana*. A similar break was found for intact cell respiration in *C. stolonifera* (YOSHIDA and TAGAWA 1979). This indicates that in *Cornus* mitochondria one or more components of the cytochrome path is negatively affected by temperatures below the chilling temperature and that electrons are then diverted to the alternative path. Breakpoints in the Arrhenius plots of the activity of the cytochrome path (YOSHIDA and TAGAWA 1979) and the capacity of the alternative path (McCAIG and HILL 1977, KIENER and BRAMLAGE 1981) versus straight lines in other species are likely to reflect differences in the lipid domain of the two pathways. Obviously, these differences in the lipid domain do not only occur between both pathways in one tissue, but also between the same pathway in different tissues.

6.2.4 Effects of Supra-Optimal Temperatures

Upon exposure of *Glycine max* seeds to high temperature (40 °C) at a saturated humidity for two or more days, respiration of the axes declined (LEOPOLD and MUSHGRAVE 1980). After 4 days at 40 °C axis respiration was decreased by 50%. Only the cytochrome path became less engaged, whilst the engagement of the alternative path increased from 0 to ca. 100% of its maximal capacity. The authors suggest that high temperatures cause deterioration of the cytochrome path. However, since the cytochrome path is itself controlled by oxidative phosphorylation (see Sect. 4.1), a lowering of cytochrome path activity might equally well be due to a decreased demand for metabolic energy in the temperature-stressed seeds. Experiments with uncouplers could provide the answer.

6.2.5 Temperature as an Ecological Factor

The rate of respiration at 20 °C of arctic and temperate species is higher than that of tropical ones; leaves of tropical trees at 30 °C respire at about the same rate as leaves of arctic species at 10 °C (STOCKER 1935, WAGER 1941). Although differences in the Q_{10} of respiration between plants of cold and warm areas have been suggested (EARNSHAW 1981), a wider comparison involving many species has clearly demonstrated that the temperature coefficient of respiration is not different in species from arctic, temperate, or tropical habitats (STOCKER 1935, WAGER 1941). The difference in respiratory intensity between plants from warm and cold climates does not necessarily reflect a difference in the biochemistry of respiration. First, respiration tends to show a "drift" with temperature (WAGER 1941, Sect. 6.2.2), diminishing the initial sharp effect upon a change in environmental temperature. The quoted differences in respiratory intensity between plants from different climates may, therefore, merely reflect the temperature during growth of plants (warm for the tropical species and cold for the arctic ones), rather than species-specific differences. Secondly, since respiration in general and the cytochrome path in particular are often regulated by the demand for metabolic energy, the differences in respiratory intensity may reflect a difference in demand of metabolic energy, i.e., arctic plants are likely to require more ATP for growth at 10 °C than tropical plants, since growth of the latter species is restricted at that temperature.

This is also indicated by a comparison of root respiration of species from Great Britain (CRAWFORD and PALIN 1981). When the plants of two northern species were stored at 10 °C (*Ligusticum scoticum*) or 5 °C (*Mertensia maritima*), their rate of root respiration in the range 5°–30 °C was higher than that of two southern species, stored at 25 °C (*Crithmum maritimum*) or 20 °C (*Limonium binervatum*). However, when stored at 20 °C, root respiration of *Ligusticum scoticum* was lower than when stored at 5° or 15 °C, when compared at the same temperature, between 10° and 30 °C.

Thus, there is no hard evidence that biochemical characteristics of respiration contribute to the occurrence of species in a particular environment.

6.3 Effects of Salinity and Water Stress

When *Plantago coronopus* was grown at 50 mM NaCl, which did not inhibit growth, neither the dark respiration of the shoots nor root respiration via the cytochrome path and the alternative path was different from that in control plants, grown at 0 mM NaCl (BLACQUIÈRE and LAMBERS 1981). When young plants (ca. 7 weeks old) of the same species were transferred from a nonsaline solution into a solution containing 50 mM NaCl, the alternative path in the roots declined and the same quantity of carbohydrates thus "saved" in respiration was used in the synthesis of sorbitol, a compatible solute in *Plantago* (LAMBERS et al. 1981a). In *P. coronopus*, the capacity to increase the sorbitol content upon transfer into a saline solution is lost with age. When such older plants were transferred into a saline solution, photosynthesis and growth were inhibited and the alternative path in the roots became to a larger extent engaged (A. BIERE and H. LAMBERS, unpublished data). Therefore, an increased rate of root respiration in *P. coronopus* is associated with the lack of the capacity to cope with an increased salinity level, and *not* with the need for extra energy to handle the influx of ions.

In apple leaves, respiration (CO_2-output) decreased by up to 17% after watering of the tree was stopped and then increased by up to 62% above the control level as the plant showed definite signs of wilting. Upon rewatering, leaf respiration decreased to its control level (SCHNEIDER and CHILDERS 1941).

BRIX (1962) found a decrease in top respiration of loblolly pine seedlings and tomato plants with increasing water stress. In loblolly pine, but not in tomato, top respiration increased again at water potentials of -14 bar and lower and reached a level ca. 40% above that of the control at -30 bar. Increased levels of leaf respiration during wilting have been observed in many other species (cf. HUBER and ZIEGLER 1960); however, it does not occur in the moss *Funaria hygrometrica*, which tolerates desiccation (KERNBACH 1960). BOYER (1970) found that the rate of dark respiration (CO_2-output) in leaves of *Zea mays*, *Glycine max* and *Helianthus annuus* decreased with decreasing water potential of the leaves (in the range of -4 to -16 bar). Respiration was inhibited by not more than 50% and did not further decrease at lower water potentials. In roots of wheat, exposed to a gradually increasing water stress, the activity of the alternative path declined and organic solutes accumulated (M.E. NICOLAS, personal communication). In *Eucalyptus camaldulensis*, both the rate of photosynthesis and the engagement of the alternative path declined upon addition of 200 mM NaCl to the root medium (L. THOMPSON, personal communication).

From the above data it transpires that a decreased rate of respiration under conditions of NaCl or water stress is an adaptive response. The decrease in respiration rate is associated with a decreased availability of respirable substrates, due to either a decreased rate of photosynthesis, to an increased demand of sugars for the accumulation of organic solutes, or to a combination of both. An increased rate of respiration appears to be associated with inhibition of growth and an increased engagement of the alternative path.

However, not all authors agree with the above statement. SCHWARZ and GALE (1981) observed an increased rate of respiration in saline environments in a number of plant species. Rates of total respiration were plotted versus photosynthetic carbon uptake, the variation in both parameters being obtained by growing plants at different light intensities. In such plots (cf. Sect. 2) the interception with the y-axis is supposed to be the rate of maintenance respiration. Resulting analysis of the respiration data for *Xanthium strumarium* indicated that the rate of maintenance respiration increased with decreasing osmotic potential down to -7 bar and then decreased again. General remarks about this method to separate a maintenance component of respiration have already been made in Section 2. With respect to this specific problem, one more point should be stressed. It is essential to know which of the two respiratory pathways is responsible for the increased respiration under saline conditions. If it is the alternative path only, increased respiration rates are likely to be due to reduced growth, whereas increased demands for maintenance energy should be reflected in an increased activity of the cytochrome path. In the absence of information on which respiratory path is increased in a saline environment, the increased respiration cannot conclusively be ascribed to salt-induced maintenance respiration.

6.4 Mineral Nutrition

Most investigations on effects of mineral nutrition on respiration provide information on the plant's C-economy, rather than on mechanistic aspects of respiration. PATE and coworkers have provided such information on legumes (e.g., PATE et al. 1980), and HANSEN has investigated growth and maintenance respiration in *Lolium multiflorum* (e.g., HANSEN 1980). HANSEN (1980) observed peaks in the rate of nitrate uptake by the roots, which coincided with peaks in the rate of root respiration. This observation is reminiscent of the anion respiration discussed in Section 5.1.

7 Respiration and Its Relation to Yield and the Plasticity of the Individual

Over 50 years ago it was commonly believed that respiration merely consumed carbohydrates which could otherwise have been used for growth. The respiration was believed to decrease the yield (see Sect. 7.1). About two decades ago respiration was considered the source of metabolic energy to be used in the various processes determining yield. Now, with the knowledge on energy conserving versus nonphosphorylating pathways, an intermediate view must be held; respiration *can* produce metabolic energy, but sometimes it proceeds with *little* energy conserved and predominantly heat being produced. More information will have to be collected before the exact new view can be defined. In this section the available information on the relations between respiration and yield will be reviewed. Moreover, some ideas will be presented on the relation of respiration and plasticity of a genotype.

7.1 The Negative Correlation Between Respiration and Yield or Growth Rate

SCHEIBE and MEYER ZU DREWER (1959) were probably the first to show that those cultivars of *Hordeum vulgare*, *Avena sativa*, *Secale cereale* and *Triticum aestivum* which were "anspruchsvoll" exhibited lower rates of root respiration than the cultivars which were less "anspruchsvoll". Anspruchsvoll, according to the authors, are those cultivars which have a better production under favorable climatic conditions and with a good supply of water and nutrients than the cultivars considered as less anspruchsvoll. Under less favorable conditions the former cultivars generally respond with a greater reduction in yield than the cultivars which are less "anspruchsvoll".

HEICHEL (1971) has provided information on photosynthetic and respiratory rates in two inbred *Zea mays* varieties. Pa83 accumulated more dry matter than Wf9. However, Wf9 had a higher respiration rate (expressed g^{-1} dry wt.) in both leaves (42% higher) and roots (35% higher) than Pa83. The net photosynthetic capacity (per unit area) was 41% higher in the variety with a low growth rate. However, at the light intensity during growth, the photosynthetic performance of both cultivars was about the same. Unfortunately, some of the methods (e.g., storing the roots overnight at 2 °C, before measuring the respiration) lacked the subtlety to allow rigorous conclusions; but they

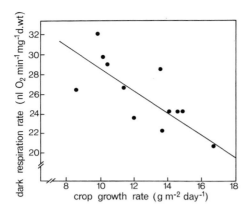

Fig. 6. Mature leaf respiration and crop growth in different genotypes of *Lolium perenne* cv. S23. A negative correlation was found between the rate of dark respiration and the rate of growth. (After WILSON 1982)

clearly provide an indication that the slower-growing variety respired a greater proportion of the photosynthates produced each day than the fast-growing one. The question now arises: Was the respiration rate high (1) because of a high demand for metabolic energy (e.g., for maintenance) and was the growth rate low because of the high demand for sugars in respiration (cf. Sect. 2), or (2) because growth required so much less carbon that an "excess" was available for respiration via nonphosphorylating pathways (cf. Sect. 3.7)?

HANSON (1971) selected lines from *Zea mays* and tested these for several parameters, including leaf respiration and photosynthesis. He concluded that net photosynthesis was not an important factor determining differences in yield. Lower rates of leaf respiration (per unit area) were associated with high yields. It is not clear from the data whether there were also differences in leaf respiration when expressed on a weight basis. WILSON (1975) observed a negative correlation between dry matter increase and mature leaf respiration per unit leaf dry weight in a comparison of six *Lolium perenne* genotypes. LAMBERS (1979) found consistently lower values for both growth and maintenance respiration in species with a high growth rate than in species with a slow growth rate. It was concluded that the fast-growing species respired more efficiently than the slow-growing ones.

More recently, WILSON (1981) selected lines from *Lolium perenne* cv. S23. The fast-growing lines had a "slow" rate of dark respiration (measured in mature leaves), whereas the slow growing ones had a "fast" respiration rate (Fig. 6). In the field, annual increases in crop productivity of 10–20% over the original cultivar were associated with a 20% reduction in mature leaf respiration (WILSON and JONES 1982). ROBSON (1982a) also found that simulated swards of the line with the slow respiration rate outyielded the fast line. The "slow" respiration line had a 22–34% lower rate of dark respiration (expressed per unit dry weight and measured on the entire community, including the roots). The differences in mature leaf respiration between the slow and the fast line was thus reflected in the behavior of the entire community. Canopy net photosynthesis was somewhat higher in the slow line, but only during the initial period following cutting of the canopy (regrowth period; ROBSON 1982a). In a comparison of the leaf respiration in seedlings of the two lines, ROBSON (1982b)

Table 5. Growth and respiration in roots of four inbred lines of *Plantago major*. The physiological plasticity of the four lines was found to decrease from line 1 to line 4 (data from KUIPER 1983; rates expressed on a dry weight basis; 35 day-old seedlings were used; numbers in brackets: respiration via the alternative path as a percentage of total root respiration)

Line	Root growth rate $mg\ g^{-1}\ day^{-1}$	Respiration via	
		Cytochrome path	Alternative path
		$mg\ O_2\ h^{-1}\ g^{-1}$	
1	90	4	4 (50)
2	96	6.5	3.5 (35)
3	141	7	4 (36)
4	165	8	4 (33)

showed that the differences in respiration were already apparent in 3-week-old plants. Two week-old seedlings did not show a difference in respiration rate (it was ca. 4 mg CO_2 g^{-1} dry wt. h^{-1}). The rate of mature leaf respiration declined with age in both lines (to ca. 2.4 and 3.2 mg CO_2 g^{-1} dry wt. h^{-1} in the slow respiration line and in the fast line, respectively, in 7-week-old seedlings). ROBSON (1982b) concluded that the differences in seedling growth between the lines were totally associated with difference in respiration, and no further differences needed to be invoked to explain the difference in carbon balance. Also, other parameters, e.g., ash content, shoot to root ratio, and leaf area index were not different for the various genotypes. The question of the cause and effect of respiration and growth, as formulated above, remains, however, unresolved!

KUIPER (1983) compared growth and respiration of the roots of four inbred lines of *Plantago major*. The lines with the highest relative growth rate showed the highest rate of root respiration (Table 5). However, the relative growth rate only showed a *positive* correlation with the activity of the cytochrome path. Growth rate was negatively correlated with the relative contribution of the alternative path (Table 5). D. KUIPER (unpublished results) also compared growth and root respiration of male sterile and hermaphrodite genotypes of *Plantago lanceolata*. The male sterile plants had a higher relative growth rate, but the rate of root respiration was about the same (ca. 6.5 mg O_2 h^{-1} g^{-1} dry wt. in 47-day-old plants). However, the activity of the cytochrome path was highest in the male sterile plants (5 vs. 3.5; same units), whereas the activity of the alternative path was lowest (1.5 vs. 3.0; same units).

These results on *Plantago* species suggest that a selection for a high growth rate should not be based only on the rate of respiration per se, but also on the relative contribution of the alternative path.

7.2 Are There Penalties, Associated with Slow Respiration Lines?

From a plant breeder's point of view, it appears rather attractive to improve yield in association with decreased respiration rates. However, do we only gain from the increased efficiency in C-economy, or are there also losses? More information is needed on the biochemical basis of fast and slow lines, before definite answers can be given. Two possibilities can be envisaged.

First, there is the possibility that growth in the fast respiration line is restricted by the shortage of carbohydrate. Any decrease in demand for carbohydrates, e.g., a decreased rate of maintenance respiration, due to a lowered protein turnover rate or decreased leakiness of membranes (cf. Sect. 2), will then result in a higher growth rate. It is hard to predict whether penalties are associated with the slow respiration line if the high respiration rate is due to a high demand for maintenance energy.

Secondly, there is the possibility that the fast line respired to a larger extent via nonphosphorylating pathways than the "slow" line. In that case selection of the slow respiration line involved primarily an increased incorporation of carbohydrates into dry matter, so that less substrates were left for respiration (cf. Sect. 3.7). It remains to be seen whether this type of selection involves penalties. LAMBERS et al. (1980) suggested that an operative alternative pathway contributes to the plasticity of the plant. This suggestion was based on experiments with *Plantago coronopus* (see Sect. 6.3). More examples in the same vein have become available since (cf. LAMBERS 1982). However, the question remains whether the engagement of the alternative path, prior to exposure to 50 mM NaCl, was really *essential* for the adaptation of *P. coronopus*. Starch levels in the roots did not change after exposure to NaCl, but does this mean that this was not an alternative source of carbon for sorbitol synthesis?

Results on inbred lines of *Plantago major* (Table 5) suggest the absence of a link between root respiration via the alternative path and the plasticity of a genotype. (Plasticity is defined here as the extent to which a number of physiological parameters were affected by a change in the nutrient supply; the lines varied with respect to the *rate* of the response). In line 1, which showed the lowest degree of physiological plasticity, the relative activity of the alternative path was higher than in any of the other lines.

Unfortunately, no answers are available for the questions raised above. This is to be regretted, since these answers might provide insight into the ultimate role of the alternative path in the metabolism of the plant and into the possibility of using information on this pathway as a tool in breeding programs.

8 Developmental Aspects

8.1 Germination

Upon the break of dormancy, during imbibition, the respiration of seeds rises rapidly but plateaus after ca. 20 min (for a review of the early literature see STILES 1960). During this early phase of imbibition the energy charge in *Lactuca sativa* (lettuce) seeds, rose from 0.2 to 0.8. This increase in energy charge requires O_2 and is inhibited by the presence of cyanide (HOURMANT and PRADET 1981).

After 5 h of soaking, this respiration of intact seeds of *Glycine max* and *Phaseolus aureus* was 100% and 70% inhibited by KCN and was not affected by inhibitors of the alternative path (SIEDOW and GIRVIN 1980). A high degree of cyanide sensitivity during inbibition has also been reported for seeds of *Lactua sativa* (HOURMANT and PRADET 1981) and *Cicer arietum* (BURGUILLO and NICOLAS 1977). Seeds of *Xanthium pensylvanicum* were ca. 25% resistant to cyanide after 24 h of soaking at 23 °C and inhibition by hydroxamic acid was also ca. 25% (ESASHI et al. 1982; note that the p plots in this paper are misleading and not in accordance with those described in Sect. 3.1).

It is concluded that the major part of respiration upon the break of dormancy is via the cytochrome path, albeit in some seeds, respiration may be up to 25% via the alternative path. There is no evidence that the presence and/or engagement of the alternative path is required for seed germination.

8.2 Root and Leaf Development

The rate of respiration (O_2-uptake) of the apical 10-mm root segment of *Hordeum vulgare* was several times higher than that of older segments along the root (MACHLIS 1944). Similarly, BERRY and NORRIS (1949) observed the highest rate of root respiration (expressed g^{-1} dry wt.) in the tip (5 mm). YEMM (1965) summarized the older literature as follows. The small undifferentiated cells of the meristem (active mitosis!) have a relatively *low* respiration rate on a cell basis. Upon vacuolation there is a marked rise in the rate per cell. Beyond this zone, respiration is maintained at a high level, as differentiation of cortical and stelar tissues continues. In maturing cells, respiration per cell is somewhat lower again, but still about five times higher than in the meristematic cells. On a protein basis there is much less variation in the rate of respiration and the rates tend to be highest in the meristematic zone. Despite the high energy requirement in the root tip, the alternative path still contributed significantly in the 5 mm tip of *Zea mays* roots (LAMBERS and POSTHUMUS 1980). At low light intensity during growth, the engagement of the alternative path in the mature root parts declined toward the end of the night, but remained high (56% of total respiration) in the root tip. This suggests that carbohydrates per se did not limit the metabolism in maize roots, even when growth occurred at low light intensity.

AZCÓN-BIETO et al. (1983c) studied respiration during leaf development in *Phaseolus vulgaris*. Respiration declined from ca. 3 in very young leaves to less than 1 μmol O_2 m^{-2} s^{-1} in 14-day-old leaves. During that period the capacity of the alternative path was constant at ca. 0.6 μmol O_2 m^{-1} s^{-1}, whereas its engagement gradually declined from ca. 30% to ca. 10%. The decline in respiration was largely due to a decreased activity of the cytochrome path.

8.3 Senescence

The information on respiration during senescence is rather conflicting. On the one hand, the data on respiration of leaves, attached to the plant, (*Triticum aestivum*; FELLER and EHRISMANN 1978) or detached immediately before the measurements (*Phaseolus vulgaris*; AZCÓN-BIETO et al. 1983c) showed a gradual decline toward senescence (in both cases plants were grown in the light). On the other hand, the senescence of detached *Avena sativa* leaves showed a rapid rise of respiration during senescence in the dark (TETLEY and THIMANN 1974). After 3 days in the dark, leaf respiration had increased 2.5 times. Inhibitors of senescence, such as kinetin (TETLEY and THIMANN 1974) and methyl jasmonate (SATLER and THIMANN 1981) prevented this respiratory rise. The difference in the effects of senescence on leaf respiration when senescence occurs with intact leaves as compared with detached leaves is likely to be due to the accumulation of respiratory substrates, e.g., amino acids or sugars (TETLEY and THIMANN 1974), in leaves which senesce after excision. The gradual decline of respiration in bean leaves with increasing age (AZCÓN-BIETO et al. 1983c) was largely due to a decreased activity of the cytochrome path. In these bean leaves the *capacity* of the cytochrome path declined with age, whereas that of the alternative path was unaffected.

8.4 Fruit Ripening

"Climacteric" fruits show a typical respiratory burst (the climacterium) during ripening. Ripening and the climacteric respiratory burst are associated with increased ethylene production, and exogenous ethylene induces both ripening and the climacteric burst (FORWARD 1965). Ripening is associated with an increased rate of protein synthesis. The pattern of changes in the glycolytic intermediates (see Sect. 4.2) in intact avocados (*Persea americana*) upon exposure to ethylene indicated an enhanced flux through glycolysis, due to an increased demand for metabolic energy (SOLOMOS and LATIES 1974). In *Daucus carota* roots ethylene treatment caused the proliferation of polyribosomes which was closely correlated to stimulation of respiration (CHRISTOFFERSEN and LATIES 1982). Ethylene might have similar effects in ripening fruits.

It is likely that the respiratory rise during the climacterium and upon exposure to ethylene is a response to the increased demand for metabolic energy. However, confirmation of this hypothesis awaits a proper analysis of the engagement of the alternative path in *intact* fruits.

9 Host-Parasite and Symbiotic Associations

When plant tissues are infected by fungi, their respiration increases (infection-induced respiration). The increased respiration is not due to the respiratory activity of the invading parasites themselves, and is located in both the infected and noninfected surrounding cells (EBERHARDT 1960). Parasitic angiosperms tend to have similar effects on the respiration of the host (SINGH and SINGH 1971). In addition to these topics the respiration of symbiotic associations will be discussed.

9.1 Host-Parasite Associations

Root respiration of both *Brassica campestris* and *Lycopersicum esculentum* was ca. 50% higher in plants infected by the angiosperm parasite *Orobanche* (broom-rape) than in uninfected healthy plants (SINGH and SINGH 1971). The stimulatory effect on respiration was stronger close to the point of infection than at a distance of a few centimeters from the infection point. Both inhibition by azide and stimulation by DNP (both expressed as percentages of the control rate) were the same in healthy and infected plants, indicating that the total respiratory capacity per unit of dry weight was increased without a further change in the biochemistry of respiration. [Unfortunately it is not clear from the paper by SINGH and SINGH (1971) if a change in the composition of host root dry matter, e.g., more storage carbohydrates in the healthy roots, could explain the higher respiration rates per unit dry weight in infected roots].

Infection of *Ipomoea batatas* (sweet potato) slices by the fungal parasite *Ceratocystis fimbriata* stimulated respiration about fourfold, 40–60 h after the start of the infection process. After infection, respiration is cyanide-resistant, similar to that of uninfected slices, but it is not known to what extent the cyanide-resistant path is engaged, or even if the cyanide resistance involves the alternative path (URITANI and ASAHI 1980). In response to parasitic infection, starch was rapidly degraded in the root storage tissue of sweet potato (KATO and URITANI 1976). The increased rate of respiration is associated with increased metabolic activity, such as protein synthesis and the production of phytoalexins (for a review, see URITANI and ASAHI 1980). Mixed function oxygenases, e.g., impomeamarone 15-hydroxylase, are induced upon infection of *Ipomoea batatas* roots by *Ceratocystis fimbriata* (FUJITA et al. 1982). These enzymes are involved in the synthesis of antifungal compounds, belonging to the phytoalexins, and will be responsible for a part of the increased oxygen uptake upon infection or damage.

It is likely that the increased respiration rate primarily allows the energy-requiring synthesis of a number of enzymes and stress metabolites necessary to combat the infecting organisms. As everyone knows who has enthusiastically started eating an apple, only to meet an insect which had started some time earlier, these stress metabolites are also synthesized upon damage by insects or wounding. The increase in respiration upon wounding, infection, and damage by insects is also similar (URITANI and ASAHI 1980).

What is needed in this field is a proper energy balance, to test if the increased demand for metabolic energy, e.g., for synthesis of phytoalexins and proteins, can explain the increased respiration rate. Only when such a balance has been made, can one fully appreciate the respiratory responses upon infection or wounding.

9.2 Symbiotic Systems

Upon infection of a *Cymbidium* hybrid and *Dactylorhiza purpurella* (two orchid species) by the endophytic mycorrhizal fungus *Rhizoctonia* sp. respiration (expressed g^{-1} fr. wt.) of the protocorms was increased (BLAKEMAN et al. 1976). Culture filtrates or mycelium extracts of the mycorrhizal fungus similarly enhanced respiration. The activity of a polyphenol oxidase, ascorbic acid oxidase, peroxidase, and catalase also increased upon infection. It is not clear if, and to what extent, these oxidases participate in the increased respiration upon infection.

A plant–symbiont system studied in great detail is the legume–*Rhizobium* system. However, since it is the C-economy which has received major attention, rather than elements of the respiratory system per se, this topic is only briefly discussed in this chapter (Sect. 2.2).

10 Concluding Remarks

A comparison of the information in the present chapter with that in the original volume on respiration in the Encyclopedia of Plant Physiology allows us to make a balance of 25 years of work in this area. This balance is remarkably positive for such topics as cyanide-resistant respiration and regulation. We can no longer consider respiration merely as the result of the energy-requiring reactions in the cell, nor as the determinant of the rate of these energy-requiring processes. We have come to consider the interdependence of energy conservation and energy utilization and also that respiration per se is not necessarily a measure for energy conservation.

With the insight into the role of a nonphosphorylating pathway as an energy overflow, questions have arisen or been revived as to the role of respiration in yield formation and in adaptation to a dynamic environment. However, we still do not know a great deal more about the relation between respiration and growth rate (Sects. 2 and 7) and these topics need a more analytical rather than a descriptive approach if further knowledge is to be gained. The problem is that we do not yet fully understand the quantitative relation between growth or yield and respiration even though there are a large number of values relating the two processes. What do we really understand of differences in rate of maintenance respiration or in respiration related to ion uptake? Another subject deserving further investigation is the regulation of the partitioning of carbohydrates between storage and respiration. In some plants or tissues significant amounts of sugars are stored, whereas these carbohydrates are grist for the mill of the alternative pathway in others, and it remains unclear why sugars can sometimes accumulate without increasing the activity of the alternative path. Compartmentation of respiratory substrates in the cell (storage in vacuoles) may be involved, but there is not sufficient information to provide definitive answers. In addition to all these problems, the question as to how the level of carbohydrates is a regulatory factor in the engagement of the alternative path needs investigation at a biochemical level.

As a final suggestion I should like to add that research on respiration in an ecophysiological context (shade leaves vs. sun leaves, shade-tolerant vs.

shade-intolerant plants, temperature effects) seems rather promising. The balance is clearly positive as far as the amount of data is concerned but the *insight* into the mechanisms behind the observations is often still small. Work with defined genotypes and their progenies is likely to provide information far beyond the strict ecophysiological context, and may increase our insight into the regulation of the whole plant's functioning.

The sum total of this chapter is therefore that we have made considerably progress in the last 25 years and that we are now able to define more clearly the problems which need to be investigated in the next decade. The excitement in this field is still ahead of us!

References

Algera L (1932) Energiemessungen bei *Aspergillus niger* mit Hilfe eines automatischen Mikro-Kompensations-Calorimeters. PhD Thesis, Groningen, De Bussy, Amsterdam

Alves LM, Heisler EG, Kissinger JC, Patterson JM III, Kalan EB (1979) Effects of controlled atmospheres on production of sesquiterpenoid stress metabolites by white potato tuber. Possible involvement of cyanide-resistant respiration. Plant Physiol 63:359–362

ap Rees T, Fuller WA, Wright BW (1977) Measurements of glycolytic intermediates during the onset of thermogenesis in the spadix of *Arum maculatum*. Biochem Biophys Acta 461:274–282

Aviram I (1981) The interaction of benzhydroxamic acid with horseradish peroxidase and its fluorescent analogs. Arch Biochem Biophys 212:483–490

Azcón-Bieto J, Osmond CB (1983) Relationship between photosynthesis and respiration. The effect of carbohydrate status on the rate of CO_2 production by respiration in darkened and illuminated wheat leaves. Plant Physiol 71:574–581

Azcón-Bieto J, Day DA, Lambers H (1983a) The regulation of respiration in the dark in wheat leaf slices. Plant Sci Lett 32:313–320

Azcón-Bieto J, Lambers H, Day DA (1983b) The effect of photosynthesis and carbohydrate status on respiratory rates and the involvement of the alternative path in leaf respiration. Plant Physiol 72:598–603

Azcón-Bieto J, Lambers H, Day DA (1983c) Respiratory properties of developing bean and pea leaves. Aust J Plant Physiol 10:237–245

Bahr JT, Bonner WD Jr (1973) Cyanide-insensitive respiration. I. The steady states of skunk cabbage spadix and bean hypocotyl mitochondria. J Biol Chem 248:3441–3445

Bauer JR, Workman M (1964) Relationship between cell permeability and respiration in ripening banana fruit tissue. Plant Physiol 39:540–543

Beevers H (1961) Respiratory metabolism in plants. Harper & Row, New York Evanston, London

Berry LJ, Norris WE (1949) Studies on onion root respiration. II: The effect of temperature on the apparent diffusion coefficient in different segments of the root tip. Biochim Biophys Acta 3:607–614

Blacquière T, Lambers H (1981) Growth, photosynthesis, and respiration in *Plantago coronopus* as affected by salinity. Physiol Plant 51:265–269

Blacquière T, Visser De R (1984) Capacity of cytochrome and alternative paths in coupled and uncoupled root respiration of *Pisum* and *Plantago*. Physiol Plant 62:427–432

Blakeman JP, Mokahel MA, Hadley G (1976) Effect of mycorrhizal infection on respiration and activity of some oxidase enzymes of orchid protocorms. New Phytol 77:697–704

Bowling DJF (1976) Uptake of ions by plant roots. Chapman & Hall, London

Boyer JS (1970) Leaf enlargement and metabolic rates of corn, soybean, and sunflower at various leaf water potentials. Plant Physiol 46:233–235

Breeze V, Elston J (1978) Some effects of temperature and substrate content upon respiration and the carbon balance of field beans (*Vicia faba* L.). Ann Bot (London) 42:863–876

Brix H (1962) The effect of water stress on the rates of photosynthesis and respiration in tomato plants and loblolly pine seedlings. Physiol Plant 15:10–20

Brunton CJ, Palmer JM (1973) Pathways for the oxidation of malate and reduced pyridine nucleotide by wheat mitochondria. Eur J Biochem 39:283–291

Burguillo PF, Nicolas G (1977) Appearance of an alternate pathway cyanide-resistant during germination of seeds of *Cicer arietinum*. Plant Physiol 60:524–527

Cammack R, Palmer JM (1973) EPR studies of iron-sulphur proteins of plant mitochondria. Ann NY Acad Sci 222:816–823

Chalkey DB, Millar RL (1982) Cyanide-insensitive respiration in relation to growth of a low-temperature basidiomycete. Plant Physiol 69:1121–1127

Challa H (1976) An analysis of the diurnal course of growth, carbon dioxide exchange and carbohydrate reserve content of cucumber. PhD Thesis, Wageningen.

Chatterton NJ, McKell CM, Strain BR (1970) Intraspecific differences in temperature-induced respiratory acclimation of desert saltbush. Ecology 51:545–547

Christoffersen RE, Laties GG (1982) Ethylene regulation of gene expression in carrots. Proc Natl Acad Sci USA 79:4060–4063

Collinge DB, Hughes MA (1982) Development and physiological studies on the cyanogenic glucosides of white clover, *Trifolium repens* L. J Exp Bot 33:154–161

Crawford RMM, Palin MA (1981) Root respiration and temperature limits to the north-south distribution of four perennial maritime plants. Flora 171:338–354

Cunningham GL, Syvertsen JP (1977) The effect of nonstructural carbohydrate levels on dark CO_2 release in creosote bush. Photosynthetica 11:291–295

Daday H (1954) Gene frequencies in wild populations of *Trifolium repens* L. I. Distribution by latitude. Heredity 8:61–78

Day DA, Lambers H (1983) The regulation of glycolysis and electron transport in roots. Physiol Plant 58:155–160

Day DA, Arron GP, Laties GG (1980) Nature and control of respiratory pathways in plants: The interaction of cyanide-resistant respiration with the cyanide-sensitive pathway. In: Stumpf PK, Conn EE (eds) The biochemistry of plants. A comprehensive treatise, vol II. Academic Press, London New York, pp 197–241

Dizengremel P (1975) La voie d'oxydation insensible au cyanure dans les mitochondries de tranches de pomme de terre (*Solanum tuberosum* L.) maintenues en survie. Physiol Veg 13:39–54

Earnshaw MJ (1981) Arrhenius plots of root respiration in some arctic plants. Arct Alp Res 13:425–430

Eberhardt F (1960) Der Einfluß von mechanischer Beanspruchung, Verletzung und Infektion auf die Atmung. In: Ruhland W (ed) Encyclopedia of plant physiology, Vol XII/2. Springer, Berlin Göttingen Heidelberg, pp 388–415

Eliasson L, Mathiesen I (1956) The effect of 2,4-dinitrophenol and some oxidase inhibitors on the oxygen uptake in different parts of wheat roots. Physiol Plant 9:265–279

Esashi Y, Komatsu H, Ushizawa R, Sakai Y (1982) Breaking of secondary dormancy in cocklebur seeds by cyanide and azide in combination with ethylene and oxygen, and their effects on cytochrome and alternative respiratory pathways. Aust J Plant Physiol 9:97–111

Fales FN (1953) The effect of sodium azide on alcoholic fermentation. J Biol Chem 202:157–167

Farrar JF (1980) The pattern of respiration rate in the vegetative barley plant. Ann Bot (London) 46:71–76

Feller K, Ehrismann KH (1978) Changes in gas exchange and in the activities of proteolytic enzymes during senescence of wheat leaves (*Triticum aestivum* L.). Z Pflanzenphysiol 90:235–244

Forward DF (1960) Effects of temperature on respiration. In: Ruhland W (ed) Encyclopedia of plant physiology, vol XII/2. Springer, Berlin Göttingen Heidelberg, pp 234–258

Forward DF (1965) The respiration of bulky organs. In: Steward FC (ed) Plant physiology. A treatise, vol IV-A. Academic Press, London New York, pp 311–376

Foulds W, Young L (1977) Effect of frosting, moisture stress and potassium cyanide on the metabolism of cyanogenic and acyanogenic phenotypes of *Lotus corniculatus* L. and *Trifolium repens* L. Heredity 38:19–24

Fritz G, Miller WG, Harris RH, Anderson L (1958) Direct incorporation of molecular oxygen into organic material by respiring corn seedlings. Plant Physiol 33:159–161

Fujita M, Ôba K, Uritani I (1982) Properties of a mixed function oxygenase catalyzing ipomeamarone 15-hydroxylation in microsomes from cut-injured and *Ceratocystis fimbriata*-infected sweet potato root tissue. Plant Physiol 70:573–578

Genevois ML (1929) Sur la fermentation et sur la respiration chez les végétaux chlorophylliens. Rev Gen Bot 41:252–271

Goldstein AH, Anderson JO, McDaniel RG (1980) Cyanide-insensitive and cyanide-sensitive O_2 uptake in wheat. I, Gradient-purified mitochondria. Plant Physiol 66:488–493

Graham D (1980) Effects of light on "dark" respiration. In: Stumpf PK, Conn EE (eds) The biochemistry of plants. A comprehensive treatise, vol II. Academic Press, London New York, pp 525–579

Graham D, Patterson BD (1982) Responses of plants to low, nonfreezing temperatures: Proteins, metabolism, and acclimation, Annu Rev Plant Physiol 33:347–372

Grime JP (1966) Shade avoidance and shade tolerance in flowering plants. In: Bainbridge R, Evand GC, Rackman O (eds) Light as an ecological factor. Blackwell, Oxford, pp 187–207

Grover SD, Laties GG (1978) Characterization of the binding properties of disulfiram, an inhibitor of cyanide-resistant respiration. In: Ducet G, Lance C (eds) Plant mitochondria. Elsevier, Amsterdam, pp 259–266

Grover SD, Laties GG (1981) Disulfiram inhibition of the alternative respiratory pathway in plant mitochondria. Plant Physiol 68:393–400

Hacket DP (1960) Respiratory inhibitors. In: Ruhland W (ed) Encyclopedia of plant physiology, vol XII/2. Springer, Heidelberg Göttingen Berlin, pp 23–41

Hall FR, Hollingworth RM, Shankland DL (1971) Cyanide tolerance in millipedes: The biochemical basis. Comp Biochem Physiol 38:723–737

Hansen GK (1980) Diurnal variation of root respiration rates and nitrate uptake as influenced by nitrogen supply. Physiol Plant 48:421–427

Hansen GK, Jensen CR (1977) Growth and maintenance respiration in whole plants, tops and roots of *Lolium multiflorum*. Physiol Plant 39:155–164

Hanson WD (1971) Selection for differential productivity among juvenile maize plants: Associated net photosynthetic rate and leaf area changes. Crop Sci 11:334–339

Heichel GH (1970) Prior illumination and the respiration of maize leaves in the dark. Plant Physiol 46:359–362

Heichel GH (1971) Confirming measurements of respiration and photosynthesis with dry matter accumulation. Photosynthetica 5:93–98

Henry MF, Nyns E-J (1975) Cyanide-insensitive respiration. An alternative mitochondrial pathway. Sub-Cell Biochem 4:1–65

Hew CS, Thio YC, Wong SY, Chin TY (1978) Rhythmic production of CO_2 by tropical orchid flowers. Physiol Plant 42:226–230

Horn ME, Mertz D (1982) Cyanide-resistant respiration in suspension cultured cells of *Nicotiana glutinosa* L. Plant Physiol 69:1439–1443

Hourmant A, Pradet A (1981) Oxidative phosphorylation in germinating lettuce seeds (*Lactuca sativa*) during the first hours of imbibition. Plant Physiol 68:631–635

Huber B, Ziegler H (1960) Atmung und Wasserhaushalt. In: Ruhland W (ed) Encyclopedia of plant physiology, vol XII/2. Springer, Berlin Göttingen Heidelberg, pp 150–169

Huq S, Palmer JM (1978) Superoxide and hydrogen peroxide production in cyanide resistant *Arum maculatum* mitochondria. Plant Sci Lett 11:351–358

Jackson PC, St. John JB (1982) Effects of 2,4-dinitrophenol on membrane lipids of roots. Plant Physiol 70:858–862

Jacobi B (1899) Über den Einfluss verschiedener Substanzen auf die Atmung und Assimilation submerser Pflanzen. Flora 86:289–327

Kandler O (1950) Untersuchungen über den Zusammenhang zwischen Atmungsstoffwechsel und Wachstumsvorgängen bei in vitro kultivierten Maiswurzeln. Z Naturforsch 5 B:203–211

Kandler O (1953) Über den „synthetischen Wirkungsgrad" in vitro kultivierter Embryonen, Wurzeln und Sprosse. Z Naturforsch 8 B:109–117

Kano H, Kageyama M (1977) Effects of cyanide on the respiration of musk-melon (Cucumis melo) roots. Plant Cell Physiol 18:1149–1154

Kato C, Uritani I (1976) Changes in carbohydrate content of sweet potato in response to cutting and infection by black rot fungus. Ann Phytopathol Soc Jpn 42:181–186

Kernbach B (1960) Wasserhaushalt, Atmung und Regeneration bei Funaria hygrometrica. Z Bot 48:415–441

Kiener CM, Bramlage WJ (1981) Temperature effects on the activity of the alternative respiratory pathway in chill-sensitive Cucumis sativus. Plant Physiol 68:1474–1478

Kimura M (1970) Analysis of production processes of an undergrowth of subalpine Abies forest, Pteridophyllum racemosum population. 2. Respiration, gross production and economy of dry matter. Bot Mag (Tokyo) 83:304–311

Kimura M, Yokoi Y, Hogetsu K (1978) Quantitative relationships between growth and respiration. II. Evaluation of constructive and maintenance respiration in growing Helianthus tuberosus leaves. Bot Mag (Tokyo) 91:43–56

Kinraide TB, Behan MJ (1975) Restoration of organic acid accumulation in sectioned leaves of Bryophyllum tubiflorum Harv. Plant Physiol 56:830–835

Kinraide TB, Marek LF (1980) Wounding stimulates cyanide-sensitive respiration in the highly cyanide-resistant leaves of Bryophyllum tubiflorum Harv. Plant Physiol 65:409–410

Klerk-Kiebert De YM, Kneppers TJA, Plas van der LHW (1981) Participation of the CN-resistant alternative oxydase pathway in the respiration of white and green soybean cells during growth in batch suspension culture. Z Pflanzenphysiol 104:149–159

Klerk-Kiebert De YM, Kneppers TJA, Plas Van der LHW (1982) Influence of chloramphenicol on growth and respiration of soybean (Glycine max L.) suspension cultures. Physiol Plant 55:98–102

Knutson RM (1974) Heat production and temperature regulation in eastern skunk cabbage. Science 186:746–747

Kolkwitz R (1899) Über den Einfluss des Lichtes auf die Atmung niederer Pilze. Jahrb Wiss Bot 33:128–165

Kowallik W, Schätzle S (1980) Enhancement of carbohydrate degradation by blue light. In: Senger H (ed) The blue light syndrome. Springer, Berlin Heidelberg New York, pp 344–360

Kuiper D (1983) Genetic differentiation in Plantago major: Growth and root respiration and their role in phenotypic adaptation. Physiol Plant 57:222–230

Lambers H (1979) Efficiency of root respiration in relation to growth, morphology and soil composition. Physiol Plant 46:194–202

Lambers H (1980) The physiological significance of cyanide-resistant respiration. Plant Cell Environ 3:293–302

Lambers H (1982) Cyanide-resistant respiration: A nonphosphorylating electron transport pathway acting as an energy overflow. Physiol Plant 55:478–485

Lambers H, Posthumus F (1980) The effect of light intensity and relative humidity on growth rate and root respiration of Plantago lanceolata and Zea mays. J Exp Bot 31:1621–1630

Lambers H, Steingröver E (1978a) Efficiency of root respiration of a flood-tolerant and a flood-intolerant Senecio species as affected by low oxygen tension. Physiol Plant 42:179–184

Lambers H, Steingröver E (1978b) Growth respiration of a flood-tolerant and a flood-intolerant Senecio species: Correlation between calculated and experimental values. Physiol Plant 43:219–224

Lambers H, Noord R, Posthumus F (1979) Respiration of *Senecio* shoots: Inhibition during photosynthesis, resistance to cyanide and relation to growth and maintenance. Physiol Plant 45:351–356

Lambers H, Layzell DB, Pate JS (1980) Efficiency and regulation of root respiration in a legume: Effects of the N source. Physiol Plant 50:319–325

Lambers H, Blacquière T, Stuiver CEE (1981a) Interactions between osmoregulation and the alternative respiratory pathway in *Plantago coronopus* as affected by salinity. Physiol Plant 51:63–68

Lambers H, Posthumus F, Stulen I, Lanting L, Dijk Van de SJ, Hofstra R (1981b) Energy metabolism of *Plantago major major* as dependent on the nutrient supply. Physiol Plant 51:245–252

Lambers H, Day DA, Azcón-Bieto J (1983a) Cyanide-resistant respiration in roots and leaves. Measurements with intact tissues and isolated mitochondria. Physiol Plant 58:148–154

Lambers H, Szaniawski RK, Visser De R (1983b) Respiration for growth, maintenance and ion uptake. An evaluation of concepts, methods, values and their significance. Physiol Plant 58:556–563

Laties GG (1982) The cyanide-resistant, alternative path in higher plant respiration. Annu Rev Plant Physiol 33:519–555

Leopold AC, Mushgrave ME (1979) Respiratory changes with chilling injury of soybeans. Plant Physiol 64:702–705

Leopold AC, Mushgrave ME (1980) Respiratory pathways in aged soybean seeds. Physiol Plant 49:49–54

Ludwig LJ, Charles-Edwards DA, Withers AC (1975) Tomato leaf photosynthesis and respiration in various light and carbon dioxide environments. In: Marcelle R (ed) Environmental and biological control of photosynthesis. Junk, The Hague, pp 29–36

Lundegårdh H (1954) Enzyme systems conducting the aerobic respiration of roots of wheat and rye. Ark Kem 7:451–478

Lundegårdh H (1955) Mechanisms of absorption, transport, accumulation, and secretion of ions. Annu Rev Plant Physiol 6:1–24

Lundegårdh H (1960) Anion respiration. In: Ruhland W (ed) Encyclopedia of plant physiology, vol XII/2. Springer, Berlin Göttingen Heidelberg, pp 185–233

Lyons JM, Raison JK (1970) Oxidative activity of mitochondria isolated from plant tissues sensitive and resistant to chilling injury. Plant Physiol 45:386–389

Machlis L (1944) The respiratory gradient in barley roots. Am J Bot 31:281–282

Marx R, Brinkmann K (1978) Characteristics of rotenone-insensitive oxidation of matrix-NADH by broad bean mitochondria. Planta 142:83–90

Massimino D, André M, Richaud C, Daguenet A, Massimino J, Vivoli J (1980) Évolution horaire au cours d'une journée normale de la respiration foliaire et racinaire et de la nutrition N.P.K. chez *Zea mays*. Physiol Plant 48:512–518

Massimino D, André M, Richaud C, Daguenet A, Massimino J, Vivoli J (1981) The effect of a day at low irradiance of a maize crop. I. Root respiration and uptake of N, P and K. Physiol Plant 51:150–155

McCaig TN, Hill RD (1977) Cyanide-insensitive respiration in wheat: cultivar differences and effects of temperature, carbon dioxide, and oxygen. Can J Bot 55:549–555

McCree KJ (1970) An equation for the rate of respiration of white clover plants grown under controlled conditions. In: Setlik I (ed) Prediction and measurements of photosynthetic productivity. Proc IBP/PP Tech Meet, Trebon, 1969, pp 221–230

McCree KJ (1974) Equations for the rate of dark respiration of white clover and grain *Sorghum*, as functions of dry weight, photosynthetic rate, and temperature. Crop Sci 14:509–514

McCree KJ, Kresovich S (1978) Growth and maintenance requirements of white clover as a function of daylength. Crop Sci 18:22–25

Meeuse BJD (1975) Thermogenic respiration in aroids. Annu Rev Plant Physiol 26:117–126

Mercier PJ, Poole RJ (1980) Electrogenic pump activity in red beet: Its relation to ATP levels and to cation influx. J Membr Biol 55:165–174

Møller IM (1978) Balance between cyanide-sensitive and -insensitive respiration in wheat root mitochondria, as influenced by salt concentration in the plant growth medium. Physiol Plant 42:157–162

Møller IM, Palmer JM (1982) Direct evidence for the presence of a rotenone-resistant NADH-dehydrogenase on the inner surface of the inner membrane of plant mitochondria. Physiol Plant 54:267–274

Passam HC (1976) Cyanide-insensitive respiration in root tubers of cassava (*Manihot esculenta* Crantz.). Plant Sci Lett 7:211–218

Passam HC, Bird MC (1978) The respiratory activity of honeydew melons during the climacteric. J Exp Bot 29:325–333

Pate JS, Layzell DB, Atkins CA (1979) Economy of carbon and nitrogen in a nodulated and nonnodulated (NO_3-grown) legume. Plant Physiol 64:1083–1088

Pate JS, Layzell DB, Atkins CA (1980) Transport exchange of carbon, nitrogen and water in the context of whole plant growth and functioning – Case history of a nodulated annual legume. Dtsch Bot Ges 93:243–255

Pearson OP (1948) Metabolism of small mammals, with remarks on the lower limit of mammalian size. Science 108:44

Penning de Vries FWT (1975) Use of assimilates in higher plants. In: Cooper J (ed) Photosynthesis and productivity in different environments. Cambridge Univ Press, Cambridge, pp 459–480

Penning de Vries FWT, Van Laar HH (1975) Substrate utilization in germinating seeds. In: Lansberg JJ, Cutting CV (eds) Environmental effects on crop physiology. Proc 5th Long Ashton Symp. Academic Press, London New York, p 217–228

Penning de Vries FWT, Brunsting AHM, Laar Van (1974) Products, requirements and efficiency of biosynthetic processes: A quantitative approach. J Theor Biol 45:339–377

Peterson RB, Zelitch I (1982) Relationship between net CO_2 assimilation and dry weight accumulation in field-grown tobacco. Plant Physiol 70:677–685

Pirt SJ (1975) Principles of microbe and cell cultivation. Blackwell, Oxford

Pitman MG, Mowat J, Nair H (1971) Interaction of processes for accumulation of salt and sugar in barley plants. Aust J Biol Sci 24:619–632

Plas Van der LHW, Wagner MJ (1983) Regulation of the activity of the alternative oxidase in callus-forming potato tuber tissue discs. Physiol Plant 58:311–317

Pomeroy MK, Andrews CJ (1975) Effect of temperature on respiration of mitochondria and shoot segments from cold-hardened and nonhardened wheat and rye seedlings. Plant Physiol 56:703–706

Popp M, Osmond CB, Summons RE (1982) Pathway of malic acid synthesis in response to ion uptake in wheat and lupin roots: Evidence for fixation of ^{13}C and ^{14}C. Plant Physiol 69:1289–1292

Prins HBA, Walsarie Wolff R (1974) Photorespiration in leaves of *Vallisneria spiralis*; the effect of oxygen on the carbon dioxide compensation point. Proc Kon Ned Acad Wetensch Amsterdam Ser C77:239–245

Rhodes MJC (1980) Respiration and senescence of plant organs. In: Stumpf PK, Conn EE (eds) The biochemistry of plants. A comprehensive treatise. Academic Press, London New York, pp 419–462

Rich PR (1978) Quinol oxidation in *Arum maculatum* mitochondria and its application to the assay, solubilization and partial purification of the alternative oxidase. FEBS Lett 96:252–256

Rich PR, Wiegand NK, Blum H, Moore AL, Bonner WD (1978) Studies on the mechanism of inhibition of redox enzymes by substituted hydroxamic acids. Biochem Biophys Acta 525:325–337

Richardson SD (1953) Studies of root growth in *Acer saccharinum* L. I. The relation between root growth and photosynthesis. Proc Kon Ned Acad Wetensch, Amsterdam, Ser C56:185–253

Riehl TE, Jaffe MJ (1982) Physiological studies on pea tendrils. XIII. Respiration is necessary for contact coiling. Physiol Plant 55:192–196

Rissler JF, Millar RL (1977) Contribution of a cyanide-insensitive alternate respiratory system to increases in formamide hydro-lyase activity and to growth in *Stemphylium loti* in vitro. Plant Physiol 60:857–861

Robson MJ (1982a) The growth and carbon economy of selection lines of *Lolium perenne* with 'fast' and 'slow' rates of dark respiration. 1. Grown as simulated swards during a regrowth period. Ann Bot (London) 49:321–329

Robson MJ (1982b) The growth and carbon economy of selection lines of *Lolium perenne* cv S23 with 'fast' and 'slow' rates of dark respiration. 2. Grown as young plants from seed. Ann Bot (London) 49:331–339

Rosenstock G, Ried A (1960) Der Einfluss sichtbarer Strahlung auf die Pflanzenatmung. In: Ruhland W (ed) Encyclopedia of plant physiology, vol XII/2. Springer, Berlin Göttingen Heidelberg, pp 259–333

Ruget F (1981) Respiration de croissance et respiration d'entretien: méthodes de mesure, comparison des résultats. Agronomie 1:601–610

Ryle GJA (1984) Respiration and plant growth. In: Palmer JM (ed) Physiology and biochemistry of plant respiration. Cambridge Univ Press, Cambridge (in press)

Saglio PH, Pradet A (1980) Soluble sugars, respiration, and energy charge during aging of excised maize root tips. Plant Physiol 66:516–519

Satler SO, Thimann KV (1981) Le jasmonate de méthyle: nouveau et puissant promoteur de la sénescence des feuilles. Physiol Veg 293:735–740

Satter RL, Galston AW (1973) Leaf movements: rosetta stone of plant behavior? BioScience 23:407–416

Satter RL, Hatch AM, Gill MK (1979) A circadian rhythm in oxygen uptake by *Samanea* pulvini. Plant Physiol 64:379–281

Scheibe A, Meyer zu Drewer H (1959) Vergleichende Untersuchungen zur Atmungsintensität der Wurzeln unterschiedlicher Genotypen bei Getreidearten. Z Acker-Pflanzenb 108:223–252

Schneider CW Childers NF (1941) Influence of soil moisture on photosynthesis, respiration, and transpiration of apple leaves. Plant Physiol 16:565–583

Schonbaum GR, Bonner WD Jr, Storey B, Bahr JT (1971) Specific inhibition of the cyanide-insensitive respiratory pathway in plant mitochondria by hydroxamic acids. Plant Physiol 47:124–128

Schwarz M, Gale J (1981) Maintenance respiration and carbon balance of plants at low levels of sodium chloride salinity. J Exp Bot 32:933–941

Siedow JN, Bickett DM (1981) Structural features required for inhibition of cyanide-insensitive electron transfer by propyl gallate. Arch Biochem Biophys 207:32–39

Siedow JN, Girvin ME (1980) Alternative respiratory pathway. Its role in seed respiration and its inhibition by propyl gallate. Plant Physiol 65:669–674

Simon EW, Minchin A, McMenahim MM, Smith JM (1976) The low temperature limit for seed germination. New Phytol 77:301–311

Singh JN, Singh JN (1971) Studies on the physiology of host–parasite relationship in Orobanche. I. Respiratory metabolism of host and parasite. Physiol Plant 24:380–386

Smakman G, Hofstra R (1982) Energy metabolism of *Plantago lanceolata* L., as affected by a change in root temperature. Physiol Plant 56:33–37

Solomos T, Laties GG (1974) Similarities between the actions of ethylene and cyanide in initiating the climacteric and ripening of avocados. Plant Physiol 54:506–511

Solomos T, Laties GG (1976) Induction by ethylene of cyanide-resistant respiration. Biochem Biophys Res Commun 2:663–671

Steingröver E (1981) The relationship between cyanide-resistant root respiration and the storage of sugars in the taproot in *Daucus carota* L. J Exp Bot 32:911–919

Stiles W (1960) The composition of the atmosphere (oxygen content of the air, water, soil, intercellular spaces, diffusion, carbon dioxide and oxygen tension). In: Ruhland W (ed) Encyclopedia of plant physiology, vol XII/2. Springer, Berlin Göttingen Heidelberg, pp 114–148

Stocker O (1935) Assimilation und Atmung westjavanischer Tropenbäume. Planta 24:402–445

Strain BR (1969) Seasonal adaptations in photosynthesis and respiration in four desert shrubs growing in situ. Ecology 50:511–513

Szaniawski RK (1981) Growth and maintenance respiration of shoots and roots in Scots pine seedlings. Z Pflanzenphysiol 101:391–398

Szaniawski RK, Adams MS (1974) Root respiration of *Tsuga canadensis* seedlings as influenced by intensity of net photosynthesis and dark respiration of shoots. Am Midl Nat 91:464–468

Szaniawski RK, Kielkiewicz M (1982) Maintenance and growth respiration in shoots and roots of sunflower plants grown at different root temperatures. Physiol Plant 55:500–504

Tànczos OG (1974) Invloed van lage temperatuur op de bladeren van *Cucumis sativus* L. Thesis, Univ Groningen, The Netherlands

Tànczos OG (1977) Influence of chilling on electrolyte permeability, oxygen uptake and 2,4-dinitrophenol stimulated oxygen uptake in leaf discs of the thermophilic *Cucumis sativus*. Physiol Plant 41:289–292

Tetley RM, Thimann KV (1974) The metabolism of oat leaves during senescence. I. Respiration, carbohydrate metabolism, and the action of cytokinins. Plant Physiol 54:294–303

Theologis A, Laties GG (1978a) Relative contribution of cytochrome-mediated and cyanide-resistant electron transport in fresh and aged potato slices. Plant Physiol 62:232–237

Theologis A, Laties GG (1978b) Cyanide-insensitive respiration in fresh and aged sweet potato slices. Plant Physiol 62:243–248

Theologis A, Laties GG (1978c) Respiratory contribution of the alternative path during various stages of ripening in avocado and banana fruits. Plant Physiol 62:249–255

Theologis A, Laties GG (1981) Alternative path-mediated chloride absorption in cyanide-resistant tissues. Plant Physiol 68:240–243

Thomas M (1951) Carbon dioxide fixation and acid synthesis in crassulacean acid metabolism. Symp Soc Exp Biol 5:72–93

Thornley JHM (1970) Respiration, growth and maintenance in plants. Nature (London) 227:304–305

Turner JF, Turner DH (1980) The regulation of glycolysis and the pentose phosphate pathway. In: Stumpf PK, Conn EE (eds) The biochemistry of plants. A comprehensive treatise, vol II. Academic Press, London New York, pp 279–316

Ulrich A (1941) Metabolism of nonvolatile organic acids in excised barley roots as related to cation/anion balance during salt accumulation. Am J Bot 28:526–537

Uritani I, Asahi T (1980) Respiration and related metabolic activity in wounded and infected tissues. In: Stumpf PK, Conn EE (eds) The biochemistry of plants. A comprehensive treatise, vol II. Academic Press, London New York, pp 463–485

Veen BW (1980) Energy cost of ion transport. In: Rains DW, Valentine RC, Hollaender A (eds) Genetic engineering of osmoregulation. Impact on plant productivity for food, chemicals, and energy. Plenum Press, New York London, pp 187–195

Visser De R, Blacquière T (1984) Inhibition and stimulation of root respiration in *Pisum* and *Plantago* by hydroxamate. Its consequences for the assessment of alternative path activity. Plant Physiol 75:813–817

Visser De R, Lambers H (1983) Growth and the efficiency of root respiration of *Pisum sativum* as dependent on the source of nitrogen. Physiol Plant 58:533–543

Wager HG (1941) On the respiration and carbon assimilation rates of some arctic plants as related to temperature. New Phytol 40:1–19

Waring AJ, Laties GG (1977a) Dependence of wound-induced respiration in potato slices on the time-restricted actinomycin-sensitive biosynthesis of phospholipid. Plant Physiol 60:5–10

Waring AJ, Laties GG (1977b) Inhibition of the development of induced respiration and cyanide-insensitive respiration in potato tuber slices by cerulenin and dimethylaminoethanol. Plant Physiol 60:11–16

Wedding RT, McReady CC, Harley JL (1973) Cyanide sensitivity of respiration during ageing of *Arum* spadix slices. New Phytol 72:15–26

Wilson D (1975) Variation in leaf respiration in relation to growth and photosynthesis of *Lolium*. Ann Appl Biol 80:323–338

Wilson D (1982) Response to selection for dark respiration rate of mature leaves in *Lolium perenne* and its effects on growth of young plants and simulated swards. Ann Bot (London) 49:303–312

Wilson D, Jones JG (1982) Effect of selection for dark respiration rate of mature leaves on crop yields of *Lolium perenne* cv. S23. Ann Bot (London) 49:313–320

Wolfe J, Bagnall DJ (1980) Arrhenius plots – curves or straight lines? Ann Bot (London) 45:485–488

Woodstock LW, Pollock BM (1965) Physiological predetermination: Imbibition, respiration, and growth of lima beans seeds. Science 150:1031–1032

Yamaguchi J (1978) Respiration and the growth efficiency in relation to crop productivity. J Fac Agric, Hokkaido Univ 59:59–129

Yemm EW (1965) The respiration of plants and their organs. In: Steward FC (ed) Plant physiology. A treatise, vol IV-A. Academic Press, London New York, pp 231–310

Yoshida S, Niki T (1979) Cell membrane permeability and respiratory activity in chilling-stressed callus. Plant Cell Physiol 20:1237–1242

Yoshida S, Tagawa F (1979) Alteration of the respiratory function in chill-sensitive callus due to low temperature stress. I. Involvement of the alternate pathway. Plant Cell Physiol 20:1243–1250

Author Index

Page numbers in *italics* refer to the references

Subject Index

Encyclopedia of Plant Physiology

New Series

Editors: A. Pirson, M. H. Zimmermann

Springer-Verlag
Berlin
Heidelberg
New York
Tokyo